GNSS Applications in Earth and Space Observations

Global Navigation Satellite Systems (GNSSs) have become an essential technology used in navigation, positioning, and timing applications in meteorology, environmental monitoring, disaster management, and space exploration. This comprehensive book explores the various applications of GNSS technology in different fields of Earth and Space observations and provides researchers, professionals, and students valuable insights into these emerging trends. It discusses the challenges that impact the performance of GNSS technology and offers solutions through several case studies on Space weather and climate disasters, opening a different dimension of approaches in various paradigms of GNSS technology.

Features:

- Covers the most up-to-date GNSS applications in three major areas related to Earth and Space observations: climate studies, disaster management, and Space weather monitoring.
- Includes case studies of best practices in climate studies and disaster management.
- Explains the impacts of Space weather events on the near-Earth environment.
- Describes limitations and future possibilities of better use of GNSS in Earth and Space observation and monitoring.
- Highlights an integrated and interdisciplinary approach valuable to a wide range of readers studying Earth and Space interactions.

This book is a valuable resource for professionals, researchers, academics, and students in Remote Sensing and GIS, Earth Science, Physics and Electronics, Climate Studies, Disaster Management, Geophysics, and Space Science.

Microsatellite Remote Sensing, GNSS, and Applications

Series Editor

Josaphat Tetuko Sri Sumantyo
Department Head, Department of Environmental Remote Sensing,
Graduate School of Integrated Science and Technology, Chiba University, Japan

GNSS Applications in Earth and Space Observations: Challenges and Prospective Approaches
Edited by Dileep Kumar Gupta and Abhay Kumar Singh

GNSS Applications in Earth and Space Observations

Challenges and Prospective Approaches

Edited by
Dileep Kumar Gupta
Abhay Kumar Singh

CRC Press
Taylor & Francis Group
Boca Raton London New York

CRC Press is an imprint of the
Taylor & Francis Group, an **informa** business

Designed cover image: Shutterstock

First edition published 2025
by CRC Press
2385 NW Executive Center Drive, Suite 320, Boca Raton FL 33431

and by CRC Press
4 Park Square, Milton Park, Abingdon, Oxon, OX14 4RN

CRC Press is an imprint of Taylor & Francis Group, LLC

Library of Congress Cataloging-in-Publication Data
Names: Gupta, Dileep Kumar, editor. | Singh, Abhay Kumar, (Professor of physics), editor.
Title: GNSS applications in Earth and space observations : challenges and prospective approaches / edited by Dileep Kumar Gupta and Abhay Kumar Singh.
Other titles: Global Navigation Satellite Systems applications in Earth and space observations
Description: First edition. | Boca Raton, FL : CRC Press, Taylor & Francis Group, 2025. | Series: Microsatellite remote sensing, GNSS, and applications | Includes bibliographical references.
Identifiers: LCCN 2024046890 (print) | LCCN 2024046891 (ebook) | ISBN 9781032712437 (hardback) | ISBN 9781032712451 (paperback) | ISBN 9781032712444 (ebook)
Subjects: LCSH: Global Positioning System. | Astronomical observatories.
Classification: LCC G109.5 .G325 2025 (print) | LCC G109.5 (ebook) | DDC 550.28/4--dc23/eng/20250203
LC record available at https://lccn.loc.gov/2024046890
LC ebook record available at https://lccn.loc.gov/2024046891

ISBN: 978-1-032-71243-7 (hbk)
ISBN: 978-1-032-71245-1 (pbk)
ISBN: 978-1-032-71244-4 (ebk)

DOI: 10.1201/9781032712444

Typeset in Times
by SPi Technologies India Pvt Ltd (Straive)

Contents

Section 1 Introduction to GNSS and Its Advances in Earth Observation

Section 2 GNSS Applications for Disaster Management

Section 3 GNSS Applications for Climate Studies

Section 4 GNSS Applications for Space Weather

Section 5 Challenges and Future Prospects

About the Editors

Dr. Dileep Kumar Gupta received the M.Sc. degree in Electronics from the University Institute of Engineering and Technology, C.S.J.M. University, Kanpur, Uttar Pradesh, India, in 2009. He has completed a Doctorate with the Department of Physics, IIT (BHU), Varanasi, India. Dr. Gupta qualified in the National Eligibility Test in the subject of Electronic Science and the Graduate Aptitude Test for Engineers in the subject of Electronics and Communication Engineering. Currently, he is working as the Assistant Professor (Grade II) at Galgotias University, Greater Noida, UP, India. With a rich research portfolio, Dr. Gupta has published more than 60 research articles in esteemed journals, conference proceedings, and book chapters. He serves as a Guest Editor for the reputable journal MPDI Sensors, overseeing the special issue *Space-borne Radar Remote Sensing of Agricultural Canopies and Soil Moisture*. Furthermore, he has edited the book, *Radar Remote Sensing: Applications and Challenges*, published by an esteemed publisher. His research interests are multi-sensor remote sensing and climate data processing, geo informatics, artificial intelligence and machine learning algorithms, disaggregation schemes for high-resolution soil moisture retrieval, agriculture remote sensing, atmospheric and aerosols studies, cryosphere studies, microwave active and passive remote sensing for soil moisture retrieval, the assessment of climate change impact on agriculture, GNSS Applications, NISAR applications, and NWP modeling, such as WRF.

Prof. Abhay Kumar Singh has made outstanding contributions to teaching and research. He is distinguished for his many crucial and outstanding contributions in the area of Space weather study of the upper atmosphere over low latitude and characteristics of aerosols over IGB. He has succeeded in resolving issues of crop residual burning aerosols and their long-range transport over the whole IGB, which impacted the scientific community thinking. His study of the nature, formation, and transport of polluted aerosols in the Asian troposphere aerosol layer (ATAL) by a series of BATAL campaigns has revealed various unsolved properties of ATAL. His study of a solar eclipse, analyzed for the first time at eight different cities, induced a significant reduction in solar irradiance, surface ozone, and dynamical changes to the meteorological parameters, as well as TIDs oscillations in ionosphere generated by atmospheric gravity waves. His work is widely cited in books, reviews, and papers. Prof. Singh has published more than 200 research papers, supervised 15 Ph.D. students, and completed 15 research projects, amounting to Rs. 4 crores to his credit. He single-handedly established an Atmospheric Research Laboratory recognized by ISRO. He worked at Umeå University, Sweden, as BOYSCOST Fellow for one year. Prof. Singh visited several countries abroad, namely The Netherlands; ICTP, Italy; Sopron, Hungry; Busan, South Korea; and New Orleans, Pasadena, and Boston, USA. He has developed excellent scientific temperament, experimental skill, high-quality research work, and distinct academic and administrative capabilities. He is also an esteemed and actively involved member of several world-renowned scientific societies.

Contributors

Shiv Prasad Aggarwal
North Eastern Space Applications Centre
India

Nitin Arora
Department of Computer Science and Engineering
University Institute of Engineering, Chandigarh
    University
Mohali, Punjab, India

Nitesh Awasthi
Department of Remote Sensing, Birla Institute
    of Technology
Mesra, Ranchi

Deepthi Ayyagari
Department of Astronomy, Astrophysics &
    Space Engineering, IIT Indore
Laboratoire GEOLOC
Université Gustave Eiffel
France

Ogunsua Babalola
Federal University of Technology Akure
Nigeria

S. Bala Subramaniyam
XR Developer
Larsen and Toubro Technology Services

Shashank C. Bangi
Assistant Professor, Department of Civil
    Engineering, KLS Gogte Institute of
    Technology
Belagavi, India

Bhuvnesh Brawar
Department of Astronomy, Astrophysics &
    Space Engineering
IIT Indore

Sumanjit Chakraborty
Department of Astronomy, Astrophysics &
    Space Engineering
IIT Indore
Indian Institute of Geomagnetism
Navi Mumbai, Maharashtra, India

Rajeev Singh Chandel
Independent Research, Raibareli (U.P.) and
    Centre for Climate Change and Water
    Research
Suresh Gyan Vihar University
Jaipur, Rajasthan, India

Sunil Kumar Chaurasiya
Department of Physics
SRM Institute of Science and Technology
Delhi-NCR Campus
Modinagar, Ghaziabad, U.P. 201204, India

Neelam Dahiya
Chitkara University Institute of Engineering
    and Technology
Chitkara University
Punjab, India

Abhirup Datta
Department of Astronomy, Astrophysics &
    Space Engineering
IIT Indore

Digvijay Dubey
Department of Earth & Planetary
    Sciences
University of Allahabad
Allahabad, India

Raj Kumar Goel
North-Eastern Hill University Tura Campus
    (a Central University)
Meghalaya

Susana Layana Guerrero
Millennium Institute on Volcanic Risk
    Research - Ckelar Volcanoes
Antofagasta, Chile

Ayushi Gupta
Remote Sensing Laboratory
Institute of Environment and Sustainable
    Development
Banaras Hindu University
Varanasi, India

Dileep Kumar Gupta
Department of Electronics
Galgotias University
Greater Noida, Gautam Buddh Nagar,
Uttar Pradesh

Raj Gusain
Graphic Era Deemed to be University

C. Jeganathan
Department of Remote Sensing
Birla Institute of Technology
Mesra, Ranchi

Abhijeet Kumar
Indian Institute of Tropical Meteorology
Pune, Maharashtra
Ministry of Earth Sciences
Government of India

Rajesh Kumar
Department of Environmental Science
School of Earth Sciences
Central University of Rajasthan
Ajmer, India

Ramesh Kumar
Marwadi University Research Centre
Marwadi University
Rajkot, Gujarat, India

Sanjay Kumar
Department of Physics
Nehru Gram Bharti (Deemed to be University)
Kotwa-Jamunipur-Dubawal, Prayagraj

Shayam Sundar Kundu
North Eastern Space Applications Centre
India

Sushil Kumar
School of Information Technology,
Engineering, Mathematics and Physics
The University of the South Pacific
Suva, Fiji

Megha Maheshwari
Space Navigation Group
U R Rao Satellite Centre-ISRO
Bengaluru, India

Sarvesh Mangla
Department of Astronomy, Astrophysics &
Space Engineering
IIT Indore
Ludwig-Maximilians-Universität
München, Germany

Raziqa Masood
Department of Computer Science &
Engineering
Integral University

Ajeet Kumar Maurya
Department of Physics
Babasaheb Bhimrao Ambedkar University
Lucknow, Uttar Pradesh, India

Roopa Maurya
Department of Earth & Planetary
Sciences
University of Allahabad
Allahabad, India

Arti Mishra
Department of Physics
Nehru Gram Bharti (Deemed to be University)
Kotwa-Jamunipur-Dubawal, Prayagraj

Rahul Kumar Misra
Department of Economics
Khwaja Moinuddin Chishti Language
University
Lucknow, India

Mukulika Mondal
Department of Physics, Institute of Science
Banaras Hindu University
Varanasi, UP, India

Damanti Murmu
Institute of Environment and Sustainable
Development
Banaras Hindu University
Varanasi, India

Kalpana Patel
Department of Physics
SRM Institute of Science and Technology
Delhi-NCR Campus
Modinagar, Ghaziabad, U.P. 201204, India

Prity Singh Pippal
Department of Environmental Science
School of Earth Sciences
Central University of Rajasthan
Ajmer, India

Rishi Prakash
Graphic Era Deemed to be University

Praveen Kumar Rai
Department of Geography
Khwaja Moinuddin Chishti Language
  University
Lucknow, India

Mini Rajput
Department of Physics, Institute of Science
Banaras Hindu University
Varanasi, UP, India

Surjeet Singh Randhawa
H.P. Council for Science, Technology and
  Environment
Shimla, H.P., India

Sardar Singh Rao
Udaipur Solar Observatory/Physical Research
  Laboratory
Udaipur, Rajasthan, India

Reshma Raskar-Phule
Department of Civil Engineering,
Bharatiya Vidya Bhavan's Sardar Patel College
  of Engineering
Bhavan's Campus
Munshi Nagar, Andheri West, Mumbai

V. S. Rathore
Department of Remote Sensing
Birla Institute of Technology
Mesra, Ranchi, Jharkhand, India

Leesh Ray
Department of Botany
Banaras Hindu University
Varanasi, India

Suniti Saharan
Department of Physics, Doon University
Dehradun, Uttarakhand, India

Sakshi
Department of Computer Science and
  Engineering
University Institute of Engineering, Chandigarh
  University
Mohali, Punjab, India

Francisca Sánchez
Millennium Institute on Volcanic Risk
  Research - Ckelar Volcanoes
Antofagasta, Chile

Gopi Krishna Seemala
Indian Institute of Geomagnetism
Navi Mumbai, India

Apoorva Sharma
Department of Computer and Engineering
University Institute of Engineering
Chandigarh University
Punjab, India

Bhawana Sharma
Department of Wildlife Sciences
University of Kota
Rajasthan, India

Gopal Sharma
North Eastern Space Applications Centre
India

Payal Sharma
Department of Environmental Science
School of Earth Sciences
Central University of Rajasthan
Ajmer, India

Mahesh N. Shrivastava
Department of Geological Sciences,
  Universidad Católica del Norte
Antofagasta, Chile
and
Millennium Institute on Volcanic Risk
  Research - Ckelar Volcanoes
Antofagasta, Chile

Ashish Kumar Shukla
Indian Space Research Organization

Abhay Kumar Singh
Atmospheric Research Lab.
Department of Physics
Banaras Hindu University
Varanasi, Uttar Pradesh, India

Ashutosh Kumar Singh
Post Graduate Department of Physics
C. M. Science College Darbhanga
Bihar, India (A constituent unit of L. N. Mithila
    University Darbhanga-Bihar, India)

Atar Singh
Center for Cryosphere and Climate
    Change Studies
National Institute of Hydrology
Roorkee, India

Gurwinder Singh
University Institute of Computing
Chandigarh University
Punjab, India

Rajesh Singh
KSK Geomagnetic Research Laboratory IIG
Prayagraj (Allahabad), India

Sartajvir Singh
Department of Computer and Engineering
University Institute of Engineering
Chandigarh University
Punjab, India

Shruti Singh
Department of Environmental Sciences, SSBSR
Sharda University
Greater Noida, UP, India

M. Somorjit Singh
North Eastern Space Applications Centre
India

Vivek Singh
Indian Institute of Tropical Meteorology,
    New Delhi Branch
Ministry of Earth Sciences, Government of India
New Delhi, India

Irjesh Sonker
Department of Earth & Planetary Sciences
University of Allahabad
Allahabad, India

Vishakha Sood
Department of Civil Engineering
Indian Institute of Technology
Ropar, India

Abhay Srivastava
North Eastern Space Applications Centre
India

Anubhava Srivastava
Computer Science & Engineering
    Department
Sharda School of Engineering Technology
Sharda University
Greater Noida, India

Swarnim
Department of Earth & Planetary Sciences
University of Allahabad
Allahabad, India

Varsha Tanu
IIHMR University
Prabhu Dayal Marg, Near Sanganer Airport
Jaipur, India

Prasad S. Thenkabail
Senior Scientist (ST)
United States Geological Survey (USGS)
Flagstaff, Arizona, United States

Gaurav Tiwari
Center for Environmental Remote Sensing
Chiba University
Japan

Jayant Nath Tripathi
Department of Earth & Planetary Sciences
University of Allahabad
Allahabad, India

Shikha Verma
Computer Science & Engineering
    Department
Sharda School of Engineering Technology
Sharda University
Greater Noida, India

Anurag Vidyarthi
Graphic Era Deemed to be University

Preface

The Global Navigation Satellite System (GNSS) technology plays a critical role in the study and monitoring of the various aspects of our planet, such as geodetic surveys, climate studies and environmental changes, monitoring of Space weather activities, and disaster management, etc., with its primary purpose to provide accurate positioning, navigation, and timing information. This book discusses the use of GNSS signals in climate change and weather monitoring, disaster monitoring and response efforts, Space weather/ionospheric monitoring, and geodetic surveys by analyzing the characteristics of the signals received from multiple satellites and reflected signals from various objects. Additionally, one of the key aspects covered in this book is the exploration of challenges associated with utilizing GNSS technologies such as atmospheric interference, multipath effects, clock modeling, data processing techniques, and integration with other sensing systems.

These challenges provide an opportunity for the identification of innovative solutions and development strategies to enhance the accuracy, reliability, and efficiency of GNSS-based applications. It also covers the ongoing efforts with rapid advancements in technology to enhance the capabilities of GNSS through the integration of other complementary technologies such as remote sensing, unmanned aerial vehicles (UAVs), the development of enhanced positioning algorithms, the deployment of resilient GNSS networks, the use of multi-constellation and multi-frequency systems, improved data fusion techniques, and the integration of real-time monitoring that pave the way for future advancements in precise positioning, disaster and climate studies, and Space weather monitoring.

The content of this book is designed by a team of experts from academia, industry, and government agencies of various disciplines brought together to share their expertise and experiences in utilizing the GNSS technology for a wide range of applications, mainly for Earth and Space weather monitoring. This book is organized into five distinct sections, each focusing on a critical area of GNSS applications. The first section introduces the fundamentals of GNSS and its evolving role in Earth observation. It covers the basics of various GNSSs, with a particular focus on the Global Positioning System (GPS), and delves into the technical intricacies of minimizing GPS measurement errors and optimizing coordinate systems.

The second section highlights GNSS applications in disaster management. From monitoring cyclones and tsunamis to assessing the risk of landslides and forest fires, the chapters in this section demonstrate how GNSS, combined with remote sensing and other geospatial technologies, can enhance our ability to predict and respond to natural disasters. These applications are essential for building resilient communities in the face of increasing climate variability and extreme weather events.

In the third section, the focus shifts to GNSS applications in climate studies. This section covers a range of topics, including the monitoring of sea level variations, glacier dynamics, and atmospheric water vapor. These studies underscore the importance of GNSS in tracking the impacts of climate change on natural systems and in developing models that can predict future changes. The integration of GNSS data with other climatic observations offers new opportunities for improving the accuracy of climate models and supporting sustainable development initiatives.

The fourth section explores the role of GNSS in Space weather research. Space weather phenomena, such as solar flares and geomagnetic storms, can significantly impact GNSS signals, leading to potential disruptions in navigation and communication systems. This section provides insights into how GNSS can be used to study these phenomena and develop mitigation strategies to protect critical infrastructure.

The final section addresses the challenges facing GNSS technology and explores future prospects. As GNSS continues to evolve, it faces technical, operational, and societal challenges that must be overcome to unlock its full potential. This section discusses these challenges, from signal interference and data management issues to ethical considerations and economic constraints. It also looks ahead to developments in GNSS, such as advancements in multi-GNSS systems, integration with emerging technologies, and contributions to sustainable development goals.

As editors, our primary objective is to inspire and inform readers about the immense potential of GNSS and its applications in Earth and Space observations. We hope this book not only serves as a reference guide but also sparks curiosity and encourages further research and collaboration. By collectively exploring the applications, addressing the challenges, and envisioning the future prospects of GNSS, we may drive the innovations, expand scientific knowledge, and contribute to a more connected and informed world.

The readers of this book will delve into the critical role of GNSS in tracking and predicting Earth's changing climate patterns, monitoring natural disasters, and facilitating precise positioning and navigation for various applications and will also explore the contributions of GNSS to the understanding of Space weather, opening up new frontiers for scientific discovery and human exploration. Really, it will serve as a valuable resource for researchers, engineers, students, and professionals seeking to delve deeper into the multidisciplinary field of GNSS and its far-reaching impact.

We extend our gratitude to the contributors who have shared their expertise and insights, making this book possible. We also thank the readers for their interest in exploring the exciting world of GNSS applications. Together, let us embark on this journey to uncover the vast potential of GNSS and shape the future of Earth and Space observations.

In conclusion, this book is a comprehensive resource for anyone seeking a deeper understanding of the immense potential and challenges associated with GNSS technology. It offers a collection of expert perspectives, cutting-edge research, and visionary insights into how GNSS is transforming the way we observe, understand, and navigate both our planet and the vast expanses of space. For the scientist, researcher, student, or industry professional, this book will undoubtedly expand horizons and inspire readers to explore the limitless possibilities that GNSS presents in the realm of Earth and Space observations.

Section 1

Introduction to GNSS and Its Advances in Earth Observation

1 Introduction to Navigation Satellite Systems (GNSSs) in a Global Perspective

Sushil Kumar

1.1 INTRODUCTION

Global Navigation Satellite Systems (GNSSs) provide services relating to positioning, timing, navigation, mapping, public safety surveillance and geographic surveys, time standards, mapping, and weather. GNSSs boast the most significant advancements in navigation and surveying during the past century. Along with the internet and mobile communications, GNSSs have become another essential component of today's technology civilization. Current smart cellphones have integrated the position, navigation, and timing (PNT) functionality into a single CMOS chip (https://www. gpsworld.com/dual-band-gnss-market-moving-from-insignificant-to-billions-in-less-than-5-years/).

The Soviet Union launched Sputnik, the first artificial satellite, on October 4, 1957, to gather data on the space environment (Kuznetsov et al., 2015). This launch was likely the first step toward space-based communication. The US military developed the first satellite systems, including Transit, Timation, and NAVSTAR. The first NAVSTAR was launched in 1989, followed by the 24th satellite in 1994, and its complete operational capacity was announced in April 1995. Currently, over 8000 satellites are in orbit, forming the different GNSSs, which are also used for research and communication.

Currently, four GNSSs and two regional navigation satellite systems (RNSSs) are operational. The Global Positioning System (GPS) of the United States of America, the European Satellite Navigation System (Galileo), the Global Navigation Satellite System (GLONASS) of the Russian Federation, and the BeiDou Navigation Satellite System (also known as COMPASS) of China are the four GNSSs. GPS and GLONASS are the two most advanced systems, each with 24 satellites.

The GPS (NAVSTAR) satellites orbit approximately 20,200 km above the Earth's surface in three orbital planes, and GLONASS satellites orbit 19,100 km above the Earth. The Galileo and the BeiDou Navigation Satellite System are being developed and deployed, and their full operational capability will greatly enhance the quality of services and increase the number of potential users and applications. The Galileo system consists of 26 satellites positioned in three circular medium Earth orbit (MEO) planes at 23,222 km above the Earth. BeiDou is a regional and global navigation system with a constellation of 30 MEO satellites at an altitude of 21,150km and five geostationary orbits (GEO) satellites at 36000 km above the Earth.

Japan's Quasi-Zenith Satellite System (QZSS) and the Indian Regional Navigation Satellite System (IRNSS, also known as NavIC) are available at the regional level. The QZSS uses three elliptical inclined geostationary orbits (GEO) spacecraft that are 120 degrees apart and orbit the same ground path. QZSS is built such that there is always at least one satellite in close proximity to zenith over Japan. It works in combination with GPS and Galileo to improve its accuracy and services. The IRNSS comprises four satellites in a GEO synchronous orbit positioned at an inclination of 29 degrees with longitude crossings at 55 and 111 degrees east and three satellites in a

DOI: 10.1201/9781032712444-2

geostationary orbit at 34, 83, and 132 degrees east longitude. IRNSS uses L5 (1191.795MHz) and S (2491.005MHz) signals and has been mainly designed to provide accurate position information service to aid in ship navigation in the Indian Ocean.

A user must be able to see (i.e., have a direct line of sight to) at least four satellites simultaneously for GNSS systems to determine their position. For more detailed information about different GNSSs and RNSSs, the reader is referred to the Interoperable Global Navigation Satellite Systems Space Service Volume, Second Edition (2021) (https://www.unoosa.org/res/oosadoc/data/documents/2021/stspace/stspace75rev_1_0_html/st_space_75rev01E.pdf) and individual chapters of a book edited by Morton et al. (2021).

1.2　TECHNICAL COMPONENTS OF A GNSS SYSTEM

The technical components of a GNSS system are: i) The Space Segment, ii) The Control Segment, and iii) The User Segment, which for GPS systems have been briefly described by Kumar (2023). The GPS is owned and operated by the US Department of Defense (DoD), which was developed mainly for US military applications; however, in 1994, it was made available for civilian communities with a wide range of applications.

1.3　GNSS ANTENNA AND RECEIVER

GNSS signals are right-hand circularly polarized (RHCP) radio waves. As GNSS signals travel from satellites to ground receivers, they experience attenuation, atmospheric (troposphere and ionosphere) delays, multipath, and interference. GNSS antennas have lower effective antenna gain, which is compensated by adding a low-noise amplifier (LNA) to the antenna. For practical GNSS applications, narrowband Helical, Spiral, and microstrip Patch Antennas are used. For further details about GNSS antennas, the reader is referred to the book by Chen et al. (2012), and in particular, the chapter on Fundamental Considerations for GNSS Antennas.

Figure 1.1 depicts the basic layout of a GNSS receiving system. Numerous GNSS receivers are employed for diverse applications, including but not limited to navigation, surveying, and ionospheric research. With reference to locally generated reference signals, GNSS receivers estimate the code phase, carrier frequency, and carrier phase. The carrier phase measurements provide positioning services. For improved positioning services, cycle ambiguities that affect the carrier phase may need to be fixed. For highly accurate millimeter-level carrier-phase-based relative positioning, real-time kinematic (RTK) receivers are employed. Nowadays, multi-constellation GNSS receivers are used for practical and scientific applications. The elimination of ionospheric errors, radio frequency interference robustness, improved multipath rejection, increased spoofing detection, and other benefits are among the many benefits of multi-frequency GNSS receivers.

1.4　RECEIVER DIFFERENTIAL CODE BIASING

The high accuracy of the GNSS can be significantly impacted by ionospheric radio wave propagation delays and signal fading (scintillations). The ionosphere is characterized by its total electron content (TEC), which is a measure of the total number of electrons per meter squared (el/m^2) along the line of sight from a transmitter on a satellite to the ground receiver. A TEC unit (TECU) is defined as 1 TECU = 1 × 10^{16} electrons m^{-2}. TEC has been studied for scientific as well as for satellite applications like positioning accuracy. The ionosphere is highly dynamic and shows diurnal, seasonal, latitudinal, day-to-day, and solar and geomagnetic activity variability. The reliable and continuous TEC data are important to explore the nature of the ionospheric variability for the improved GNSS services as our reliance on trans-ionospheric signals has increased. In addition to propagational errors (refraction and multipath), other sources of errors that affect PNT services include receiver noise, ephemeris errors, and satellite clock errors (Kumar et al., 2021).

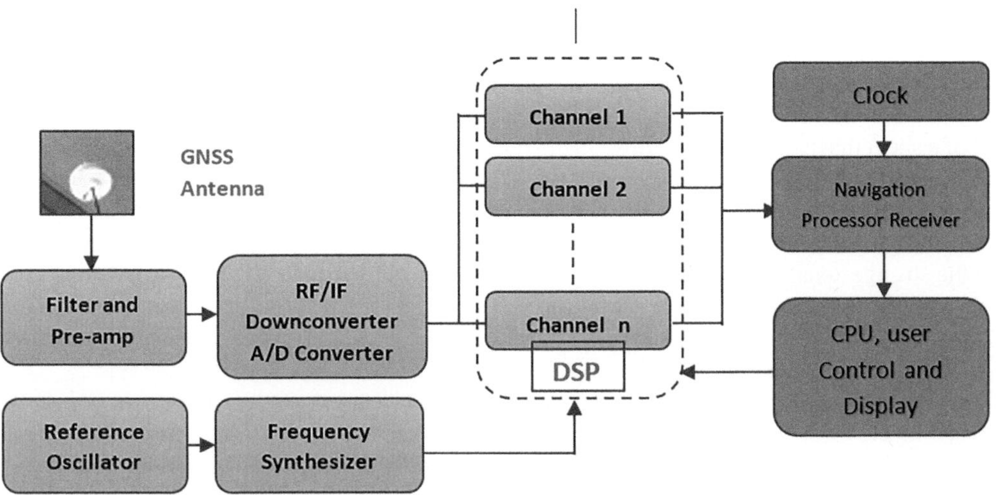

FIGURE 1.1 Block diagram of a typical GNSS receiver [concept adapted from Chen et al. (2012), Antennas for Global Navigation Satellite Systems, Chapter 2, Figure 2.9, https://onlinelibrary.wiley.com/doi/book/10.1002/9781119969518.

The satellite and receiver differential code biases (DCBs) can also induce significant errors in the GNSS-derived TEC of the ionosphere. To obtain absolute TEC, these biases need to be removed by using reliable methods (Themens et al., 2013). To eliminate these biases for individual receiver stations, three main methods have been developed: i) linear regression fit to compare GPS TEC with another set of data (Themens et al., 2013); ii) the minimization of standard deviation method (Ma & Maruyama, 2003); and iii) least-square fitting to a polynomial method (Coco, 1991; Coco et al., 1991).

The first method is quite straightforward, in which a linear regression fit is used to compare the GPS TEC to TEC obtained from any trustworthy model. In this method, the extrapolated y-intercept reveals the receiver biases. The second method was developed assuming that, once the receiver and satellite biases are appropriately eliminated, the vertical total electron content (vTEC) from every satellite is the same (Ma & Maruyama, 2003). In the least-squares fitting to a polynomial method, the GPS vTEC is fitted to a second order bivariate polynomial at the ionospheric pierce point (IPP) of the respective station. Prasad et al. (2016) did a statistical estimation of receiver DCBs of GSV4004B receiver at a low latitude station, Suva (lat. 18.15°S and long. 178.45°E), Fiji, using GPS TEC and CODE TEC and IRI-2012 model TEC data. To substantiate the results, they also used the minimization of the standard deviation (SD) method. They estimated DCB of $\sim 63 \pm 2$ TECU using GPS TEC 2010 with IRI-2012 TEC, while the minimization of standard deviation technique with GPS TEC 2010 data with CODE TEC gave a DCB of $\sim 62 \pm 2$ TECU.

1.5 IONOSPHERIC IRREGULARITIES

Ionospheric electron density irregularities can cause serious GNSS PNT errors for a few mintes to a few hours and even a loss of signal lock on the receivers due to strong irregulatities mainly associated with strong geomagnetic storms. The plasma instabilities are responsible for irregularities with scale sizes varying from centimeters to kilometers. According to their scale sizes, ionospheric irregularities have also been classified into several categories: planetary (> 1000 km), medium scale (10–100 km), intermediate (0.1–10 km), transitional (10–100 m), and short wavelength (< 10 m) (Livingston et al., 1981). The ionospheric irregularities are divided into E- and F-region irregularities.

The E-region irregularities are known as Sporadic-E layer or Es. At E-region altitudes of 100–120 km, where ion motion is mostly governed by collisions with neutrals, the Es is distinguished by a thin reflective layer in the ionosphere that appears and disappears normally for shorter durations (Bhattacharya, 1991). Es often occurs in the form of a patch but occasionally it takes the form of a sheet that can entirely enclose the F-region above it. Es is caused by the wind shear theory mechanism (Axford, 1963; Haldoupis, 2011, 2012; Kunduri et al., 2023 and references therein; Whitehead, 1961) in which East-West winds in the E-region induce a vertical movement that compresses the ions into thin dense layers. Although the Es can occur at every latitude, the mid-latitudes have the strongest and most common occurrence. Equatorial Sporadic-E (Esq) is the name given to Es developed in the equatorial zone. It has been reported that Esq forms as a result of plasma instability caused by the high electron drift velocity associated with Equatorial Electrojet (EEJ) (Rastogi, 1972).

F-region irregularities, also termed Plasma bubbles or plumes, were initially reported in the 1950s (Booker, 1958) and associated with equatorial spread-F (ESF) (Briggs, 1964). F-region irregularities have been studied using a variety of methods, including radio-star scintillations, VHF-backscatter radar, Langmuir probes onboard satellites, topside and bottomside ionosondes, night airglow measurements, GPS scintillations, etc.

It is generally accepted that at equatorial and low latitudes, ESF occurs at the bottom of the F-region due to the Gravitational Rayleigh-Taylor (GRT) instability and Pre-Reversal Enhancement (PRE) (Dasgupta et al., 1985; Eccles, 1998; Fejer et al., 1979; Kelly, 2009; Santos et al., 2021; Sultan et al., 1996). The requirement for GRT fulfills after sunset at the geomagnetic equator when the bottom-side of the F-layer recombines with the dense neutral atmosphere (Akala et al., 2016; Beer, 1974). Concurrently, the pre-reversal enhancement raises the F-region to a higher altitude, generating a sharp vertical gradient in the electron density.

The equatorial plasma bubbles (EPBs) thus generated rise to higher altitudes and latitudes due to non-linear evolution of the **E×B** drift (e.g., Fejer et al., 1999; Kumar Su & Kumar Sa, 2020). EPBs, on reaching the maximum height in the F-region, diffuse at locations off the magnetic equator at approximately ±20° to the geomagnetic equator, which is known as the Equatorial Ionization Anomaly (EIA) (Aarons, 1982; Basu et al., 1978; Dabas et al., 2000; Fejer et al., 1979; Kumar & Gwal, 2000; Gladek et al., 2019).

The high and low latitudes regions are of particular interest in GNSS applications because of the signals' sensitivity to scintillation events associated with ionospheric irregularities (Aarons, 1982; Aarons et al., 1982; Basu & Basu, 1985; Fejer & Kelley, 1980; Kumar Su & Kumar Sa, 2020). At nighttime, associated with EPBs (ESF or F-region irregularities), the ionospheric scintillations occur in post-sunset hours, evolve to midnight hours and can extend to post-midnight hours at stations located off the magnetic equator with rare extension into occurrence during daytime (Wang et al., 2019; Jiao & Morton, 2015). The scintillation activity begins around 19:00 LT and reaches its peak occurrence between 20:00–21:00 LT (Kumar & Gwal, 2000; Sunda et al., 2015). Daytime scintillations are associated with E-region irregularities, often referred to as Sporadic-E (Es) (or Equatorial Sporadic-E (Esq) (D'ujanga & Taabu, 2017; Prasad & Kumar, 2017). Esq occurs mainly during the day, with a higher chance after sunrise with some seeding by gravity waves (Gupta & Kersley, 1976; Prasad & Kumar, 2017; Zeng & Sokolovskiy, 2010). At the low and equatorial latitudes, daytime scintillations randomly occur due to the sporadic E-layer, and the nighttime scintillations predominantly occur due to the ESF.

The intensity of the amplitude scintillation is measured by scintillation indices; S_1, S_2, S_3, and S_4 given as (Bartusek & Felgate, 1966; Fremouw & Bates, 1971):

$$S_1 = \frac{\left\langle \left| SI - \left\langle SI \right\rangle \right| \right\rangle}{\left\langle SI \right\rangle} \tag{1.1}$$

$$S_2 = \frac{\sqrt{\left\langle \left(SI - \langle SI \rangle \right)^2 \right\rangle}}{\langle SI \rangle} \tag{1.2}$$

$$S_3 = \frac{< SI^2 - \langle SI >^2 \rangle}{< SI >^2} \tag{1.3}$$

$$S_4 = \sqrt{\frac{\langle SI^2 \rangle - < SI >^2}{< SI >^2}} \tag{1.4}$$

where SI and SI^2 are the signal intensity and power, respectively, and $< >$ represents the time averages.

The S_1 gives the normalized deviation of amplitude, and S_2 is the normalized root mean square deviation of signal amplitude. The S_3 is the normalized deviation of signal power, and the S_4 gives the normalized root mean square deviation of signal power, which is most widely used for scintillation studies.

1.6 GNSS PERFORMANCE AND LIMITATIONS

The position accuracy, carrier-to-noise C/No (dBHz), percentage of signal availability (%), and maximum outage duration (MOD) are the four important parameters of any GNSS system. The position accuracy and carrier-to-noise C/No (dBHz) ratio are important parameters defining any GNSS's performance. The Civilian Position accuracy for four GNSSs (GPS, GLONASS, Galileo, BeiDou) and a regional GNSS system (QZSS) is about 7.1 meters for GPS (NAVSTAR), 7.5 meters for GLONASS, 4.0 meters for Galileo, 10.0 meters for BeiDou, and 1.6 for QZSS. The position measurements must be adjusted for the biases (or errors) associated with each satellite measurement, which can include orbital errors, receiver noise and clock offset, RF interference, multipath (reflections, diffraction, and scattering of the incident radio waves from the structure of the surrounding environments), and ionospheric and tropospheric delay errors, etc. One of the main causes of error in GNSSs and RNSSs for the signal for a static receiver is ionospheric delay.

The C/No is the ratio of carrier power to the noise power mixed with the signal in a 1Hz bandwidth. The C/No gives the GNSS receiver sensitivity, which varies with satellite elevation. In addition, it provides the user with information regarding signal quality, carrier lock, and code tracking loops. The signals from various satellites are weighted in a typical GNSS receiver based on their C/No ratio, so that the signal can have the best C/No, which informs the user about the sufficient signal strength that can be measured with a given instrument to form a usable measurement. The C/No in dBHz for four GNSSs (GPS, GLONASS, Galileo, BeiDou) and two regional GNSS systems (QZSS, IRNSS) is about 37–45 for GPS (NAVSTAR), 20–45 meters for GLONASS, 30–55 for Galileo, 31–48 on B1 signal for BeiDou, 35–53 for QZSS, and 44.79 on L5 for IRNSS. While position accuracy is relatively less variable, the C/NO exhibits high variability.

The percentage of signal availability (%) and MOD) tells the user about the usability of a particular GNSS system. MOD, which is expressed in minutes, is the maximum amount of time a user can anticipate going without a signal. For GNSSs (GPS, GLONASS, Galileo, BeiDou, and a regional QZSS), the percentage of signal availability for at least one signal (L1/E1/B1) is approximately 100% and 99.6% for QZSS, respectively; for four or more signals, it varies from 79.4% to 99.8% for GALILEO. The MOD for a single signal is 197 minutes for QZSS, 11 minutes for GALILEO, and 0 minutes for GPS, GLONASS, and BeiDou. The reader is directed to The Interoperable Global Navigation Satellite Systems Space Service Volume, 2018 for additional information regarding the percentage of time for five GNSS signals (GPS, GLONASS, Galileo BeiDou,

QZSS, and MOD) and their global performance estimates (page 20, Table 5.2) and full volume on https://www.unoosa.org/res/oosadoc/data/documents/2021/stspace/stspace75rev_1_0_html/st_space_75rev01E.pdf).

Ionospheric scintillations can degrade the performance of GNSSs under severe Space weather conditions, which happen more frequently during periods of high solar activity (Stankov et al., 2006; Rama Rao et al., 2009). Ionospheric scintillations, both on the amplitude and phase of radio signals, are produced while passing through small-scale ionospheric irregularities (Yeh & Liu, 1982). Ionospheric scintillation, which mostly happens in equatorial, auroral, and polar regions (Aarons, 1982; Jiao & Morton, 2015; Bumrungkit et al., 2022), refers to the random amplitude and phase fluctuations happening in radio signals traveling through ionospheric electron density irregularities. The occurrence and intensity of scintillations depend on the local time, seasons, radio signal frequency, latitude and longitude, the elevation angle of the line-of-sight to the satellite, zenith angle, solar cycle, and geomagnetic activities (e.g., Aarons, 1982; Briggs & Parkin, 1963; Jiao & Morton, 2015; Kumar & Gwal, 2000; Kumar Su & Kumar Sa, 2020; Salles et al., 2021; Spogli et al., 2013). Goswami et al. (2018) studied the relative performance of different navigational satellite constellations (GPS, GLONASS, GALILEO) under unfavorable ionospheric conditions during the equinoctial months of 2014 and March 2016 at Calcutta (22.58°N 88.38°E geographic; 32°N magnetic dip) and for September 2016 at Siliguri (26.72°N, 88.39°E; 39.49°N magnetic dip), located near the northern crest of equatorial ionization anomaly in the Indian longitude sector. They reported C/No ratio fluctuations of 36.6 to 60% more over limited ionospheric conditions at both stations. Jiao and Morton (2015) compared the occurrence levels of high-latitude and equatorial ionospheric scintillation on GPS signals during the maximum of solar cycle 24. They found that scintillations at high latitudes are moderate and generally dominated by phase fluctuations, whereas scintillation activity in the equatorial region is generally more severe and long-lasting.

An effort is being made by the United Nations International Committee on GNSS (ICG) to guarantee that GNSS signals in the Space Service Volume (SSV) are accessible and compatible with all international global constellations and regional augmentations (https://www.unoosa.org/res/oosadoc/data/documents/2021/stspace/stspace75rev_1_0_html/st_space_75rev01E.pdf). The ICG Working Group B (WG-B) on "Enhancement of GNSS Performance, New Services and Capabilities" is responsible for carrying out this program. The book "Position, Navigation, and Timing Technologies in the 21st Century" Volume 1, IEEE Press, 445 Hoes Lane, Piscataway, NJ 08854, edited by Morton et al. (2021), is recommended reading for details and on applications of GNSSs and RNSSs.

The accuracy of GNSSs can be significantly impacted by the atmospheric delay, which is made up of tropospheric and ionospheric propagation delays. The ionospheric delay changes approximately with the inverse square of the signal frequency, but tropospheric delay is independent of the signal frequency. The ionospheric delay also depends upon the location of the user in the equatorial, mid-latitude, and high latitude regions.

Ionospheric refraction is the main factor limiting GNSS accuracy. Due to the satellite trajectories and the surroundings of the receiving GNSS antenna, multipath is another important source of error. Other sources of errors that affect PNT services besides propagational errors include receiver noise, ephemeris error, satellite clock error, and others (Kumar et al., 2021).

The Ground-Based Augmentation System (GBAS) and the Satellite-Based Augmentation System (SBAS) have been developed to enhance GNSS service to meet air navigation requirements related to approach, takeoff, ascent, and en route. When designing a navigation system, the four parameters—accuracy, integrity, continuity, and availability—are considered. Navigation systems intended for use in aviation are evaluated based on these parameters. GBAS systems have been installed at airports in several countries, such as the United States, Europe, Australia, India, Brazil, and Russia. For further details about GBAS and SBAS, the reader is referred to chapters 12 and 13 of the book on "Position, Navigation, and Timing Technologies in the 21st Century."

1.7 GNSS APPLICATIONS

Four GNSSs (GPS, GLONASS, Galileo, Bidou) and two regional GNSSs (QZSS, IRNSS) are used globally and regionally for practical and scientific applications. However, GPS is used most widely for a variety of applications and can be affected by the ionosphere (Klobuchar, 1991). GPS is used for Intelligent Vehicle Highway Systems (IVHS); land, sea, air, and space navigation; search and rescue operations; and recreational surveying. Standard Positioning Service (SPS) and Precise Positioning Service (PPS) are the two navigation and positioning tiers that GPS provides. The SPS gives 100 m horizontal accuracy, 156 m vertical accuracy, and 167 ns time accuracy. According to French (1996), PPS improves accuracy to 17.8 m in the horizontal direction, 27.7 m in the vertical direction, and 100 ns in the time domain. Under both normal and Space weather situations, the dual-frequency GPS receivers modified with a Kalman Filter have been utilized to record ionospheric total electron content (TEC) and L band scintillations (Prasad & Kumar, 2017; Kumar Su & Kumar Sa, 2020). A comprehensive review of GNSS contributions to studies of extreme weather and climate events has been provided by Bonafoni et al. (2019). Water vapor is the major constituent of the atmosphere, which responds to a rise in tropospheric temperature. Based on multi-instrument (ground-based microwave radiometer in Bern, Fourier transform infrared, FTIR, spectrometer at Jungfraujoch, Swiss ground-based GNSS stations) and reanalysis of ERA5 and MERRA-2 data, Bernet et al. (2020) reported that IWV is increasing by 2% per decade to 5% per decade, which is consistent with global warming on a regional scale and stresses the importance of water vapor feedback in climate change. GNSSs are also used in weather forecasting, natural hazards (terrestrial and Space weather) and disaster management, and climate and environment studies (e.g., Crowell et al., 2012; Guergana et al., 2016; Segall & Davis, 1997; Tregoning & Dam, 2005; Kumar, 2023). For detailed GNSS Applications for Disaster Management, Space Weather, and Climate Studies, the reader is referred to Chapters 2, 3, and 4 of this book, respectively.

REFERENCES

Aarons, J. 1982. Global Morphology of ionospheric scintillations. *Proceedings of IEEE*, 70, 360–373.

Aarons, J., J. Klobuchar, H. Whitney, J. Austen, A. Johnson, and C. Rino. (1982), Gigahertz scintillations associated with equatorial patches, *Radio Science*, 18(03), 421–434, doi:10.1029/RS018i003p00421

Akala, A. O., R. Idolor, F. M. D'ujanga, and P. H. Doherty. 2016. GPS Amplitude Scintillations over Kampala, Uganda, During 2010–2011, *Acta Geophysica*, 64(5), 1903–1915, doi:10.1515/acgeo-2016-0052

Axford, W. I. 1963. The formation and vertical movement of dense ionized layers in the ionosphere due to neutral wind shears. *Journal of Geophysical Research (1896-1977)*, 68(3), 769–779. doi:10.1029/JZ068i003p00769

Bartusek, K., and D. Felgate. 1966. Some aspects of irregular diffraction studied by means of ultrasonic waves, *Australian Journal of Physics*, 19(4), 545–548.

Basu, S., S. Basu, J. Aarons, J. McClure, and M. Cousins. 1978. On the coexistence of kilometer-and meter-scale irregularities in the nighttime equatorial F region, *Journal of Geophysical Research: Space Physics*, 83(A9), 4219–4226, doi:10.1029/JA083iA09p04219

Basu, S., and S. Basu. 1985. Equatorial scintillations: Advances since ISEA-6, *Journal of Atmospheric and Terrestrial Physics*, 47(8–10), 753–768, doi:10.1016/0021-9169(85)90052-2

Beer, T. 1974. On the Dynamics of Equatorial Spread F. *Australian Journal of Physics*, 27, 391–400.

Bernet, L., E. Brockmann, T. von Clarmann, N. Kämpfer, E. Mahieu, C. Mätzler, G. Stober, and K. Hocke. 2020. Trends of atmospheric water vapour in Switzerland from ground-based radiometry, FTIR and GNSS data, *Atmospheric Chemistry and Physics*, 20, 11223–11244.

Bhattacharya, A. 1991. Phase and amplitude scintillations due to electrojet irregularities. *Journal of Atmospheric and Solar - Terrestrial Physics*, 53 (5), 379–387.

Bonafonia, S., R. Biondib, H. Brenotc, R. Anthes. 2019. Radio occultation and ground-based GNSS products for observing, understanding and predicting extreme events: A review, *Atmospheric Research*, 230, 104624. doi:10.1016/j.atmosres.2019.104624

Booker, H. G.1958. The use of radio stars to study irregular refraction of radio waves in the ionosphere. *Proceedings of the IRE*, 46, 298–314.

Briggs, B. H. 1964. Observations of radio star scintillations and spread-F echoes over a solar cycle. *Journal of Atmospheric and Solar - Terrestrial Physics*, 26, 1–23.

Briggs, B. H., and J. A. Parkin. 1963. On the variation of radio star and satellite scintillations with zenith angle. *Journal of Atmospheric and Solar-Terrestrial Physics*, 25(6), 339–366, doi:10.1016/0021-9169(63)90150-8

Bumrungkit, A., P. Supnithi, S. Saito, L. M. M. Myint. 2022. A study of equatorial plasma bubble structure using VHF radar and GNSS scintillations over the low-latitude regions, *GPS Solutions*, 26, 148, doi:10.1007/s10291-022-01321-4

Chen, X., C. G. Parini, B. Collins, Y. Yao, M. Ur Rehman. 2012. *Antennas for Global Navigation Satellite Systems*, John Wiley & Sons Ltd.

Coco, D. 1991. GPS- Satellite of opportunity for ionospheric monitoring. *Innovation - GPS World*, 2, No. 9, pp. 47–50.

Coco, D.S., C. Coker, S.R. Dahlke, J. R. Clynch. 1991. Variability of GPS satellite differential group delay biases. *IEEE Transactions on Aerospace Electronic Systems* 27, 931–938.

Crowell, B.W., Y. Bock, D. Melgar. 2012. Real-time inversion of GPS data for finite fault modeling and rapid hazard assessment. *Geophysical Research Letters*, 39. doi:10.1029/2012GL051318

Dabas, R., B. M. Reddy, D. Lakshmi, and K. Oyama. 2000. Study of anomalous electron temperature variations in the topside ionosphere using HINOTORI satellite data, *Journal of Atmospheric and Solar - Terrestrial Physics*, 62(15), 1351–1359, doi:10.1016/s1364-6826(00)00126-7

DasGupta, A., A. Maitra, and S. Das. 1985. Post-midnight equatorial scintillation activity in relation to geomagnetic disturbances, *Journal of Atmospheric and Terrestrial Physics*, 47(8–10), 911–916, doi:10.1016/0021-9169(85)90067-4

D'ujanga, F. M., and S. D. Taabu. 2017. Study on the occurrence characteristics of VHF and L-band ionospheric scintillations over East Africa, *Indian Journal of Radio and Space Physics*, 43(4–5), 263–273.

Eccles, J. V. 1998. Modeling investigation of the evening prereversal enhancement of the zonal electric field in the equatorial ionosphere, *Journal of Geophysical Research: Space Physics*, *103*(A11), 26709–26719, doi:10.1029/98JA02656

Fejer, B. G., C. Gonzales, D. Farley, M. Kelley, and R. Woodman. 1979. Equatorial electric fields during magnetically disturbed conditions 1. The effect of the interplanetary magnetic field, *Journal of Geophysical Research: Space Physics*, *84*(A10), 5797–5802, doi:10.1029/JA084iA10p05797

Fejer, B. G., and M. Kelley. 1980. Ionospheric irregularities, *Reviews of Geophysics*, 18(2), 401–454, doi:10.1029/RG018i002p00401

Fejer, B. G., L. Scherliess, E. R. de Paula, 1999. Effects of the vertical plasma drift velocity on the generation and evolution of equatorial spread F. *Journal of Geophysical Research*, 104 (A9), 19859–19869, doi: 10.1029/1999JA900271

Fremouw, E. J., and H. F. Bates. 1971. Worldwide behavior of average VHF and UHF scintillations. *Radio Science*, 6 (10), 863–869.

French, G. T. 1996. *Understanding The GPS*. Bethesta, Maryland: GeoResearch Inc.

Gladek, Y. C., J. Sousasantos, L. Salles, V. C. Lima Filho, B. Vani, and A. De O. Moraes. 2019. GPS Amplitude Fading Due to Ionospheric Scintillation Near the Equatorial Ionospheric Anomaly, in AIAA Scitech 2019 Forum, edited, p. 0057, *Journal of Engineering, Science and Management Education*, doi:10.2514/6.2019-0057

Guergana, G.,, J. Jones, J. Douša, G. Dick, S. de Haan, E. Pottiaux, O. Bock, R. Pacione, G. Elgered, H. Vedel, and M. Bende. 2016. Review of the state of the art and future prospects of the ground-based GNSS meteorology in Europe, *Atmospheric Measurement Techniques*, 9, 5385–5406.

Goswami, S., A. Paul, and S. Haldar. 2018. Study of relative performance of different navigational satellite constellations under adverse ionospheric conditions. *Space Weather*, *16*, 667–675. doi:10.1029/2017SW001762

Gupta, A. D., and L. Kersley. 1976. Summer daytime scintillation and sporadic-E, *Journal of Atmospheric and Solar - Terrestrial Physics*, 38(6), 615–618, doi:10.1016/0021-9169(76)90157-4

Haldoupis, C. 2011. *A tutorial review on sporadic E layers* (pp. 381–394). Springer Netherlands. doi:10.1007/978-94-007-0326-1_29

Haldoupis, C. 2012. Midlatitude sporadic e. a typical paradigm of atmosphere-ionosphere coupling. *Space Science Reviews*, 168(1), 441–461 doi:10.1007/s11214-011-9786-8

Jiao, Y., and Y. T. Morton. 2015. Comparison of the effect of high-latitude and equatorial ionospheric scintillation on GPS signals during the maximum of solar cycle 24, *Radio Science*, *50*(9), 886–903, doi:10.1002/2015RS005719

Kelly, M. C. 2009. The earth's ionosphere: plasma physics and electrodynamics, 2nd edn. Academic Press, San Diego.

Klobuchar, J. A. 1991. Ionospheric effects on GPS. In *Innovation: GPS world*, Vol. 2, pp. 48–51. Air Force Geophysics Lab.

Kumar, S., & Gwal, A. K. 2000. VHF ionospheric scintillations near the equatorial anomaly crest: solar and magnetic activity effects. *Journal of Atmospheric and Solar - Terrestrial Physics*, 62, 157–167.

Kumar, S., and Sarvesh Kumar. 2020. Equatorial ionospheric TEC and scintillations under the space weather events of 4–9 September 2017: M-class solar flares and a G4 geomagnetic storm, *Journal of Atmospheric and Solar - Terrestrial Physics*, 209, 105421, doi:10.1016/j.jastp.2020.105421

Kumar, S., A. K. Singh, R.P. Singh, Chapter 6 - Probing the tropospheric water vapor using GPS, *GPS and GNSS Technology in Geosciences* 2021, pp. 119–134.

Kumar, S. 2023. Global Navigation Satellite Systems and their Applications in Remote Sensing of Atmosphere: Atmospheric Remote Sensing-Principles and Applications.ed. A. K. Singh and S. Tiwari, 39–62. Elsevier (ISBN 978-0-323-99262-6) doi:10.1016/B978-0-323-99262-6.00006-7

Kunduri, B. S. R., Erickson, P. J., Baker, J. B. H., Ruohoniemi, J. M., Galkin, I. A., & Sterne, K. T. 2023. Dynamics of mid-latitude sporadic-E and its impact on HF propagation in the North American sector. *Journal of Geophysical Research: Space Physics*, 128, e2023JA031455. doi:10.1029/2023JA031455

Kuznetsov, V., V. Sinelnikov, and S. Alpert. 2015. Yakov Alpert: Sputnik-1 and the first satellite ionospheric experiment, *Advances in Space Research*, 55(12), 2833–2839, doi:10.1016/j.asr.2015.02.033

Ma, G., T. Maruyama. 2003. Derivation of TEC and estimation of instrumental biases from GEONET in Japan. *Annales de Geophysique* 21, 2083–2093.

Morton et al. 2021. *Position, Navigation, and Timing Technologies in the 21st Century"* Volume 1, IEEE Press 445 Hoes Lane Piscataway, NJ 08854, edited by Morton et al. (2021), *IEEE Press*, 445 Hoes Lane Piscataway, NJ 08854.

Livingston, R. C., C. L. Rino, J. P. McClure, W. B. Hanson. 1981. Spectral characteristics of medium-scale equatorial F region irregularities. *Journal of Geophysical Research*, 86, 2421–2432.

Prasad, R., S. Kumar, P.T. Jayachandran. 2016. Receiver DCB estimation and GPS vTEC study at a low latitude station in the South Pacific, *Journal of Atmospheric and Solar - Terrestrial Physics*, 149, 120–130.

Prasad, R., and S. Kumar. 2017. Day and nighttime L-Band amplitude scintillations during low solar activity at a low latitude station in the South Pacific region, *Journal of Atmospheric and Solar - Terrestrial Physics*, 165, 54–66, doi:10.1016/j.jastp.2017.11.005

Rama Rao, P.V.S., S. G. Krishna, J. V. Prasad, S. N. V. S. Prasad, D. S. V. V. D. Prasad, and K. Niranjan. 2009. Geomagnetic storm effects on GPS based navigation, Annales Geophysicae, 27, 2101–2110, www.ann-geophys.net/27/2101/2009/

Rastogi, R. G. 1972. Equatorial Sporadic E and crossfield instability. *The proceedings of the Indian Academy of Sciences*, pp. 181–194.

Salles L.A., B.C. Vani, A. Moraes, E. Costa E. R. De Paula 2021. Investigating ionospheric scintillation effects on multifrequency GPS signals. *Surveys in Geophysics*, 42(4):999–1025. doi:10.1007/s10712-021-09643-7

Santos, A. M., Batista, I. S., Brum, C. G. M., Sobral, J.H.A., Abdu, M. A., & Souza, J. R. (2021). F region electric field effects on the intermediate layer dynamics during the evening prereversal enhancement at equatorial region over Brazil. *Journal of Geophysical Research: Space Physics*,e2020JA028429. doi:10.1029/2020JA028429

Segall, P. and J. L. Davis. 1997. GPS applications for geodynamics and earthquake studies. *Annual Review of Earth and Planetary Sciences*, 25: 301–336.

Spogli, L., L. Alfonsi, P.J. Cilliers et al., 2013. GPS scintillations and total electron content climatology in the southern low, middle and high latitude regions, *Annals of Geophysics*, 56, 2, 2013, R0220; doi:10.4401/ag-6240

Stankov, S. M., N. Jakowski, K. Tsybulya, and V. Wilken. 2006. Monitoring the generation and propagation of ionospheric disturbances and effects on Global Navigation Satellite System positioning, *Radio Science*, 41, RS6S09, doi:10.1029/2005RS003327

Sultan P. J. 1996. Linear theory and modeling of the Rayleigh-Taylor instability leading to the occurrence of equatorial spread F *Journal of Geophysical Research: Space Physics*, 101(A12):26875–26891. doi:10.1029/96JA00682

Sunda, S., R. Sridharan, B. Vyas, P. Khekale, K. Parikh, A. Ganeshan, C. Sudhir, S. Satish, and M. S. Bagiya. 2015. Satellite-based augmentation systems: A novel and cost-effective tool for ionospheric and space weather studies, *Space Weather*, 13(1), 6–15, doi:10.1002/2014SW001103

Themens, D.R., Jayachandran, P.T., Langley, R.B., MacDougall, J.W., Nicolls, M.J., 2013. Determining receiver biases in GPS-derived total electron content in the auroral oval and polar cap regions using iono-sonde measurements. *GPS Solutions* 17 (3), 357–369. doi:10.1007/s10291-012-0284-6

Tregoning, P. and T. van Dam. 2005. Atmospheric pressure loading corrections applied to GPS data at the observation level. *Geophysical Research Letters* 32. 153.

The Interoperable Global Navigation Satellite Systems Space Service Volume, (2018), *United Nations Publications*, Sales No. E.19.IV.1, ISBN 978-92-1-130355-1eISBN 978-92-1-047440-5.

Wang, N., L. Guo, Z. Zhao, Z. Ding, T. Xu, and S. Sun (2019), A comparative study of ionospheric spread-F and scintillation at low-and mid-latitudes in China during the 24th solar cycle, *Advances in Space Research*, *63*(2), 986–998, doi:10.1016/j.asr.2018.10.010

Whitehead, J. D. 1961. The formation of sporadic-E layer in the temperate zone. *Journal of Atmospheric and Solar - Terrestrial Physics*, 20, 49.

Yeh K.C., C. H. Liu. 1982. Radio wave scintillations in the ionosphere. *Proceedings of IEEE* 70(4):324–360. doi:10.1109/PROC.1982.12313

Zeng, Z., and S. Sokolovskiy (2010), Effect of sporadic E clouds on GPS radio occultation signals, *Geophysical Research Letters*, 37(18), doi:10.1029/2010GL044561

2 Quantitative Analysis of Variance and Optimization of Coordinate Systems and Temporal Conditions for Minimizing GPS Measurement Errors

Nitesh Awasthi, V. S. Rathore, Jayant Nath Tripathi, and C. Jeganathan

2.1 INTRODUCTION

The Global Positioning System (GPS) is a transformative technology providing precise location and timing information across the globe. Developed by the United States Department of Defense, GPS leverages a constellation of satellites to deliver accurate positional data critical for navigation, mapping, and various scientific applications. The technology's fundamental operation involves satellites transmitting signals containing their positions and the precise time of transmission, which are then received by GPS receivers on Earth. By measuring the time delay between transmission and reception, these receivers calculate their exact location (Hofmann-Wellenhof, Lichtenegger, & Collins, 2008).

GPS data collection relies on different coordinate systems, primarily the Geographic Coordinate System (GCS) and the Cartesian Coordinate System (CCS). The GCS employs angular measurements—latitude and longitude—to specify positions on the Earth's surface. This system is widely used in geographic and cartographic applications due to its alignment with conventional mapping practices and intuitive representation of locations (Hofmann-Wellenhof et al., 2008). In contrast, the CCS uses linear coordinates (x, y, z) to represent points in a three-dimensional space, with the origin typically at Earth's center. The CCS is preferred in engineering and 3D modeling contexts due to its straightforward approach to spatial calculations (Hofmann-Wellenhof et al., 2008; Teunissen & Montenbruck, 2017).

Despite its extensive applications, GPS technology is susceptible to various errors that can affect data accuracy. These errors are broadly categorized into several types.

2.1.1 EPHEMERIS ERRORS

These errors stem from inaccuracies in the satellite's orbital data. Discrepancies between the predicted and actual satellite positions can lead to significant deviations in the calculated location (Misra & Enge, 2011; Parkinson & Spilker, 1996).

DOI: 10.1201/9781032712444-3

2.1.2 SATELLITE CLOCK ERRORS

Minor discrepancies in the atomic clocks aboard the satellites can lead to errors in time measurement, subsequently affecting positional accuracy (Misra & Enge, 2011).

2.1.3 ATMOSPHERIC DELAYS

The Earth's ionosphere and troposphere can alter GPS signal propagation, leading to signal speed variations and ranging errors. Ionospheric delays are influenced by solar activity and geomagnetic conditions, while tropospheric delays are affected by factors such as moisture content and atmospheric pressure (Klobuchar, 1996; Saastamoinen, 1972; Ciraolo et al., 2015).

2.1.4 MULTIPATH EFFECTS

GPS signals that reflect off surfaces before reaching the receiver can cause multipath errors, impacting the accuracy of the measurements (Van Sickle, 2001).

2.1.5 RECEIVER NOISE

Noise introduced by the GPS receiver itself can affect measurement precision, contributing to overall positioning errors (Misra & Enge, 2011; Van Sickle, 2001).

Modeling and mitigating these errors are crucial for enhancing GPS accuracy. Techniques for ionospheric modeling, such as the Klobuchar model and the NeQuick model, estimate ionospheric delays based on solar and geomagnetic activity (Klobuchar, 1996; Radicella & Leitinger, 2001). Tropospheric models, including the Saastamoinen and Hopfield models, provide corrections based on atmospheric parameters (Saastamoinen, 1972; Hopfield, 1969). Post-processing methods like Differential GPS (DGPS) and Real-Time Kinematic (RTK) positioning enhance accuracy by analyzing data from multiple receivers (Teunissen & Montenbruck, 2017; Borre et al., 2007).

Understanding and mitigating these errors are critical for improving GPS accuracy. Various models and techniques are employed to address atmospheric delays. For ionospheric delays, models such as the Klobuchar model and the NeQuick model estimate delays based on solar and geomagnetic activity (Klobuchar, 1996; Radicella & Leitinger, 2001). Tropospheric models, including the Saastamoinen and Hopfield models, correct for delays based on atmospheric conditions (Saastamoinen, 1972; Hopfield, 1969). Post-processing methods like Differential GPS (DGPS) and Real-Time Kinematic (RTK) positioning enhance accuracy by analyzing data from multiple receivers (Teunissen & Montenbruck, 2017; Borre et al., 2007).

This study aims to explore the impact of atmospheric errors on GPS data by comparing measurements collected using the GCS and CCS. By analyzing how these errors influence positional accuracy in different coordinate systems, the research seeks to enhance the precision of GPS technology and provide insights into the temporal and spatial variations of GPS errors.

2.2 STUDY AREA AND EXPERIMENTAL FRAMEWORK

2.2.1 DATA ACQUISITION

Point positioning data was obtained using Leica GPS receivers in the open space in front of BIT Main Building (the entrance to the Upper Lawn) at different times of the day. The data was collected in the third week of November (from November 15 to November 20) at two times of the day (Morning and Afternoon).

For the first two days (November 15 and November 16), the morning data was collected at 5:30 a.m. and, subsequently, for the next pair of days at 6:30 a.m. (November 17 and November 18) and

7:30 a.m. (November 19 and November 20). Similarly, afternoon data was collected at 11:30 a.m., 12:30 p.m., and 1:30 p.m., respectively, for the three pairs of days. The data was collected using two Leica handheld GPS receivers, one in the Geographic Coordinate System and the other in the Cartesian Coordinate System.

For accuracy estimation of the datasets, the location of the point under observation was determined using the static DGPS technique, which would obviously result in a much more accurate value than the handheld single receiver approach. The location of the observation point using the static DGPS technique was found to be 23.41285302 N, 85.43974244 E.

2.2.2 METHODOLOGY

In this study, data collection was conducted using two separate GPS receivers to gather positioning information. The objective was to evaluate and compare the accuracy of data collected in the Geographic Coordinate System (GCS) and the Cartesian Coordinate System (CCS). To ensure a valid comparison, it was essential to control for factors that could influence the accuracy of the GPS measurements.

Firstly, it was verified that the satellite geometry remained consistent across both receivers. This was crucial because the Dilution of Precision (DOP)—a measure of satellite geometry and its effect on the accuracy of GPS measurements—could otherwise vary, potentially skewing the comparison. By ensuring that data collection occurred simultaneously in both receivers, any variations due to satellite geometry were minimized, thereby maintaining a consistent DOP throughout the study.

The next step involved addressing the differences between the coordinate systems used for data collection. One dataset was collected using the Geographic Coordinate System (GCS), which provides positional information in terms of latitude and longitude. The other dataset was recorded using the Cartesian Coordinate System (CCS), which uses linear coordinates (x, y, z) to define positions in three-dimensional space. To facilitate a meaningful comparison, it was necessary to convert the CCS data into the GCS framework.

The conversion of coordinate systems was performed using the Coordinate Calculator Tool in Erdas Imagine 2014, a specialized software for geospatial data analysis and visualization. This tool facilitated the transformation of Cartesian coordinates into Geographic Coordinates, aligning both datasets within a unified reference framework. This step was crucial, as it allowed for a direct comparison of data accuracy and variability between the two coordinate systems by ensuring that both were represented in the same spatial context.

Once the conversion was completed, the analysis involved plotting the coordinate values from both datasets against the number of samples collected. This graphical representation was essential for visualizing and assessing the accuracy and variability of the GPS measurements. Differential GPS (DGPS) measurements, known for their high precision, were used as a benchmark to evaluate the accuracy of the data from both the GCS and CCS. By comparing these plotted values to the DGPS measurements, the study could determine the extent of variability and the relative accuracy of each coordinate system.

Further, the study aimed to identify optimal data collection times to reduce the impact of GPS errors. This was achieved by analyzing temporal variations in mean coordinate values across different times of the day. To this end, the mean values for each time period were plotted, and fluctuations were examined to pinpoint times when GPS errors were minimized. This temporal analysis provided insights into periods when GPS data collection could yield the highest accuracy, thus helping in selecting optimal times for future data collection.

The methodological approach of using coordinate conversion combined with detailed graphical and temporal analysis ensured a thorough comparison of GPS data accuracy between the GCS and CCS. This comprehensive methodology facilitated a robust evaluation of data variability and accuracy, ensuring that the findings were both accurate and reliable.

2.3 RESULTS AND ANALYSIS

The variability of GPS coordinate values (latitude and longitude) was analyzed by plotting these measurements against the number of samples collected. In the Geographic Coordinate System (GCS), latitude and longitude values exhibited lower variability as sample size increased, indicating greater stability and consistency in positional accuracy. This trend highlights the GCS's effectiveness in accurately representing geographic locations.

In contrast, the Cartesian Coordinate System (CCS) showed higher variability in coordinate values with an increasing number of samples. This fluctuation suggests that CCS measurements were more prone to inconsistencies compared to GCS, likely due to the CCS's linear coordinate framework, which does not account for the Earth's curvature as effectively as GCS. Overall, the GCS demonstrated superior accuracy and reliability, with reduced variability in measurements, making it more suitable for applications requiring high positional precision.

The variance of the latitude values with respect to the static DGPS latitude value in the Geographic and Cartesian Coordinate Systems was found to be 1.98278×10^{s9} and 4.99431×10^{-9}, respectively.

In this chapter, the variance of longitude values in the datasets collected using the Geographic Coordinate System (GCS) and Cartesian Coordinate System (CCS) was analyzed and compared to the static Differential GPS (DGPS) longitude values. The variance for the longitude values in the GCS dataset was found to be 2.80047×10^{-9}, while the variance in the CCS dataset was 4.23331×10^{-9}. This analysis reveals that the longitude values obtained from the Cartesian Coordinate System exhibit a higher variance compared to those from the Geographic Coordinate System.

The increased variance in the Cartesian Coordinate System dataset suggests that the data collected may be less consistent or accurate compared to the GCS dataset. Several factors could contribute to this observation. One possible reason is the inherent inaccuracies associated with data collected in the Cartesian Coordinate System. Such inaccuracies might arise from the conversion process or from limitations inherent to the CCS itself.

Another potential factor is the error introduced during the conversion of Cartesian coordinates to Geographic coordinates. This conversion process, though necessary for comparison, can introduce additional inaccuracies or distortions. These errors might amplify variances in the data, particularly when the coordinate systems handle positional data differently. To mitigate the effects of coordinate system conversion and to ensure a more accurate comparison, an alternative approach could be employed. This would involve converting the Geographic Coordinate System dataset into Cartesian coordinates and then comparing the variance of these converted coordinates with the variance observed in the data collected using the Cartesian Coordinate System. By performing this reverse conversion, any discrepancies caused by the initial conversion process can be better understood and quantified.

Repeating the analysis in this manner would provide a clearer picture of whether the observed differences in variance are due to inherent characteristics of the coordinate systems or to errors introduced during the conversion process. This approach would help isolate the impact of coordinate system conversion from the inherent accuracy of the coordinate systems themselves. In summary, the higher variance observed in the Cartesian Coordinate System dataset compared to the Geographic Coordinate System dataset may be attributed to inaccuracies in the data collected or errors associated with the conversion process. Further analysis, including the conversion of GCS data to CCS and comparing it with native CCS data, could provide deeper insights into these variances and help improve the accuracy of GPS measurements across different coordinate systems.

The variance of the Northing values with respect to the static DGPS Northing value in the Cartesian and Geographic Coordinate Systems was found to be 40.14013333 and 35.66268804, respectively.

In this chapter, the variance of Easting values relative to static Differential GPS (DGPS) Easting values was compared between datasets collected using the Cartesian Coordinate System (CCS) and the Geographic Coordinate System (GCS). The variance for Easting values in the CCS dataset was found to be 44.795475, while in the GCS dataset it was 29.49244548. These results indicate that the GCS dataset exhibits a lower variance compared to the CCS dataset when compared to the static DGPS coordinates. Consequently, the GCS dataset is observed to be more accurate and reliable in terms of its alignment with the static DGPS measurements. To assess the impact of time on GPS errors, a comprehensive analysis of the temporal variation in mean coordinate values was conducted. This analysis involved plotting the mean coordinates—both in the Geographic and Cartesian Coordinate Systems—against the time of data collection. By examining these plots, the study aimed to determine periods when GPS errors are most pronounced.

Accuracy estimation was performed using statistical techniques. Specifically, the variance of each mean coordinate value was calculated for different times of observation and compared with the static DGPS coordinates. This approach allowed for an assessment of how the accuracy of the datasets varied with time. The resulting temporal variations of mean coordinate values in the Geographic Coordinate System, expressed in terms of latitude and longitude, were carefully analyzed. By plotting these values over different times of the day, patterns of accuracy and error were identified. This analysis helped in pinpointing specific times when GPS errors were maximized, providing insights into the optimal times for accurate data collection.

The analysis of variance in Easting values and the temporal variation of mean coordinates highlight the importance of considering both the coordinate system used and the timing of data collection. The lower variance observed in the GCS dataset underscores its superior accuracy compared to the CCS dataset. Additionally, understanding the time-related variations in GPS accuracy enables more informed decisions regarding the timing of data collection to minimize errors.

The analysis of variance for latitude and longitude values from the Geographic Coordinate System (GCS) datasets, relative to the static Differential GPS (DGPS) values, reveals notable insights into GPS measurement accuracy. The total variance for latitude values was calculated to be 4.43297×10^{-10}, while for longitude values it was 2.43791×10^{-9}. These variances highlight the discrepancies between the GCS dataset and the highly accurate static DGPS measurements, indicating that GPS errors, particularly atmospheric delays, play a significant role in reducing the resolution and accuracy of positioning data.

To further investigate temporal effects on GPS accuracy, the variances of latitude and longitude values were examined for different times of day. The morning and afternoon variances in latitude values for the GCS dataset, compared to the static DGPS latitude values, were 1.17818×10^{-10} and 7.68777×10^{-10}, respectively. This indicates that the variance in latitude values is notably higher in the afternoon compared to the morning. This increase in variance during the afternoon may be attributed to factors such as increased atmospheric activity or signal degradation as the day progresses.

Similarly, the morning and afternoon variances in longitude values for the GCS dataset, relative to the static DGPS longitude values, were found to be 2.35218×10^{-11} and 4.8523×10^{-9}, respectively. This substantial increase in the variance of longitude values from morning to afternoon underscores a significant drop in accuracy during the latter part of the day. The large variance in the afternoon could be due to enhanced atmospheric disturbances or other environmental factors that affect GPS signal quality as the day progresses.

These findings suggest that atmospheric delays and other error sources contribute significantly to the variances observed in GPS measurements. The differences in variance between morning and afternoon times highlight the importance of considering temporal factors when assessing GPS data accuracy. The increased variances in both latitude and longitude values during the afternoon point to a possible degradation in GPS signal quality or increased atmospheric interference during this period. Overall, the analysis indicates that GPS measurement accuracy can be considerably impacted by atmospheric conditions and time of day. These factors must be accounted for to improve the reliability of GPS data. Future studies could focus on minimizing these variances by optimizing data collection times and mitigating the effects of atmospheric delays to enhance GPS measurement precision.

2.4 DISCUSSIONS

From the analysis of Figures 2.1 through 2.8, it is evident that the data collected using the Geographic Coordinate System (GCS) consistently aligns more closely with the static Differential GPS (DGPS) values compared to the data from the Cartesian Coordinate System (CCS). This alignment is further supported by the lower variance values observed in the GCS dataset. The reduced variance indicates that the GCS dataset provides higher resolution and greater accuracy in positioning data. This enhanced accuracy is attributed to the lower temporal variability observed in the GCS dataset, which suggests that it is less prone to errors compared to the CCS dataset. Consequently, the Geographic Coordinate System is deemed superior for precise point location tasks when compared to the Cartesian Coordinate System.

Figures 2.9 through 2.14 illustrate a notable variation in the values of the GCS dataset during the afternoon compared to the morning. This variation results in increased positioning errors in the afternoon. These observations highlight that GPS data is significantly more affected by errors during the afternoon than in the morning. This phenomenon is likely due to ionospheric delays, which have a substantial impact on GPS signal accuracy.

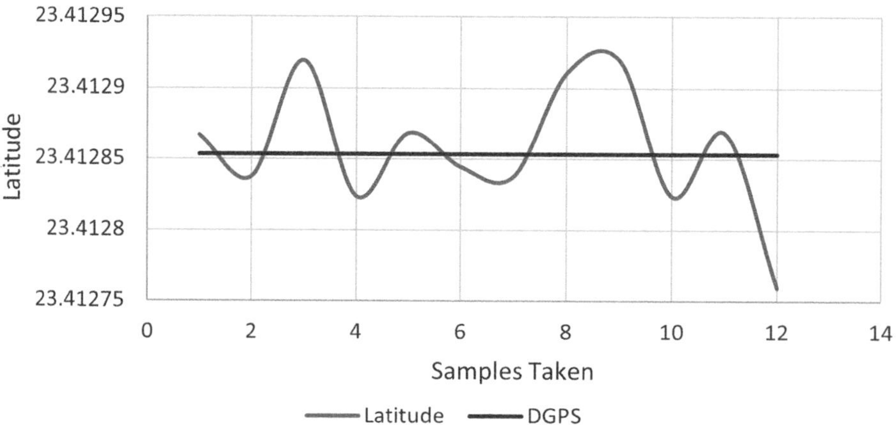

FIGURE 2.1 Variation of latitude of GCS data.

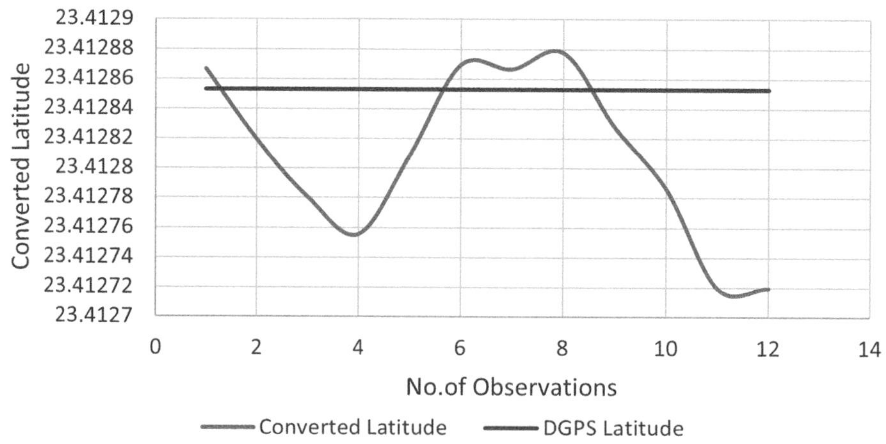

FIGURE 2.2 Variation of converted latitude of CCS data.

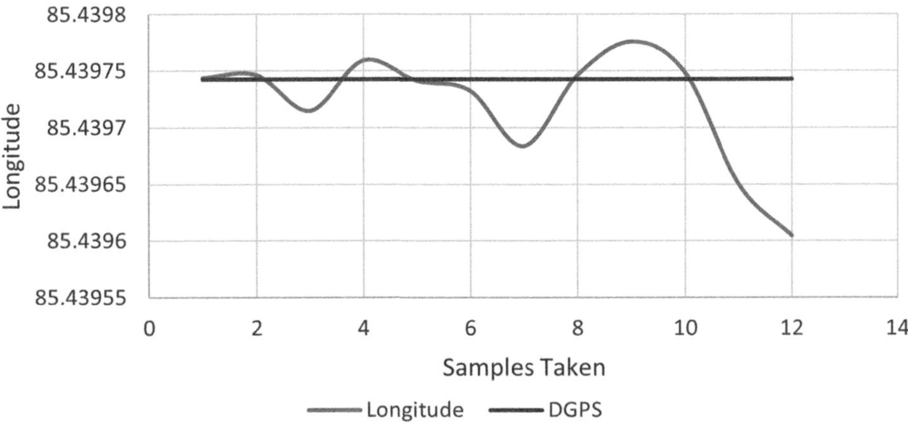

FIGURE 2.3   Variation of longitude of GCS data.

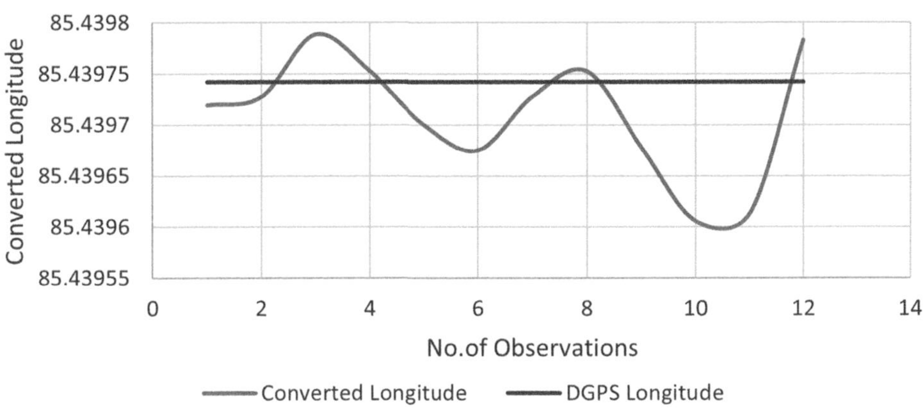

FIGURE 2.4   Variation of converted longitude of CCS data.

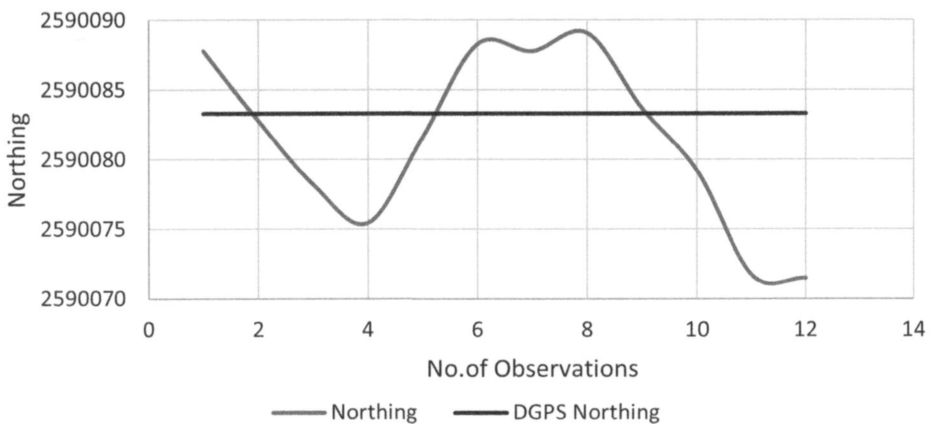

FIGURE 2.5   Variation of northing of CCS data.

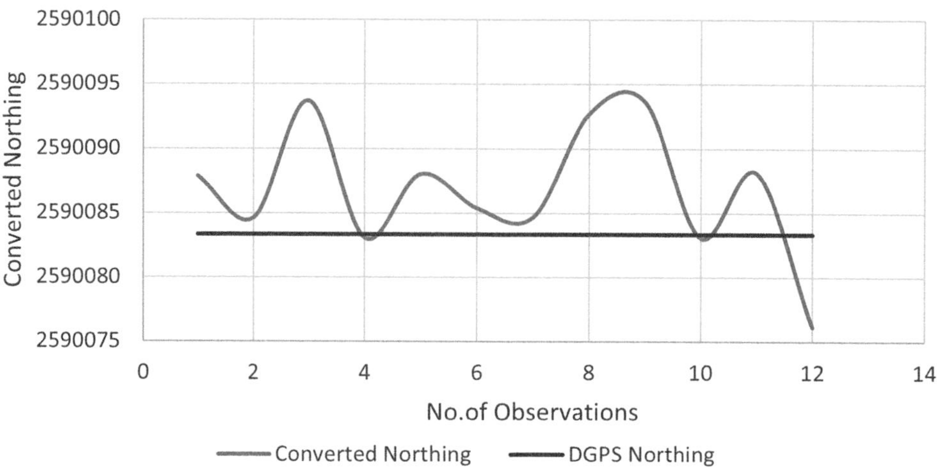

FIGURE 2.6   Variation of converted northing of GCS data.

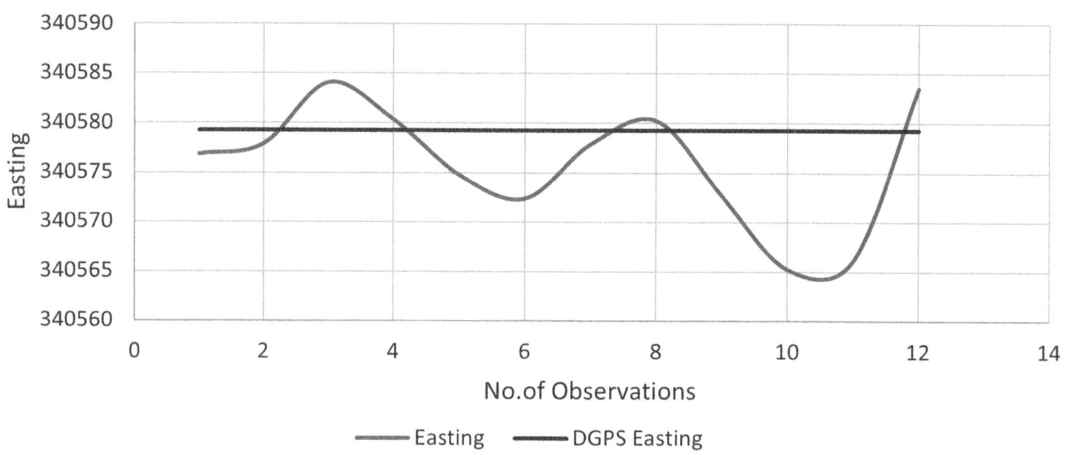

FIGURE 2.7   Variation of easting of CCS data.

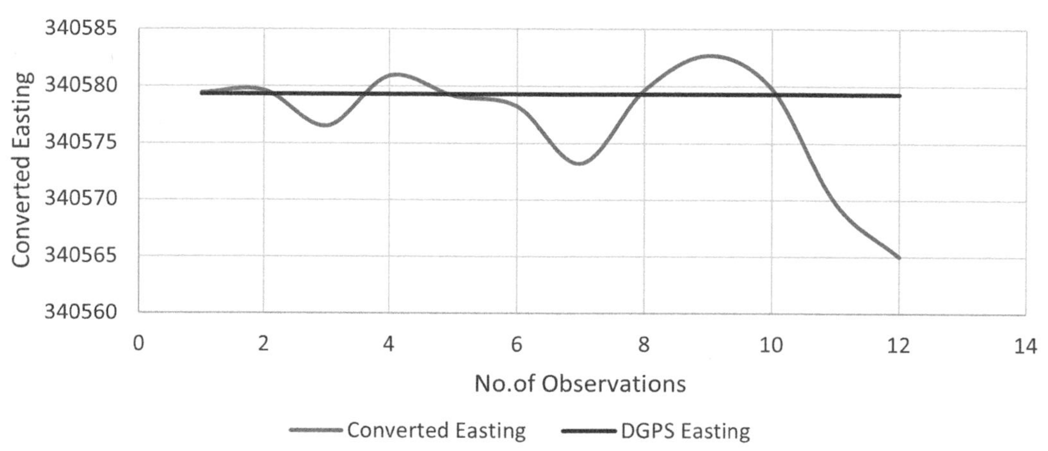

FIGURE 2.8   Variation of converted easting of GCS data.

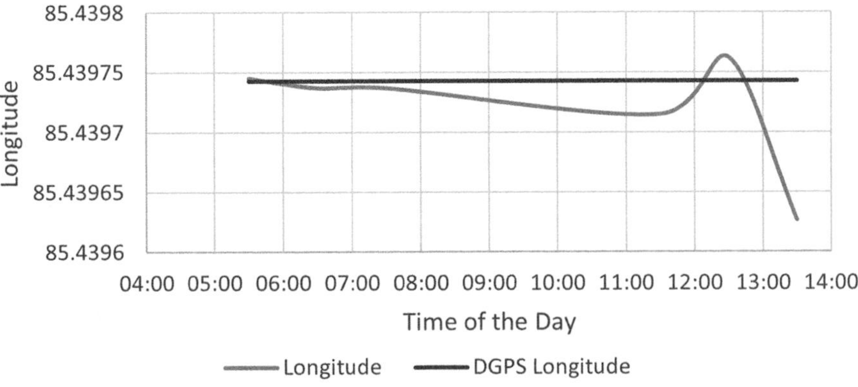

FIGURE 2.9   Temporal variation of longitude.

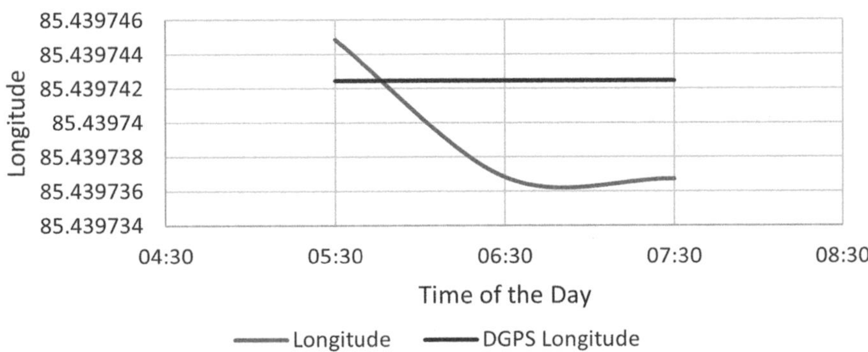

FIGURE 2.10   Morning variation of longitude.

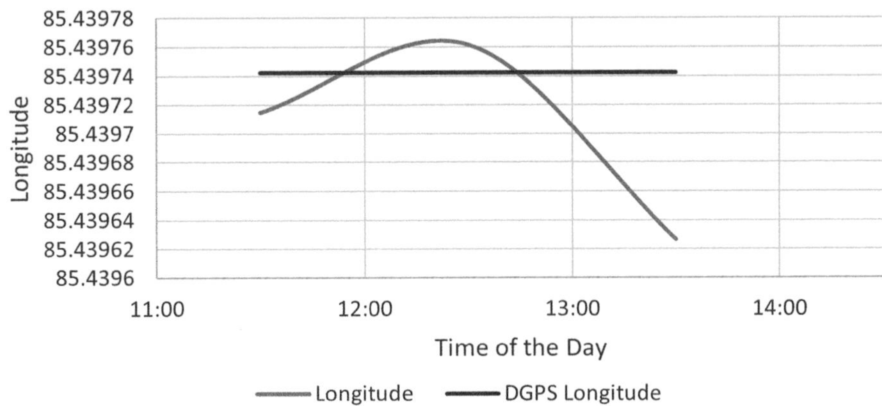

FIGURE 2.11   Afternoon variation of longitude.

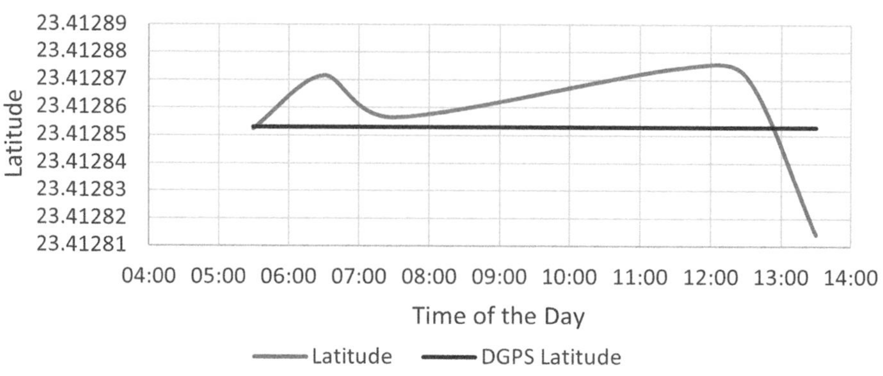

FIGURE 2.12   Temporal variation of latitude.

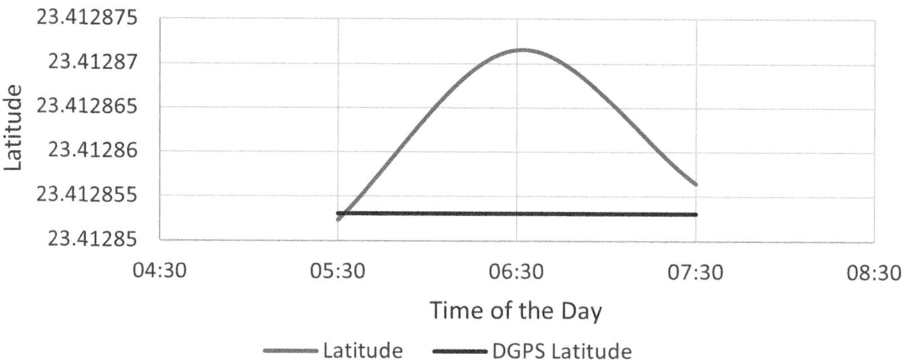

FIGURE 2.13   Morning variation of latitude.

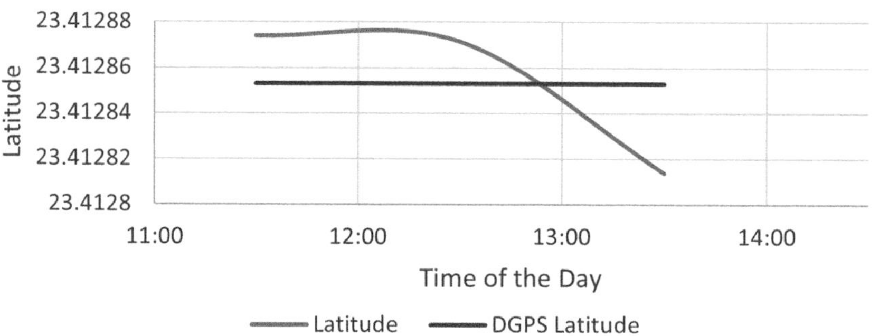

FIGURE 2.14   Afternoon variation of latitude.

The ionospheric delay tends to cause a northward shift in coordinates for locations in higher latitudes, leading to an increase in coordinate values, or a southward shift for locations in equatorial latitudes, resulting in a decrease in coordinate values. The magnitude of this shift is generally greater for southward shifts than for northward shifts (Weiping Jiang, Liansheng Deng, Zhao Li). Additionally, the Total Electron Content (TEC) of the ionosphere is highest in the afternoon, contributing to the most significant ionospheric delays during this time. This effect is evident from Figures 2.9, 2.11, 2.12, and 2.14, where the reduction in latitude and longitude values in the afternoon aligns with the increased ionospheric delay.

The variance in latitude and longitude values for the GCS dataset can therefore be attributed to ionospheric delays. The standard deviations of these values provide an estimate of the error magnitude induced by such delays. In the afternoon, the error magnitude in latitude and longitude was found to be 2.77268×10^{-5} degrees and 6.96584×10^{-5} degrees, respectively. Using the empirical conversion factor (1 degree = 110 km), these errors correspond to positional inaccuracies of approximately 3.0499 meters in latitude and 7.6654 meters in longitude during the afternoon. This confirms that ionospheric delays can induce significant errors in GPS measurements, particularly during periods of high TEC.

In summary, the results highlight that the Geographic Coordinate System dataset is more accurate and reliable compared to the Cartesian Coordinate System dataset. Furthermore, temporal variations, particularly in the afternoon, significantly impact GPS accuracy due to ionospheric delays. Understanding these factors is crucial for optimizing GPS data collection and improving overall positional accuracy.

2.5 CONCLUSIONS

This chapter investigated whether there was a statistically significant difference between GPS data collected using different coordinate systems, specifically the Geographic Coordinate System (GCS) and the Cartesian Coordinate System (CCS). The findings indicate that such differences are indeed present and vary with the type of coordinate system used, as well as with the time of day. The analysis revealed that the accuracy of GPS measurements is influenced by the type of coordinate system employed. Specifically, the Geographic Coordinate System demonstrated superior accuracy compared to the Cartesian Coordinate System. This conclusion is supported by the observation that the GCS dataset exhibited lower variance and greater consistency in comparison to the CCS dataset. The reduced variance in the GCS data underscores its reliability and precision in representing positional information.

Additionally, the study highlighted that the time of day plays a crucial role in GPS accuracy. Through statistical techniques and analysis, it was determined that signal variations and resultant accuracy were significantly affected by the time of observation. The results showed that greater deviations from actual coordinate values occurred in the afternoon compared to other times of the day. This is attributed to increased ionospheric delays during the afternoon, which have a pronounced effect on GPS signal quality.

Overall, the research confirms that the Geographic Coordinate System provides better accuracy and consistency than the Cartesian Coordinate System. Moreover, it emphasizes the importance of considering the time of day when collecting GPS data to account for variations due to ionospheric delays. Understanding these factors can help in improving the precision of GPS measurements and in selecting optimal times for data collection to minimize errors.

REFERENCES

Borre, K., Dalgaard, P., & Madsen, E. (2007). *GPS: Theory, Algorithms, and Applications*. Springer.

Ciraolo, L., Zuffada, C., & Xie, L. (2015). The Impact of Ionospheric and Tropospheric Delays on GPS Measurements. *IEEE Transactions on Geoscience and Remote Sensing*, 53(5), 2513–2524.

Hofmann-Wellenhof, B., Lichtenegger, H., & Collins, J. (2008). *GPS: Theory and Practice*. Springer.

Hopfield, H. S. (1969). Tropospheric Time Delay for Geodetic Measurement. *Journal of Geophysical Research*, 74(18), 4487–4499.

Klobuchar, J. A. (1996). Ionospheric Effects on GPS. In *Global Positioning System: Theory and Applications* (Vol. 1, pp. 359–378). American Institute of Aeronautics and Astronautics.

Misra, P., & Enge, P. (2011). *Global Positioning System: Signals, Measurements, and Performance*. Ganga-Jamuna Press.

Parkinson, B. W., & Spilker, J. J. (1996). *Global Positioning System: Theory and Applications*. American Institute of Aeronautics and Astronautics.

Radicella, S. M., & Leitinger, R. (2001). The Ne Quick Model of the Ionosphere. In *The Handbook of Atmospheric Electrodynamics* (Vol. 1). CRC Press.

Saastamoinen, J. (1972). Atmospheric Correction for the Geodetic Use of Satellite Data. In *The Use of Artificial Satellites for Geodesy* (pp. 247–251). American Geophysical Union.

Teunissen, P. J. G., & Montenbruck, O. (2017). *Springer Handbook of Global Navigation Satellite Systems.* Springer.

Van Sickle, J. (2001). *GPS for Land Surveyors.* CRC Press.

3 Application of GNSS in Earth System Sciences

Bhawana Sharma, Leesh Ray, Damanti Murmu,
Ayushi Gupta, and Dileep Kumar Gupta

3.1 INTRODUCTION

Earth System Science (ESS) has been a continuously emerging interdisciplinary branch of science since the 1980s (Steffen et al., 2020). The confluence of hydrosphere, atmosphere, lithosphere, and biosphere with their natural processing and anthropogenic influences is mainly explored in ESS. It was described that the matter dynamic of Earth is a closed system, whereas the same is not true for the energy dynamics due to the prominent role of solar radiation in the Earth system (Jacobson et al., 2000). Major elements move from one geosphere (i.e., atmosphere, lithosphere, hydrosphere, pedosphere) to another due to their physical and chemical transformation. It is evident that the biosphere has a significant impact on these transformations. Thus, the foundation of ESS was to recognize the changes—that happened in the past, currently happening in the present, or might happen in the future—with their relevant explanations (Jacobson et al., 2000). GNSS plays a crucial role in various Earth observation-related research studies.

GNSS is composed of a globally as well as regionally available collection of satellites and their augmentation systems to simply navigate, compute accurate locations, and predict the time required to reach a certain destination. Satellite navigation systems, such as GPS, GLONASS, Galileo, and BeiDou, provide continuous location data with high accuracy at the global level. NavIC and QZSS, along with BeiDou, are regional navigation satellite systems, limiting their service to specific regions of the world. Global satellites are placed in the Medium Earth Orbit at an altitude ranging from approximately 2000 to 35000 km, whereas the regional satellites revolve in a Geostationary Orbit at an altitude >35000 km (Riebeek, 2009; Kaplan and Hegarty, 2017).

The application of GNSS in geoscience covers fields of geology, oceanography, seismology, meteorology, volcanology, paleontology, hydrology, etc. A researcher (Jin et al., 2022) emphasized the role of GNSS in numerous sub-fields of ESS such as meteorology, ionosphere, Space weather, earthquake assessment, reflectometry, tracking of natural hazards (Figure 3.1), and use of unified technology to monitor lands and other structural health. Another study by (Kumar et al., 2021) outlined the importance of this technology in ever-changing Earth surface monitoring, enhanced transmission time, measuring speed, agriculture practices, etc. Some real-time applications of GNSSs in various disciplines of ESS are discussed below (Section 3.3).

3.2 WORKING/PRINCIPLE OF GNSS

A constellation of satellites known as GNSS circles the planet and provides precise Positioning, Navigation, and Timing (PNT) information (Lavrakas, 2020) to receivers on the ground. This makes a wide range of applications in various domains, including Earth system science, possible. The following are the main attributes of GNSS's PNT services:

- Positioning: With an accuracy of only a few meters, this method pinpoints the exact location of a receiver, including latitude, longitude, and altitude.

DOI: 10.1201/9781032712444-4

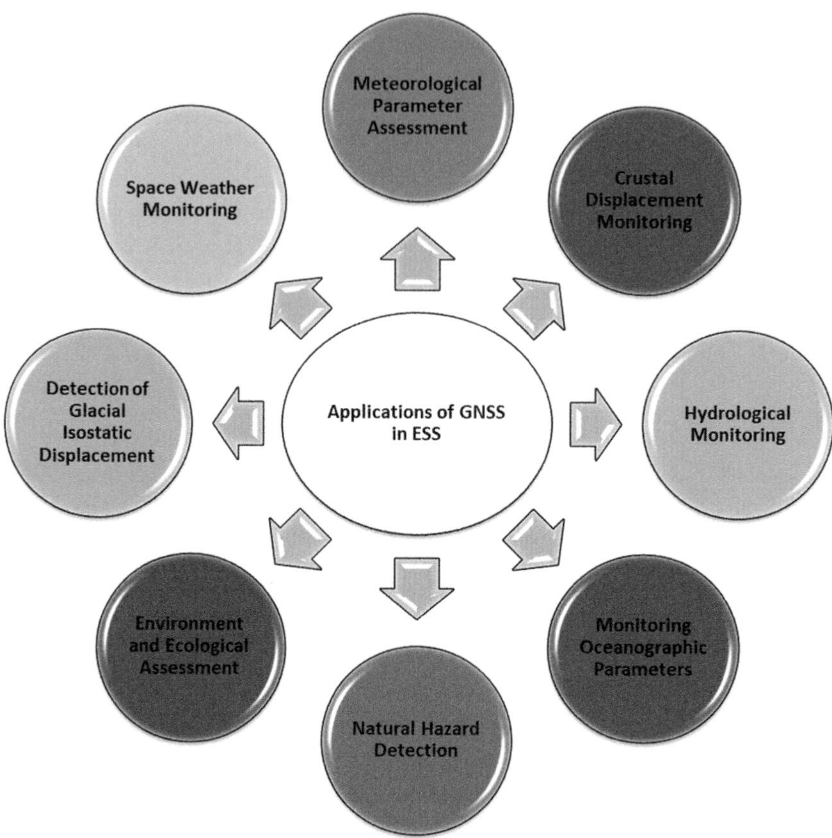

FIGURE 3.1 Diagrammatic representation of different GNSS applications in Earth System Sciences.

- Navigation: This feature allows users to plan routes and follow their movement in real time by providing velocity and direction data.
- Timing: Provides precise time synchronization, which is necessary for several applications, including banking, science, and telecommunication.

Figure 3.2 Illustrates more secondary services that are available in addition to primary PNT services.

The GNSS offers Positioning, Navigation, and Timing (PNT) services based on the following methods and procedures:

i. *Trilateration*: It determines a receiver's location based on its separation from many satellites, which is how GNSS works to provide positioning services. This is accomplished using a triangulation (Partsinevelos et al., 2020) procedure in which the receiver determines its position, velocity, and time by utilizing coded signals sent by many satellites at specific intervals. The receiver uses computations involving the time lag between a signal's transmission and reception to precisely determine its location by merging data from several satellites.

ii. *Signal Transmission and Signal Reception*: Turn on the navigation service so that receivers can determine their direction and speed of travel. This is accomplished by receiving signals from many satellites, each of which transmits two different types of signals

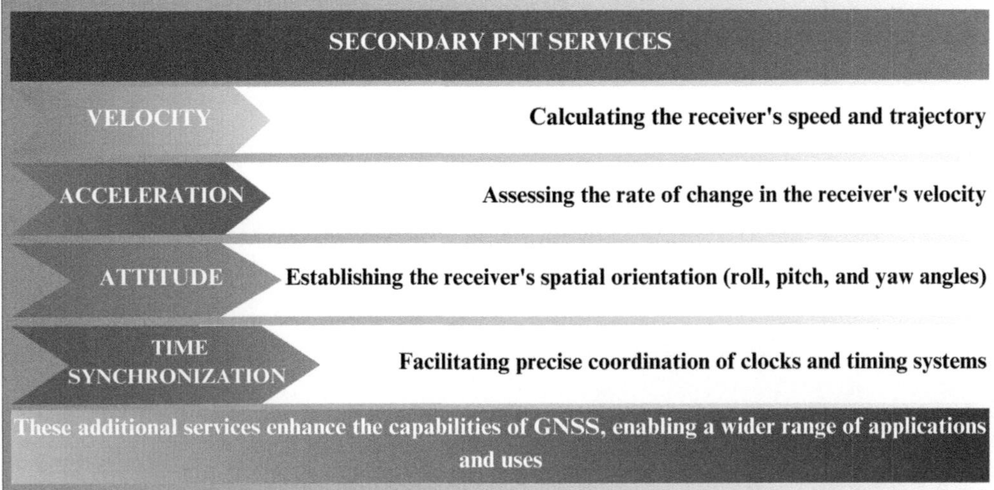

FIGURE 3.2 Secondary PNT services.

(Dardari et al., 2011; Bacci et al., 2012): L1 (1575.42 MHz) and L2 (1227.60 MHz), which carry exact code along with navigation data and coarse/acquisition code, respectively. A GNSS receiver detects and decodes broadcast signals from at least four visible satellites, utilizing the information to estimate its movement and calculate its velocity and direction.

iii. *Time Delay and Pseudorange Calculation*: It enables the timing service, allowing receivers to synchronize their clocks with satellite time. This is achieved by measuring the time it takes for a signal to travel from the satellite to the receiver and calculating the pseudorange (Lu and Yao, 2020), or apparent distance, using the speed of light. By determining the time delay between signal transmission and reception, the receiver can calculate the exact distance from the satellite, enabling precise time synchronization with the satellite's clock (Gonzalez-Garrido et al., 2022).

iv. *Error Correction*: Improves attitude (Bisnath, 2020) determination by refining position and velocity calculations, enabling receivers to accurately determine their orientation (roll, pitch, yaw). This is achieved by incorporating additional data, such as atmospheric delay and satellite orbit corrections, to enhance the precision of position calculations and ultimately determine the receiver's attitude.

The ability to provide exact position and timing information to ground-based receivers is madepossible by an understanding of the basic GNSS concepts. PNT services and GNSS principles are integrated to serve a wide range of applications that need precise positioning, navigation, and timing information (Figure 3.3). Three components make up a GNSS system: the user segment, which consists of various entities such as the military, government, commercial, and private sectors; the space segment, which consists of a constellation of 24 orbiting satellites; and the control segment, which consists of global monitoring stations that maintain satellite health. These principles will be applied in a variety of ways in Earth System Science, such as tracking sea level rise and glacier movements to monitor climate change, analyzing ground deformation to study earthquakes and volcanic eruptions, comprehending ocean currents and circulation patterns, managing natural resources by keeping an eye on changes in land use and deforestation, forecasting weather patterns and storms by analyzing atmospheric conditions, etc.

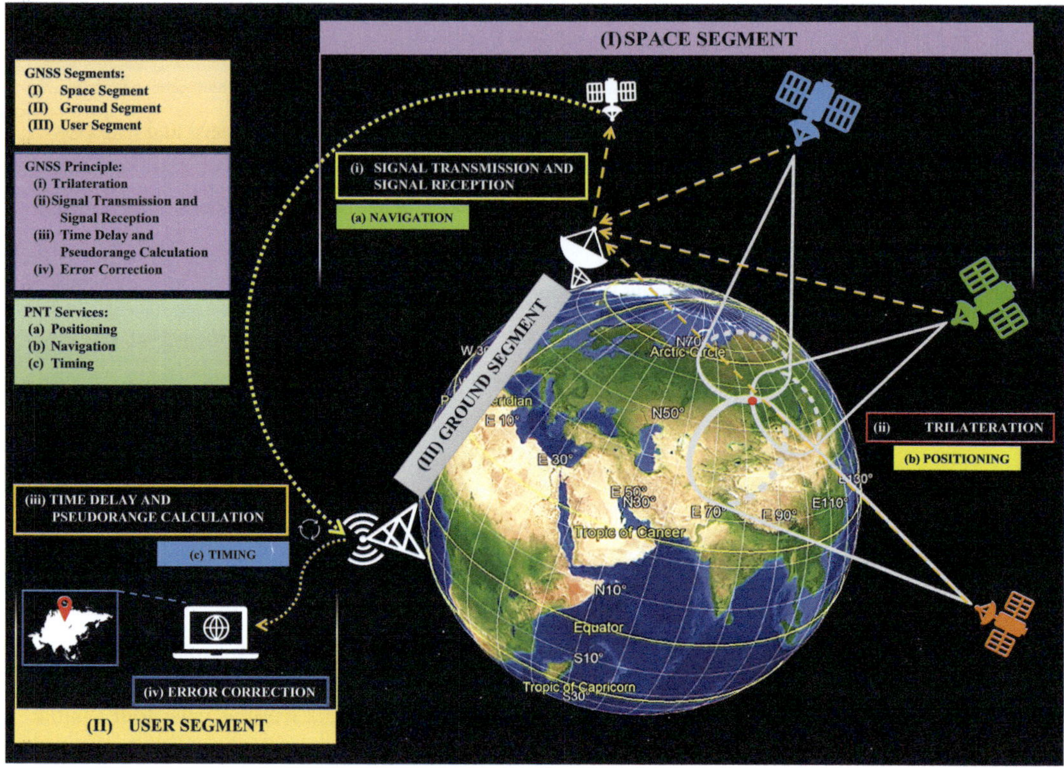

FIGURE 3.3   GNSS segments, principles, and services are interconnected and graphically represented.

3.3   APPLICATIONS OF GNSS IN EARTH SYSTEM SCIENCES

3.3.1   APPLICATION IN METEOROLOGY

Atmospheric water vapor can be estimated using GNSS technology. Tropospheric delays are primarily estimated to reduce error in GNSS positioning features, but its further analysis contributes to direct water vapor estimation. Estimation of wet delay out of zenith total delays yields this crucial and abundant parameter. It is a highly important parameter to assess climatic changes. It is also a prominent parameter for numerical weather modeling (Männel et al., 2021). Tomography is an emerging technique facilitating the 3D distribution of water vapor using multi-GNSS observations with increased signals. Many studies have regarded the combination of GNSS with other techniques or data sources to study water vapor dynamics in the atmosphere. Satellite data and algorithms are also combined for enhanced prediction of this parameter (Vaquero-Martínez and Antón, 2021).

Radio occultation (RO) technology in GNSSs prominently contributes to weather and other atmosphere-related research. GNSS-RO proved to give higher accuracy, global coverage, higher resolution, and all-time meteorological data to monitor various atmospheric parameters when compared with radiosonde and Lidar data. A recent study (Bai et al., 2020) demonstrated Numerical Weather Prediction and tropical Cyclone forecasting using the RO technique. The bending angle of the GNSS signal due to atmospheric refraction is mainly considered for atmospheric parameter analysis in the GNSS-RO method; these signals are received by low-earth orbiting satellites placed opposite to the GNSS satellite. It provides more accurate atmospheric profiles above the oceanic covers of the southern hemisphere as the RO signal remains least deterred by the clouds and the rains. The higher number of RO observations marginally increases the accuracy of Numerical Weather Prediction. Thus, large data directly ensures better predictions of weather using the GNSS-RO technique.

Tropical cyclone predictions are made using Numerical Weather Prediction. Low wind shear, warm sea surface, humidity, and high vorticity with deep convection are ideal conditions for a tropical cyclone. Therefore, meticulous predictions of tropical cyclones are made when parameters like temperature, wind, and water vapor are accurately assessed. The useful feature of propagation through clouds in RO signals facilitates water vapor analysis over the ocean for tropical cyclone forecasting. It was found that higher frequency and increased RO sounding significantly enhance these predictions, initiating various approaches to achieve this goal. New features—polarimetric RO and multi-feature GNSS receiver integrating GNSS-Reflectometry (GNSS-R) with RO functions—are explored to detect heavy precipitation with its surrounding water vapor profile, wind field monitoring over the ocean, and measurement of the ionosphere and sea waves. These features largely improve Numerical Weather Prediction and tropical cyclone forecasting (Bai et al., 2020).

A research study (Wang et al., 2023) was conducted to detect tropical cyclones using a space-borne GNSS-R full-delay Doppler map and monitor wind speed using a scene-classified model. An alternate study (Al-Khaldi et al., 2019) estimated the wind speed of hurricanes using Cyclone GNSS (CYGNSS) measurements. CYGNSS was commissioned in 2016, consisting of eight micro-satellites with a Delay Doppler Mapping Instrument as a payload. It is a constellation of low-earth orbiting satellites placed at ~520 km altitude. Apart from wind speed estimation above the ocean surface, geophysical parameters such as soil moisture, biomass, and surface water can also be retrieved using CYGNSS data (Carreno-Luengo et al., 2021).

3.3.2 Applications in Hydrology

Analyzing the spatial and temporal distribution of land-based water is crucial for understanding hydrological processes, climate fluctuations, and the management of water resources (Figure 3.4). Ground-based GNSS equipment is capable of precisely measuring small deformations of the Earth's

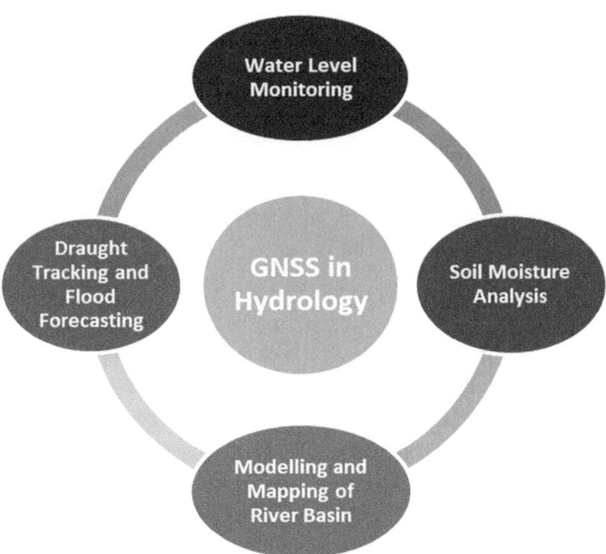

FIGURE 3.4 Applications of GNSS in hydrology.

surface with an accuracy of millimeters. These instruments have been effectively used to monitor changes in water mass at various spatial and temporal scales (Jiang et al., 2021).

The basis of GNSS-based hydrogeodesy was laid by early hydrogeologic research that identified a correlation between vertical displacement measured by GNSS and changes in terrestrial water (White et al., 2022). In addition, two researchers (Heki and Letters, 2003) estimated snow depth across Japan using vertical displacements at GNSS stations, accurately reproducing the spatial variability of snow depths measured by in situ sensors and demonstrating the power of inverse modeling with GNSS, in which hydrologic loads can be estimated from the observed Earth deformation. His direct comparison of snow depths to vertical surface displacements observed at GNSS stations across Japan suggested that snow loads produce most of the annual variations in the GNSS vertical time series there.

GNSS measurements can determine fine-scale spatial fluctuations in terrestrial water storage in a region heavily instrumented (Argus et al., 2014; Argus et al., 2017; Hsu et al., 2020), enabling the recording of regional loading changes spanning up to 300 km as well as local hydrologic changes at watershed scales of tens of kilometers (Knappe et al., 2019). Primarily, a dense GNSS array offers a novel means of tracking surface movements linked to meteorological extremes that are marked by notable variations in water mass with extremely short latency times (less than one day). This has enormous potential for operational hydrological monitoring and helps with early warning systems for disasters linked to hydrometeorological extremes, like storm surges (Geng et al., 2021).

GNSSs can be used to estimate the water level of different water bodies. GNSS receivers can be mounted on buoys or permanent structures to monitor the water levels in rivers, lakes, and reservoirs. This offers up-to-date information for managing water resources and anticipating floods (Tabibi and Francis, 2020). By examining GNSS signal reflections, GNSS interferometric reflectometry (GNSS-IR) approaches can identify variations in the water table's height. Changes in soil moisture can be identified by examining the GNSS signals reflected off the ground. This information is essential for agricultural purposes, drought tracking, and flood forecasting (Edokossi et al., 2020).

GNSS data provides accurate location and measurements of various hydrological variables and thus can enhance the accuracy of hydrological models, which contributes to constructing accurate models in water resource management (Michel et al., 2021). GNSSs are used in tide gauges to monitor sea level fluctuations. This data is vital for coastal management, flood risk assessment, and investigating the impacts of climate change on sea levels (Santamaría-Gómez et al., 2015). Depending on the present situation regarding hazards in various river basin areas, it is important to monitor different river basin areas and analyze future possibilities. Data acquired from GNSSs can be indispensable in building different models and mapping watersheds and river basins (Zhang and Lu, 2024).

3.3.3 Monitoring Crustal Deformation

In the past, crustal deformation was detected using triangulation and leveling techniques. NASA initiated further development in studying plate tectonic movements through space geodetic methods including Satellite Laser Ranging (SLR) and Very Long Baseline Interferometry (VLBI). This method demands highly skilled individuals in significant numbers. Consequently, it was replaced to an extent with GPS technology, as SLR still provides global coverage for the same (Larson, 1995).

A study (Wei et al., 2010) carried out an integrated approach to compute inter-seismic crustal deformations. This study merged GPS measurements with InSAR data to assess surface deformation that occurs between consecutive earthquakes and the procedural front of seismic zone deformations. The emerging new technique of GNSS reflectometry (GNSS-R) is employed in monitoring slope deformation (Yang et al., 2019). It is a remote approach to assess deformation based on signal reflection. Deformation at even sub-centimeter level was detected using this technique. Another study by (Yokota et al., 2019) discussed the inversion method applied to obtain the gradient effect and structural details of the seafloor. This study corroborates the feasibility of the GNSS-Acoustic ranging (GNSS-A) method in monitoring crustal deformation at the seafloor.

3.3.4 APPLICATIONS IN HAZARD DETECTION

3.3.4.1 Earthquake

A recent study (Maciuk, 2021) highlighted the rapidly increasing utilization of GNSSs in nature-induced phenomena, such as crustal deformation, resulting earthquake events, volcanic eruptions, and landslides (Figure 3.5), as it produces high-end results when integrated with other measurement techniques. The kinematic positioning method of the GNSS point determination technique works more efficiently in studying deformations caused due to earthquake events. Using a Continuously Operating Reference Station (CORS) (Corsa et al., 2022), movement in the earth's surface due to earthquakes can be quickly discerned. The system detects movement before, after, and during the occurrence of the earthquake at 1-mm scale (Maciuk, 2021). GNSSs also facilitate the study of post- and co-seismic ionospheric anomalies through Total Electron Content or TEC monitoring in the close range of the earthquake epicenter.

Fault displacement generates compressed acoustic gravity waves. These resulting waves cause ionospheric anomalies that are easily detected through TEC measurement. A researcher (Srivastava et al., 2021) used ground-based GPS data to measure TEC during the two consecutive earthquakes that occurred in Wharton Basin of Sumatra in 2012. The wave signature of TEC indicated the intensity and span of earthquakes, while the extensive number of GPS stations accommodated the directivity of ionospheric disturbances.

3.3.4.2 Landslide

Collaborative landslide assessment is conducted in real-time cases using 3D location data and geodetic methods of GNSSs. It mainly incorporated digital photogrammetry with remote sensing technologies. GNSS data validates the landslide-predicting models. These models are generated using InSAR- or LIDAR-based satellite data as they continuously monitor the landslide-prone areas (Maciuk, 2021). In a study by (Chadwick et al., 2005), a 12-m shift was estimated at the Salmon Falls landslide of Idaho by incorporating precise GPS data with a time-series archive of remote sensing. He studied the horizontal and vertical motion of the ground since 1900 to gain a

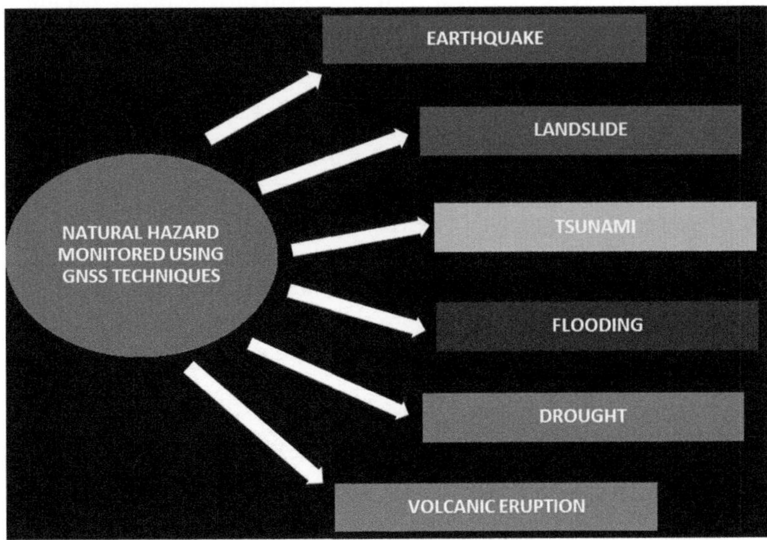

FIGURE 3.5 Multiple natural hazards are monitored using various GNSS techniques.

better understanding of the potential risk landslides carry in terms of the hydrological dynamics of the area.

3.3.4.3 Flood

Flooding instances can be witnessed very often due to changing climatic phenomena. Satellite-based flood monitoring includes optical and microwave approaches. Optical sensing is hindered by clouds and vegetation, while microwave sensing delivers coarse temporal and spatial resolution; these drawbacks are omitted when the GNSS technique of reflectometry is used to map flooded areas. Comparative assessment of CYGNSS data with Sentinel-1 SAR data indicated better performance of the GNSS method with higher accuracy in the detection of surface water, whereas the Visible Infrared Imaging Radiometer Suite (VIIRS) and MODIS flood data were underestimated (Downs et al., 2023). It is necessary to detect early signs of flooding in the future so that effective management and emergency arrangements are made in time.

A research study (Yang et al., 2021) combined the GNSS-R technique with remote sensing to conduct high-resolution flood monitoring in the Henan province of China. Surface reflectivity data of days after and before the flooding incident were acquired from CYGNSS. Changes in reflectivity were mapped for each day. Then it was further evaluated by analyzing Moderate Resolution Imaging Spectroradiometer (MODIS) data and compared using Soil Moisture Active Passive (SMAP) observations for a similar duration. The author concluded that CYGNSS data gives better accuracy in flood monitoring using the GNSS-R technique and will prove to be beneficial for future flood-related studies.

3.3.4.4 Tsunami

Causes of tsunamis include seismic events in the ocean crust with up to 100 km depth, submarine landslides, volcanic eruptions with seismic tremors, submarine eruptions, and asteroidal or meteoritic collision (Bryant, 2009). In a recent study (Ravanelli et al., 2021), GNSS technology was used to detect early signs of tsunamis. TEC measurement was integrated with ground displacement data to conduct this study. The total Variometric Approach was created by merging the Variometric Approach for Displacement Analysis Stand-alone Engine (VADASE) and Variometric Approach for Real-Time Ionosphere Observation (VARION) algorithms to quickly identify the source parameters and to predict tsunami origin. This approach enhanced the pre-existing alerting method for upcoming tsunamis.

Researchers (Yu, 2014) used the GNSS-R technique to approximate the parameters of tsunami-induced waves. GNSS-R uses altimetry to estimate the current height of the sea surface. Tsunami wave data, captured using buoy sensors, was utilized as input to generate a GNSS-R-based sea surface height measurement model. The propagation direction, propagation speed, and wavelength of tsunami waves were basically estimated in this study.

3.3.4.5 Drought

Events of drought have a profound impact on every living being, from natural vegetation and wildlife to the human population. A group of researchers (Edokossi et al., 2024) investigated drought using soil moisture data from SMAP and CYGNSS. Results were validated using soil moisture data from the Global Land Data Assimilation System (GLDAS). A good correlation with low RMSE strengthens the reliability of this approach. According to the study, the estimation of soil moisture is enough to assess drought conditions.

Regional drought monitoring using GNSS-derived zenith troposphere delay (ZTD) and precipitation was used by (Zhao et al., 2023). Instead of using pre-generated drought monitoring indices—Standardized Precipitation Evapotranspiration Index (SPEI) and Standardized Precipitation Conversion Index (SPCI)—a new index, regional precipitation and ZTD index (RPZI), was generated. It proved to successfully detect changes in drought conditions at the regional level.

3.3.4.6 Volcanic Activity

Volcanic eruptions can have a huge impact on climatic conditions. Substantial amounts of sulfur gases are released into the stratosphere due to volcanic eruptions. Oxidation of these gases produces sulfuric acid that absorbs and reflects solar insolation in the stratospheric layer causing stratospheric warming and subsequent cooling of the troposphere layer. Depending on location, it affects climatic conditions at either global or hemispherical levels (Zielinski, 2002).

Some researchers (Manta et al., 2021) used GNSS measurements of TEC to analyze disturbances caused by volcanic activities in the ionospheric layer. The new index, the Ionospheric Volcanic Power Index (IVPI), was created to compute the energy released from volcanic eruption and its propagation to the ionosphere. Results were validated using the previously generated index, the Volcanic Explosivity Index (VEI), and the ash plume height parameter. The application of GNSSs proved to be efficient in monitoring inaccessible volcanic arcs in different regions with less favorable weather conditions for satellite measurements. Geodetic time series measurements were used by (Corsa et al., 2022) to observe and detect deformation inciting volcanic eruptions. Time series of satellite-based Differential Interferometric SAR (DInSAR) combined with GNSS measurements to observe ground displacement at higher accuracy in three-dimensional maps. This approach helps in the detection of early ground motion leading to volcanic eruptions, reducing its greater potential to destroy humans and wildlife to an agreeable extent.

3.3.5 Applications in Oceanography

The megathrust earthquake mechanism and interplate subduction is very important to analyze earthquake incidents, and to understand that, seafloor crustal deformation must be studied. The GNSS-A technique, by combining the GNSS and acoustic ranging, is appropriate for detecting absolute seafloor crustal deformation with detailed information (Yokota et al., 2019). This observation has led to key scientific discoveries in the fields of geodesy and seismology, e.g., the discoveries of the giant crustal deformations due to the 2011 Tōhoku-oki Earthquake (Kido et al., 2011; Sato et al., 2011), the postseismic mechanisms following this earthquake (Watanabe et al., 2014; Tomita et al., 2016), and the interplate coupling condition along the Nankai Trough megathrust zone (Tadokoro et al., 2012; Yasuda et al., 2017; Yokota and Ishikawa, 2020). GNSSs possess many essential uses in oceanography, giving precise data that helps our understanding of marine processes.

3.3.5.1 Sea Level Monitoring

Highly accurate measurements of changes in sea level are made using GNSS-equipped tidal gauges (Dawidowicz, 2014). Understanding long-term patterns in sea level, managing coastal areas, and comprehending the effects of climate change all depend on this data. Sea surface heights are measured using satellite altimeters, which are calibrated using the GNSS. As a result, measurements of sea levels around the world are more accurate (Yokota et al., 2019).

3.3.5.2 Oceanic Circulation Studies

GNSSs are used to monitor floating instruments that give salinity, temperature, and ocean current data. Large-scale ocean circulation patterns are crucial for climate studies, and these patterns can be modeled and understood with the use of this data (Saynisch et al., 2015; Erena et al., 2020).

3.3.5.3 Bathymetric Mapping

Fine-grained bathymetric maps that depict the ocean floor's topography can be constructed through GNSS in conjunction with sonar. These maps are crucial for environmental research, maritime building, and navigation (Hatcher et al., 2023).

3.3.5.4 Marine Navigation and Safety

GNSSs are extensively used to track ships in real time, providing safe navigation and effective route planning, and plays a crucial role in maritime search and rescue operations by giving precise location information for lost or disabled vessels and individuals (Grant et al., 2011).

3.3.5.5 Oceanographic Data Buoys

Temperature, salinity, wave height, and other real-time oceanographic data are gathered and transmitted using GNSS-enabled data buoys. Programs for ocean monitoring, climate research, and weather forecasting are supported by this data (Jiang et al., 2021).

3.3.6 Applications in Environment and Ecology

The collection of data is one of the most crucial stages in environmental biology research (Franklin, 2009). Although they have several drawbacks, conventional techniques like measuring tapes, theodolites, compasses, and paper maps can be rather precise (Beever, 2006). Researchers had to rely on these techniques to describe individual plants, entire plant habitats, biotopes, and interactions between them before the GNSS and GNSS-receiving devices were more widely available in affordable units (Sonti, 2015).

The necessity to alter fieldwork techniques arose from the rapid development and increased accessibility of Geographical Information Systems (GIS) and GNSS technology in spatial analysis of environmental and animal data (Sonti et al., 2022). Many new gradations are happening in the field of GNSS; for example, Mapit GIS Ltd (2020) created the mobile application Mapit Spatial for Android-powered devices. It is the replacement for the Mapit GIS and Mappad programs, adding new features and utilizing an entirely new data management strategy (Nowak et al., 2020). QField is another such app developed by OPENGIS. It is available for free use and modification under the terms of the GNU Public License (GPL) Version 2 or above (Dalton et al., 2021).

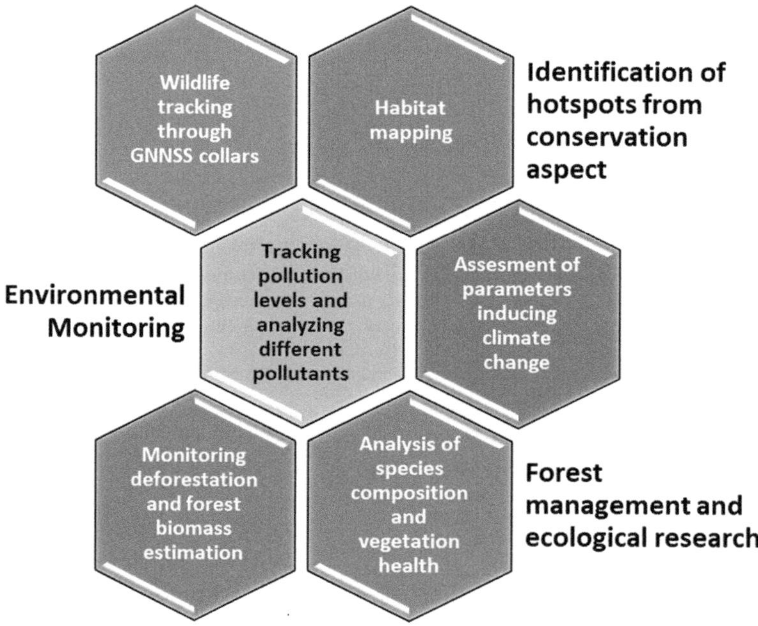

FIGURE 3.6 Different applications of GNSS in environmental monitoring and ecological research and management.

GNSS is contributing to aspects in the field of environment monitoring and monitoring and conservation of ecology throughout the globe (Figure 3.6). Those areas of benefit are:

3.3.6.1 Wildlife Tracking and Conservation

Tracking the migrations of animals, such as birds, mammals, and marine creatures, is done by GNSS collars and tags (Garrido-Carretero et al., 2023). Understanding migratory patterns and the effects of environmental changes on animals are all made easier with GNSS. Data from GNSS technology monitoring endangered species is used to support conservation plans, anti-poaching initiatives, and habitat preservation (Gor et al., 2017).

3.3.6.2 Habitat Mapping and Land Use Analysis

Mapping of various land cover types, including forests, marshes, and grasslands, is being constructed using GNSS in conjunction with remote sensing (Chung, 2017). The management of habitats, evaluation of changes in land use, and application of conservation regulations all depend on this knowledge (Smaby, 2021). Using GNSS data, precise ecosystem mapping makes it easier to identify hotspots for biodiversity or regions with a high concentration of species that need to be protected (Yang et al., 2023).

3.3.6.3 Environmental Monitoring

Oil spills, air pollution, and agricultural runoff are just a few examples of the pollutants that GNSS is used to track (Perez-Portero et al., 2021). This information is essential to understand the effects of different contaminants on the ecosystem. Accurate measurements of environmental factors like sea level rise, glacier movement, and vegetation changes are provided by GNSS data, which aids in the research of climate change (Awange and Awange, 2018).

3.3.6.4 Forest Management

GNSSs help monitor forest degradation by providing accurate data on forest borders and changes over time. Forest biomass is mapped using GNSS technology to evaluate carbon stocks and understand the function that forests play in sequestering carbon (Ferrazzoli et al., 2011).

3.3.6.5 Ecological Research

GNSSs model species distributions across a range of environmental conditions and may predict future changes in those distributions by giving accurate position data for species observations (Keefe et al., 2019). GNSSs are used to track the species composition, growth trends, and overall health of the vegetation. The management of habitats and ecological studies both benefit from this data (Murgaš et al., 2018).

Besides all these mentioned areas, GNSSs are widely used in different environmental and ecological monitoring sectors as well, such as wetlands and riparian zones mapping, hydrological studies like water level monitoring, analysis of flow rates, and the spatial distribution of water bodies, flood mapping and monitoring, landslide and erosion monitoring, tracking and monitoring of ecologically harmful invasive species, selection of sites for green energy production, etc. (Faka et al., 2021).

In addition, citizens can take part in environmental monitoring by providing data on species sightings, pollution levels, and habitat changes using GNSS-enabled devices, including cell phones (Zangenehnejad and Gao, 2021).

3.3.7 Applications in Glacial Isostatic Adjustment

Glacial isostatic adjustment (GIA) is the response of the planet to the growth and decay of massive ice sheets. It takes more than thousands of years to detect changes of up to 100 m in sea level and solid surface deformation (Whitehouse, 2018; Peltier et al., 2022).

The GIA model was generated using Holocene relative sea level and GNSS vertical land motion data of coastal regions in the North Sea by (Simon et al., 2021). The study shows discrepancies in predicted regional and global GIA models, therefore indicating the need for geological data to estimate changes in relative sea levels that are validated using tide gauge data.

Similarly, in another study conducted at 73 GPS stations in the Antarctic region, the uplift velocity of GPS stations was computed using GAMIT/GLOBK software on the data observed from 1996 to 2014. Previously generated GIA models for the region show inconsistency and were less reliable. In order to assess the vertical velocities calculated by different sources, the Antarctic regions were divided into seven parts. Results showed high variations between GIA models and estimated GPS data, concluding the need for further enhancement in GIA modeling methods and geodetic measurements in the Antarctic region (Li et al., 2020).

3.3.8 Monitoring Space Weather

Solar radio bursts can negatively affect the tracking ability of GNSS receivers while the ionospheric plasma irregularities cause scintillation, which is known to downgrade the location accuracy of the GNSS receiver. Therefore, monitoring of Space weather conditions is necessary to mitigate adverse effects and improve the performance of GNSS receivers simultaneously (Sreeja, 2016). Integration of ground, as well as space-borne GNSS measurements, characterizes continuous monitoring of ionosphere and plasmasphere systems. Ground-based tracking data provided by the International GPS Service (IGS) and GPS data of the CHAMP instrument mounted on the LEO satellite is used (Jakowski et al., 2005) to study the different ionospheric perturbations and their impact on the density and distribution of ionospheric plasma. TEC measurements taken by ground-based GNSSs correspond to the horizontal plasma redistribution during stormy events, whereas the space-borne RO and navigation data are used to detect vertical plasma redistribution in a few selected events.

3.4 CONCLUSION

GNSS merged with multiple technologies to emanate extremely reliable results in various ESS disciplines. Frequent integration of GNSS and remote sensing can be seen in meteorological dynamic assessment, surface deformation monitoring, high resolution surface water and flood monitoring, drought detection, ground displacement detection prior to a volcanic eruption, land cover, as well as ecosystem mapping and forest change monitoring.

GNSSs in combination with radio occultation, reflectometry, and acoustic ranging techniques is often used for different purposes such as NWP and cyclone forecastings, wind speed estimation during cyclones, soil moisture and surface water mapping, biomass estimation, slope deformation detection, crustal deformation at sea floor, tsunami-induced sea surface height measurement, and space weather monitoring. It was observed that Cyclone GNSS data with reflectometry is very crucial in many attributes of ESS.

In meteorology, a multi-GNSS approach was taken to map 3D water vapor distribution, whereas GNSS measurement of tropospheric delay produces direct water vapor estimates. Ground-based GNSS data is important to detect highly accurate surface deformation, TEC measurement, and many other ESS parameters. The TEC measurement forms the base for earthquake intensity and span approximation near its epicenter. CORS, a reference station of GNSS, provides real-time data of the Earth's surface movement before, after, and during an earthquake event.

It is evident that GNSS data largely improves modeling accuracy in many disciplines. However, some fields still need further development in GNSS technologies to enhance its prediction accuracy and provide highly reliable information.

Advancement in GNSSs broadens its scope and applicability in interdisciplinary fields of Earth System Sciences. The state-of-the-art techniques of GNSSs are used by many researchers due to its undeterred signal quality, high accuracy of ground- and space-based receivers and real-time data availability to effectively pursue remote approaches in large or inaccessible areas with uncertain weather conditions. GNSS measurements are often integrated with subject-specific models and techniques to obtain enhanced outputs, thus facilitating a better understanding of continuous change and its impact on the physical and biological dynamics of the Earth System. It has valuable contributions in making effective resource and risk management strategies. Expansion of this approach will be beneficial for futuristic highly detailed and accuracy-oriented studies.

REFERENCES

Al-Khaldi, M.M., Johnson, J.T., Kang, Y., Katzberg, S.J., Bringer, A., Kubatko, E. and Wood, D., 2019. Track-based cyclone maximum wind retrievals using the cyclone global navigation satellite system (CYGNSS) mission full DDMs. IEEE Journal of Selected Topics in Applied Earth Observations and Remote Sensing, 13, pp. 21–29.

Argus, D.F., Fu, Y. and Landerer, F.W., 2014. Seasonal variation in total water storage in California inferred from GPS observations of vertical land motion. Geophysical Research Letters, 41(6), pp. 1971–1980.

Argus, D.F., Landerer, F.W., Wiese, D.N., Martens, H.R., Fu, Y., Famiglietti, J.S., Thomas, B.F., Farr, T.G., Moore, A.W. and Watkins, M.M., 2017. Sustained water loss in California's mountain ranges during severe drought from 2012 to 2015 inferred from GPS. Journal of Geophysical Research: Solid Earth, 122(12), pp. 10,559–10,585.

Awange, J. and Awange, J., 2018. Environmental monitoring. GNSS Environmental Sensing: Revolutionizing Environmental Monitoring, pp. 1–13.

Bacci, G., Falletti, E., Fernández-Prades, C., Luise, M., Margaria, D. and Zanier, F., 2012. Satellite-Based Navigation Systems. In Satellite and Terrestrial Radio Positioning Techniques (pp. 25–74). Academic Press.

Bai, W., Deng, N., Sun, Y., Du, Q., Xia, J., Wang, X., Meng, X., Zhao, D., Liu, C., Tan, G. and Liu, Z., 2020. Applications of GNSS-RO to numerical weather prediction and tropical cyclone forecast. Atmosphere, 11(11), p. 1204.

Beever, E.A., 2006. Monitoring biological diversity: strategies, tools, limitations, and challenges. Northwestern Naturalist, 87(1), pp.66–79.

Bisnath, S., 2020, April. PPP: Perhaps the natural processing mode for precise GNSS PNT. In 2020 IEEE/ION Position, Location and Navigation Symposium (PLANS) (pp. 419–425). IEEE.

Bryant, E., 2009. Tsunami. A&E Television Networks.

Carreno-Luengo, H., Crespo, J.A., Akbar, R., Bringer, A., Warnock, A., Morris, M. and Ruf, C., 2021. The CYGNSS mission: On-going science team investigations. Remote Sensing, 13(9), p. 1814.

Chadwick, J., Dorsch, S., Glenn, N., Thackray, G. and Shilling, K., 2005. Application of multi-temporal high-resolution imagery and GPS in a study of the motion of a canyon rim landslide. ISPRS Journal of Photogrammetry and Remote Sensing, 59(4), pp. 212–221.

Chung, J., 2017. An accuracy study of rtk gnss positioning applied to comparing maps for bentgrass habitat modeling.

Corsa, B., Barba-Sevilla, M., Tiampo, K. and Meertens, C., 2022. Integration of DInSAR time series and GNSS data for continuous volcanic deformation monitoring and eruption early warning applications. Remote Sensing, 14(3), p. 784.

Dalton, D.T., Pascher, K., Berger, V., Steinbauer, K. and Jungmeier, M., 2021. Novel technologies and their application for protected area management: A supporting approach in biodiversity monitoring. Protected Area Management-Recent Advances.

Dardari, D., Luise, M. and Falletti, E. eds., 2011. Satellite and terrestrial radio positioning techniques: a signal processing perspective. Academic Press.

Dawidowicz, K., 2014. Sea level changes monitoring using GNSS technology–a review of recent efforts. Acta Adriatica, 55(2), pp. 145–162.

Downs, B., Kettner, A.J., Chapman, B.D., Brakenridge, G.R., O'Brien, A.J. and Zuffada, C., 2023. Assessing the relative performance of GNSS-R flood extent observations: Case study in south Sudan. IEEE Transactions on Geoscience and Remote Sensing, 61, pp. 1–13.

Edokossi, K., Calabia, A., Jin, S. and Molina, I., 2020. GNSS-reflectometry and remote sensing of soil moisture: A review of measurement techniques, methods, and applications. Remote Sensing, 12(4), p. 614.

Edokossi, K., Jin, S., Mazhar, U., Molina, I., Calabia, A. and Ullah, I., 2024. Monitoring the drought in Southern Africa from space-borne GNSS-R and SMAP data. Natural Hazards, 120(8), pp. 7947–7967.

Erena, M., Domínguez, J.A., Atenza, J.F., García-Galiano, S., Soria, J. and Pérez-Ruzafa, Á., 2020. Bathymetry time series using high spatial resolution satellite images. Water, 12(2), p. 531.

Faka, A., Tserpes, K. and Chalkias, C., 2021. Environmental sensing: a review of approaches using GPS/GNSS. GPS and GNSS Technology in Geosciences, pp. 199–220.

Ferrazzoli, P., Guerriero, L., Pierdicca, N. and Rahmoune, R., 2011. Forest biomass monitoring with GNSS-R: Theoretical simulations. Advances in Space Research, 47(10), pp. 1823–1832.

Franklin, J., 2009. Mapping species distributions: spatial inference and prediction. Cambridge University Press.

Freymueller, J., 2017. Geodynamics. Springer handbook of global navigation satellite systems, pp. 1063–1106.

Garrido-Carretero, M.S., Azorit, C., de Lacy-Pérez de los Cobos, M.C., Valderrama-Zafra, J.M., Carrasco, R. and Gil-Cruz, A.J., 2023. Improving the precision and accuracy of wildlife monitoring with multi-constellation, multi-frequency GNSS collars. The Journal of Wildlife Management, 87(4), p. e22378.

Geng, M., Wang, K., Yang, N., Li, F., Zou, Y., Chen, X., Deng, Z. and Xie, Y., 2021. Spatiotemporal water quality variations and their relationship with hydrological conditions in Dongting Lake after the operation of the Three Gorges Dam, China. Journal of Cleaner Production, 283, p. 124644.

Gonzalez-Garrido, A., Querol, J. and Chatzinotas, S., 2022, September. Hybridization of GNSS and 5G Measurements for Assured Positioning, Navigation and Timing. In *Proceedings of the 35th International Technical Meeting of the Satellite Division of The Institute of Navigation (ION GNSS+ 2022)* (pp. 2377–2384).

Gor, M., Vora, J., Tanwar, S., Tyagi, S., Kumar, N., Obaidat, M.S. and Sadoun, B., 2017, July. GATA: GPS-Arduino based Tracking and Alarm system for protection of wildlife animals. In 2017 international conference on computer, information and telecommunication systems (CITS) (pp. 166–170). IEEE.

Grant, A., Williams, P., Shaw, G., De Voy, M. and Ward, N., 2011, January. Understanding GNSS availability and how it impacts maritime safety. In *Proceedings of the 2011 International Technical Meeting of the Institute of Navigation* (pp. 687–695).

Hatcher, G.A., Warrick, J.A., Kranenburg, C.J. and Ritchie, A.C., 2023. Accurate maps of reef-scale bathymetry with synchronized underwater cameras and GNSS. Remote Sensing, 15(15), p. 3727.

Heki, K., 2003. Snow load and seasonal variation of earthquake occurrence in Japan. Earth and Planetary Science Letters, 207(1–4), pp. 159–164.

Hsu, Y.J., Fu, Y., Bürgmann, R., Hsu, S.Y., Lin, C.C., Tang, C.H. and Wu, Y.M., 2020. Assessing seasonal and interannual water storage variations in Taiwan using geodetic and hydrological data. Earth and Planetary Science Letters, 550, p. 116532.

Jacobson, M., Charlson, R.J., Rodhe, H. and Orians, G.H., 2000. Earth System Science: from biogeochemical cycles to global changes. Academic Press.

Jakowski, N., Wilken, V., Schlueter, S., Stankov, S.M. and Heise, S., 2005. Ionospheric space weather effects monitored by simultaneous ground and space based GNSS signals. Journal of Atmospheric and Solar-Terrestrial Physics, 67(12), pp. 1074–1084.

Jiang, Z., Hsu, Y.J., Yuan, L., Cheng, S., Li, Q. and Li, M., 2021. Estimation of daily hydrological mass changes using continuous GNSS measurements in mainland China. Journal of Hydrology, 598, p. 126349.

Jin, S., Wang, Q. and Dardanelli, G., 2022. A review on multi-GNSS for earth observation and emerging applications. Remote Sensing, 14(16), p. 3930.

Kaplan, E.D. and Hegarty, C. eds., 2017. Understanding GPS/GNSS: principles and applications. Artech house.

Keefe, R.F., Wempe, A.M., Becker, R.M., Zimbelman, E.G., Nagler, E.S., Gilbert, S.L. and Caudill, C.C., 2019. Positioning methods and the use of location and activity data in forests. Forests, 10(5), p. 458.

Kido, M., Osada, Y., Fujimoto, H., Hino, R. and Ito, Y., 2011. Trench-normal variation in observed seafloor displacements associated with the 2011 Tohoku-Oki earthquake. Geophysical Research Letters, 38(24).

Knappe, E., Bendick, R., Martens, H.R., Argus, D.F. and Gardner, W.P., 2019. Downscaling vertical GPS observations to derive watershed-scale hydrologic loading in the northern Rockies. Water Resources Research, 55(1), pp. 391–401.

Kumar, P., Srivastava, P.K., Tiwari, P. and Mall, R.K., 2021. Application of GPS and GNSS technology in geosciences. In GPS and GNSS Technology in Geosciences (pp. 415–427). Elsevier.

Larson, K.M., 1995. Crustal deformation. Reviews of Geophysics, 33(S1), pp. 371–377.

Lavrakas, J.W., 2020, September. GNSS Performance Standards: How are They Holding Up? In *Proceedings of the 33rd International Technical Meeting of the Satellite Division of The Institute of Navigation (ION GNSS+ 2020)* (pp. 1261–1267).

Li, F., Ma, C., Zhang, S., Lei, J., Hao, W., Zhang, Q. and Li, W., 2020. Evaluation of the glacial isostatic adjustment (GIA) models for Antarctica based on GPS vertical velocities. Science China Earth Sciences, 63, pp. 575–590.

Liu, J., Hyyppä, J., Yu, X., Jaakkola, A., Kukko, A., Kaartinen, H., Zhu, L., Liang, X., Wang, Y. and Hyyppä, H., 2017. A novel GNSS technique for predicting boreal forest attributes at low cost. IEEE Transactions on Geoscience and Remote Sensing, 55(9), pp. 4855–4867.

Lu, M. and Yao, Z., 2020. Position, navigation, and timing technologies in the 21st century: Integrated satellite navigation, sensor systems, and civil applications. In BeiDou navigation satellite system. 1, pp. 143–170. Wiley-IEEE Press.

Maciuk, K., 2021. GNSS monitoring natural and anthropogenic phenomena. In GPS and GNSS Technology in Geosciences (pp. 177–197). Elsevier.

Männel, B., Zus, F., Dick, G., Glaser, S., Semmling, M., Balidakis, K., Wickert, J., Maturilli, M., Dahlke, S. and Schuh, H., 2021. GNSS-based water vapor estimation and validation during the MOSAiC expedition. Atmospheric Measurement Techniques, 14(7), pp. 5127–5138.

Manta, F., Occhipinti, G., Hill, E.M., Perttu, A., Assink, J. and Taisne, B., 2021. Correlation between GNSS-TEC and eruption magnitude supports the use of ionospheric sensing to complement volcanic hazard assessment. Journal of Geophysical Research: Solid Earth, 126(2), p. e2020JB020726.

Michel, A., Santamaría-Gómez, A., Boy, J.P., Perosanz, F. and Loyer, S., 2021. Analysis of GNSS displacements in Europe and their comparison with hydrological loading models. Remote Sensing, 13(22), p. 4523.

Murgaš, V., Sačkov, I., Sedliak, M., Tunák, D. and Chudý, F., 2018. Assessing horizontal accuracy of inventory plots in forests with different mix of tree species composition and development stage.

Nowak, M.M., Dziób, K., Ludwisiak, Ł. and Chmiel, J., 2020. Mobile GIS applications for environmental field surveys: A state of the art. Global Ecology and Conservation, 23, p. e01089.

Partsinevelos, P., Chatziparaschis, D., Trigkakis, D. and Tripolitsiotis, A., 2020. A novel UAV-assisted positioning system for GNSS-denied environments. Remote Sensing, 12(7), p. 1080.

Peltier, W.R., Wu, P.P.C., Argus, D.F., Li, T. and Velay-Vitow, J., 2022. Glacial isostatic adjustment: physical models and observational constraints. Reports on Progress in Physics, 85(9), p. 096801.

Perez-Portero, A., Munoz-Martin, J.F., Park, H. and Camps, A., 2021. Airborne gnss-r: A key enabling technology for environmental monitoring. IEEE Journal of Selected Topics in Applied Earth Observations and Remote Sensing, 14, pp. 6652–6661.

Ravanelli, M., Occhipinti, G., Savastano, G., Komjathy, A., Shume, E.B. and Crespi, M., 2021. GNSS total variometric approach: first demonstration of a tool for real-time tsunami genesis estimation. Scientific Reports, 11(1), p. 3114.

Riebeek, H., 2009. Catalog of earth satellite orbits. Earth Observatory: NASA. Recuperado de: http://earthobservatory.nasa.gov/Features/OrbitsCatalog/page1.php

Santamaría-Gómez, A., Watson, C., Gravelle, M., King, M. and Wöppelmann, G., 2015. Levelling co-located GNSS and tide gauge stations using GNSS reflectometry. Journal of Geodesy, 89(3), pp. 241–258.

Sato, M., Ishikawa, T., Ujihara, N., Yoshida, S., Fujita, M., Mochizuki, M. and Asada, A., 2011. Displacement above the hypocenter of the 2011 Tohoku-Oki earthquake. Science, 332(6036), pp. 1395–1395.

Saynisch, J., Semmling, M., Wickert, J. and Thomas, M., 2015. Potential of space-borne GNSS reflectometry to constrain simulations of the ocean circulation: A case study for the South African current system. Ocean Dynamics, 65, pp. 1441–1460.

Simon, K.M., Riva, R.E.M. and Vermeersen, L.L.A., 2021. Constraint of glacial isostatic adjustment in the North Sea with geological relative sea level and GNSS vertical land motion data. Geophysical Journal International, 227(2), pp. 1168–1180.

Smaby, R., 2021. Vegetation change analysis from 2010-2018 using aerial photography and RTK-GNSS to assist Lake Mattamuskeet Restoration Efforts in North Carolina, USA. East Carolina University.

Sonti, S., Tyagi, K., Pande, A., Daniel, R., Sharma, A.L. and Tyagi, M., 2022. Crossroads of drug abuse and HIV infection: neurotoxicity and CNS reservoir. Vaccine, 10(2), p. 202.

Sonti, S.H., 2015. Application of geographic information system (GIS) in forest management. Journal of Geography & Natural Disasters, 5(3), p. 1000145.

Sreeja, V., 2016. Impact and mitigation of space weather effects on GNSS receiver performance. Geoscience Letters, 3(1), p. 24.

Srivastava, S., Chandran, A., Manta, F. and Taisne, B., 2021. GNSS TEC-based detection and analysis of acoustic-gravity waves from the 2012 Sumatra double earthquake sequence. Journal of Geophysical Research: Space Physics, 126(6), p. e2020JA028507.

Steffen, W., Richardson, K., Rockström, J., Schellnhuber, H.J., Dube, O.P., Dutreuil, S., Lenton, T.M. and Lubchenco, J., 2020. The emergence and evolution of Earth System Science. Nature Reviews Earth and Environment, 1(1), pp. 54–63.

Tabibi, S. and Francis, O., 2020. Can GNSS-R detect abrupt water level changes? Remote Sensing, 12(21), p. 3614.

Tadokoro, K., Ikuta, R., Watanabe, T., Ando, M., Okuda, T., Nagai, S., Yasuda, K. and Sakata, T., 2012. Interseismic seafloor crustal deformation immediately above the source region of anticipated megathrust earthquake along the Nankai Trough, Japan. Geophysical Research Letters, 39(10).

Tomita, T., Arikawa, T., Takagawa, T., Honda, K., Chida, Y., Sase, K. and Olivares, R.A.O., 2016. Results of post-field survey on the Mw 8.3 Illapel earthquake tsunami in 2015. Coastal Engineering Journal, 58(02), p. 1650003.

Vaquero-Martínez, J. and Antón, M., 2021. Review on the role of GNSS meteorology in monitoring water vapor for atmospheric physics. Remote Sensing, 13(12), p. 2287.

Wang, F., Zhang, G., Yang, D. and Kuang, H., 2023. Single-pass tropical cyclone detector and scene-classified wind speed retrieval model for spaceborne GNSS reflectometry. IEEE Transactions on Geoscience and Remote Sensing, 61, pp. 1–16.

Watanabe, S.I., Sato, M., Fujita, M., Ishikawa, T., Yokota, Y., Ujihara, N. and Asada, A., 2014. Evidence of viscoelastic deformation following the 2011 Tohoku-Oki earthquake revealed from seafloor geodetic observation. Geophysical Research Letters, 41(16), pp. 5789–5796.

Wei, M., Sandwell, D. and Smith-Konter, B., 2010. Optimal combination of InSAR and GPS for measuring interseismic crustal deformation. Advances in Space Research, 46(2), pp. 236–249.

White, A.M., Gardner, W.P., Borsa, A.A., Argus, D.F. and Martens, H.R., 2022. A review of GNSS/GPS in hydrogeodesy: Hydrologic loading applications and their implications for water resource research. Water Resources Research, 58(7), p. e2022WR032078.

Whitehouse, P.L., 2018. Glacial isostatic adjustment modelling: historical perspectives, recent advances, and future directions. Earth Surface Dynamics, 6(2), pp. 401–429.

Yang, N., Dai, X., Wang, B., Wen, M., Gan, Z., Li, Z. and Duffy, K.J., 2023. Mapping potential human-elephant conflict hotspots with UAV monitoring data. Global Ecology and Conservation, 43, p. e02451.

Yang, W., Gao, F., Xu, T., Wang, N., Tu, J., Jing, L. and Kong, Y., 2021. Daily flood monitoring based on space-borne GNSS-R data: A case study on Henan, China. Remote Sensing, 13(22), p. 4561.

Yang, Y., Zheng, Y., Yu, W., Chen, W. and Weng, D., 2019. Deformation monitoring using GNSS-R technology. Advances in Space Research, 63(10), pp. 3303–3314.

Yasuda, K., Tadokoro, K., Taniguchi, S., Kimura, H. and Matsuhiro, K., 2017. Interplate locking condition derived from seafloor geodetic observation in the shallowest subduction segment at the Central Nankai Trough, Japan. Geophysical Research Letters, 44(8), pp. 3572–3579.

Yokota, Y., Ishikawa, T. and Watanabe, S.I., 2019. Gradient field of undersea sound speed structure extracted from the GNSS-A oceanography. Marine Geophysical Research, 40(4), pp. 493–504.

Yokota, Y. and Ishikawa, T., 2020. Shallow slow slip events along the Nankai Trough detected by GNSS-A. Science Advances, 6(3), p. eaay5786.

Yu, K., 2014. Tsunami-wave parameter estimation using GNSS-based sea surface height measurement. IEEE Transactions on Geoscience and Remote Sensing, 53(5), pp. 2603–2611.

Zangenehnejad, F. and Gao, Y., 2021. GNSS smartphones positioning: Advances, challenges, opportunities, and future perspectives. Satellite Navigation, 2, pp. 1–23.

Zhang, W. and Lu, X., 2024. Inversion Method for Monitoring Daily Variations in Terrestrial Water Storage Changes in the Yellow River Basin Based on GNSS. Water, 16(13), p. 1919.

Zhao, Q., Liu, K., Sun, T., Yao, Y. and Li, Z., 2023. A novel regional drought monitoring method using GNSS-derived ZTD and precipitation. Remote Sensing of Environment, 297, p. 113778.

Zielinski, G.A., 2002. Climatic impact of volcanic eruptions. The Scientific World Journal, 2(1), pp. 869–884.

4 Introduction to the Measurement Techniques of GNSS Geodesy in Surveying
Enhancing Precision and Efficiency

Shashank C. Bangi

4.1 INTRODUCTION

Global Navigation Satellite Systems (GNSSs) have transformed the field of geodesy and surveying by providing precise positioning data essential for a wide range of applications. The measurement techniques employed in GNSS geodesy are fundamental in determining accurate coordinates, crucial for tasks such as land surveying, mapping, and infrastructure development (Hegarty and Chatre, 2008). These techniques rely on a constellation of satellites, such as those in the GPS, GLONASS, Galileo, and BeiDou systems, which continuously transmit signals containing orbital and timing information (Pan et al., 2019). Receivers on the ground use these signals to calculate their exact location through triangulation, measuring the time it takes for signals to travel from the satellites to the receiver.

The primary techniques in GNSS geodesy include static, kinematic, Real-Time Kinematic (RTK), and Precise Point Positioning (PPP) (Alkan et al., 2020). Static GNSS surveying involves long observation periods at fixed points, providing high-precision data ideal for establishing control networks (Abdallah and Schwieger, 2016). In contrast, kinematic surveying allows for data collection while in motion, beneficial for applications such as mapping and mobile data gathering. RTK enhances this by using a base station that sends real-time corrections to a rover, achieving centimeter-level accuracy crucial for tasks like construction and land surveying (Luo et al., 2021). PPP, on the other hand, uses satellite clock and orbit corrections, enabling high precision without the need for a local base station, which is particularly advantageous in remote areas.

Each of these techniques has specific applications and benefits, personalized to the varying requirements of different surveying tasks. For instance, static methods are preferred for projects needing the highest accuracy, while kinematic methods are ideal for dynamic environments (Elaksher et al., 2020). The ability to choose the appropriate technique based on project needs highlights the versatility of GNSS in geodesy (Li et al., 2023). Despite its many advantages, GNSS geodesy faces challenges, such as signal multipath, atmospheric interference, and obstructions in densely built environments. These factors can impact the accuracy and reliability of the data. However, advancements in error correction, multi-frequency receivers, and the use of augmentation systems like DGNSS and RTK have significantly mitigated these issues, enhancing the overall performance of GNSSs in surveying applications (Zangenehnejad and Gao, 2021).

As technology continues to advance, future developments in GNSSs are expected to further improve accuracy, reliability, and efficiency. These enhancements will likely include increased satellite constellations, improved signal processing, and integration with other geospatial technologies, ensuring that GNSSs remain an indispensable tool in geodesy and surveying (Gentle, Gledhill, and Blick, 2016). The measurement techniques of GNSS geodesy are foundational to modern surveying

DOI: 10.1201/9781032712444-5

practices, providing the precision and efficiency necessary for a wide range of applications in our increasingly interconnected world.

The primary objective of this chapter is to assess the accuracy and efficiency of land measurement using different GNSS techniques. By applying these techniques across various types of surveying fields, the chapter aims to identify the most effective methods for precise land measurement and to understand the advantages and limitations of each GNSS approach. This comprehensive analysis will provide valuable insights for improving surveying practices and optimizing the use of GNSS technology in diverse applications.

4.2 STUDY AREA

The campus of K.L.S. Gogte Institute of Technology (GIT), located in Udyambag, Belagavi, Karnataka, India, (Figure 4.1) has been selected as the study area for this research. GIT, established in 1979, is the flagship institute of the Karnataka Law Society. The institute occupies a sprawling 20-acre mango meadow and has been recognized as an autonomous institution permanently affiliated with Visvesvaraya Technological University. The geographic coordinates of the campus range from 15°48' 49.51"N to 15°49' 00.09"N latitude and 74°29' 08.72"E to 74°29' 21.32"E longitude. This extensive and well-defined area provides an ideal setting for conducting detailed surveying studies using various GNSS techniques.

4.3 METHODOLOGY

In this study, GNSS data over the study area were collected using various measurement techniques such as Absolute positioning with one GNSS receiver, Static GNSS surveying, Kinematic GNSS surveying, and Network real-time kinematics. Finally, accuracy assessment was done by comparing the data obtained from GNSS measurement to data obtained from Total station measurement. The procedure of GNSS measurement techniques (as shown in Figure 4.2) was explained as follows.

FIGURE 4.1   KLS GIT Campus.

Source: Google Earth image.

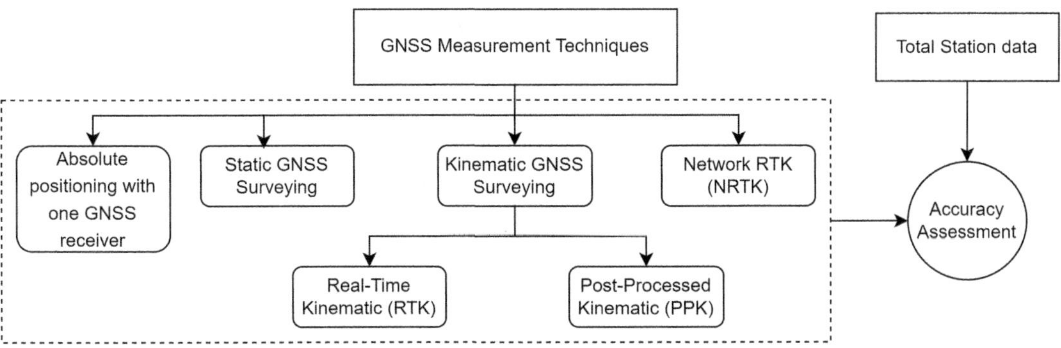

FIGURE 4.2   Methodology flow chart.

4.3.1   Absolute Positioning with One GNSS Receiver

Absolute positioning with one GNSS receiver involves determining the receiver's position relative to the GNSS satellites without relying on additional reference stations. This technique, also known as single-point positioning or stand-alone GNSS positioning, relies on signals received from multiple satellites to compute the receiver's location in three-dimensional space (Gentle, Gledhill, and Blick, 2016). In this study, to obtain absolute positioning using a single GNSS receiver, a GNSS receiver (SATLAB SL900) was selected, and the antenna was positioned in an open area with an unobstructed view of the sky to minimize errors. After powering on the receiver, it is allowed to initialize and acquire signals from at least four GNSS satellites, which typically took a few minutes. The receiver collected raw data, including pseudo-ranges and satellite positions.

Using these pseudo-range measurements, the receiver calculated the distances to each satellite and determined the antenna's coordinates through trilateration, applying necessary corrections for ionospheric and tropospheric delays to enhance accuracy. The resulting position, expressed in latitude, longitude, and altitude, was displayed on the receiver and logged for further analysis. Quality indicators such as Dilution of Precision (DOP) were monitored to ensure the reliability of the measurements, and the collected position data was stored. The procedure was conducted under optimal environmental conditions to avoid obstructions and multipath effects, ensuring the reliability of the results.

4.3.2   Static GNSS Surveying

Static measurement with post-processing is a highly precise GNSS surveying technique, often employed in geodetic and high-accuracy applications (Bakula, 2012). The process begins by setting up a GNSS receiver over a specific point of interest, ensuring the setup is stable and precisely cantered using a tripod and plummet. The receiver then records satellite signals over an extended period, often several hours, to collect comprehensive data. During this period, simultaneous measurements are taken at one reference station with known positions. The key to static measurement lies in the subsequent post-processing. The recorded raw GNSS data, typically stored in a manufacturer-specific format, is converted to RINEX (Receiver Independent Exchange format), a standard format compatible with most post-processing software (Gałdyn, Zajdel, and Sośnica, 2023). This data is then processed using post processing software (Trimble Business Centre).

Post-processing involves comparing the measurements from the survey point with those from the reference station. By analyzing the differences in the data, it is possible to correct for various errors such as satellite clock errors, orbital errors, and atmospheric disturbances. Advanced models and algorithms are used to reduce these errors further, utilizing high-precision satellite orbit and clock data, as well as ionospheric and tropospheric models. This method allows for extremely accurate position determinations, often achieving centimeter- or even millimeter-level precision (Elaksher et al., 2020). The extended observation times improve the ability to model and mitigate errors, leading to highly reliable results. Static measurement with post-processing is therefore the preferred choice for applications requiring the highest accuracy, such as establishing or densifying geodetic control networks, scientific research, and critical infrastructure projects. Although it requires more time and effort compared to real-time methods like RTK, the superior accuracy makes it indispensable for specific high-precision needs.

4.3.3 Kinematic GNSS Surveying

Kinematic GNSS surveying is a technique used to determine precise positions of moving objects by employing GNSS receivers. This method involves the use of a base station and one or more roving receivers, which collect GNSS data while in motion. There are two primary types of kinematic GNSS surveying: Real-Time Kinematic (RTK) and Post-Processed Kinematic (PPK).

Real-Time Kinematic (RTK) is a GNSS positioning technique that provides high-precision location data in real time by utilizing a base station and one or more rover receivers (Lotfy, Abdelfatah, and Hosny, 2020). In RTK, a base station is set up at a known, fixed location and continuously broadcasts correction data to rover receivers located at various points within the base station's coverage area. The rover receivers use this correction data to adjust their measurements of satellite signals, significantly reducing errors caused by factors such as atmospheric delays and satellite clock inaccuracies. This process enables the rover to calculate its position with centimeter-level accuracy in real time. RTK is commonly used in applications requiring precise positioning, such as land surveying, construction, and autonomous vehicle navigation. The system's effectiveness relies on a clear line of sight to satellites and a stable communication link between the base station and rover receivers.

Post-Processed Kinematic (PPK) is a GNSS positioning technique that enhances accuracy by using data collected during a survey and processing it after the fact (Martínez-Carricondo, Agüera-Vega, and Carvajal-Ramírez, 2023). In PPK, a base station and rover receivers are set up with the base station at a known location and the rover collecting data as it moves through the survey area. Both the base station and rover record their GNSS data independently. After the survey is completed, the collected data is post-processed using specialized software (Trimble Business Centre), which applies corrections for various errors such as ionospheric delays, satellite clock errors, and atmospheric conditions. This post-processing step refines the rover's position data by correcting the measurements based on the base station's data, resulting in high-precision position estimates typically accurate to within a few centimeters. PPK is especially useful in environments where real-time corrections are not feasible or where survey data needs to be analyzed and refined after collection.

4.3.4 Network RTK

Network RTK (Real-Time Kinematic) is an advanced GNSS positioning technique that enhances accuracy and coverage by utilizing a network of base stations rather than a single base station. This method improves the precision of GNSS measurements over a larger area and is particularly useful for applications requiring consistent and reliable positioning data (Wielgosz, Kashani, and Grejner-Brzezinska, 2005). In Network RTK, multiple base stations are distributed across a region, each continuously collecting GNSS data and broadcasting correction information. These base stations work together to create a correction network, which is then used to generate highly accurate real-time corrections for rover receivers within the network's coverage area (Berber and Arslan, 2013).

The network processes the data from all base stations to provide a unified set of corrections that account for regional variations in atmospheric conditions, satellite positions, and other factors that can affect GNSS accuracy. Rover receivers within the network receive these corrections in real time, allowing them to compute their positions with centimeter-level accuracy. In this study Continuously Operating Reference Stations (CORS) managed by a survey of India (https://cors.surveyofindia.gov.in/) have been used as Network RTK to get corrections for rovers. Network RTK significantly extends the coverage area compared to traditional single-base RTK, making it ideal for large-scale applications such as urban mapping, agricultural monitoring, and construction projects. This system offers the advantage of enhanced reliability and consistency in positioning, even in areas where individual base stations might not provide sufficient coverage.

4.4　RESULTS AND DISCUSSION

In this study, precise coordinates in the North (N), East (E), and Vertical (Z) directions were gathered from 30 different locations around the campus. The data was collected using a combination of GNSS measurement techniques and a total station, which provided a comprehensive overview of positional accuracy across diverse surveying methods. To ensure a rigorous accuracy assessment, the study focused on evaluating only the 10 most reliable coordinates from the initial 30. These selected points were carefully chosen based on their optimal conditions, specifically ensuring they were free from potential sources of interference such as tree cover and other physical obstructions. This approach aimed to minimize measurement errors and enhance the validity of the accuracy evaluation. The error obtained from the GNSS receiver in N, E, and Z directions were tabulated in the Table 4.1, 4.2, and 4.3.

The results presented in Table 4.1 highlight the errors observed in the North (N) direction across various GNSS measurement techniques. Absolute Positioning with One GNSS Receiver exhibited the highest average error (1.230 meters), indicating lower accuracy compared to other methods. This method does not incorporate real-time corrections or post-processing, leading to larger positional errors. Static GNSS Surveying demonstrated the lowest average error (0.005 meters), showcasing its high accuracy. The extended observation periods and post-processing of static surveys significantly reduce positional errors, making this method highly reliable for precise measurements.

Real-Time Kinematic (RTK) and Post-Processed Kinematic (PPK) methods both provided errors in the centimeter range, with average errors of 0.011 meters and 0.009 meters, respectively. RTK

TABLE 4.1

Errors in GNSS Measurement Data in North (N) Direction (Meters)

	Errors in Meters			
Pt No.	Absolute positioning with one GNSS receiver	Static GNSS Surveying	Real-Time Kinematic (RTK)	Post-Processed Kinematic (PPK)	Network RTK (NRTK)
2	1.2	0.003	0.009	0.008	0.015
3	0.8	0.004	0.011	0.009	0.018
4	0.9	0.004	0.012	0.008	0.021
7	1.8	0.003	0.008	0.007	0.020
10	2.1	0.005	0.013	0.011	0.018
15	1.2	0.008	0.015	0.012	0.016
16	0.7	0.006	0.011	0.011	0.014
18	0.9	0.004	0.009	0.008	0.015
25	1.6	0.005	0.009	0.007	0.018
29	1.1	0.004	0.012	0.005	0.013
Average	1.230	0.005	0.011	0.009	0.017

TABLE 4.2

Errors in GNSS Measurement Data in East (E) direction (Meters)

	Errors in Meters				
Pt No.	Absolute positioning with one GNSS receiver	Static GNSS Surveying	Real-Time Kinematic (RTK)	Post-Processed Kinematic (PPK)	Network RTK (NRTK)
2	0.7	0.005	0.008	0.007	0.009
3	0.5	0.003	0.012	0.005	0.011
4	0.8	0.003	0.008	0.009	0.021
7	0.5	0.005	0.015	0.007	0.020
10	1.1	0.005	0.008	0.005	0.018
15	0.8	0.004	0.009	0.008	0.015
16	1.2	0.003	0.008	0.009	0.009
18	1.5	0.003	0.012	0.011	0.015
25	0.9	0.004	0.011	0.015	0.008
29	0.8	0.005	0.008	0.011	0.018
Average	0.880	0.004	0.010	0.009	0.014

TABLE 4.3

Errors in GNSS Measurement Data in Vertical (Z) Direction (Meters)

	Errors in Meters				
Pt No.	Absolute positioning with one GNSS receiver	Static GNSS Surveying	Real-Time Kinematic (RTK)	Post-Processed Kinematic (PPK)	Network RTK (NRTK)
2	2.4	0.003	0.012	0.015	0.019
3	5.5	0.004	0.018	0.014	0.018
4	8.7	0.004	0.015	0.018	0.023
7	5.8	0.003	0.018	0.014	0.021
10	3.1	0.005	0.019	0.015	0.025
15	4.7	0.008	0.016	0.016	0.026
16	3.5	0.006	0.017	0.015	0.024
18	2.8	0.004	0.018	0.018	0.017
25	6.6	0.005	0.021	0.018	0.028
29	3.7	0.004	0.013	0.016	0.016
Average	4.680	0.005	0.017	0.016	0.022

offers immediate accuracy improvements through real-time corrections, whereas PPK achieves similar precision through post-survey data processing. Network RTK (NRTK) also showed good accuracy with an average error of 0.017 meters. Although slightly higher than RTK and PPK, NRTK benefits from a network of base stations providing corrections, extending its effective range and consistency over larger areas.

The results presented in Table 4.2 illustrate the errors observed in the East (E) direction using various GNSS measurement techniques. Absolute Positioning with One GNSS Receiver exhibited the highest average error (0.880 meters), reflecting the lower accuracy of this method due to the lack of real-time corrections and post-processing capabilities. Static GNSS Surveying showed the lowest average error (0.004 meters), demonstrating its superior precision. This method benefits from prolonged observation periods and extensive post-processing, which significantly reduce positional errors.

Real-Time Kinematic (RTK) and Post-Processed Kinematic (PPK) techniques both delivered errors in the centimeter range, with average errors of 0.010 meters and 0.009 meters, respectively. RTK offers real-time corrections, allowing for immediate accuracy improvements, while PPK achieves similar precision through thorough post-survey data processing. Network RTK (NRTK) had an average error of 0.014 meters, slightly higher than RTK and PPK. Despite this, NRTK provides reliable positioning over larger areas by leveraging a network of base stations, which compensates for regional variations and extends effective coverage.

The data in Table 4.3 highlights the errors observed in the vertical (Z) direction using various GNSS measurement techniques. The average value of the vertical GNSS values have shown highest values over north and east GNSS observations. The comparative analysis across the methods reveals significant differences in the accuracy of vertical measurements as compared to north and east directions.

4.5 CONCLUSION

The average errors for the North direction were lower across all techniques compared to the vertical direction. Static GNSS Surveying had the lowest error (0.005 meters), similar to the vertical results, whereas Absolute Positioning showed the highest error (1.230 meters). In east direction similar trends were observed, with Static GNSS Surveying again showing the lowest error (0.004 meters) and Absolute Positioning the highest (0.880 meters). Errors in the East direction were generally lower than in the vertical direction. The vertical (Z) direction consistently shows higher errors compared to the North (N) and East (E) directions for all GNSS techniques.

This trend highlights the inherent challenges in achieving high vertical accuracy due to factors such as atmospheric conditions and satellite geometry. The study reveals that while Absolute Positioning with one GNSS receiver is straightforward, it lacks the precision required for accurate vertical measurements. Static GNSS Surveying remains the most precise method across all directions, albeit more resource-intensive. RTK and PPK offer practical and accurate solutions for both horizontal and vertical measurements, balancing accuracy and efficiency. NRTK, while slightly less precise vertically, provides extensive coverage and reliable accuracy, making it suitable for broader surveying applications.

The measurement techniques of GNSS geodesy have revolutionized the field of surveying, offering unparalleled precision and efficiency. By leveraging sophisticated technologies such as code and carrier phase measurements, and implementing advanced methods like RTK, network RTK, and static measurement with post-processing, surveyors can achieve centimeter-level accuracy in real time or through meticulous post-survey analysis. Each technique has its specific applications, advantages, and limitations, enabling surveyors to choose the most appropriate method for their needs.

The continuous advancements in GNSS technology, including developments in multi-frequency receivers, enhanced correction algorithms, and robust communication networks, further reduce errors and improve the reliability of measurements. This technological progress is critical for various applications, from establishing geodetic control networks to supporting dynamic industries like construction, agriculture, and autonomous navigation. Understanding and effectively applying these GNSS measurement techniques is essential for modern surveyors aiming to meet the growing demands for accuracy and efficiency in geospatial data collection. As the technology evolves, so too will the capabilities and methodologies available, ensuring that GNSS geodesy remains at the forefront of precision surveying.

REFERENCES

Abdallah, A., and V. Schwieger. 2016. "Static GNSS Precise Point Positioning Using Free Online Services for Africa." *Survey Review* 48 (346): 61–77. https://doi.org/10.1080/00396265.2015.1097595

Alkan, Reha Metin, Serdar Erol, Veli İlçi, and Murat Ozulu. 2020. "Comparative Analysis of Real-Time Kinematic and PPP Techniques in Dynamic Environment." *Measurement: Journal of the International Measurement Confederation* 163 (October). https://doi.org/10.1016/j.measurement.2020.107995

Bakula, M. 2012. "An Approach to Reliable Rapid Static GNSS Surveying." *Survey Review* 44 (327): 265–271. https://doi.org/10.1179/1752270611Y.0000000038

Berber, Mustafa, and Niyazi Arslan. 2013. "Network RTK: A Case Study in Florida." *Measurement: Journal of the International Measurement Confederation* 46 (8): 2798–2806. https://doi.org/10.1016/j.measurement.2013.04.078

Elaksher, Ahmed, Tarig Ali, Franck Kamtchang, Christian Wegmann, and Adalberto Guerrero. 2020. "Performance Analysis of Multi-GNSS Static and RTK Techniques in Estimating Height Differences." *International Journal of Digital Earth* 13 (5): 586–601. https://doi.org/10.1080/1753894 7.2018.1550118

Gałdyn, Filip, Radosław Zajdel, and Krzysztof Sośnica. 2023. "RINEXAV: GNSS Global Network Selection Open-Source Software Based on Qualitative Analysis of RINEX Files." *SoftwareX* 22:101372. https://doi.org/10.1016/j.softx.2023.101372

Gentle, P., K. Gledhill, and G. Blick. 2016. "The Development and Evolution of the GeoNet and PositioNZ GNSS Continuously Operating Network in New Zealand." *New Zealand Journal of Geology and Geophysics* 59 (1): 33–42. https://doi.org/10.1080/00288306.2015.1127821

Hegarty, Christopher J., and Eric Chatre. 2008. "This Growing Civil Aviation System Is Expected to Replace a Significant Number of Ground Based Navigation Systems and Allow for More Efficient Use of the World Wide Airspace." *Proceedings of the IEEE* 96 (12): 1902–1917. https://doi.org/10.1109/JPROC.2008.2006090

Li, Xianjie, Jean Pierre Barriot, Yidong Lou, Weixing Zhang, Pengbo Li, and Chuang Shi. 2023. "Towards Millimeter-Level Accuracy in GNSS-Based Space Geodesy: A Review of Error Budget for GNSS Precise Point Positioning." *Surveys in Geophysics.* Springer Science and Business Media B.V. https://doi.org/10.1007/s10712-023-09785-w

Lotfy, A., M. A. Abdelfatah, and H. Hosny. 2020. "Improving the Performance of GNSS Kinematic Point Positioning Technique by Simulated Data." *NRIAG Journal of Astronomy and Geophysics* 9 (1): 491–498. https://doi.org/10.1080/20909977.2020.1783633

Luo, X., S. Schaufler, M. Branzanti, and J. Chen. 2021. "Assessing the Benefits of Galileo to High-Precision GNSS Positioning – RTK, PPP and Post-Processing." *Advances in Space Research* 68 (12): 4916–4931. https://doi.org/10.1016/j.asr.2020.08.022

Martínez-Carricondo, Patricio, Francisco Agüera-Vega, and Fernando Carvajal-Ramírez. 2023. "Accuracy Assessment of RTK/PPK UAV-Photogrammetry Projects Using Differential Corrections from Multiple GNSS Fixed Base Stations." *Geocarto International* 38 (1). https://doi.org/10.1080/1010604 9.2023.2197507

Pan, Lin, Xiaohong Zhang, Xingxing Li, Xin Li, Cuixian Lu, Jingnan Liu, and Qianxin Wang. 2019. "Satellite Availability and Point Positioning Accuracy Evaluation on a Global Scale for Integration of GPS, GLONASS, BeiDou and Galileo." *Advances in Space Research* 63 (9): 2696–2710. https://doi.org/10.1016/j.asr.2017.07.029

Wielgosz, P., I. Kashani, and D. Grejner-Brzezinska. 2005. "Analysis of Long-Range Network RTK during a Severe Ionospheric Storm." *Journal of Geodesy* 79 (9): 524–531. https://doi.org/10.1007/s00190-005-0003-y

Zangenehnejad, Farzaneh, and Yang Gao. 2021. "GNSS Smartphones Positioning: Advances, Challenges, Opportunities, and Future Perspectives." *Satellite Navigation.* Springer. https://doi.org/10.1186/s43020-021-00054-y

Section 2

GNSS Applications for Disaster Management

5 Cyclone Global Navigation Satellite System (CYGNSS)

Insights into the Tropical Cyclone Mocha (2023) in the Bay of Bengal

Vivek Singh, Gaurav Tiwari, Megha Maheshwari, and Abhijeet Kumar

5.1 INTRODUCTION

Tropical cyclones (TCs) are among the most destructive natural phenomena, causing significant loss of life, economic damage, and environmental degradation, particularly in coastal regions. The Indian coastal landmass is highly vulnerable to TCs, with an average of five to six TCs affecting the region annually (Gupta et al., 2019). The North Indian Ocean (NIO), consisting of the Bay of Bengal (BoB) and the Arabian Sea (ARB), experiences significant cyclone activity (Xiao-Ting et al., 2020; Singh et al., 2021; Singh et al., 2023), with the BoB being particularly more active than the ARB in terms of cyclone formation and intensity (Deshpande et al., 2021; Sahoo & Bhaskaran, 2016; Tiwari et al., 2022a; Tiwari et al., 2022b). These areas' unique geophysical location and warm tropical waters provide an ideal setting for cyclogenesis. Accurate monitoring and prediction of these TCs are crucial for disaster preparedness and mitigation efforts. Traditional satellite-based monitoring systems, such as geostationary and polar-orbiting satellites, have been instrumental in tracking and forecasting cyclones (Nam & Park, 2018; Velden & Hawkins, 2010). However, they have temporal and spatial resolution limitations, which can affect the accuracy of predictions, especially in the rapidly changing atmospheric conditions associated with TCs (Katsaros et al., 2002; Olander & Velden, 2007).

The Cyclone Global Navigation Satellite System (CYGNSS), launched by the National Aeronautics and Space Administration (NASA), United States of America, in December 2016, presents a novel approach to TC monitoring by utilizing a constellation of eight microsatellites (Figure 5.1a) (Carreno-Luengo et al., 2021). These satellites receive reflected signals from Global Navigation Satellite System (GNSS) satellites, primarily the Global Positioning System (GPS), to measure ocean surface wind speeds with a high temporal and spatial resolution (Figure 5.1b; Clarizia and Ruf, 2016; Morris and Ruf, 2017a; Morris and Ruf, 2017b). CYGNSS offers several advantages over traditional remote sensing methods, including the ability to penetrate heavy precipitation (near the eye wall and inner rain bands region of TCs) and provide continuous coverage over tropical regions due to the very short revisit time of microsatellites (~ 3 hrs 'median' and ~ 7 hrs 'mean'), thereby enhancing the capability to monitor the evolution and further development of cyclones (Wang et al., 2018).

This extremely short revisit time of the CYGNSS constellation enables more frequent observations, which are crucial for capturing rapid changes during the intensification stages of TCs. Integrating CYGNSS data into cyclone monitoring and prediction frameworks can significantly improve forecasts' accuracy and lead time. Keeping this in mind, this chapter explores the potential of leveraging CYGNSS to enhance cyclone monitoring and prediction over the Indian region.

DOI: 10.1201/9781032712444-7

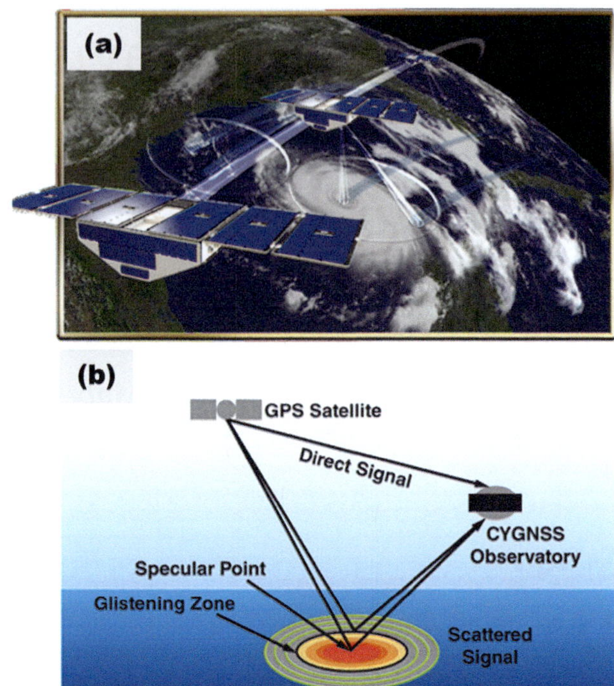

FIGURE 5.1 (a) Graphical Illustration of the CYGNSS satellite system (Image Credit: Southwest Research Institute, San Antonio, Texas), (b) Basic working mechanism where each CYGNSS observatory receives both a direct signal from a GPS satellite and a signal reflected from the ocean surface (Image Credit: University of Michigan, Ann Arbor, Michigan, United States).

We have analyzed a case study of the Extremely Severe Cyclonic Storm (ESCS) Mocha (9th–15th May 2023) over the BoB to illustrate the practical applications and benefits of utilizing CYGNSS data during cyclonic events. The analysis aims to demonstrate that CYGNSS is a valuable tool for enhancing the resilience of vulnerable communities in the Indian subcontinent against the impacts of TCs.

5.2 DETAILED OVERVIEW OF GNSS DATA PRODUCTS

5.2.1 GNSS Radio Occultation (GNSS-RO)

GNSS is a collective term for satellite systems, including Global Positioning System (GPS; United States of America), Global Navigation Satellite System (GLONASS; Russia), Galileo (European Union), NavIC (India), and BeiDou (China), that provide autonomous geospatial positioning within their service areas. Based on satellite signals, these satellite systems allow ground-based users to determine their exact location (latitude, longitude, and altitude). Beyond navigation, GNSSs have applications in numerous fields, including meteorology. The precise time and frequency shift information provided by GNSSs can be used for atmospheric sounding, offering valuable insights into humidity profiles, temperature, and ionospheric data, all critical for meteorological applications and cyclone predictions (Cardellach & Xie, 2014; Jin et al., 2011; Steiner et al., 2001).

GPS-RO and GNSS-R are two key techniques that utilize GNSS data for atmospheric and environmental monitoring, respectively (Bai et al., 2020; Bonafoni et al., 2019; Peng & Jin, 2019). GPS, the most well-known subset of the GNSS system, transmits signals on two main frequencies. The signals travel through the atmosphere before reaching ground-based receivers. As the signal traverses the atmosphere, it undergoes deceleration and refraction, experiencing temporal delays due to the presence of electrons in the ionosphere and water vapor in the troposphere (Kursinski et al., 1997). These refractions and delays provide a unique vantage point for atmospheric investigation when signals are meticulously processed and interpreted (Pottiaux, 2010).

Real-time GNSS processing has emerged as an invaluable tool, especially in meteorological contexts. It allows for immediate atmospheric estimates post-observation, proving particularly beneficial for weather forecasting and issuing severe weather warnings. Key among these real-time processes is the estimation of 'Precipitable Water Vapor' using GNSS-derived 'zenith hydrostatic delay' and 'zenith wet delay' (Li et al., 2021; Su et al., 2021). This method is pivotal in nowcasting scenarios (Benevides et al., 2015; Guerova et al., 2022). Furthermore, real-time-derived troposphere gradients, pinpointing localized atmospheric variations, have become indispensable for monitoring the emergence and development of thunderstorms and hurricanes (Kačmařík et al., 2019). Ground-based GNSS receivers can also assess the overall quantity of 'water vapor' (also known as 'Integrated Precipitable Water Vapor') present in the atmosphere through the analysis of the delay caused by the troposphere, referred to as 'Zenith Total Delay' (Pacione & Vespe, 2003). Despite its slight delay, near real-time processing is advantageous due to its enhanced data quality checks.

This form of processing bolsters the accuracy of precipitable water vapor estimates, especially when complemented with data from neighboring stations (Rózsa et al., 2021; Yoon et al., 2017; Bosy et al., 2012; Dymarska et al., 2017). A particularly notable output from near real-time processing is the creation of regional Integrated Water Vapor maps, which offer comprehensive insights into the spatial distribution of atmospheric moisture (Calori et al., 2016). Post-processing, with its inherent advantage of integrating many datasets and meticulous quality control, stands out for its applicability in climate research. One can harness post-processed GNSS data for discerning and attributing long-term climatic changes (Chen et al., 2012; Goncalves et al., 2023). Further advancements in this domain allow for the extraction of vertical atmospheric profiles, encompassing temperature and humidity, thereby shedding light on intricate climate dynamics (Syndergaard, 1999; Li et al., 2019).

5.2.2 GNSS-REFLECTOMETRY (GNSS-R)

The fundamental principle of GNSS-R is that GNSS signals can reflect off surfaces such as the ocean, land, ice, and snow and return to the various types of receivers (Figure 5.2). When GNSS signals bounce off surfaces such as the ocean, ice, or land, the characteristics of the reflected signals can provide information about 'Soil moisture levels,' 'Ocean surface winds and wave heights,' and 'Sea ice extent and thicknesses,' respectively. Thus, GNSS-R is useful for environmental monitoring, as it allows for continuous and global data collection. GNSS signals in the form of 'reflectometry' are considered to supplement the 'Scatterometer' data during cyclonic conditions due to the low attenuation of GNSS signals during rain.

In GNSS-R, GNSS signals, which are reflected from the Earth's surface (Rodriguez-Alvarez et al., 2023), carry the information on the surface roughness. This property is used for the retrieval of soil moisture content (Jia & Savi, 2017; Malik et al., 2017; Edokossi et al., 2020). Soil moisture is a crucial parameter in weather and climate modeling, affecting the energy balance between the land surface and the atmosphere. Traditional methods for measuring soil moisture, such as ground-based sensors or satellite-based radiometers, have spatial and temporal resolution limitations. GNSS-R offers the potential for high-resolution, global coverage, and continuous soil moisture monitoring.

GNSS-R can further be used to measure 'ocean surface salinity,' which is an important parameter for understanding ocean circulation, water cycle, and climate change (Sabia et al., 2007; Munoz-Martin & Camps, 2021). The information on salinity retrieved from GNSS-R measurements can complement the information provided by other sensing techniques, such as 'salinity sensors' onboard satellites or Argo floats. Another potential application of GNSS-R in meteorology is the retrieval of 'ocean surface wind speed and direction' (Komjathy et al., 2004; Li et al., 2023). Ocean

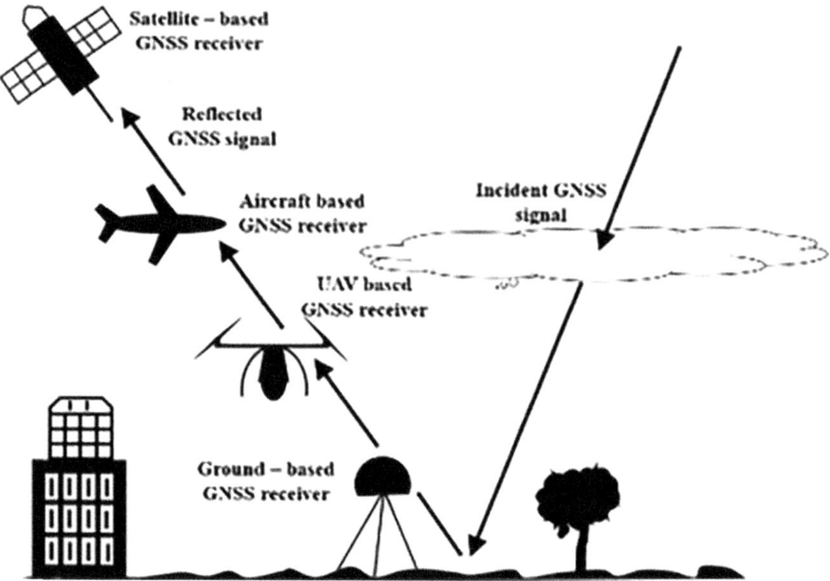

FIGURE 5.2 Utilization of GNSS-R on diverse platforms, encompassing ground-based setups, UAVs, aircraft, and satellites (figure concept adapted from Rodriguez-Alvarez et al., 2023 and replotted).

surface wind speed and direction are crucial parameters for weather and climate modeling, as they affect ocean circulation and heat exchange between the ocean and the atmosphere. Traditional methods for measuring ocean surface wind, such as scatterometers or buoys, have spatial and temporal resolution limitations. GNSS-R offers the potential for high-resolution global coverage and continuous monitoring of ocean surface wind. GNSS-R exhibits sensitivity to variations in 'surface roughnesses' induced by the interaction of wind and ocean currents (Hoseini, & Nahavandchi, 2022). This sensitivity arises from the modulation of GNSS signals reflected off the Earth's surface.

In the context of ocean surface current measurement, GNSS-R can be harnessed to infer information about these currents through the 'roughness sensitivity,' 'signal modulation,' and 'inverse problem solutions.' Further, GNSS-R can be used to measure the 'significant wave height'/ sea surface height (Peng & Jin, 2019; Qiu and Jin, 2020), which is a key parameter for understanding ocean circulation, sea level rise, and climate change and tsunamis. For example, Kegen Yu (2015) investigated the detection of weak tsunamis using noisy sea surface height (SSH) measurement data, particularly those acquired by a satellite-borne receiver employing the GNSS-R technique.

The sea surface height information retrieved from GNSS-R measurements can complement the information provided by other sensing techniques, such as radar altimetry. For example, Zhijin Li et al. (2016) analyzed the effectiveness of GNSS-R altimetry in mapping mesoscale SSH fields. The analysis utilized synthetic measurements derived from a high-resolution (1/10°) numerical model of the North Pacific. GNSS-R technique is also used to measure the dynamics of coastal zones and estuaries, including tide heights, river flows, wetland monitoring, and inland water bodies (Semmling et al., 2011; Ban et al., 2022; Wang & Jade Morton, 2021; Bai et al., 2015; Nghiem et al., 2017).

GNSS-R has proven to be an effective technique for measuring ocean surface winds in tropical cyclones, as it can penetrate through heavy rain and cloud cover that often accompanies these storms (Li et al., 2023). CYGNSS was specifically designed to take advantage of this capability of GNSS-R. While GNSS-R is a promising technique with potential applications in meteorology, there are also potential limitations and challenges associated with its use.

One challenge is related to the accuracy of the GNSS-R measurements. The accuracy of GNSS-R measurements can be affected by a range of factors, including the incidence angle of the satellite signal, the quality of the GNSS signal, and the characteristics of the surface being measured. This can result in errors or uncertainties in the retrieved parameters, such as 'soil moisture content' or 'ocean surface wind speed and direction.' Therefore, it is crucial to carefully validate the accuracy of the retrieved parameters, particularly in comparison with ground-based measurements or in-situ data.

Another challenge is related to the spatial and temporal resolution of GNSS-R measurements. The spatial and temporal resolution of GNSS-R measurements can be affected by the number and distribution of GNSS-R receivers, as well as the orbit characteristics of the GNSS satellites. This can result in gaps or inconsistencies in the data, particularly in areas with limited GNSS-R coverage. In addition, because GNSS-R relies on GNSS signals, it can be affected by signal interference or jamming, which can degrade the quality of the measured data. Moreover, GNSS-R measurements are impacted by the atmosphere, which can introduce errors in the retrieved parameters. Thus, GNSS-R provides a valuable contribution to ocean remote sensing by providing information on key parameters such as ocean surface wind speed and direction, salinity, surface currents, and sea surface height.

5.2.3 CYGNSS

Integrating GNSS and CYGNSS data into cyclone monitoring and forecasting models has greatly improved our ability to understand, track, and predict cyclones (Mueller et al., 2021; Maheshwari et al., 2023; McNoldy et al., 2017). These advancements have far-reaching implications, from enhancing early warning systems to facilitating better-informed disaster preparedness and response, ultimately reducing the impact of cyclones on coastal communities. The innovative use of GNSS technology combined with specialized satellite missions such as CYGNSS represents a significant step forward in mitigating the devastating effects of cyclones. The CYGNSS observatories conduct singular observations of circularly polarized reflectivity at the GPS frequencies, specifically at 1.5 GHz (Boutin et al., 2023).

GNSS technology aids in accurately determining the position and motion of cyclones. Meteorologists and scientists can monitor tropical cyclone events by precisely tracking the storm's movement through satellite-based "GNSS Zenith Tropospheric Delay" data (Lian et al., 2023). This improved monitoring and tracking capability is crucial for issuing early warnings, allowing communities in the cyclone's track to make informed decisions and take necessary precautions, thus potentially saving lives and reducing property damage.

One of the most challenging aspects of cyclone forecasting is accurately predicting its intensity, as it depends on complex interactions between the storm and the surrounding environment. GNSS and CYGNSS data provide valuable insights into the inner core of cyclones, where conventional sensors often fail due to heavy rainfall (Figure 5.3).

By measuring ocean 'surface roughness' and wind speeds, CYGNSS data significantly improve intensity forecasting (Ruf et al., 2017). The improved understanding of cyclone intensification processes allows for more precise and timely warnings, reducing the risk of underestimating a cyclone's destructive potential. Cyclones are associated with heavy rainfall, which can lead to devastating flooding. GNSS technology assists in assessing the atmospheric water vapor content within and around cyclones, providing valuable data for understanding and predicting these extreme weather events. This information is essential for more accurate rainfall predictions, enabling better anticipation of flood risks and the allocation of resources for disaster response and recovery efforts (Bevis et al., 1992).

The combination of GNSS and CYGNSS data improves cyclone forecasts' accuracy and enhances risk assessment and preparedness. By providing detailed information about cyclone characteristics, including wind speed, structure, and intensification processes, these technologies empower disaster management agencies and local authorities to develop more effective evacuation plans and allocate resources where they are needed most. This is particularly crucial in densely populated and vulnerable coastal regions.

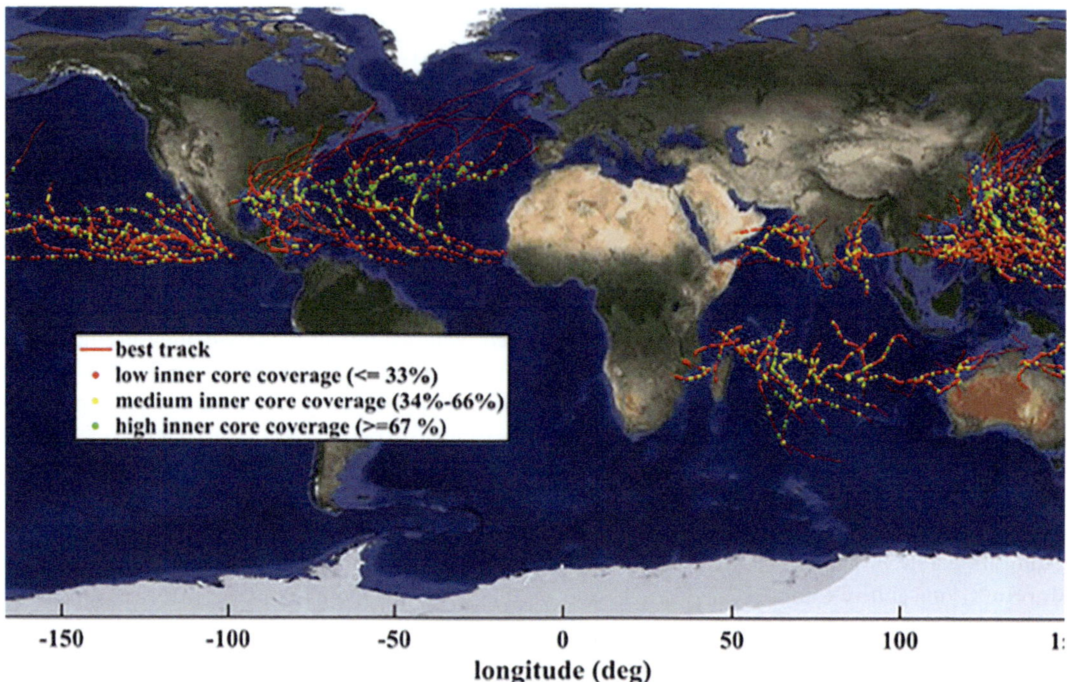

FIGURE 5.3 CYGNSS overpasses of all significant storms from 2018 to 2020, with inner core coverage quality indicated. The red curve represents the observed tracks of the storms across various world basins during this period. Red dots denote low inner core coverage (≤ 33%), yellow dots denote medium inner core coverage (34–66%), and green dots denote high inner core coverage (≥ 67%). The figure was downloaded from NASA's website at https://earthobservatory.nasa.gov/blogs/fromthefield/category/cygnss/.

The CYGNSS mission comprises a fleet of eight satellites, each with L-band receivers specifically calibrated to detect and analyze the signals emitted by the GPS. These receivers are precisely tuned to capture the GPS signals that have been reflected from the ocean's surface, specifically in the forward direction (Asharaf et al., 2021). Positioned at approximately 35° orbital inclination from the equator, CYGNSS ensures nearly continuous global coverage, primarily focusing on ocean surface winds between 38°N and 38°S latitudes critical range for studying tropical cyclone formation and movement. This contrasts radar scatterometers like ISS-RapidScat, QuikSCAT, and Advanced Scatterometer (ASCAT), which emit and receive microwave radar pulses. CYGNSS operates passively, receiving signals from surface-reflected GPS pulses, providing a critical advantage in measuring high wind speeds in hurricane eye-wall region areas, typically challenging for traditional microwave scatterometry due to signal degradation from intense rainfall.

In this study, the ESCS 'Mocha' is analyzed using CYGNSS data. The data was accessed through NASA's Physical Oceanography Distributed Active Archive Center (PODAAC) data portal (https://podaac.jpl.nasa.gov/dataset/CYGNSS_L2_V2.1#). We retrieved Level-2 (CYGNSS_L2_V2.1) data, which includes wind speed measurements. CYGNSS data is typically available in NetCDF-4 format. We extracted geolocation information, including latitude and longitude, to accurately map wind speed measurements.

5.3 SYNOPTIC FEATURES OF ESCS MOCHA (2023)

In this chapter, an examination of the Extremely Severe Cyclonic Storm (ESCS) 'Mocha' is conducted through the analysis of CYGNSS data in the form of a case study. The first cyclonic storm

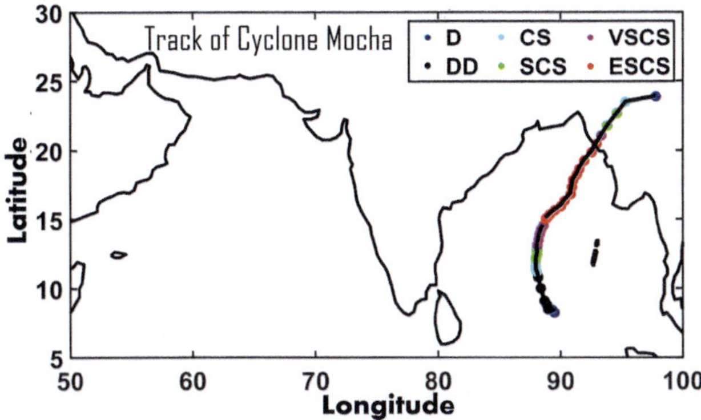

FIGURE 5.4 The observed track of ESCS 'Mocha' during its lifetime (9–15 May 2023).

of 2023, ESCS 'Mocha,' occurred over the BoB from 9–15 May 2023 (Figure 5.4 shows the observed best-track provided by IMD, New Delhi, India) during displayed several curvatures along its path. On May 6, 2023, in the early hours of 0830 IST/0300 UTC, a cyclonic circulation developed over the Southeast BoB and its environments. On May 8, 2023, at 0830 hours IST/0300 UTC, it evolved further into a low-pressure area over the southeast BoB and the adjacent south Andaman Sea. It then matured into a well-marked low-pressure area on May 9, 2023, and transformed into a Depression on the evening of May 9. This Depression further intensified into a Deep Depression on May 10 and eventually evolved into a Cyclonic Storm named 'Mocha' on May 11. Mocha continued to intensify, reaching Very Severe Cyclonic Storm status on May 12. Further, it peaked with wind speeds of 210–220 kmph, gusting to 240 kmph, over the east-central Bay of Bengal from May 13 to May 14. Subsequently, Mocha weakened and made landfall between Myanmar and Bangladesh on May 14 as an Extremely Severe Cyclonic Storm (ESCS). After landfall, the cyclone rapidly weakened into a Depression over northwest Myanmar on May 15.

5.4 MONITORING OF ESCS MOCHA USING CYGNSS DATA

The descending passage time refers to the time at which a CYGNSS satellite crosses the equator from north to south. Figure 5.5 depicts the CYGNSS-derived passage times for May 13th and 14th, 2023, respectively. The hollow circle curves in Figure 5.5a and b represent the track of cyclone Mocha. On May 13th, the descending passage times of CYGNSS satellites were spaced roughly 90 minutes apart. On the 14th of May, the passage times continued in a similar pattern, shifting slightly due to the Earth's rotation and the satellite's orbital progression. By overlaying the descending passage times of the CYGNSS satellites on the cyclone Mocha's track, we can determine when the satellites observed the storm (Figure 5.5a and b).

Further, the investigation of CYGNSS-derived wind speed on May 13th and 14th of May 2023, (Figure 5.6a and b, respectively) provides crucial insights into the atmospheric dynamics of the Mocha. The variations in wind speed throughout each day are graphically represented (shaded lines), presenting a comprehensive view of the storm's dynamic behavior. The utilization of CYGNSS data, known for its capability to measure reflected satellite signals with exceptional accuracy, offers a detailed examination of the wind speeds associated with ESCS Mocha during these relevant dates (Cyclone was rapidly intensified on these days).

National Oceanic and Atmospheric Administration (NOAA) CYGNSS winds from ascending and descending passes for 13th and 14th May are depicted in Figure 5.7. When a satellite moves from the Southern Hemisphere towards the Northern Hemisphere, it is on an ascending pass. The color bar shows the wind speed (in knots). On 13th May (Figure 5.7, first row), the color bar shows

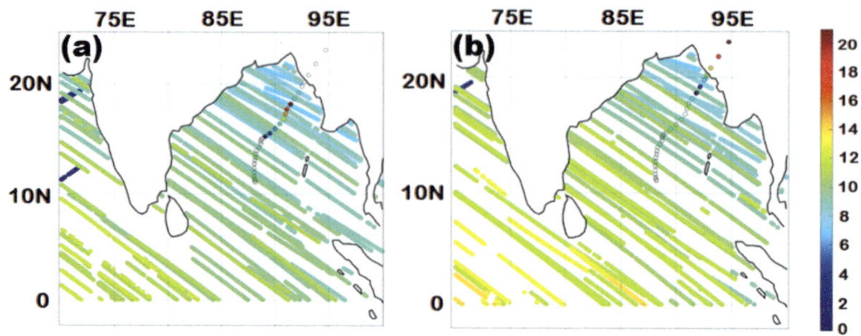

FIGURE 5.5 CYGNSS-derived descending passage time (of the day) for 13th May (a) and 14th May (b). 2023, respectively. The hollow circle curve represents the ESCS Mocha track during its lifetime.

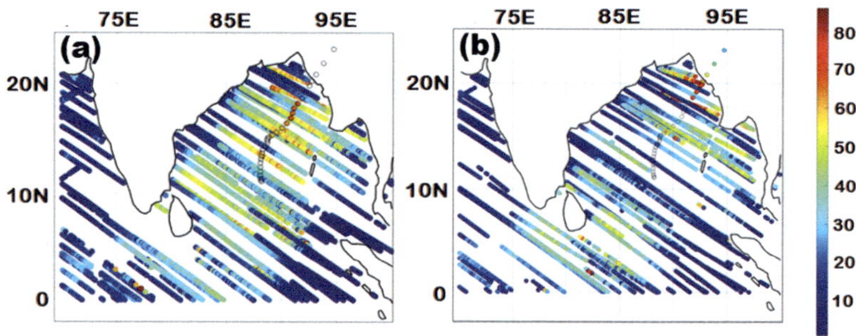

FIGURE 5.6 CYGNSS-derived wind speed for (a) 13th May and (b) 14th May 2023, respectively. The maximum intensity of ESCS "Mocha" is represented by the black circle.

stronger winds from both passes over the region of the cyclone Mocha. Referring to Figure 5.4, on the next day the cyclone was relatively weaker, and such a weakening of the cyclone can be noticed in Figure 5.7c–d.

CYGNSS is a constellation of eight microsatellites designed to measure surface wind speeds over the oceans, especially in and near the eye of tropical cyclones, hurricanes, and typhoons. Each of the eight satellites in the CYGNSS constellation is referred to as a Flight Model (FM). These are labeled FM01, FM02, FM03, …, FM08. Each FM satellite collects reflected GPS signals from the ocean surface, which are used to derive various measurements such as wind speed and surface roughness. Having multiple FMs increases the spatial and temporal coverage, enhancing the reliability and resolution of the collected data. A zoomed-in view, targeted over the BoB, of Figure 5.7 is shown in Figure 5.8.

Measured on the ocean surface, the mean square slope (MSS) represents the average slope of the small-scale waves. Its mathematical definition is the mean of the squares of the components of the surface slope in two perpendicular directions. The roughness of the ocean surface, directly correlated with wind direction and speed over the ocean, is thus described by the MSS, an important parameter derived from CYGNSS data. By examining GPS signals reflected off the sea surface, CYGNSS calculates the ocean's surface roughness. The local wind conditions have an impact on the small-scale waves on the ocean's surface, which in turn affect these reflections. In Figure 5.9a and b, relatively high MSS values over the BoB indicate a rougher surface, typically caused by stronger winds. In this way, the intensity of the cyclone Mocha can be addressed using the CYGNSS MSS product.

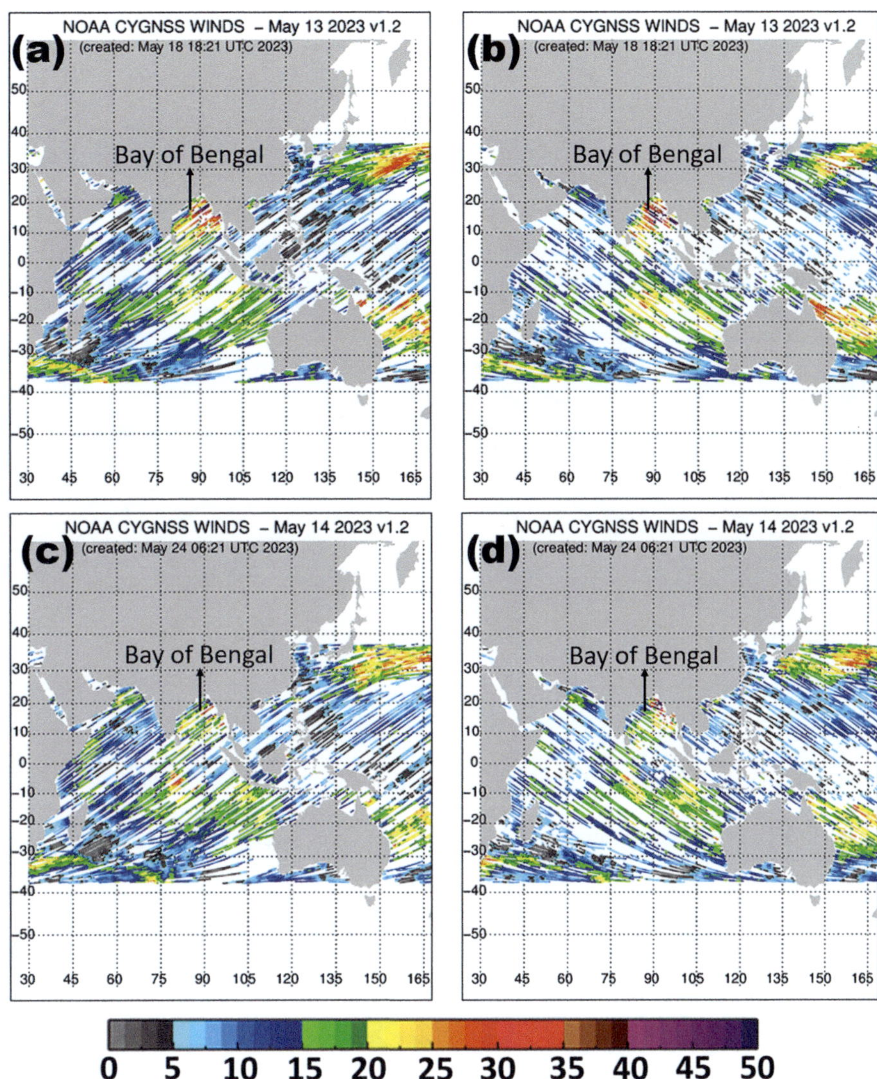

FIGURE 5.7 NOAA CYGNSS winds (knots) for 13th May (first row) and 14th May 2023 (second row) from ascending pass (first column), and descending pass (second column) (Image Credit: CYGNSS).

5.5 SUMMARY

This chapter delved into the transformative potential of Global Navigation Satellite System (GNSS) technology, with a focus on the Cyclone Global Navigation Satellite System (CYGNSS), to enhance cyclone monitoring and prediction in the Indian region. The discussion commenced with a comprehensive exploration of GNSS Reflectometry (GNSS-R) and GNSS Radio Occultation (GNSS-RO)—fundamental principles and their applications in meteorology, establishing a theoretical foundation for leveraging GNSSs in cyclone studies. The subsequent in-depth analysis of CYGNSS underscored its instrumental role in advancing cyclone monitoring and prediction capabilities, offering unique insights into critical parameters such as wind speed. The case study on Extremely Severe Cyclonic Storm 'Mocha' provided a practical application of CYGNSS data analysis, showcasing the temporal evolution of the storm's wind speed and the associated hourly variations on 13th and 14th May 2023. By integrating theoretical knowledge with practical applications, the present chapter

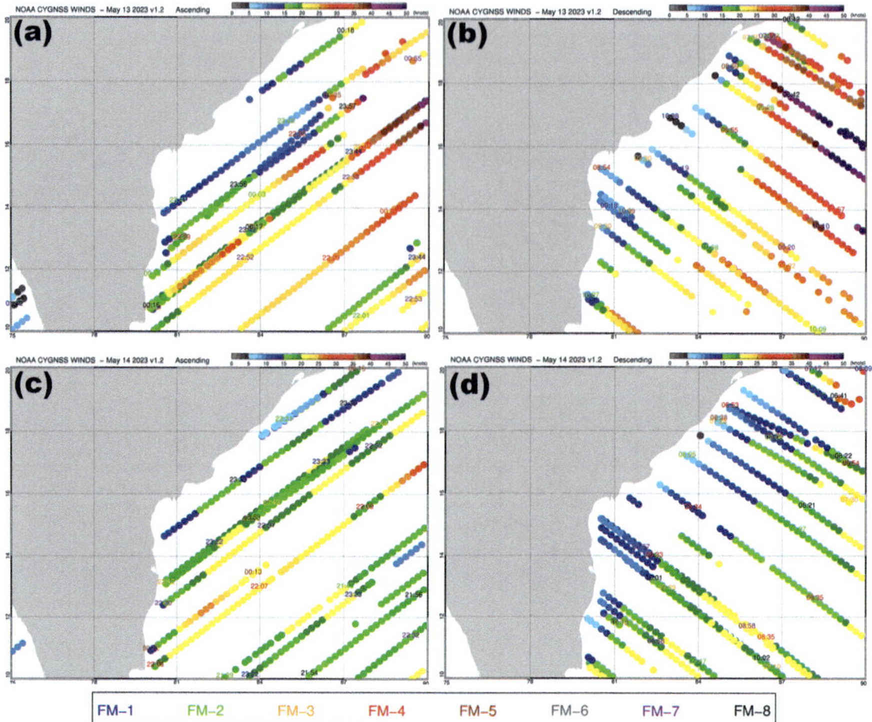

FIGURE 5.8   A zoomed-in snapshot of Figure 5.7 focused on the Bay of Bengal (Image Credit: CYGNSS).

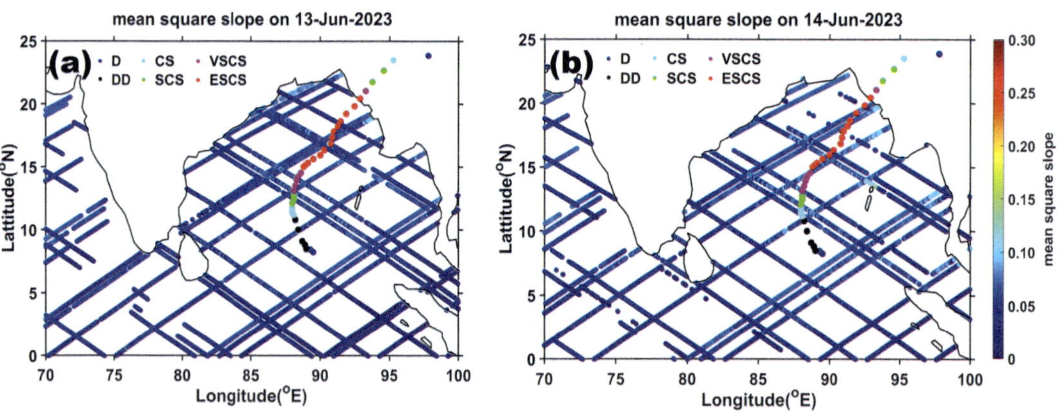

FIGURE 5.9   CYGNSS-derived mean square slope (MSS) for (a) 13th May and (b) 14th May 2023, respectively.

highlights the significant contributions of GNSS and CYGNSS to the field of meteorology, ultimately paving the way for improved cyclone forecasting methodologies and disaster preparedness in the Indian region. The insights gained from the present study may contribute to the broader discourse on leveraging advanced satellite technologies for enhanced weather monitoring and prediction.

ACKNOWLEDGEMENT

The authors express their sincere gratitude to the Directors of IITM, Pune, Ministry of Earth Sciences, Government of India, and the Space Navigation Group at ISRO, Bengaluru for their unwavering support. We acknowledge NASA for providing CYGNSS data crucial to our analysis. Lastly, our appreciation extends to the India Meteorological Department (IMD) for providing the best track data for Extremely Severe Cyclonic Storm 'Mocha' (2023).

REFERENCES

Asharaf, Shakeel, Duane E. Waliser, Derek J. Posselt, Christopher S. Ruf, Chidong Zhang, and Agie W. Putra. "CYGNSS ocean surface wind validation in the tropics." *Journal of Atmospheric and Oceanic Technology* 38, no. 4 (2021): 711–724.

Chen, J., Wu, B., Hu, X., Li, H (2012) SHA: The GNSS Analysis Center at SHAO. In: Sun J, Liu J, Yang Y, Fan S (eds) China Satellite Navigation Conference (CSNC) 2012 Proceedings. Springer, Berlin, Heidelberg, pp. 213–221

Maheshwari, M., Chakraborty, A., Kumar, A. and Nirmala, S., 2023. Uncertainty assessment and an improved CYGNSS cyclonic wind speed retrieval model for cyclones over North Indian Ocean. *Journal of Earth System Science*, *132*(1), p. 18.

McNoldy, B., Annane, B., Majumdar, S., Delgado, J., Bucci, L. and Atlas, R., 2017. Impact of assimilating CYGNSS data on tropical cyclone analyses and forecasts in a regional OSSE framework. *Marine Technology Society Journal*, *51*(1), pp. 7–15.

Morris, M. and Ruf, C.S., 2017a. Determining tropical cyclone surface wind speed structure and intensity with the CYGNSS satellite constellation. *Journal of Applied Meteorology and Climatology*, *56*(7), pp. 1847–1865.

Mueller, M.J., Annane, B., Leidner, S.M. and Cucurull, L., 2021. Impact of CYGNSS-derived winds on tropical cyclone forecasts in a global and regional model. *Monthly Weather Review*, *149*(10), pp. 3433–3447.

Morris, M. and Ruf, C.S., 2017b. Estimating tropical cyclone integrated kinetic energy with the CYGNSS satellite constellation. *Journal of Applied Meteorology and Climatology*, *56*(1), pp. 235–245.

Cardellach, Estel, and Feiqin Xie. *GNSS remote sensing*. Ed. Shuanggen Jin. Vol. 16. Dordrecht: Springer, 2014.

Sabia, R., Caparrini, M., Camps, A. and Ruffini, G., 2007. Potential synergetic use of GNSS-R signals to improve the sea-state correction in the sea surface salinity estimation: Application to the SMOS mission. *IEEE Transactions on Geoscience and Remote Sensing*, *45*(7), pp. 2088–2097.

Clarizia, M.P. and Ruf, C.S., 2016. Wind speed retrieval algorithm for the Cyclone Global Navigation Satellite System (CYGNSS) mission. *IEEE Transactions on Geoscience and Remote Sensing*, *54*(8), pp. 4419–4432.

Deshpande, M., Singh, V.K., Ganadhi, M.K., Roxy, M.K., Emmanuel, R. and Kumar, U., 2021. Changing status of tropical cyclones over the north Indian Ocean. *Climate Dynamics*, *57*, pp. 3545–3567.

Jin, S., Feng, G.P. and Gleason, S., 2011. Remote sensing using GNSS signals: Current status and future directions. *Advances in Space Research*, *47*(10), pp. 1645–1653.

Bai, Weihua, Junming Xia, Wei Wan, Limin Zhao, Yueqiang Sun, Xiangguang Meng, Congliang Liu et al. "A first comprehensive evaluation of China's GNSS-R airborne campaign: part II—river remote sensing." *Science Bulletin* 60, no. 17 (2015), pp. 1527–1534.

Steiner, A.K., Kirchengast, G., Foelsche, U., Kornblueh, L., Manzini, E. and Bengtsson, L., 2001. GNSS occultation sounding for climate monitoring. *Physics and Chemistry of the Earth, Part A: Solid Earth and Geodesy*, *26*(3), pp. 113–124.

Ban, Wei, Kefei Zhang, Kegen Yu, Nanshan Zheng, and Shuo Chen. "Detection of red tide over sea surface using GNSS-R spaceborne observations." *IEEE Transactions on Geoscience and Remote Sensing* 60 (2022), pp. 1–11.

Benevides, P., J. Catalao, and P. M. A. Miranda. "On the inclusion of GPS precipitable water vapour in the nowcasting of rainfall." *Natural Hazards and Earth System Sciences* 15(12) (2015), pp. 2605–2616.

Bevis, M., Businger, S., Herring, T. A., Rocken, C., Anthes, R. A., & Ware, R. H. (1992). GPS meteorology: Remote sensing of atmospheric water vapor using the global positioning system. *Journal of Geophysical Research: Atmospheres*, *97*(D14), pp. 15787–15801.

Bonafoni, Stefania, Riccardo Biondi, Hugues Brenot, and Richard Anthes. "Radio occultation and ground-based GNSS products for observing, understanding and predicting extreme events: A review." *Atmospheric Research* 230 (2019), p. 104624.

Bosy, Jarosław, J. Kaplon, Witold Rohm, Jan Sierny, and T. Hadas. "Near real-time estimation of water vapour in the troposphere using ground GNSS and the meteorological data." In *Annales Geophysicae*, vol. 30, no. 9, pp. 1379–1391. Göttingen, Germany: Copernicus Publications, 2012.

Boutin, Jacqueline, S. Yueh, R. Bindlish, S. Chan, D. Entekhabi, Y. Kerr, Nicolas Kolodziejczyk, T. Lee, Nicolas Reul, and M. Zribi. "Soil moisture and sea surface salinity derived from satellite-borne sensors." *Surveys in Geophysics* (2023), pp. 1–39.

Calori, A., Santos, J. R., Blanco, M., Pessano, H., Llamedo, P., Alexander, P., & de la Torre, A. (2016). Ground-based GNSS network and integrated water vapor mapping during the development of severe storms at the Cuyo region (Argentina). *Atmospheric Research, 176*, pp. 267–275.

Li, Z., Guo, F., Chen, F., Zhang, Z. and Zhang, X., 2023. Wind speed retrieval using GNSS-R technique with geographic partitioning. *Satellite Navigation, 4*(1), p. 4.

Carreno-Luengo, H., Crespo, J.A., Akbar, R., Bringer, A., Warnock, A., Morris, M. and Ruf, C., 2021. The CYGNSS mission: On-going science team investigations. *Remote Sensing, 13*(9), p. 1814.

Dymarska, Natalia, Witold Rohm, Jan Sierny, Jan Kapłon, Tomasz Kubik, Maciej Kryza, Jerzy Jutarski, Jacek Gierczak, and Ryszard Kosierb. "An assessment of the quality of near-real time GNSS observations as a potential data source for meteorology." *Meteorology Hydrology and Water Management. Research and Operational Applications* 5(1) (2017), pp. 3–13.

Gupta, S., Jain, I., Johari, P. and Lal, M., 2019. Impact of climate change on tropical cyclones frequency and intensity on Indian coasts. In *Proceedings of international conference on remote sensing for disaster management: Issues and challenges in disaster management* (pp. 359–365). Springer International Publishing.

Hoseini, M., & Nahavandchi, H. (2022). The potential of spaceborne GNSS reflectometry for detecting ocean surface currents. *Remote Sensing of Environment, 282*, p. 113256.

Ruf, Christopher, Scott Gleason, Aaron Ridley, Randall Rose, and John Scherrer. "The nasa cygnss mission: Overview and status update." In *2017 IEEE International Geoscience and Remote Sensing Symposium (IGARSS)* (pp. 2641–2643). IEEE, 2017.

Tiwari, Gaurav, Arathi Rameshan, Pankaj Kumar, Aaquib Javed, and Alok K. Mishra. "Understanding the post-monsoon tropical cyclone variability and trend over the Bay of Bengal during the satellite era." *Quarterly Journal of the Royal Meteorological Society* 148(742) (2022a), pp. 1–14.

Tiwari, Gaurav, Pankaj Kumar, Aaquib Javed, Alok Kumar Mishra, and Ashish Routray. "Assessing tropical cyclones characteristics over the Arabian Sea and Bay of Bengal in the recent decades." *Meteorology and Atmospheric Physics* 134(3) (2022b), p. 44.

Peng, Q. and Jin, S., 2019. Significant wave height estimation from space-borne cyclone-GNSS reflectometry. *Remote Sensing, 11*(5), p. 584.

Singh, V., Konduru, R. T., Srivastava, A. K., Momin, I. M., Kumar, S., Singh, A. K., … & Sinha, A. K. (2021). Predicting the rapid intensification and dynamics of pre-monsoon extremely severe cyclonic storm 'Fani'(2019) over the Bay of Bengal in a 12–km global model. *Atmospheric Research, 247*, p. 105222.

Sahoo, Bishnupriya, and Prasad K. Bhaskaran. "Assessment on historical cyclone tracks in the Bay of Bengal, east coast of India." *International Journal of Climatology 36*(1) (2016), pp. 95–109.

Velden, C. and Hawkins, J., 2010. Satellite observations of tropical cyclones. In *Global Perspectives on Tropical Cyclones: From Science to Mitigation* (pp. 201–226).

Nghiem, Son V., Cinzia Zuffada, Rashmi Shah, Clara Chew, Stephen T. Lowe, Anthony J. Mannucci, Estel Cardellach, G. Robert Brakenridge, Gary Geller, and Ake Rosenqvist. "Wetland monitoring with global navigation satellite system reflectometry." *Earth and Space Science* 4(1) (2017), pp. 16–39.

Yu, Kegen. "Weak tsunami detection using GNSS-R-based sea surface height measurement." *IEEE Transactions on Geoscience and Remote Sensing* 54, no. 3 (2015): 1363–1375.

Wang, T., Ruf, C.S., Block, B., McKague, D.S. and Gleason, S., 2018. Design and performance of a GPS constellation power monitor system for improved CYGNSS L1B calibration. *IEEE Journal of Selected Topics in Applied Earth Observations and Remote Sensing, 12*(1), pp. 26–36.

Li, Zhijin, Cinzia Zuffada, Stephen T. Lowe, Tong Lee, and Victor Zlotnicki. "Analysis of GNSS-R altimetry for mapping ocean mesoscale sea surface heights using high-resolution model simulations." *IEEE Journal of Selected Topics in Applied Earth Observations and Remote Sensing* 9(10) (2016), pp. 4631–4642.

Munoz-Martin, Joan Francesc, and Adriano Camps. "Sea surface salinity and wind speed retrievals using GNSS-R and L-band microwave radiometry data from FMPL-2 onboard the FSSCat mission." *Remote Sensing* 13(16) (2021), p. 3224.

Komjathy, Attila, Michael Armatys, Dallas Masters, Penina Axelrad, Valery Zavorotny, and Steven Katzberg. "Retrieval of ocean surface wind speed and wind direction using reflected GPS signals." *Journal of Atmospheric and Oceanic Technology 21*(3) (2004), pp. 515–526.

Semmling, A. M., Georg Beyerle, Ralf Stosius, Galina Dick, Jens Wickert, F. Fabra, Estel Cardellach et al. "Detection of Arctic Ocean tides using interferometric GNSS-R signals." *Geophysical Research Letters 38*(4) (2011).

Jia, Y., & Savi, P. (2017). Sensing soil moisture and vegetation using GNSS-R polarimetric measurement. *Advances in space research, 59*(3), pp. 858–869.

Bai, Weihua, Nan Deng, Yueqiang Sun, Qifei Du, Junming Xia, Xianyi Wang, Xiangguang Meng, et al. "Applications of GNSS-RO to numerical weather prediction and tropical cyclone forecast." *Atmosphere 11*(11) (2020): 1204.

Katsaros, K.B., Vachon, P.W., Liu, W.T. and Black, P.G., 2002. Microwave remote sensing of tropical cyclones from space. *Journal of Oceanography, 58*, pp. 137–151.

Li, Y., Kirchengast, G., Scherllin-Pirscher, B., Schwaerz, M., Nielsen, J. K., Ho, S. P., & Yuan, Y. B. (2019). A new algorithm for the retrieval of atmospheric profiles from GNSS radio occultation data in moist air and comparison to 1DVar retrievals. *Remote Sensing, 11*(23), p. 2729.

Malik, J. S., Jingrui, Z., & Naqvi, N. A. (2017, November). Soil moisture content estimation using GNSS reflectometry (GNSS-R). In *2017 Fifth International Conference on Aerospace Science & Engineering (ICASE)* (pp. 1–9). IEEE.

Nam, S. and Park, K., 2018. Status and prospects of marine wind observations from geostationary and polar-orbiting satellites for tropical cyclone studies. *Journal of the Korean earth science society, 39*(4), pp. 305–316.

Rodriguez-Alvarez, N., Munoz-Martin, J. F., & Morris, M. (2023). Latest Advances in the Global Navigation Satellite System—Reflectometry (GNSS-R) Field. *Remote Sensing, 15*(8), pp. 2157.

Syndergaard, S. (1999). *Retrieval analysis and methodologies in atmospheric limb sounding using the GNSS radio occultation technique* (pp. 31–35). Copenhagen, Denmark: Danish Meteorological Institute.

Wang, Yang, and Y. Jade Morton. "River slope observation from spaceborne GNSS-R carrier phase measurements: a case study." *IEEE Geoscience and Remote Sensing Letters 19* (2021), pp. 1–5.

Qiu, H. and Jin, S., 2020. Global mean sea surface height estimated from spaceborne cyclone-GNSS reflectometry. *Remote Sensing, 12*(3), p. 356.

Goncalves, Rodrigo Mikosz, Júlia Isabel Pontes, Flávia Helena Manhães Vasconcellos, Lígia Albuquerque de Alcântara Ferreira, Heithor Alexandre de Araújo Queiroz, and Paulo Henrique Gomes de Oliveira Sousa. "High spatial resolution data obtained by GNSS and RPAS to assess islets flood-prone scenarios for 2100." *Applied Geography 150* (2023), pp. 102817.

Edokossi, Komi, Andres Calabia, Shuanggen Jin, and Iñigo Molina. "GNSS-reflectometry and remote sensing of soil moisture: A review of measurement techniques, methods, and applications." *Remote Sensing 12*(4) (2020), p. 614.

Pacione, R., & Vespe, F. (2003). GPS zenith total delay estimation in the Mediterranean area for climatological and meteorological applications. *Journal of Atmospheric and Oceanic Technology, 20*(7), pp. 1034–1042.

Kursinski, E. R., G. A. Hajj, J. T. Schofield, R. P. Linfield, and K. Rer Hardy. "Observing Earth's atmosphere with radio occultation measurements using the Global Positioning System." *Journal of Geophysical Research: Atmospheres 102*(D19) (1997), pp. 23429–23465.

Lian, Dajun, Qimin He, Li Li, Kefei Zhang, Erjiang Fu, Guangyan Li, Rui Wang, Biqing Gao, and Kangming Song. "A Novel Method for Monitoring Tropical Cyclones' Movement Using GNSS Zenith Tropospheric Delay." *Remote Sensing 15*(13) (2023), pp. 3247.

Olander, T.L. and Velden, C.S., 2007. The advanced Dvorak technique: Continued development of an objective scheme to estimate tropical cyclone intensity using geostationary infrared satellite imagery. *Weather and Forecasting, 22*(2), pp. 287–298.

Guerova, Guergana, Jan Douša, Tsvetelina Dimitrova, Anastasiya Stoycheva, Pavel Václavovic, and Nikolay Penov. "GNSS storm nowcasting demonstrator for Bulgaria." *Remote Sensing 14*(15) (2022), pp. 3746.

Kačmařík, Michal, Jan Douša, Florian Zus, Pavel Václavovic, Kyriakos Balidakis, Galina Dick, and Jens Wickert. "Sensitivity of GNSS tropospheric gradients to processing options." In *Annales Geophysicae*, vol. 37, no. 3, pp. 429–446. Copernicus GmbH, 2019.

Su, Hang, Tao Yang, Kan Wang, Baoqi Sun, and Xuhai Yang. "Evaluation of Precipitable Water Vapor Retrieval from Homogeneously Reprocessed Long-Term GNSS Tropospheric Zenith Wet Delay, and Multi-Technique." *Remote Sensing 13*(21) (2021), p. 4490.

Pottiaux, E., 2010. Sounding the Earth's atmospheric water vapour using signals emitted by global navigation satellite systems (Doctoral dissertation, PhD thesis, Department of Physics, Earth and Life Institute, Catholic University of Louvain).

Li, Longjiang, Suqin Wu, Kefei Zhang, Xiaoming Wang, Wang Li, Zhen Shen, Dantong Zhu, Qimin He, and Moufeng Wan. "A new zenith hydrostatic delay model for real-time retrievals of GNSS-PWV." *Atmospheric Measurement Techniques 14*, (10) (2021), pp. 6379–6394.

Rózsa, Szabolcs, Abir Khaldi, Ágnes Ács, and Bence Turák. "Multi-GNSS near real-time precipitable water vapour estimation for severe weather prediction." *Scientific Bulletin Series D: Mining, Mineral Processing, Non-Ferrous Metallurgy, Geology and Environmental Engineering 35*(2) (2021), pp. 777–786.

Yoon, Ha Su, Jung Ho Cho, Han Earl Park, and Sung Moon Yoo. "Development of Near Real Time GNSS Precipitable Water Vapor System Using Precise Point Positioning." *Journal of the Korean Society of Surveying, Geodesy, Photogrammetry and Cartography 35*(6) (2017), pp. 471–484.

Xiao-Ting, F., Ying, L.I., Ai-Min, L. and Long-Sheng, L., 2020. Statistical and comparative analysis of tropical cyclone activity over the Arabian Sea and Bay of Bengal (1977-2018). *Journal of Tropical Meteorology*, 26(4), pp. 441–452.

Singh, Vivek, Atul Kumar Srivastava, Rajeeb Samanta, Amarendra Singh, Arun Kumar, and Abhay Kumar Singh. "Investigating Physics Behind the Rapid Intensification and Catastrophic Landfall of Cyclone 'Titli' (2018) in the Bay of Bengal." *Indian Journal of Pure & Applied Physics (IJPAP) 61*(3) (2023), pp. 175–181.

6 Lightning Locations and Thunderstorm-induced Ionospheric Perturbations
Applications of GNSS Technology

Abhay Srivastava, Ogunsua Babalola, Gopal Sharma, Shayam Sundar Kundu, M. Somorjit Singh, and Shiv Prasad Aggarwal

6.1 INTRODUCTION

Lightning is a crucial atmospheric event that leads to loss of life and infrastructure. However, proper dissemination and prevention methods can reduce these losses. Dissemination is possible by suitable forecast and for which the understanding of lightning characteristics is necessary. One way to understand lightning characteristics is to identify and locate the location of the lightning event. In general, lightning can be located using ground-based lightning location networks, satellite-based observations, ground-based cameras, and electric field mills, etc.

Ground-based lightning location networks use radiation signals from the lightning that range from very low frequency (VLF) networks like the worldwide lightning location network (WWLLN), low frequency (LF) networks like the Earth Network Total Lightning Network (ENTLN), and the Beijing Lightning Network (BLNET) (Jacobson et al., 2006; Mallick et al., 2015; Srivastava et al., 2017). However, to understand the propagation mechanism of lightning leader and its initiation, a very high frequency (VHF) lightning mapping array is used worldwide (Thomas et al., 2004, Sun et al., 2022). All these above-mentioned networks are either using time of arrival or magnetic direction finder algorithms to detect and locate lightning (Rakov and Uman 2003).

In recent years, the application of the space technology in the field of severe weather forecasting, meteorological research, and severe thunderstorm tracking by lightning have been enhanced. Worldwide, the space technology has advanced and the low Earth orbit-based lightning detection on the International Space Station (ISS) have been replaced by geostationary meteorological satellites like GOES and FY-4A to detect lightning (Goodman et al., 2013; Liu et al., 2021a). In India, INSAT3D/ 3DR are the core satellites for meteorological services and weather forecasting (Mohapatra et al., 2021) that don't have lightning sensors. This opens a scope to include lightning sensors in future meteorological satellites.

In such cases, one of the necessities of these ground- and space-based detection methods is to start recording of signals in nanoseconds time precision. In other words, all the sensors should be precisely synchronized. In addition, the occurrence of lightning produces a huge impact to the ionosphere and disturbs satellite communication. The detection of lightning and study perturbation of Ionosphere due to lightning are possible by Global Navigation Satellite System (GNSS) technologies, specially by using Global positioning system (GPS) technology. Section 6.2 discusses the basics of lightning physics, Section 6.3 presents an overview of GNSS, Sections 6.4 and 6.5 present the applications of

DOI: 10.1201/9781032712444-8

GNSS/GPS systems for lightning and thunderstorms studies, and lightning hazard-related issues, prevention, and warning systems are discussed in the last section of this chapter.

6.2 LIGHTNING INITIATION AND PHYSICAL PROCESSES

Lightning is initiated inside thunderstorms by development of charges separation inside the thunderstorms. These charges could be balanced by the atmospheric electric field, which plays a significant role in global electric circuit (Williams, 2006).

6.2.1 THUNDERSTORM ELECTRIFICATION

Thunderstorms originate in an area where high moisture and instability occur in the atmosphere. These thunderstorms move in the direction of wind and may achieve several categories like squall line, multi-cell, or super-cell. Inside these thunderstorms the hydrometers (ice crystal, graupel) play a crucial role in charging these clouds. These thunderstorms preserve enough charge to reach a certain threshold level of atmospheric electric field to break down by acquiring huge charge separation (Williams, 1985, 1989). The charging causes the cloud to become a dipole, tri-pole, or even more complex charge structures that generate different types of lightning events (Qie et al., 2020; Rakov & Uman, 2003). These lightning events are the electrical breakdowns that occur inside the cloud, cloud to air or between cloud to ground.

6.2.2 TYPES OF LIGHTNING

Some lightning events occur inside the cloud, and some strike the ground. Therefore, lightning is categorized based on its termination and discharge (Srivastava et al., 2019; Yuan et al., 2020). Figure 6.1 illustrates several terms of lightning where the blue lines are negative lightning leader and red lines are positive lightning leader. These lightning leader processes establish lightning flashes, which form due to successive breakdown in the atmosphere. Intra-cloud (IC) lightning are events that occur in the clouds and may be seen from the ground. In Cloud to Ground (CG) lightning, some of the lightning channels travel from the cloud towards the ground, known as downward leader (Figure 6.1). In response to these channels, several upward leaders travel towards downward leaders. One of them makes a connection in the air called an attachment point. At the attachment stage, a strong current flows from the ground towards the cloud. This current neutralizes accumulated charge

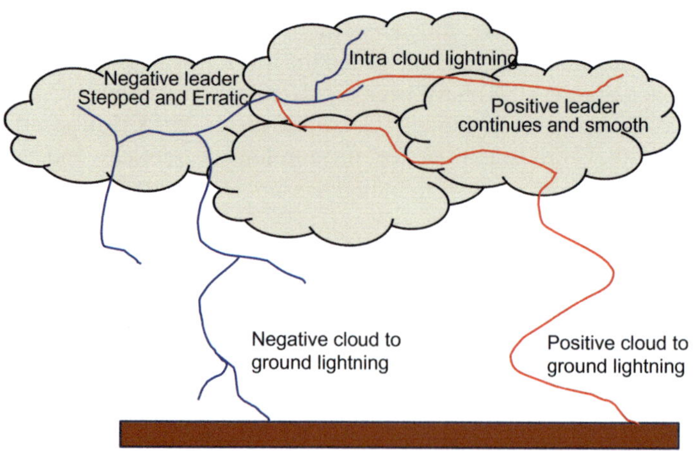

FIGURE 6.1 Illustration of lightning leaders, cloud to ground and intra cloud lightning.

in the cloud that is called return strokes (Jiang et al., 2021). The return stroke can be followed by 'dart leader' and subsequent return strokes.

Based on the flow of charges, CG lightning is categorized in terms of upward negative, upward positive, downward positive, and downward negative lightning (Rakov and Uman, 2003). By optical observation, scientific communities can identify the polarity of the lightning leaders by analyzing its pattern (stepped and erratic or continued and smooth). The propagation characteristics and physical mechanism of polarity asymmetry are the reason for the macro effects of lightning. The macro effects are lightning intensity, flash ratio, polarity of the thunderstorm and further effects on the middle and upper atmosphere, global electric circuit, and ionospheric parameters (Williams, 2006).

6.2.3 LIGHTNING HAZARDS

The lightning leaders inside the cloud or toward the ground are not the main culprit but the return stroke, which releases an enormous amount of energy in the form of light, heat, sound, and current. Therefore, the return stroke poses significant dangers, both in terms of direct and indirect threats to human safety and property loss. It should be noted that not all lightning flashes are harmful to humans, as many occur within clouds without reaching the Earth's surface, usually mentioned as IC lightning flashes. Conversely, some lightning strikes do reach the ground, called CG lightning flashes, which pose substantial risks to the public, telecommunications, and outdoor activities (Yair, 2018; Islam and Schmidlin, 2020). Therefore, to mitigate the hazards associated with CG lightning, one of the most effective methods is to suspend outdoor activities and seek shelter in a secure and protected environment, accompanied by appropriate advance warnings that are discussed in the last section of this chapter.

There are several objectives to investigate lightning physics even knowing the facts of lightning. Presently, research is concentrated to understand lightning leader propagation mechanism, lightning attachment process, thunderstorm electrification, and its associated phenomena at middle and upper atmosphere. Additionally, rigorous study of lightning locations, daily and monthly variations, and annual lightning density are required to aid policymakers in developing robust safety preparations to reduce potential losses. A ground-based lightning location network, satellite-based lightning sensors, high-speed video camera, electric field mill, and physical models are tools for the above suggested investigations. The investigations depend directly or indirectly on the time synchronization, positioning, navigation. These synchronizations are only possible by the utilization of GPS/GNSS system to detect and locate lightning events.

6.3   GLOBAL NAVIGATION SATELLITE SYSTEM (GNSS): OVERVIEW

GNSS is used to determine positions, time, and navigation services on global and regional levels by using multiple satellite constellations, receivers, and monitoring stations (Pisova and Chod, 2015). It includes the constellations of Global coverage (operated by the United States, Russia, Europe, and China) and numerous regional coverages and augmentation systems by diverse nations across the globe. *Augmentation* of a GNSS is a method of improving the integrity, continuity, and availability of GNSS signals, adopted by various nations. In general, GNSSs may be classified according to their coverage and the nation operating them (Sharma, 2019). Global coverage systems are (a) Global Positioning System (GPS), US; (b) GLObalnaya NAvigatsionnaya Sputnikovaya Sistema (GLONASS), Russia; (c) Galileo, Europe; (d) BDS (BeiDou Navigation Satellite System), China. Local/regional coverage systems are (a) IRNSS (Indian Regional Navigational Satellite System), India; (b) QZSS (Quasi-Zenith Satellite System), Japan. GNSS augmentation Systems are (a) EGNOS (European Geostationary Navigation Overlay Service), Europe; (b) WAAS (Wide Area Augmentation System), US; (c) GPS Aided GEO Augmented Navigation (GAGAN), India; (d) CWAAS (Canadian Wide Area Augmentation System), Canada; and (e) MSAS (Multi-functional Satellite Augmentation System), Japan.

6.3.1 GNSS Signals

Each GNSS constellation consists of different numbers of satellites and operates in different frequencies. The GNSS frequency plan is as per the radio regulations of the International Telecommunications Union (ITU). In accordance with the available spectrum, most GNSS satellites operate in the upper L band (1561–1610 MHz) and the bottom of the Lower L Band (1176–1278 MHz). IRNSS operates in S-band (2492.028) (GALILEO, 2005; Wang Johnson, 2012). It should be noted that future GNSS antennas will strive to cover more bands and constellations (Liu et al., 2010, Stansell et al., 2011).

6.3.2 Components of GNSS and Ranging Technique

GNSSs consist of three segments that are similar in all the constellations (GPS, GLONASS, GALILEO, etc.), and together make up the GNSS system, which is the user segment, control segment, and space segment that helps in ranging of an unknown position on the Earth and satellite (Hofmann-Wellenhof et al., 2001). It deals with the measurement of the time the signals from satellites take to reach the receivers on the ground at an unknown position. By using the multiplication relationship between the time and speed of light (electromagnetic signals), the range between the satellite and the receiver can be obtained (Hofmann-Wellenhof et al., 2001; Leick, 2003). The measurement achieved by the receiver is made by comparing received signal at the antenna and reference signal by the receiver at a certain time. The signals, called pseudorange, are affected by the clock's errors, in which the obtained range are not accurate. The pseudorange is measured by two methods: Code pseudorange and Phase pseudorange measurements.

6.3.3 GNSS Errors

GNSS signals produced with weaker power are likely to get several noises and errors from different sources (Karaim et al., 2018). The sources of errors in GNSS can be categorized as below.

6.3.3.1 Clock-related Errors

Errors related to time in the satellite or in receiver are categorized as clock-related errors. Normally, GNSS receivers are prepared with crystal clocks, whose accuracy is lower in comparison to satellite clocks in which precise atomic clocks are used (Farrell, 2008). These introduce errors in the GNSS signal and can be fixed in two ways—by the use of external precise clocks or through differencing techniques (Aboelmagd et al., 2013). With the availability of multi-constellation, GNSS data has its own timing system, and introduces intersystem clock biases in a multi-constellation system (Jin, 2012).

6.3.3.2 Signal Propagation Errors

Errors related to the propagation of GNSS signals can be categorized as Sagnac errors, Ionosphere errors, Troposphere errors, and multipath errors and are corrected by various mechanisms.

6.3.3.2.1 Sagnac Error

Earth's rotation creates a relativistic error during signal propagation between the satellite and the receiver called Sagnac error (Kaplan and Hegarty, 2006). This error may be rectified by adding the Sagnac correction factor during the pseudorange measurement. Overall, the Sangac effect needs to be addressed by omitting it, which may cause around 30 m position error (Ashby and Weiss, 1999).

6.3.3.2.2 Ionosphere Errors

The Earth's ionosphere contains various types of gases that become ionized due to continuous radiation from the sun (Misra and Enge, 2010). The ionosphere absorbs the signals and causes a delay in

GNSS frequency primarily due to the existence of free electrons. These delays are proportional to ionosphere Total Electron Content (TEC) and can be estimated using dual-frequency GNSS satellite signals with 99.99% confidence (Juan et al., 2012; Davies and Hartmann, 1997; Norsuzila et al., 2010). In the case of single-frequency GNSS receivers, the ionospheric delay must be modeled such as the GPS Klobuchar model and the Galileo NeQuick model (Klobuchar, 1987; Nava et al., 2008). The ionosphere alone accounts for about ± 5 m errors in GNSS ranging when not addressed carefully. TEC itself is used to study the thunderstorm effects and lightning that are discussed in Section 6.5.

6.3.3.2.3　Troposphere Errors

During the GNSS propagation, it next moves through the troposphere, which is composed of dry gases and water vapor (Hobiger and Norbert, 2017). The troposphere is a refractive layer that also delays GNSS signals (Jin, 2012). The tropospheric delay has wet and dry delay components. The dry component is responsible for most of the GNSS delay and can be modeled precisely using models such as the Hopfield and the Saastamoinen model (Schuler, 2001). The wet delay accounts for only 10% and is difficult to model. Satellite elevation can increase the tropospheric delay, which may introduce the errors of 2.5 m to 25 m (Misra and Enge, 2010).

6.3.3.2.4　Multipath Errors

GNSS signals may be received at the receiver's antenna on the ground in two ways. Firstly, the received signals would be the direct line-of-sight (LOS) signal, the second being the signal that reaches the receiver's antenna via more than one path because of reflections from surrounding structures or the ground (Meguro et al., 2009). The second GNSS signal that is received from surrounding structures or the ground through reflection is known as multipath. In most situations, the multipath error may introduce up to 100 m in pseudorange measurements (Borre et al., 2007).

6.3.3.3　System Errors

GNSS errors resulting from the overall nature of the system are categorized as below.

6.3.3.3.1　Orbital Errors

Satellite positions are calculated with the help of ephemeris, which is estimated at the control segment (Aboelmagd et al., 2013). These are projected using a curve fit technique to guess the satellite orbit, which introduces residual errors and may correspond to about 2 m errors in range measurements (Misra and Enge, 2010).

6.3.3.3.2　Receiver Noise

Receiver noise is commonly produced by microwave radiations sensed or generated by the antenna, cables, and amplifiers; and the signal quantization noise, etc., that is called white noise and almost impossible to totally compensate for or remove (Tsui, 2000). These errors can be minimized with modern technology receivers. Additionally, these errors can be minimized by observation of differencing techniques to cancel the common errors in receivers and satellites.

6.3.3.4　Dilution of Precision (DOP)

Dilution of precision (DOP) is an indicator of the geometrical strength of the satellites being tracked at the time of measurement. The DOP number is unitless, and its value typically ranges from 1–50. 1 is ideal, 2–3 excellent, 4–6 good, 7–8 moderate, 9–20 fair, and 21–50 poor (Misra and Enge, 2010).

6.3.4　GNSS Data Acquisition Techniques

High-precision GPS data can be collected using various techniques that depend on several factors, including survey objectives and applications, desired precision, available equipment, and field logistics. For example, to locate the lightning, a continuous survey style is used that provides < 0.5 cm accuracy in position determination. In addition to continuous survey method, Static, Rapid static, and Kinematic methods are commonly used, which have accuracy levels from 0.5–2.5 cm, 1–3 cm, and 1–5 cm (source: UNAVCO Campaign GPS/GNSS handbook, 2010). These methods have typical applications in crustal deformation, geophysics, ionosphere perturbation, mapping, navigation, etc.

6.4 LIGHTNING DETECTION AND APPLICATION OF GPS SYSTEMS

GNSS/GPS precisely locate the lightning by considering the lightning physics as lightning radiates a broadband range of electromagnetic signals that reaches far from the lightning location (Cummins et al., 1998; Betz et al., 2009; Taszarek et al., 2015; Nag et al., 2011; Srivastava et al., 2017). Once these signals reach a particular sensor, the associated GPS is triggered and identifies the received time of the signal in nano-second accuracy, as shown in Figure 6.2. The lightning location method and role of GNSS is discussed in this section.

6.4.1 Time of Arrival (TOA) Method to Locate Lightning

The time of arrival method uses a minimum of four sensors to locate the lightning (Nag et al., 2015). For example, let us consider that the lightning occurred at a position 'x, y, z' at time 't' (see Figure 6.2). To locate the lightning events, four lightning sensors are installed at different places of the region at positions:

$$\begin{aligned}
&\text{Sensor 1 with GPS receiver located at } x1, y1, z1 \\
&\text{Sensor 2 with GPS receiver located at } x2, y2, z2 \\
&\text{Sensor 3 with GPS receiver located at } x3, y3, z3 \\
&\text{Sensor 4 with GPS receiver located at } x4, y4, z4
\end{aligned} \qquad (6.1)$$

Lightning initiation and propagation are random phenomena, and it can occur anywhere within or around the thunderstorms. Therefore, the distances of these sensors are not equal from the lightning events. Here, the distances of these sensors are considered at d1, d2, d3, and d4 from the lightning

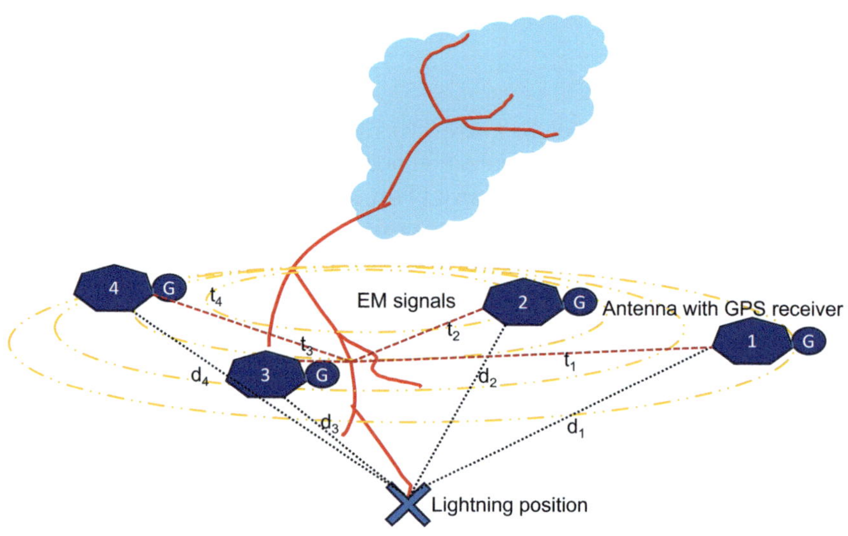

FIGURE 6.2 Schematic diagram of lightning location network and role of GPS.

event, respectively, as shown in Figure 6.2. By using distance formula with speed and time, the distance between two points can be represented by the following equations:

$$\left(x_1 - x\right)^2 + \left(y_1 - y\right)^2 + \left(z_1 - z\right)^2 = c^*\left(t_1 - t\right)$$
$$\left(x_2 - x\right)^2 + \left(y_2 - y\right)^2 + \left(z_1 - z\right)^2 = c^*\left(t_2 - t\right)$$
$$\left(x_3 - x\right)^2 + \left(y_3 - y\right)^2 + \left(z_3 - z\right)^2 = c^*\left(t_3 - t\right)$$
$$\left(x_4 - x\right)^2 + \left(y_4 - y\right)^2 + \left(z_4 - z\right)^2 = c^*\left(t_4 - t\right)$$

$$(6.2)$$

In the above equation, four unknowns (x, y, z, t) are the position and time of the lightning event. However, the position of the sensors is known by GPS receiver, and the accurate time is utilized. The accurate time is considered when the signal is received to a particular sensor (Rakov and Uman, 2003). The signals reach a peak level and begin to decay, which are associated with time stamps. The received signals are processed and used to identify the peak from the signals and their time stamps. Once the time and peak of the signals are identified, a cross-correlation method is applied to identify the common peak from different signals, and finally, equation 6.2 is used to locate the lightning (Srivastava et al., 2017). Here, the point of intersection is considered the location of lightning obtained by the solutions of equation (6.2). Further, these locations are validated by several mechanisms like relative detection efficiency and location accuracy using high-speed camera observations.

6.4.2 Lightning Mapping Arrays

One of the keys to map the lightning is the propagation of negative and positive leaders. The negative leaders propagate in a positive charge region and radiate a strong electromagnetic signal, although positive leaders propagate a negative charge region and weaker radiation signals. This phenomenon helps in lightning detection, mapping, and even identifying the thunderstorm charge regions (Sun et al., 2022). A lightning mapping array is commonly used to map the lightning that works on the same principle on VHF frequencies as the lightning location network and utilizes the GPS system for continuous time stamping.

6.4.3 Electric Field Meter to Detect Charge Transfer During Lightning

One of the basic instruments is the electric field meter, commonly called an electric field mill, or EFM, used to detect lightning by measuring the atmospheric electric field (Bennett and Harrison, 2007; Srivastava et al., 2015). To estimate the charge transfer, a multipoint measurement of the electric field is performed (Yamashita et al., 2024). In the electric field mills, exposing the sensing plate and shielding plate needs to be synchronized to know the charge transfer in the atmosphere by lightning discharge at several locations (Yamashita et al., 2024). The synchronization of exposing and shielding of these sensing plates is made possible from GPS technologies by controlling the speed of rotating motors.

In general, all these lightning sensors and a few EFM networks are associated with the GPS receivers that work globally. Although future development of other types of navigation system, receivers like IRNSS, GLONASS, etc., may be utilized for regional lightning location networks that may advance the dependency of lightning location networks on the GPS systems to GNSS technology.

6.5 THE ROLE OF GNSS AND LIGHTNING LOCATION IN UNDERSTANDING THE IONOSPHERIC IMPACT OF THUNDERSTORMS

The GNSS system gets interference from lightning propagation, and this section discusses the perturbation of the ionosphere due to lightning and uses of GNSS systems.

6.5.1 Thunderstorm and Lightning Effect on the Ionosphere

Lightning events are usually associated with thunderstorms, which is characterized by two components (Ogunsua et al., 2020; Ogunsua et al., 2023). The two components of thunderstorms are the electromagnetic component, which is visually represented by lightning in most cases, and the mechanical component, which includes the acoustic-gravity wave transportation (Vadas et al., 2012; Ogunsua et al., 2020; Ogunsua et al., 2023) and other mechanical transportation processes (Qin et al., 2023; Liu et al., 2021b). Lightning activity can be seen as the physical manifestation of the electromagnetic component of thunderstorm events that can occur as transient luminous events (TLEs) with different energy spectra and characteristics (Yang et al., 2018). These TLEs, which include cloud to ground, cloud to cloud lightning, sprites, halos, elves, etc., have been reported to have different effects on the ionosphere (Fukunishi et al., 1996; Liu et al., 2015; Blaes et al., 2016; Kolmašová et al., 2021). These effects include ionospheric potential breakdown, localized ionization lower ionospheric quasi-electrostatic heating, electron precipitation, and more (Pasko et al., 1995; Rycroft and Odzimek, 2010; Inan et al., 2010; Blaes et al., 2016; Li et al., 2019; Gordillo-Vázquez and Pérez-Invernón, 2021; Kolmašová et al., 2021; Tomicic et al., 2023).

Besides the ionospheric heating, high energy lightning activities from the same thunderstorms could affect the ionospheric current system through the global electric circuit. The mechanical component could transport energy in the form of velocity gradients, which results in vertical and horizontal gravity waves up to ionospheric heights. In the ionosphere, the mechanical component of thunderstorms includes the momentum and energy transfer, which could manifest in terms of acoustic and gravity waves and variation in the ionospheric electron density. The impact of either of these components usually depends on the magnitude of the event in most cases.

6.5.2 Lightning Detection for Ionospheric Investigation

There is a possibility of an electro-acoustic resonance effect during some extreme events involving lightning and acoustic-gravity wave disturbance, such as thunderstorms and other heavy storm systems. This might be attributed to the movement of clouds from which the lightning is produced. Simultaneous production of lightning and acoustic/gravity wave can result in electro-acoustic or electromechanical resonance.

Lightning data is used to study and understand the mechanical perturbations in the ionosphere considering the Total Electron Content (TEC) measurements from GNSS stations. It is important to track or locate the actual position and direction of the storm by tracking the lightning cell system. This can be done using the World-Wide Lightning Location Network (WWLLN), Sferics detection systems, and the GLD 360 system, which uses sensor networks to track storms. The imaging approach could also be used to locate the position of the lightning using measurements from a Lightning Imaging Sensor (LIS), which was on the International Space Station (ISS), or the TRMM mission satellites. It is very important to carry out the lightning location for effective mapping of the area of the storm impact. The GNSS estimation of the perturbation of storm impact could be carried out more accurately for a properly mapped area of storm impact based on the measurement conducted within the mapped region of the event.

6.5.3 GNSS and the Ionospheric Effects of Thunderstorms

GNSS, which is utilized for the study of ionospheric density variation, can be used for thunderstorms effects application because of the presence of ionospheric delay in genesis satellite transmission. The ability to correct for this delay results from the trans ionospheric radio propagation-based dual frequencies. L1 and L2 frequency bands have enabled the evaluation of the electron density and the path of the GNSS radio signal. Generally, this electron density is evaluated from GNSS method as the TEC. This is computed based on frequencies f_1 and f_2, such that the carrier phase measurements obtained from GNSS are used to estimate the slant TEC (sTEC) along the signals path from the satellite to the receiver as follows:

$$sTEC = -\frac{f_1^2 f_2^2}{40.3\left(f_1^2 - f_1^2\right)}\left[\left(\lambda_2 L_2 - \lambda_1 L_1\right) - \left(\lambda_1 A_1 + \lambda_2 A_2 + \varepsilon_L\right)\right] \tag{6.3}$$

where L_1 is the carrier phase frequency band measurement on f_1 (cycles), L_2 is the carrier phase frequency band measurement on f_2 (cycles), A_1 and A_2 are the ambiguity integer for the carrier phases on L_1 and L_2 frequency (cycles) respectively, ε_L is the multipath noise associated with the frequency (cycles), λ_1 (19.05 cm) and λ_2 (24.5 cm) are the corresponding wavelengths (m) to f_1 (1575.42 MHz) and f_2 (1227.60 MHz), respectively (Ciraolo et al., 2007).

Theoretically, TEC can be defined as the total number of electrons in 1 m^2 column along the path of the GNSS signal from the satellite to the receiver. This can be expressed mathematically as follows:

$$N_{eT} = \int_S^R N_e.ds \tag{6.4}$$

Where N_{eT} is the total number of electrons from satellite to receiver, N_e is the total number of electrons within a surface area.

$$vTEC = sTEC \times MF \tag{6.5}$$

where $MF = \left[1 - \left(\frac{R_e \cos\cos(\theta)}{R_e + h_{\max}}\right)^2\right]^{\frac{1}{2}}$ is the mapping function considering ionosphere as single

layer, $M(\theta) = \left[1 - \left(\frac{\cos(\theta)}{1 + \dfrac{h}{R_e}}\right)^2\right]^{\frac{1}{2}}$ Re is the mean radius of the Earth, θ is the satellite elevation angle,

$h_{\max}$ is the mean ionospheric shell height at ionospheric pierce point (IPP) at approximately 350 km altitude, for computational convenience (Mannucci et al., 1998; Norsuzila et al., 2009).

In most applications, the slant Total Electron Content (sTEC) will be converted to vertical Total Electron Content (vTEC). Considering that sTEC is dependent on the ray path geometry of the ionospheric transmission, it is vital to compute the corresponding vertical TEC (vTEC) value that will be independent of the ray path's angle of elevation. To achieve this, the sTEC is converted to vTEC using the mapping function (MF) eq 5, as described by Klobuchar (1987) and Mannucci et al. (1998).

Considering that the mechanical perturbations in the ionosphere due to thunderstorms involves the variations in influx and outflow of neutral particles and ionic variations due to acoustic-gravity wave movements. GNSSs can be used to compute the deviations in these densities during thunderstorms. In different instances, some authors may want to attribute the TEC deviations and observe the ionospheric TEC perturbations to lightning; however, these responses are purely from the mechanical component of thunderstorms like the acoustic-gravity wave disturbances.

Ogunsua et al., (2023), in their earlier investigations, used lightning location data to track the position of thunderstorms and to study the influence of acoustic and gravity waves using GNSS measurements. Initial research exploring the impact of thunderstorms on ionospheric TEC obtained from GNSS data repositories highlighted responsive TEC deviations attributable to associated gravity wave effects (Lay et al. 2013; Lay et al., 2015). Ogunsua et al. (2020) elucidated the dynamics of the equatorial ionosphere, revealing significant daytime GNSS TEC perturbations.

Of additional note are gravity wave effects throughout thunderstorms and lightning events. These observed perturbations, induced by thunderstorm-associated gravity waves, modified the ionosphere above the event region during the day. However, the nighttime gravity wave remained undetectable due to the irregularities of the equatorial ionosphere (Ogunsua et al., 2020). In recent studies, Chowdhury et al. (2023) studied the thunderstorm effects on the ionosphere, considering different locations around the globe using TEC measured from GNSSs from selected locations. Ogunsua et al. (2023) studied the mechanical signatures of the ionosphere during thunderstorms in the mid-latitude region around China and Korea.

6.5.4 GNSS SIGNAL SCINTILLATION

The ionospheric behavior during thunderstorm can studied by combining GNSS data with the lightning location data. The impact of the thunderstorm component on the ionosphere can be evaluated. Another way to understand the systemic or localized response of the ionosphere to a thunderstorm is to study the scintillation index (S4) or the Rate of change of TEC index (ROTI). The ionospheric index ROTI can be computed and used as a proxy for the S4 index to understand the variations in the ionospheric irregularities resulting from GNSS signal scintillation (Kumar et al., 2017; Chowdhury et al., 2023). The ROTI can be computed from Rate of change of TEC ROT as follows (Li et al., 2022):

$$ROT = \frac{\delta}{\delta t}\left(VTEC\left(t_k\right)\right) = \frac{VTEC\left(t_{k+1}\right) - VTEC\left(t_k\right)}{t_{k-1} - t_k} \tag{6.6}$$

such that,

$$ROTI = \sqrt{\langle ROT^2 \rangle - \langle ROT \rangle^2} \tag{6.7}$$

Lightning can also induce localized ionization in the ionosphere through electron precipitation or due to a change in chemical composition in the lower atmosphere resulting from D-layer heating. Further, it could possibly result in variations in ionospheric irregularities including the ionospheric sporadic E-layer enhancement (Shao et al., 2013; Yu et al., 2015, 2017; Gordillo-Vázquez and Pérez-Invernón, 2021). Different investigators have reported a correlation between the variations in ionospheric indices (S4) and lightning intensity (Kumar et al., 2017; Chowdhury et al., 2023; Osei-Poku et al., 2022).

Sudden changes in ionospheric micro-density fluctuations and or changes in ROTI have been found to correlate with high-intensity lightning discharges (Kumar et al., 2017; Chowdhury et al. (2023)). Even though the effect of lightning on ionospheric irregularities might be plausible, the possibility bulk change in TEC remains unrealistic. Usually, the ionization of the ionosphere due to lightning discharge will be negligible due to the bulk variation of TEC (Showen and Slingeland, 1998). These studies can help monitor the extreme events related to lightning using the perturbations in the ionosphere as proxies to study the dynamical impact and evolution of lightning events. This can help us better understand the dynamics of natural hazards and for improved modeling of the response of the ionosphere to these events.

6.6 LIGHTNING PREVENTION, PROTECTION, AND WARNING SYSTEM

In recent years, lightning has shown an increasing trend in south Asian countries, and it is possible that intensified thunderstorm frequency has increased. Lightning looks like a random phenomenon to the common public; however, it is a systematic event that occurs in thunderstorms. To reduce the risk from lightning and to take preventive measures, there are two common mechanisms—one is to install protection systems to reduce the indoor risk, and another is to generate warnings in advance to mitigate the losses of outdoor activities and field workers.

6.6.1　Lightning Protection Systems

Losses due to lightning can be reduced using safety protocols. On buildings and high towers, lightning arresters should be installed with proper grounding. There are several mechanisms to protect the buildings and complex structures and several mechanisms to reduce lightning risks in buildings like Faraday cages, lightning arresters using the rolling sphere method, and protection angle methods (Velazquez et al., 1982; D'Alessandro, 2003, 2007). This mechanism can be used by physical modeling to improve the protection systems (Srivastava and Mishra, 2013, 2015). The GNSS receiver itself needs a lightning protection system/surge protection system; however, the protection systems on instruments are out of scope from this chapter.

6.6.2　Lightning Prevention and Warning System

The structures could be protected from the lightning; the losses in open places cannot be reduced by these methods for workers and farmers. The best method to reduce loss in open places is to make the public aware of the do's and don'ts during thunderstorms and lightning. Further, they should be warned well in advance about the occurrence of lightning so as to take precautionary measures. Overall, the prevention of lightning is possible with suitable warning systems. Warnings of thunderstorms and lightning are generated by various mechanisms, and in all methods the time synchronization can utilize the GNSS systems.

Real-time lightning location with high detection efficiency works as quality data for thunderstorm tracking. In this, GPS systems play a crucial role, as discussed in Sections 6.4 and 6.5, to generate high accuracy nowcasts (Betz et al., 2008; Dixon and Wiener, 1993; Srivastava et al., 2022). Further recent developments of GPS synchronized electric field meters would help improve the accuracy of lightning nowcasts in areas of concern (Srivastava et al., 2015; Bennett, 2018). These techniques are only capable of nowcasting lightning one to two hours in advance, and the ability to forecast thunderstorm and lightning several hours before a numerical weather prediction-based model are beneficial. Srivastava et al. (2023) have shown an operational application of Weather Research Forecast (WRF) coupled additional electricity module that was developed by Fierro et al. (2013). In this type model, the lightning location information has been assimilated to improve the forecast accuracy. These models are highly dependent on initial conditions by Earth observation using ground and satellite data that need a precise time from GNSS technologies.

Overall, the chapter gives a new theme to utilize GNSS technologies and shows that accommodating multilayer of forecast and GNSS technologies together can help reduce the losses from lightning and improve our understanding of lightning phenomena.

REFERENCES

Aboelmagd, N., Karmat, T., Georgy, J. (2013). *Fundamentals of Inertial Navigation, Satellite-Based Positioning and their Integration.* 1st ed. New York: Springer-Verlag Berlin Heidelberg.

D'Alessandro, F. Striking distance factors and practical lightning rod installations: a quantitative study, *J. Electrost.* 59 (2003) 25–41.

D'Alessandro, F. On the optimum rod geometry for practical lightning protection systems, J. Electrost. 65 (2007) 113–121.

Ashby, N., Weiss, M. (1999). *Global Positioning System Receivers and Relativity.* Boulder, CO, USA: National Institute of Standards and Technology (NIST);1385:1–46.

Bennett, A.J. (2018) Warning of imminent lightning using single-site meteorological observations. *Weather* 73(6):187–193. https://doi.org/10.1002/wea.2782

Bennett, A.J., Harrison, R.G. (2007) Atmospheric electricity in different weather conditions. *Weather* 62(10):277–283.

Betz, H.D., Schmidt, K., Oettinger, W.P., Montag, B. (2008) Cell-tracking with lightning data from LINET. *Adv. Geosci.* 17:55–61. https://doi.org/10.5194/adgeo-17-55-2008

Betz, H.D., Schmidt, K., Laroche, P., Blanchet, P., Oettinger, W.P., Defer, E., Dziewit, Z., Konarski, J., 2009. LINET - an international lightning detection network in Europe. *Atmos. Res.* 91 (2), 564–573. http://dx.doi.org/10.1016/j.atmosres.2008.06.012

Blaes, P. R., Marshall R. A., and U. S. Inan (2016), Global occurrence rate of elves and ionospheric heating due to cloud-to-ground lightning, *J. Geophys. Res. Space Physics*, 121, 699–712, https://doi.org/10.1002/2015JA021916

Borre, K., Akos, D.M., Bertelsen, N., Rinder, P., Jensen, S.H. (2007). *A Software-Defined GPS and Galileo Receiver a Single-Frequency Approach*. Boston: Birkha̋user Boston.

Chowdhury, S., Kundu, S., Ghosh, S., Sasmal, S., Brundell, J., & Chakrabarti, S. K. (2023). Statistical study of global lightning activity and thunderstorm-induced gravity waves in the ionosphere using WWLLN and GNSS-TEC. *J. Geophys. Res. Space Physics*, 128, e2022JA030516. https://doi.org/10.1029/2022JA030516

Ciraolo, L., Azpilicueta, F., Brunini, C., Meza, A., Radicella, S.M., 2007. Calibration errors on experimental slant total electron content (TEC) determined with GPS. *J. Geod.* 81, 111–120, doi: https://doi.org/10.1007/s00190-006-0093-1

Cummins, K.L., Murphy, M.J., Bardo, E.A., Hiscox, W.L., Pyle, R.B., Pifer, A.E., 1998. A combined TOA/MDF technology upgrade of the U.S. National Lightning Detection Network. *J. Geophys. Res. Atmos.* 103 (D8), 9035–9044. http://dx.doi.org/10.1029/98JD00153

Davies K, Hartmann GK (1997) Studying the ionosphere with the global positioning system. *Radio Sci.* 32(4):1695–2170.

Dixon, M., Wiener, G. (1993) TITAN: thunderstorm identification, tracking, analysis, and nowcasting-A radar-based methodology. *J. Atmos. Ocean. Technol.* 10(6):785–797.

Farrell, J. (2008). *Aided Navigation GPS with High Rate Sensors*. New York: McGraw-Hill.

Fierro, A. O., Mansell, E. R., MacGorman, D. R., Ziegler, C. L., 2013. The implementation of an explicit charging and discharge lightning scheme within the WRF-ARW model: benchmark simulations of a continental squall line, a tropical cyclone, and a winter storm. Mon. Weather Rev. 141(7):2390–2415.

Fukunishi, H. Y., Takahashi, Y., Kubota, M. et al. (1996) Elves: Lightning-induced transient luminous events in the lower ionosphere. *Geophys. Res. Lett.* 23(16):2157–2160. https://doi.org/10.1029/96GL01979

GALILEO (2005). Mission High Level Definition (HLD) (2002), European Commission Communication Document, W. Doc. 2002/05 - Version 3, 23. September 2002, http://www.galileoju.com, http://www.esa.int/esaNA/index.html

Goodman, S. J., Blakeslee, R. J., Koshak, W. J., Mach, D., Bailey, J., Buechler, D., ... & Stano, G. (2013). The GOES-R geostationary lightning mapper (GLM). *Atmos. Res.*, *125*, 34–49.

Gordillo-Vázquez, F. J., Pérez-Invernón, F.J., A review of the impact of transient luminous events on the atmospheric chemistry: Past, present, and future, *Atmos. Res.*, 252, 2021, 105432, https://doi.org/10.1016/j.atmosres.2020.105432

Hobiger, T., Norbert, J. (2017). Atmospheric signal propagation. In: Teunissen, P., Montenbruck, O., editors. Springer Handbook of Global Navigation Satellite Systems. Cham: Springer; pp. 165–193.

Hofmann-Wellenhof, B., Lichtenegger, H., and Collins, J. (2001), *Global Positioning System: Theory and Practice*, 5th ed. New York: Springer Verlag Wien.

IGS products. [Online]. [cited 2017 December 3. Available from: http://www.igs.org/Products

Inan, U. S., Cummer, S. A., and Marshall, R. A. (2010), A survey of ELF and VLF research on lightning-ionosphere interactions and causative discharges, *J. Geophys. Res.*, 115, A00E36, https://doi.org/10.1029/2009JA014775

Islam, M. S., & Schmidlin, T. W. (2020). Lightning hazard safety measures and awareness in Bangladesh. *Nat. Hazards*, *101*(1), 103–124.

Jacobson, A.R., Holzworth, R., Harlin, J., Dowden, R., Lay, E., 2006. Performance assessment of the World-Wide Lightning Location Network (WWLLN), using the Los Alamos Sferic Array (LASA) as ground truth. J. Atmos. Ocean. Technol. 23, 1082–1092. http://dx.doi.org/10.1175/JTECH1902.1

Jiang, R., Srivastava, A., Qie, X., Yuan, S., Zhang, H., Sun, Z., et al. (2021). Fine structure of the breakthrough phase of the attachment process in a natural lightning flash. *Geophys. Res. Lett.*, *48*, e2020GL091608. https://doi.org/10.1029/2020GL091608

Jin, S. (2012). *Global Navigation Satellite Systems Signal*. InTech: Theory and Applications Croatia.

Wang Johnson J. H. (2012) Antennas for Global Navigation Satellite System (GNSS). *Proceedings of the IEEE 13 January 2012. Engineering, Physics.*

Juan, J., Hernández-Pajares, M., Sanz, J., Ramos-Bosch, P., Aragón-Àngel, A., Orús, R., (2012). Enhanced precise point positioning for GNSS users. IEEE Trans. Geosci. Remote Sens.. 50(10), 4213–4222.

Kaplan, E., Hegarty, C. (2006). *Understanding GPS Principles and Applications*. Norwood, MA, USA: ARTECH HOUSE, Inc.

Klobuchar, J. (1987). Ionospheric time-delay algorithm for single-frequency GPS users. IEEE Trans. Aerosp. Electron. Syst.. 3, 325–331.

Kolmašová, I., Santolík, O., Kašpar, P., Popek, M., Pizzuti, A., Spurný, P., et al. (2021). First observations of elves and their causative very strong lightning discharges in an unusual small-scale continental spring-time thunderstorm. *J. Geophys. Res. Atmos.*, 126, e2020JD032825. https://doi.org/10.1029/2020JD032825

Kumar, S., W. Chen, M. Chen, Z. Liu, and R. P. Singh (2017), Thunderstorm-/lightning-induced ionospheric perturbation: An observation from equatorial and low-latitude stations around Hong Kong, *J. Geophys. Res. Space Physics*, 122, https://doi.org/10.1002/2017JA023914

Lay, E. H., Shao, X. M., and Carrano C. S., 2013. Variation in total electron content above large thunderstorms, *Geophys. Res. Lett.*, 40, 1945–1949, https://doi.org/10.1002/grl.50499

Lay, E. H., Shao, X. M., Kendrick, A. K., and Carrano, C. S., 2015. Ionospheric acoustic and gravity waves associated with midlatitude thunderstorms, *J. Geophys. Res. Space Physics*, 120, 6010–6020. https://doi.org/10.1002/2015JA021334

Leick, A. (2003) *GPS Satellite Surveying*, 3rd ed. New York: John Wiley and Sons.

Li, W., Song, S., & Jin, X. (2022). Ionospheric scintillation monitoring with ROTI from geodetic receiver: Limitations and performance evaluation. *Radio Sci.*, 57, e2021RS007420. https://doi.org/10.1029/2021RS007420

Li, D., Luque, A., Rachidi, F., Rubinstein, M., Azadifar, M., Diendorfer, G., & Pichler, H. (2019).The propagation effects of lightning electromagnetic fields over mountainous terrain in the Earth-ionosphere waveguide. *J. Geophys. Res.* https://doi.org/1029/2018JD030014

Liu, N., Dwyer, J. R., Stenbaek-Nielsen, H. C., and McHarg, M. G., 2015. Sprite streamer initiation from natural mesospheric structures. Nat. Commun., 6, 7540. https://doi.org/10.1038/ncomms8540

Liu, T., Yu, Z., Ding, Z., Nie, W., Xu, G., 2021a. Observation of Ionospheric Gravity Waves Introduced by Thunderstorms in Low Latitudes China by GNSS. *Remote Sens.*, 13, 4131. https://doi.org/10.3390/rs13204131

Liu, W., X. Zhan, L. Liu, and M. Niu (2010). GNSS RF compatibility assessment V Interference among GPS, galileo, and compass, [GPS World, Dec. 2010. [Online].

Liu, Y., Wang, H., Li, Z., & Wang, Z. (2021b). A verification of the lightning detection data from FY-4A LMI as compared with ADTD-2. *Atmos. Res.*, *248*, 105163.

Mallick S, Rakov VA, Hill JD, Ngin T et al (2015) Performance characteristics of the ENTLN evaluated using rocket-triggered lightning data. *Electr. Power Syst. Res.* 118:15–28.

Mannucci, A. J., Wilson, B. D., Yuan, D. N., Ho, C.H., Lindqvister, U. J., Runge, T. F., (1998). A global mapping technique for GPS-derived ionospheric total electron content measurements. *Radio Sci.*, 33(3):565–582.

Karaim Malek, Elsheikh Mohamed and Noureldin Aboelmagd (2018). Multifunctional Operation and Application of GPS. Edited by Rustam B. Rustamov and Arif M. Hashimov. BoD–Books on Demand.

Meguro, J., Murata, T., Takiguchi, J., Amano, Y., Hashizume, T. (2009). GPS multipath mitigation for urban area using omnidirectional infrared camera. *EEE Trans. Intell. Transp. Syst.*.10(1), 22–30.

Misra, P., Enge, P. (2010). *Global Positioning System: Signals, Measurements and Performance*. Massachusetts: Ganga-Jamuna.

Mohapatra, M., Mitra, A. K., Singh, V., Mukherjee, S. K., Navria, K., Prashar, V., ... & Kumar, R. (2021). *INSAT-3DR-rapid scan operations for weather monitoring over India. Current Science*, 1026–1034.

Nag, A., Mallick, S., Rakov, V.A., Howard, J.S., Biagi, C.J., Hill, J.D., Uman, M.A., Jordan, D.M., Rambo, K.J., Jerauld, J.E., DeCarlo, B.A., Cummins, K.L., Cramer, J.A., 2011. Evaluation of U.S. National Lightning Detection Network performance characteristics using rocket-triggered lightning data acquired in 2004–2009. *J. Geophys. Res. Atmos.* 116 (D2), 1–8. http://dx.doi.org/10.1029/2010JD014929

Nag, A., Murphy, M.J., Schulz, W., Cummins, K.L., 2015. Lightning locating systems: characteristics and validation techniques. *Earth Space Sci.* 2 (4), 65–93. http://dx.doi.org/10.1002/2014EA000051

Nava, B., Coisson, P., Radicella, S. (2008). A new version of the NeQuick ionosphere electron density model. *J. Atmos. Sol. Terr. Phys.* 70(15), 1856–1862.

Norsuzila, Y., Abdullah, M., Ismail, M., Ibrahim, M., Zakaria, Z. (2010) Total electron content (TEC) and estimation of positioning error using Malaysia data. *Proceedings of the World Congress on Engineering, June 30 - July 2, London, U.K.* 715–719.

Norsuzila, Y., Abdullah, M., Ismail, M., and Zaharim, A.: (2009). Model validation for GPS total electron content using 10th polynomial function technique at an equatorial region, *WSEAS Trans Comp (Portugal)*, 8(9), 1533–1542.

Ogunsua, B. O., Qie, X., Srivastava, A., et al. (2023). Ionospheric Perturbations Due to Large Thunderstorms and the Resulting Mechanical and Acoustic Signatures. *Remote Sens..* 15(10):2572. https://doi.org/10.3390/rs15102572

Ogunsua, B. O., Srivatava, A., Bian, J., Qie, X., Wang, D., Jiang, R., Yang, J., (2020): Significant Day-time Ionospheric Perturbation by Thunderstorms along the West African and Congo Sector of Equatorial Region. *Nat. Sci. Rep.* 10, 8466. https://doi.org/10.1038/s41598-020-65315-3

Osei-Poku, L.; Tang, L.; Chen, W.; Chen, M.; Acheampong, A. A. Comparative Study of Predominantly Daytime and Nighttime Lightning Occurrences and Their Impact on Ionospheric Disturbances. Remote Sens. 2022, 14, 3209. https://doi.org/10.3390/rs14133209

Pasko, V. P., Inan, U. S., Taranenko, Y. N., and Bell, T. F., 1995. Heating, ionization and upward discharges in the mesosphere due to intense Quasi electrostatic thundercloud fields, *Geophys. Res. Lett.*, 22, 365–368. https://doi.org/10.1029/95GL00008

Pisova, P., & Chod, J. (2015). Detection of GNSS Signals Propagation in Urban Canyos Using 3D City Models. *Information and Communication Technnologies and Services*, 22–29.

Noureldin, A., Karamat, T. B., Georgy, J. (2013). Fundamentals of Inertial Navigation, Satellite-based Positioning and their Integration. New York: Springer-Verlag Berlin Heidelber, XVIII, pages 314.

Qie, X., S. Yuan, Z. Chen, D. Wang, D. Liu, M. Sun, Z. Sun, A. Srivastava, H. Zhang, J. Lu, H. Xiao, Y. Bi, L. Feng, Y. Tian, Y. Xu, R. Jiang, M. Liu, X. Xiao, S. Duan, D. Su, C. Sun, W. Xu, Y. Zhang, G. Lu, D. Zhang, Y. Yin, Y. Yu, 2020. Understanding the dynamical–microphysical and lightning processes associated with severe thunderstorms over the Beijing metropolitan region. *Sci. China Earth Sci.*, 64(1):10–26, https://doi.org/10.1007/s11430-020-9656-8

Qin Z., Yang J., Yang T., Jing X., Lu C., Wang Y., Yin Y., Zhang Q., Chen B. 2023. Vertical transport of water in isolated convective clouds in the interior western United States as observed using airborne in-situ measurements, *Atmos. Res.*, 285, 106629, https://doi.org/10.1016/j.atmosres.2023.106629

Rakov, V. A., & Uman, M. A. (2003). *Lightning: physics and effects.* Cambridge University Press.

Robert L. Showen and Alexander Slingeland 1998. Measuring lightning-induced ionospheric effects with incoherent scatter radar or with cross-modulation, *J. Atmos. Sol. Terr. Phys.*, 60(7–9), 951–956, https://doi.org/10.1016/S1364-6826(98)00016-9

Rycroft, M. J., and Odzimek A.: (2010), Effects of lightning and sprites on the ionospheric potential, and threshold effects on sprite initiation, obtained using an analog model of the global atmospheric electric circuit, *J. Geophys. Res.*, 115, A00E37, https://doi.org/10.1029/2009JA014758

Schuler T (2001). *On Ground-Based GPS Tropospheric Delay Estimation: Univ.* der Bundeswehr Munchen.

Shao, X. M., Lay, E. H., and Jacobson, A. R., 2013. Reduction of electron density in the night-time lower ionosphere in response to a thunderstorm, *Nat. Geosci.*, 6, 29–33. https://doi.org/10.1038/ngeo1668

Sharma, G (2019). Global Navigation Satellite System: Field procedures and Applications. *Chapter in a book entitled "A glimpse of Geospatial Technologies and Applications".* ISBN: 9789388881180.

Srivastava A., Mishra, M., Lightning modeling and protection zone of conducting rod using Monte Carlo technique, *Appl. Math. Model. Elsevier* 37 (24) (2013) 9858e9864.

Srivastava, A., Mishra, M.: Positioning of lightning rods using Monte Carlo technique. *J. Electrost.* 76, 201–207 (2015).

Srivastava A, Mishra M, Kumar M (2015) Lightning alarm system using stochastic modelling. *Nat. Hazards* 75(1):1–11. https://doi.org/10.1007/s11069-014-1247-8x

Srivastava A, Tian Y, Qie X et al (2017) Performance assessment of Beijing Lightning Network (BLNET) and comparison with other lightning location networks across Beijing. *Atmos. Res.* 197:76–83.

Srivastava A, Liu D, Xu C, Yuan S, Wang D, Ogunsua B, Sun Z, Chen Z, Zhang H (2022) Lightning nowcasting with an algorithm of thunderstorm tracking based on lightning location data over the Beijing area. *Adv. Atmos. Sci.* 39(1):178–188. https://doi.org/10.1007/s00376-021-0398-2

Srivastava, A., Kundu, S.S., Pawar, S.D. et al. Evaluation of WRF-ELEC model to forecast lightning over the North Eastern region of India. *Meteorog. Atmos. Phys.* 135, 39 (2023). https://doi.org/10.1007/s00703-023-00977-y

Srivastava, A., Jiang, R., Yuan, S., Qie, X., Wang, D., Zhang, H., et al. (2019). Intermittent propagation of upward positive leader connecting a downward negative leader in a negative cloud to ground lightning. *J. Geophys. Res. Atmos.*, 124, 13,763–13,776. https://doi.org/10.1029/2019JD031148

Stansell, T. A., K. W. Hudnut, and R. G. Keegan, (2011). Future wave L1C signal performance and receiver design, [GPS World, Apr. [Online]. Available: http://www.gpsworld.com/gnss-system/gps-modernization/future-wave-11401

Sun, Z., Qie, X., Liu, M., Jiang, R., & Zhang, H. (2022). Three-dimensional mapping on lightning discharge processes using two VHF broadband interferometers. *Remote Sens.*, 14(24), 6378.

Taszarek, M., Czernecki, B., Kozioł, A., 2015. A cloud-to-ground lightning climatology for Poland. *Mon. Weather Rev.* 143 (11), 4285–4304. http://dx.doi.org/10.1175/MWRD-15-0206.1

Thomas, R. J., Krehbiel, P. R., Rison, W., Hunyady, S. J., Winn, W. P., Hamlin, T., & Harlin, J. (2004). Accuracy of the lightning mapping array. *J. Geophys. Res. Atmos.*, *109*, D14207. https://doi.org/10.1029/2004JD004549

Tomicic, M., Chanrion, O., Farges, T., Mlynarczyk, J., Kolmašová, I., Soula, S., Lapierre, J., Köhn, C., & Neubert, T. (2023). Observations of Elves and Radio Wave Perturbations by Intense Lightning. *J. Geophys. Res. Atmos.*, 128(10), Article e2022JD036541. https://doi.org/10.1029/2022JD036541

Tsui J (2000). *Fundamentals of Global Positioning System Receivers: A Software Approach.* New York: John Wiley & Sons, Inc.

UNAVCO Campaign GPS Handbook, January 2010.

Vadas, S., J. Yue, and T. Nakamura (2012), Mesospheric concentric gravity waves generated by multiple convective storms over the North American Great Plain, *J. Geophys. Res.*, 117, D07113, https://doi.org/10.1029/2011JD017025

Velazquez, R.S., V. Gerez, D. Mukhedkar, Y. Gervais, Probabilistic calculations of lightning protection for tall buildings, *IEEE Trans. Ind. Appl.* IA-18 (3) (1982) 252e259.

Williams, E. R. (2006). Problems in lightning physics—The role of polarity asymmetry. *Plasma Sources Sci. Technol.*, 15(2), S91–S108. https://doi.org/10.1088/0963-0252/15/2/S12

Williams, E. R. (1989). The tripole structure of thunderstorms. *J. Geophys. Res. Atmos.*, 94(D11), 13151–13167.

Williams, E. R., 1985: Large scale charge separation in thunderclouds. *J. Geophys. Res.*, 90, 6013–6025.

Yamashita, K., Fujisaka, H., Wang, D. H., Iwasaki, H., Yamamoto, K., Michimoto, K., and Hayakawa, M. (2024). A new electric field mill array with each of the mill's rotor controlled precisely by a GPS module: Equipment and initial results. *Earth Planet. Phys.*, 8(2), 423–435. https://doi.org/10.26464/epp2024009

Yang, J., Sato, M., Liu, N., Lu, G., Wang, Y., & Wang, Z., 2018. A gigantic jet observed over a mesoscale convective system in midlatitude region. *J. Geophys. Res. Atmos.*, 123, 977–996, https://doi.org/10.1002/2017JD026878

Yu, B., Xue, X., Lu, G., Ma, M., et al., 2015. Evidence for lightning-associated enhancement of the ionospheric sporadic E layer dependent on lightning stroke energy, *J. Geophys. Res. Space Physics*, 120, 9202–9212. https://doi.org/10.1002/2015JA021575

Yu, B., Xue, X., Lu, G., Kuo, C., Dou, X., Gao, Q.,Tang, Y., 2017. The enhancement of neutral metal Na layer above thunderstorms. *Geophys. Res. Lett.*, 44, 9555–9563. https://doi.org/10.1002/2017GL074977

Yuan, S., Qie, X., Jiang, R., Wang, D., Sun, Z., Srivastava, A., & Williams, E. (2020). Origin of an uncommon multiple-stroke positive cloud to ground lightning flash with different terminations. *J. Geophys. Res. Atmos.*, *125*, e2019JD032098. https://doi.org/10.1029/2019JD032098

Yair, Y. (2018). Lightning hazards to human societies in a changing climate. *Environ. Res. Lett.*, *13*(12), 123002.

7 Advancing Sustainability
A Comprehensive Review of Geospatial Technology in Monitoring Climate Change-driven Landslide Hazards

Rajeev Singh Chandel, Praveen Kumar Rai, and Rahul Kumar Misra

7.1 INTRODUCTION

Monitoring landslide hazards driven by climate change is of utmost importance due to the increasing frequency and intensity of extreme precipitation-induced climate change. These incidents make slopes more susceptible, increasing the likelihood of landslides (Suh et al., 2011). In addition, the rapid population growth and urbanization in areas prone to landslides emphasize the need for monitoring, as it allows for the implementation of early warning systems and the mitigation of potential loss of life and infrastructure damage (Lacasse & Nadim, 2009). Furthermore, landslides have critical environmental consequences, such as ecosystem disruption and soil erosion, which may be efficiently managed by extensive monitoring and land management measures (Tiwari, 2000). Considering the economic implications and the necessity for climate change adaptation, monitoring provides valuable data for research, modeling, and developing precise risk assessment methodologies. By constantly tracking and comprehending climate change-driven landslide threats (Jiang et al., 2022), we can efficiently and effectively safeguard communities (Siriwardana et al., 2018), reduce environmental harm, and improve resilience in the face of a changing climate (Easton-Gomez et al., 2022).

Geospatial information technology plays a crucial and diverse role in landslide monitoring (Bello & Aina, 2014; M. Sharma et al., 2023). Its importance lies in its wide range of benefits and contributions to managing and understanding landslide hazards (Dai et al., 2002). Geospatial technology involves a wide range of tools and techniques, including satellite remote sensing, LiDAR, radar-based systems, aerial imagery, and GIS software, which collectively provide beneficial insights and data for landslide monitoring.

The fundamental contribution of geospatial technology is the collection of spatial data. Satellite remote sensing and aerial imagery allow for the acquisition of high-resolution data (Sawaya et al., 2003) over large areas, enabling the identification and mapping of landslide-prone zones (Pourghasemi & Rossi, 2017). These technologies provide detailed information on land cover, topography, and surface characteristics, essential for assessing landslide risks. Early warning systems for landslides rely heavily on geospatial technology (Ahmed et al., 2020; Singh et. al., 2021; Sur et al., 2021; Tripathi et al., 2023). Continuous monitoring of land surfaces using satellite imagery, radar, or LiDAR-based techniques enables the detection of changes in terrain conditions (Tarchi et al., 2003). Ground movement, slope deformation, and alterations in vegetation patterns can all be monitored on a vast scale (Pánek & Klimeš, 2016), providing critical indicators of potential

DOI: 10.1201/9781032712444-9

landslides. Early warnings can be issued by promptly detecting these signs, allowing for evacuation measures to be implemented, and mitigating the potential loss of life and infrastructure damage.

Geospatial technology is also crucial in risk assessment and zoning (Dewan et al., 2007). Comprehensive risk assessments can be conducted by integrating various data sources, such as topographic data, geological information, and historical landslide records. This information helps identify areas with high susceptibility to landslides, evaluate the potential impact, and inform land-use planning and zoning regulations. This proactive approach aids in minimizing vulnerability (Adinehvand et al., 2018) and reducing exposure to landslide hazards.

Monitoring and modeling of landslide-prone areas are significantly enhanced by geospatial technology. Remote sensing techniques, including satellite imagery and LiDAR, capture subtle changes in land surfaces, such as ground movement, slope creep, erosion patterns, and vegetation dynamics. These data provide a wealth of information that contributes to a better understanding of landslide dynamics (S. K. Singh & Pandey, 2014; Stumpf et al., 2013), improves predictive modeling, and enhances risk assessment methodologies. GIS software allows for the integration, analysis, and visualization of geospatial data (Huang et al., 2001), facilitating identifying high-risk areas and assessing landslide susceptibility (Reichenbach et al., 2018). During landslide events, geospatial technology enables rapid response and efficient disaster management (Abdalla & Li, 2010). Real-time monitoring combined with geospatial data helps emergency responders and authorities assess the extent of a disaster quickly, identify affected areas, prioritize response efforts, and allocate resources effectively.

Geospatial tools also assist in post-disaster damage assessment (Fayaz et al., 2022; Mejri et al., 2017), facilitating the planning and coordination of recovery and reconstruction efforts. Furthermore, geospatial technology supports climate change adaptation in landslide monitoring (A. Sharma et al., 2020; B. Singh, 2014). By integrating geospatial data with climate data, researchers can analyze long-term trends, identify climate change-induced factors contributing to landslide risks, and develop effective adaptation strategies. This integration provides insights into the evolving dynamics of landslide hazards under changing climatic conditions, helping communities and decision-makers prepare for and mitigate the impacts of climate-related landslides.

7.2 REVIEW OF PREVIOUS LITERATURE BASED ON CLIMATE CHANGE-DRIVEN LANDSLIDE HAZARDS

The relation between landslides and climate change has become a crucial topic of research, particularly in regions prone to geo-hydrological hazards. This literature review aims to explore various studies conducted on the effect of climate change on landslide hazards and risks involved in different parts of the world. The selected papers delve into the impact of changing rainfall patterns, temperature variations, and other climate-related factors on slope stability and landslides. Additionally, the review investigates the effectiveness of preventive measures and risk management strategies in mitigating landslide hazards. By examining these studies collectively, we can gain a comprehensive knowledge of the implications of climate change on landslide occurrences and how societies can better prepare and adapt to reduce vulnerability to these hazardous events.

a. (Alvioli et al., 2018) Investigates the relation between landslide occurrence and fine-scale climate projections in Central Italy. Using rainfall measurements and downscaled synthetic rainfall data, the study reveals that although rainfall thresholds for triggering landslides are projected to shift in the future, the landslide area probability distribution remains constant. This suggests that landslide hazards in the region is likely to be affected by projected variations in rainfall conditions, contributing to the understanding of climate change's impact on landslide hazards and slope stability.

b. (Picarelli et al., 2021) Discusses the effect of climate change on hazards and the risk of landslides. They highlight that climate change can result in a rise in geo-hydrological hazards, including landslides. They summarize the state of slope safety preparedness around the world and propose steps for enhanced landslide risk management.

c. (Andersson-Sköld et al., 2013) Examines landslide risk management, highlighting that preventive measures are preferred but rarely implemented, especially in developing countries, due to a lack of evidence that they pay off. The research in Sweden and Norway reveals that climate change is now considered in municipal spatial planning, and the use of landslide susceptibility maps has been influential in the Göta älv river valley for preventive measures. Better documentation and communication among stakeholders could enhance landslide management efforts.

d. (Van Beek & Van Asch, 2004) Explains how to utilize physically based models to analyze how changing land use affects the risk of landslides. The models are essential tools for hazard assessment and planning because they are more objective and need less data than conventional, statistically based methodologies. To assess the impacts of land use change on landslide activity, the physically based model utilized in the study was applied to a 15-km^2 watershed in the Alcoy region (SE Spain). Under the new conditions of land use, the results show a slight drop in the spatial frequency of land-sliding but a significant decrease in the temporal activity.

e. (Ciabatta et al., 2016) Assesses the effect of climate change scenarios on the incidence of landslides in Italy's Umbria Region. The study used a warning mechanism that utilized soil saturation levels to establish rainfall thresholds, as well as five distinct Global Circulation Models to predict the number of landslide events each year. The findings show a rise of landslides in the future, primarily during the winter season, as a result of the tight interplay between rainfall magnitude/intensity, temperature, and soil moisture.

f. (Kim et al., 2015) Discusses the landslide risk evaluation in Gangwondo, South Korea, using RCP 4.5 and 8.5 scenarios of climate change and the MaxEnt model. It explains the methods and variables employed to construct the optimal landslide model based on landslide occurrence data and GIS analysis. It also presents the results and implications of the model, highlighting potential landslide hazard areas and their relationship with land use types.

g. (Niculiţă, 2020) Analyzes the historical rainfall data and its correlation with landslide events in northeastern Romania. It also uses the EURO-CORDEX regional climate models to project future rainfall scenarios and their implications for landslide hazards. The study concludes that climate change will likely increase the frequency and intensity of rainfall events, leading to more landslides, especially in areas with existing landslides and high urbanization pressure. The article suggests some adaptation measures to reduce vulnerability and increase resilience to landslide hazards.

h. (Johnston et al., 2021) Discusses how the Pacific Coast area of the United States is affected by precipitation accumulation in terms of landslide concentration. Panel regression with fixed effects is used in the study to account for both observable and unobserved time-variant and time-invariant impacts. The findings suggest that landslide hazards are particularly susceptible to precipitation fluctuations in metropolitan regions. This information can be helpful in understanding the effect of urbanization on landslide hazards.

i. (Uzielli et al., 2018) Evaluates the impact of climate change on the hazard of flow-like landslides in a coastal area in southern Italy. The study combines climate projections, Bayesian analysis, and empirical modeling to estimate the probability of landslide occurrence and runout under different scenarios of greenhouse gas emissions and rainfall patterns. The findings suggest that climate change will increase the frequency and intensity of rainfall events that can trigger flow-like landslides, resulting in higher hazard levels for the area and its infrastructure. The study also discusses the sources and implications of uncertainty in the hazard assessment.

j. (Pun et al., 2020) Discusses the Slope Safety System in Hong Kong, a comprehensive approach to managing landslide risk in a landslide-prone region. It highlights the Geotechnical Engineering Office's (GEO) achievements in reducing risk through strategies like land use planning, maintenance, and public education. They also address challenges posed by climate change and present technical developments to enhance the system's resilience, including risk assessment, robust construction, landslide detection, public education, and emergency information sharing.

k. (Gariano & Guzzetti, 2016) Discusses the impact of climate change on landslides and the challenges in predicting the changes in stability conditions and frequency of landslides. It also recommends filling the geographical gap in landslide-climate studies and quantifying uncertainties in landslide projections for effective decision-making.

l. (Jurchescu et al., 2018) Talks about a method to assess landslide hazards in Romania under climate change. They use historical and projected data to analyze the link between rainfall and landslides, considering two climate scenarios. The study creates hazard scenarios for current and future climates, mapping out the risks. They also address climate change's impact on landslides, emphasizing the importance of their findings for national and regional risk management.

m. (Fowze et al., 2012) Discusses the increase in rain-triggered landslides in Thailand due to climate change and the measures taken to mitigate their impact, including early warning systems and structural solutions using geosynthetics. It also highlights the sensitivity of reinforced slope structures to moisture content.

n. (Mateos et al., 2020) Analyzes the legislation across Europe that regulates the integration of landslide hazards into urban planning. The study identifies the strengths and weaknesses of current laws and proposes key actions to improve the situation. They also highlight the need for a common regulatory framework to deal with this geohazard appropriately.

o. (Sammonds et al., 2021) Bangladesh is highly vulnerable to climate change and natural disasters. The paper discusses slow-onset sea-level rise and rapid-onset hazards like cyclones, floods, and landslides. It notes progress in disaster resilience but highlights challenges from climate, refugees, pandemics, and geopolitics. Suggestions include ethical standards, transparency, and public awareness to reduce vulnerability, concluding with recommendations for building resilience to climate-driven hazards.

p. (Emberson et al., 2021) Investigates the global effects of El Niño on landslide effect, utilizing satellite rainfall data and a landslide exposure model. It reveals that El Niño has significant and far-reaching influences on the exposure of people and infrastructure to landslides, particularly in Southeast Asia and Latin America. The study also demonstrates that the magnitude and direction of the exposure changes depend on whether El Niño affects total or extreme rainfall and whether El Niño or La Niña conditions result in more landslides. The findings can enhance our understanding of landslide variability and support disaster mitigation efforts on seasonal timescales.

This chapter has shed light on the critical nexus between climate change and landslide hazards in various regions across the globe. The studies presented here have highlighted the potential alterations in rainfall patterns and their influence on slope stability, leading to increased landslide risks in several areas. While preventive measures have shown effectiveness in certain regions, there remains a geographical gap in landslide-climate studies, necessitating further research and quantification of uncertainties to inform robust decision-making. Moreover, the incorporation of landslide hazards into urban planning legislation and the implementation of comprehensive risk management strategies are crucial steps toward building resilience to disaster risks driven by climate change. As we continue to confront the challenges posed by the climate crisis, it is imperative to prioritize cooperation and information exchange among stakeholders, researchers, and policymakers to enhance preparedness and response efforts, ultimately safeguarding lives and infrastructure from escalating landslide hazards induced by a changing climate.

7.3 RESEARCH GAPS

Despite significant progress made in the field of climate change-driven landslide hazards, there are still several significant research gaps that need to be addressed. *Long-term monitoring and data availability, uncertainty in climate change projections, and integrating climate data with landslide models are vital challenges that need attention. Additionally, improving the scale and resolution of data, developing reliable early warning systems, and conducting comprehensive risk assessments considering socioeconomic and policy dimensions are essential.* Understanding the impact of climate change on different landslide types and enhancing the predictive capabilities of landslide models are also critical areas for research. Standardized guidelines and interdisciplinary collaboration will be instrumental in filling these gaps and preparing communities for the increasing risks associated with climate change-induced landslides.

7.4 LANDSLIDE TYPES AND MATERIALS

Landslides can be categorized based on the composition of materials involved and the attributes of the slopes. In general, landslides are primarily driven by the movement of soil or rock masses. If the landslide comprises particles with a sand-sized or smaller grain, it is called earth material. On the other hand, if the landslide contains larger, coarser fragments, it is termed debris. Classification of landslides can also be made based on the movement type and the content associated.

Landslides are commonly observed in mountainous regions, particularly in areas with relatively lower relief, as in Table 7.1. In regions with less pronounced relief, landslides can occur in various forms, such as cut and fill failures, which are associated with building excavations and roadways; failures along riverbanks (depending on the silt and moisture content); flows that develop on gentle slopes and exhibit rapid fluid-like movement; failures in mine-waste (mainly related to coal); and a wide range of slope deterioration observed in areas with open-cut mines.

TABLE 7.1
Classification of Landslides Based on Movement

Type of Movement		Bedrock	Type of Material	
			Engineering Soils	
			Predominantly Course	Predominantly Fine
Fall		Rock Fall	Debris Fall	Earth Fall
Topple		Rock Topple	Debris Topple	Earth Topple
Slide	Rotational (Slump)		Rotational Debris Slide	Rotational Earth Slide
	Translational	Translational Rock Slide	Translational Debris Slide	Translational Earth Slide
Lateral Spread		Rock Spread	Debris Spread	Earth Spread
Flow		Rock Flow (Deep Creep)	Debris Flow (Soil Creep)	Earth Flow (Soil Creep)
Complex		Combination of two or more principal types of movement		

Source: British Geological Survey (BGS)

Rockfalls: They occur when individual rocks or blocks detach from a steep slope and rapidly fall or roll down. They can be triggered by weathering, seismic activity, or other destabilizing factors.

Topple: Toppling occurs when a mass of rock or soil rotates forward or backward around an axis near the base, causing it to overturn and fall. This type of movement is often observed in steep or vertical rock slopes where a weak plane or joint acts as a pivot point for the rotational motion.

Slides: Slides occur when an accumulation of dirt or rock moves over a clearly defined surface, such as a bedding plane or a weak layer. Slides can be further divided into *Rotational slides* (where the movement occurs along a curved surface) and *Translational slides* (where the movement occurs along a relatively flat surface).

Spread: This refers to a mechanism in which the mass of rock, soil, or debris moves in a lateral or horizontal direction along a relatively flat surface. It often occurs in cohesive or fine-grained soils, such as clay or silt, where the material behaves like a viscous fluid under certain conditions. The movement is typically slow and can be influenced by soil moisture content, slope steepness, and underlying weak layers or geological structures.

Flows: It is characterized by the fluid-like movement of rock, soil, or debris down a slope. It is driven by gravity, influenced by slope steepness and the presence of water. Flow landslides include debris flows, mudflows, and earth flows.

Complex Landslides: Complex landslides involve a combination of different movement types and mechanisms. They can exhibit characteristics of slides, flows, or other types of landslides.

These categories represent common types of landslides, and in reality, landslides can exhibit various combinations and complexities. The specific type of landslide observed in an area relies on various factors like geological conditions, slope characteristics, and triggering events.

7.5 LANDSLIDE INFLUENCING FACTORS

Landslides can occur due to various conditions and factors, encompassing natural phenomena and human activities. These events transpire when a slope with a significant gradient becomes imbalanced (Kumar Rai et al., 2014; Onagh et al., 2012a). Natural factors contributing to landslides encompass excess water, which acts as a lubricant and leads to rock expansion or slope destabilization over time due to weight absorption (Anbalagan et al., 2015; P. Singh et al., 2021). Conversely, the presence of trees and vegetation aids in slope stabilization through their root systems and their impact on the stability of fine-grained soil (Guan et al., 2023). Earthquakes, however, often have an adverse effect by generating ground vibrations that can destabilize slopes. Both natural occurrences and human activities substantially remove stability and significantly contribute to landslide occurrences (Sur et al., 2021). Natural factors include ground vibrations caused by earthquakes, volcanic eruptions, destabilization of slopes due to groundwater pressure, erosion of the slope's toe, and wildfires that destroy vegetation and disrupt soil structure, as noted in Table 7.2.

On the other hand, human activities that can trigger landslides encompass rock or soil displacement (Lorentz et al., 2016) during excavation, mining, or quarrying operations; the use of explosives near slopes; leakage from pipes carrying liquids; and human construction on hills, such as roads and buildings. Additionally, continual heavy traffic can destabilize slopes (Mountjoy et al., 2014).

7.6 LANDSLIDE DETECTION TECHNIQUES

Effective landslide detection techniques are crucial for identifying and mitigating the risks associated with landslides. These techniques can be categorized as conventional and modern methods. Conventional techniques include geotechnical investigations, ground-based monitoring, and seismic

TABLE 7.2

Factors Influencing Landslides

Sr. No.	Natural Factors	Human Factors
1.	Extended periods of intense rainfall accompanied by unmanaged surface water runoff.	Human actions
2.	Underground water level	Errors in design
3.	Slope geometry	Errors in construction
4.	Geomorphology	Negligence
5.	Geotechnical and geological factors	Incompetence
6.	Hydrology	Poor slope maintenance
7.	Climate	Unethical practices
8.	Vegetation	Negative attitudes

monitoring. Modern techniques involve remote sensing, ground-based radar, early warning systems, and LiDAR technology. By utilizing sensors such as inclinometers, seismometers, optical sensors, and radar systems, these techniques provide valuable data for assessing slope stability and detecting signs of potential landslides. Effective landslide detection is essential for risk assessment, early warning systems, and informed decision-making in land management and infrastructure development.

Conventional techniques

Sr. No.	Techniques	Description
1.	Geotechnical Investigations	Involves field surveys, subsurface exploration, and laboratory testing to evaluate soil and rock properties.
2.	Ground-based Monitoring	Uses instruments and sensors to monitor ground movement, including inclinometers, piezometers, and strain gauges.
3.	Seismic Monitoring	It involves the use of seismometers to detect ground vibrations caused by landslides.

Modern techniques

Sr. No.	Techniques	Description
1.	Remote Sensing	Utilizes satellite imagery and aerial photography to analyze land surface features for identifying potential landslides.
2.	Ground-based Radar	Utilizes interferometric synthetic aperture radar (InSAR) and ground-penetrating radar (GPR) for ground deformation monitoring.
3.	Early Warning Systems	Integrates data from various sensors to provide real-time monitoring and alerts for potential landslides.
4.	LiDAR Technology	Uses laser scanning to collect high-resolution elevation data for detailed mapping of terrain features.

7.7 ROLE OF SATELLITES AND GEOSPATIAL TECHNOLOGY IN MONITORING CLIMATE CHANGE-DRIVEN LANDSLIDE HAZARDS

Satellites are remote sensing platforms equipped with sensors capable of acquiring pictures of the Earth's surface in different spectral bands, including visible, shortwave infrared, and near-infrared.

These satellites play a significant role in gathering valuable data and information for understanding, detecting, and monitoring landslide hazards in the context of climate change. Here are some critical aspects of their role:

   a. **Remote Sensing and Data Collection**: Satellites equipped with various sensors capture high-resolution images and data throughout the world, allowing continuous monitoring of regions prone to landslides. These data provide valuable insights into changing environmental conditions.
   b. **Climate Change Monitoring**: Satellites help monitor climate change indicators, such as temperature rise, glacier melt, and changing precipitation patterns. These indicators can influence the occurrence and frequency of landslides in vulnerable areas.
   c. **Early Warning Systems**: Geospatial technology, combined with satellite data, facilitates the development of early warning systems. By detecting changes in land cover, slope stability, and weather patterns, authorities can issue timely alerts to mitigate potential landslide disasters.
   d. **Landslide Detection and Mapping**: Satellite imagery aids in the detection and mapping of landslides. It enables experts to identify areas at risk, assess the extent of landslides, and track their progression over time.
   e. **Vulnerability Assessment**: Geospatial technology allows for comprehensive vulnerability assessments of regions exposed to climate change-driven landslide hazards. This assessment considers factors such as terrain, land cover, and population density.
   f. **Hazard Zonation and Risk Mapping**: Using geospatial data, researchers can create hazard zonation maps and risk assessments. These maps help urban planners, policymakers, and disaster management authorities make informed decisions regarding land use and infrastructure development in landslide-prone areas.
   g. **Modeling and Simulation**: Satellites and geospatial technology support the development of numerical models and simulations to predict potential landslide scenarios. These simulations aid in understanding the impact of climate change on landslide dynamics.
   h. **Post-disaster Assessment**: After a landslide event, satellites can provide real-time imagery to assess the extent of damage, locate affected areas, and coordinate rescue and relief efforts.
   i. **Long-term Monitoring and Research**: Continuous satellite monitoring allows for long-term research on landslide patterns and trends, aiding in the identification of emerging hazards and the evaluation of the effectiveness of mitigation measures.
   j. **International Collaboration**: Satellite data is often part of international initiatives, enabling global cooperation in monitoring climate change and landslide hazards. Such collaboration facilitates knowledge sharing and coordinated response to transboundary landslide risks.

Geospatial technology and satellites play a pivotal role in monitoring climate change-driven landslide hazards by providing essential data for analysis, mapping, early warning systems, and informed decision-making. Integrating these technologies in landslide monitoring efforts can significantly contribute to mitigating the impacts of landslides on communities and the environment in a changing climate. Some important satellites used in various types of hazard analyses are shown in Table 7.3.

7.8 GEOLOGY OF INDIA IN PROSPECT OF LANDSLIDE

India's geological landscape plays a crucial role in determining the susceptibility of its regions to landslides. The northern Himalayan region, characterized by young and unconsolidated sediments, is particularly vulnerable to landslides owing to the ongoing tectonic collision between the Indian

TABLE 7.3

Satellites Used in Monitoring Hazards (Bijan, 2017)

Sr. No.	Optical Satellite	Year Launch	Sensor	Spatial Resolution (meters) and (bands)					Swath (Km)	Repeat Cycle
				PAN	VNIR	SWIR	MWIR	TIR		
1.	CBERS 4	2014	PAN	5					60	1–26
			ISR		40	40 (2)		80	120	26
			WFI		73 (4)				866	5
2.	RazakSat[e]	2009	MAC	2.5	5 (4)				20	13–15[f]
3.	RapidEye A-E[d]	2008	REIS	6.5	6.5 (5)				78	1
4.	Pleiades[c] -1 and 2	2008–2009	HiRI	0.7	2.8 (4)				20	26 to 4
5.	CBERS 3	2008	MUX		20 (4)				120	26
6.	Resurs DK-1[g]	2006	ESI	1	3 (3)				28	N/A
7.	ALOS	2006	PRISM,	2.5					35 (70)	46 (2)
			AVNIR-2		10 (4)				70	
8.	KOMPSAT - 2[b]	2006	MSC	1	4 (4)				15	28
9.	China DMC+4 (Tsinghua-1)	2005	MS DMC	4	32 (3)					600
10.	EROS B-C	2005–2008	PIC	0.7	2.8				11	
11.	TopSat[b]	2005	RALCam1	2.5	5 (3)				25	4
12.	IRS - P5 (CartoSat-1)	2005	PAN-F	2.5					30	5
13.	ROCSat -2/ FormoSat-2[b]	2004	RSI	2	8 (4)				24	14
14.	OrbView - 3[b]	2003	OHRIS	1	4 (4)				8	3
15.	UK - DMC[a]	2003	ESIS		32 (3)				600	4
16.	DMC2 - NigeriaSat 1[a]	2003	ESIS		32 (3)				600	4
17.	DMC2 - BILSAT-1[a]	2003	PanCam	12					25 (300)	4
			MSIS		26 (2)				55 (300)	
			COBAN		120 (4)					
18.	IRS - P6 (ResourceSat-1)	2003	LISS-4	6	6 (3)				23.9 (70)	5
			LISS-3		23.5 (3)	23.5 (1)			141	24
			AWiFS		56 (3)	56 (1)			740	
19.	CBERS 2	2003	IR-MSS	80		80 (2)		160 (1)	120	26
			WFI		260 (2)				890	3 to 5
20.	DMC2 - AlSat1[a]	2002	ESIS		32 (3)				600	4
21.	SPOT 5	2002	HRG	2.5 - 5	10 (3)	20 (1)			60	26 (5)
			HRS	10					120	26
			Vegetation 2		1000 (3)	1000 (1)			2200	1
22.	Quickbird - 2	2001	BGIS 2000	0.6	2.5 (4)				16	3
23.	MTI	2000	MTI		5 (4), 20 (3)	20 (3)	20 (2)	20 (3)	12	
24.	EROS A1**	2000	PIC	1.9					14	2.5–4.5
25.	KOMPSAT - 1**	1999	EOC	6.6					17	28
			OSMI		1000 (6)					

(Continued)

TABLE 7.3 (Continued)

Sr. No.	Optical Satellite	Year Launch	Sensor	Spatial Resolution (meters) and (bands)					Swath (Km)	Repeat Cycle
				PAN	VNIR	SWIR	MWIR	TIR		
26.	Terra	1999	ASTER		15 (3)	30 (6)		90 (5)	60	16
27.	Ikonos 2	1999	OSA	1	4 (4)				11	3
28.	CBERS 1	1999	HRCC	20	20 (4)				113	26
29.	Landsat 7	1999	ETM+	15	30 (4)	30 (2)		60 (1)	185	16
30.	SPOT 4	1998	HRVIR	10	20 (3)	20 (1)			60 (80)	26 (4)
			Vegetation		1000 (3)	1000 (1)				
31.	IRS - 1D	1997	PAN	5.8					70	24
			WiFS		188 (2)	188 (1)			774	5
32.	IRS - P3	1996	WiFS		188 (2)	188 (1)			774	5
33.	IRS - 1C	1995	LISS-III		23.5 (3)	70.5 (1)			142	24
34.	IRS - P2	1994	LISS-II		36.4 (4)				74	22
35.	SPOT 2	1990	HRV	10	20 (3)				60 (80)	26 (4)
36.	Landsat 5	1984	MSS		80 (4)			120 (1)	185	16
			TM		30 (4)	30 (2)				

***Note**: [a] Disaster Monitoring Constellation (DMC) of four satellites of circular sun-synchronic orbit and revisit cycle daily, [b] Sun-synchronic circular orbit, [c] Two-spacecraft constellation of CNES (Space Agency of France) with provision of stereo images, [d] Five-satellite constellation, [e] Near equatorial low Earth orbit (NEO), [f] Passes/day, [g] Close to non-sun synchronous circular orbit.

Plate and the Eurasian Plate, which leads to seismic activity and the loose nature of the rock and soil. The Western and Eastern Ghats, composed of various rock types, are prone to landslides due to weathering processes that can destabilize slopes, compounded by heavy monsoon rains in the Western Ghats. The Deccan Plateau, primarily consisting of basaltic rocks, experiences fewer landslides, although hilly regions with loose soils are susceptible, especially during intense rainfall events. Coastal plains, composed of sedimentary rocks and alluvial deposits, experience fewer landslides but are vulnerable to coastal and soil erosion in areas with poor land management practices (Figure 7.1).

India's geological diversity is a critical factor, and it interacts with climatic, vegetative, and human elements to shape landslide risks across the country. There are various landslide tragedies in the past, present, and looming in the future. The Kedarnath tragedy is a notable example, resulting from geological factors and their interactions with climatic conditions, illustrating the devastating consequences of such intricate relationships.

7.9 LANDSLIDE-PRONE AREAS IN INDIA

In the Indian context, approximately 12.6% or 0.42 million square kilometers of the land area, except snow-covered regions, are vulnerable to landslide hazards. The North East Himalaya Region (NEHR), encompassing Sikkim and Darjeeling Himalaya, accounts for 0.18 million square kilometers. The North West Himalaya Region (NWHR)—including Jammu and Kashmir, Himachal Pradesh, and Uttarakhand—spans approximately 0.14 million square kilometers. Furthermore, Konkan Hills and the Western Ghats, comprising Maharashtra, Goa, Karnataka, Kerala, and Tamil Nadu, extend over 0.09 million square kilometers. Lastly, the Araku region in Andhra Pradesh covers 0.01 million square kilometers of Eastern Ghats. The Himalayan region, known for its susceptibility to landslides (Figure 7.2), is situated in the highest earthquake-prone

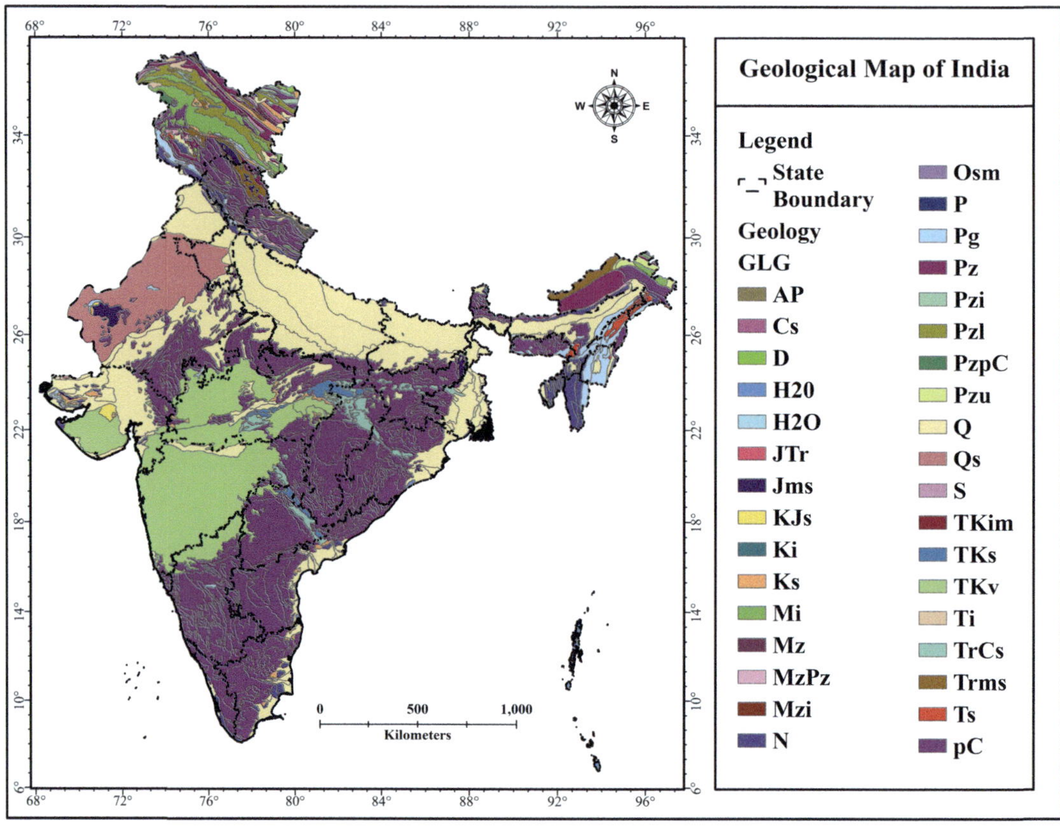

FIGURE 7.1   Geological map of India.

areas (Zone-IV and V; as per BIS, 2002), where earthquakes with a modified Mercalli intensity ranging from VIII to IX are possible, significantly increasing the risk of landslides triggered by seismic activity (Figure 7.3). The heightened risk of landslides stems from the ever-increasing population in hilly regions and the accelerated development of hydropower projects and associated infrastructure.

Moreover, the 21st century has witnessed a pivotal phase of infrastructure expansion with a focus on connecting remote and topographically challenging areas through the construction of new roads, bridges, railway lines, tunnels, and other critical facilities, both in the mainland and ecologically fragile regions. While these development endeavors are vital for meeting growing demands, they inadvertently introduce ecological imbalances, particularly in the form of debris flows and landslides, which, if not adequately addressed, can result in significant loss of property and life (Figure 7.4).

Recognizing the inherent challenges in averting landslides, the scientific community, planners, and administrators face the task of identifying hazard-prone areas and implementing comprehensive zoning strategies at various scales to enhance preparedness and mitigate risks. Earth scientists play a pivotal role in evaluating and understanding the complex dynamics of landslide hazards, while engineers, planners, public officials, and social scientists focus on assessing and managing the associated risks (Figure 7.5). Effectively addressing and minimizing the impact of landslide-related disasters necessitates a rational and objective comprehension of the underlying causes and processes within the domain of Earth system science. Such a comprehensive approach is crucial for promoting sustainable development and ensuring the resilience of communities and infrastructure in landslide-prone regions of India.

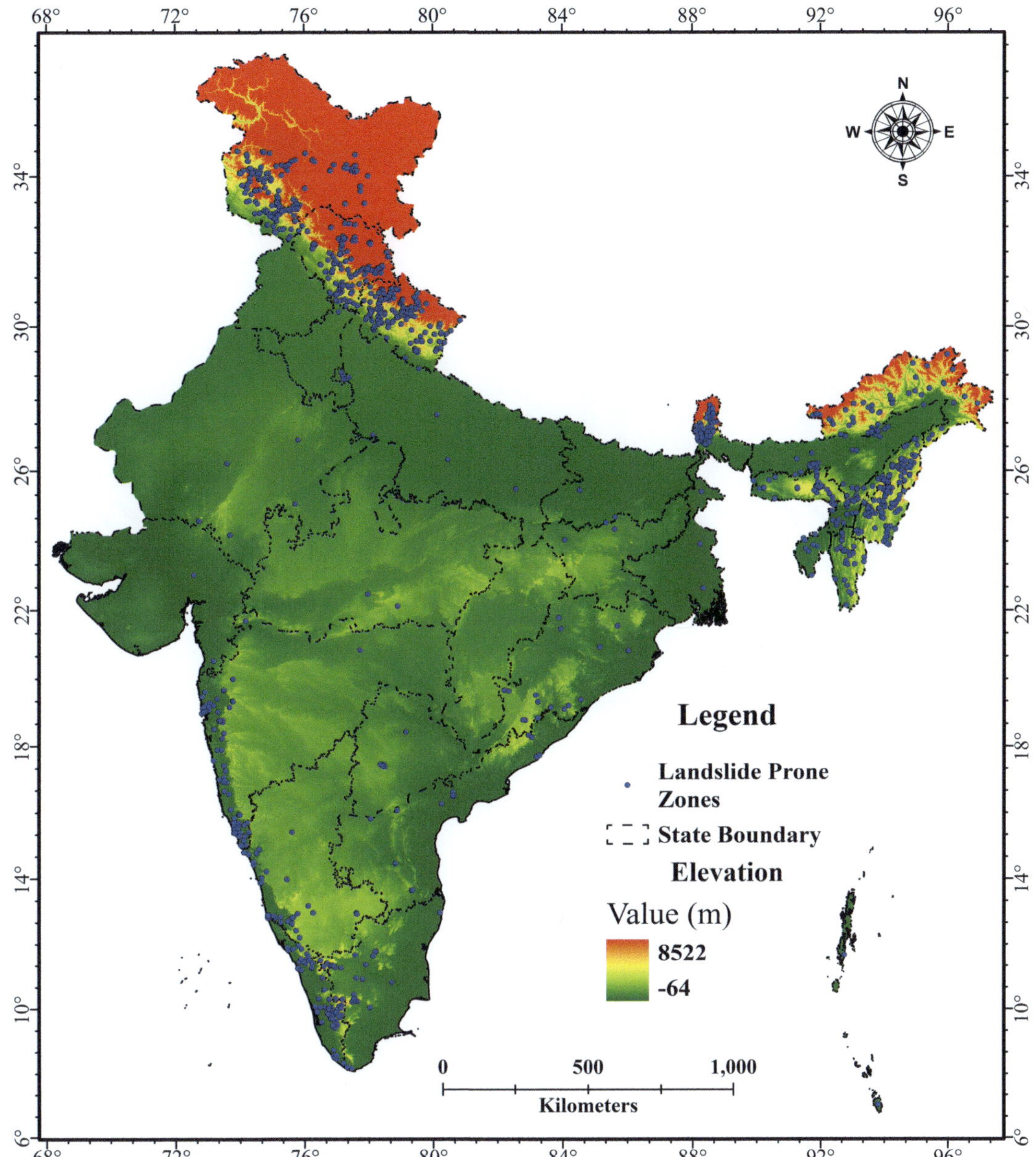

FIGURE 7.2 Representation of landslide-prone locations in India as of 2022.

7.9.1 "The Kedarnath Tragedy": A Case Study

The Kedarnath tragedy in 2013 serves as a harrowing example of the profound effect of India's intricate geology on landslide events. Nestled in the Himalayan state of Uttarakhand, Kedarnath was situated in a region profoundly influenced by geological factors. The ongoing collision of the Eurasian Plate and the Indian Plate, responsible for the development of the Himalayan Mountain range, led to heightened seismic activity, rendering the area susceptible to earthquakes, a significant trigger for landslides. Furthermore, the Himalayan geology consists of fragile, poorly consolidated rocks, and the region's glaciers, influenced by climate change, contribute to slope

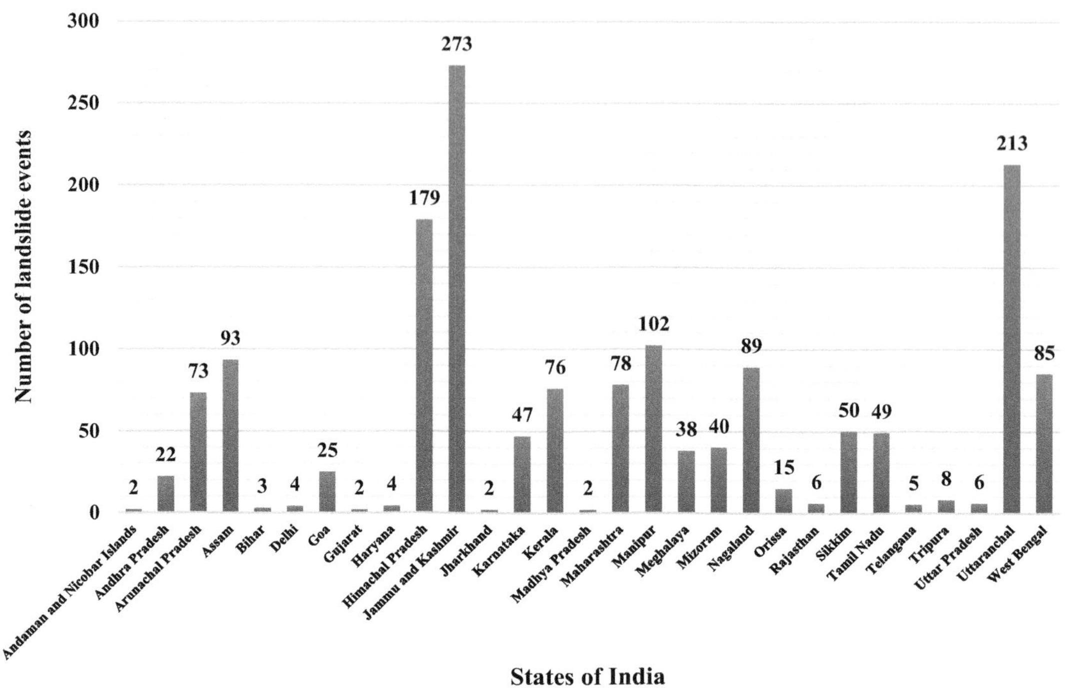

FIGURE 7.3　Landslide events caused in Indian states during 2007–22.

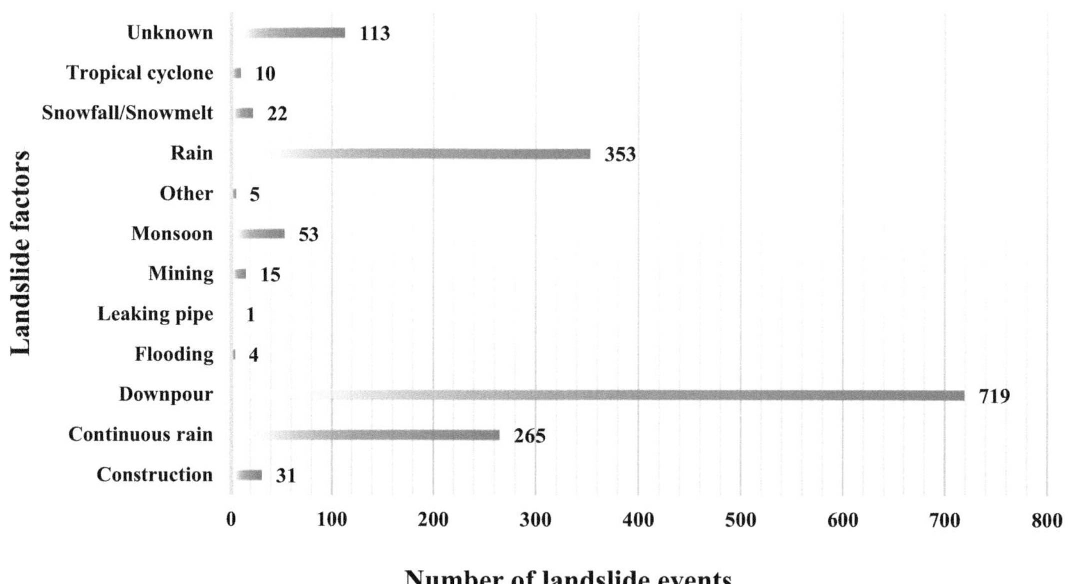

FIGURE 7.4　Factors contributing to landslide events in India during 2007–22.

instability by discharging vast quantities of water and debris. When combined with the heavy monsoon rains typical of the area, the already fragile slopes became saturated, culminating in landslides and flash floods. The Kedarnath tragedy, as depicted in Figure 7.5 and its profound repercussions, emphasized the critical need to grasp the intricate interplay between geological conditions, climate dynamics, and human activities in landslide-prone regions. This underscores

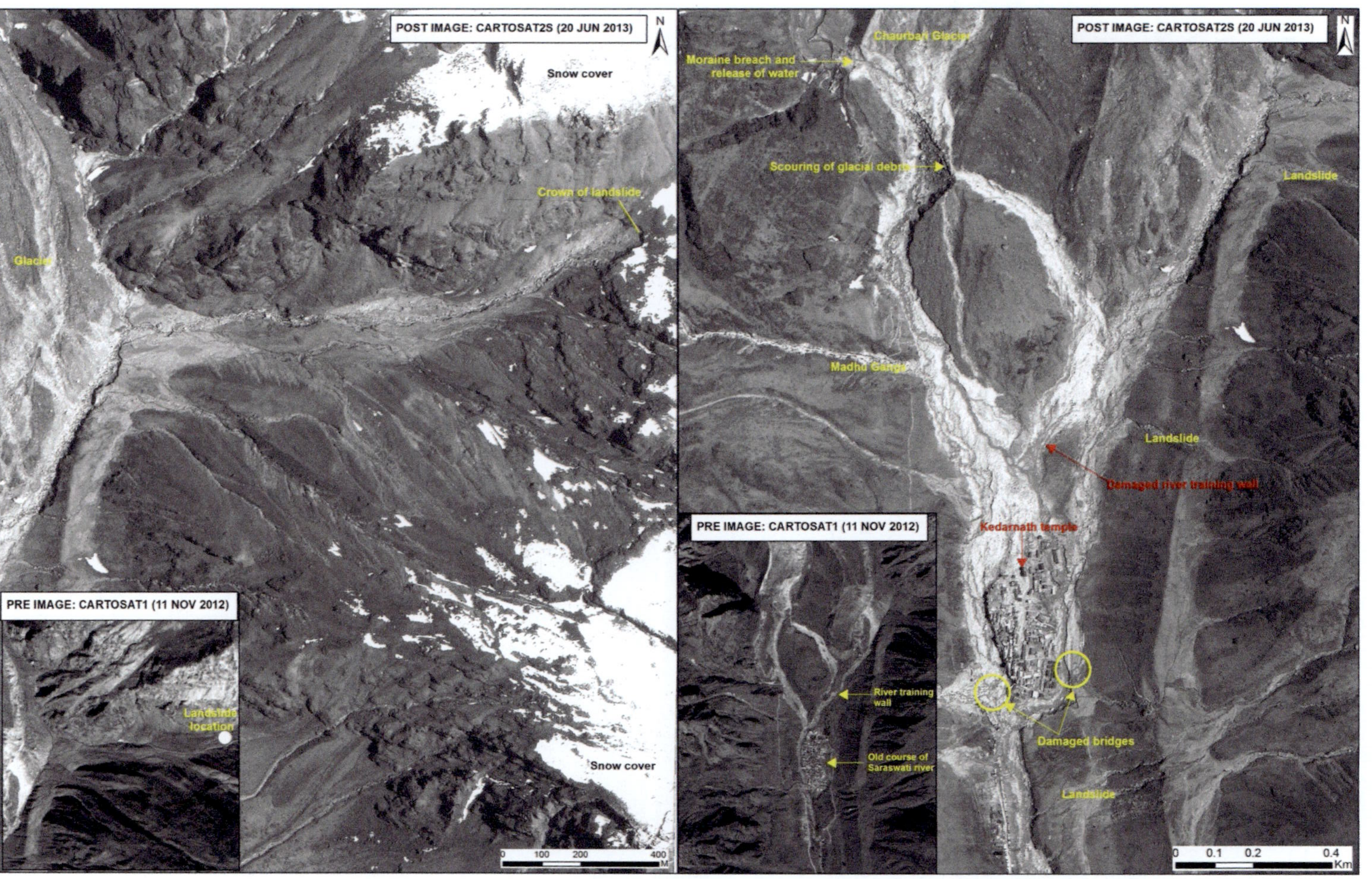

FIGURE 7.5   Geological changes leading to the disaster around Kedarnath, Uttarakhand, 2013.

Source: National Remote Sensing Centre, India

the urgency for implementing advanced monitoring systems and adopting sustainable land-use practices to mitigate future risks effectively.

7.10 RECENT DEVELOPMENTS IN MONITORING LANDSLIDES IN INDIA

Multitemporal SAR interferometry for Landslide Prediction: This technique is used in remote sensing and geodesy to monitor changes on the Earth's surface over time (Dwivedi et al., 2016). It involves the use of multiple SAR images acquired at different times to create interferograms, which are essentially maps showing the displacement of the terrain between the two acquisition dates (Zhao et al., 2018). This technique is beneficial for monitoring areas prone to natural hazards, such as landslides (Bovenga et al., 2022). The main advantage of SAR interferometry lies in its ability to penetrate cloud cover and work regardless of daylight, making it an all-weather, day-and-night monitoring method. "Multitemporal SAR interferometry for Landslide prediction" is a comprehensive technique that involves various steps and computations. The general overview is shown in Figure 7.6.

Debris Flow Modeling: It is a computational approach used to simulate and predict the behavior of debris flows, which are fast and destructive mass movements of a mixture of sediment, water, and debris down a slope (Kafle et al., 2023). Debris flows are highly hazardous natural events that can cause significant impacts on infrastructure, property, and human lives (Fuchs et al., 2007). The data required for debris flow modeling includes topographic information, soil properties, rainfall data, and initial conditions of the slope or potential debris flow area. These data inputs are essential for

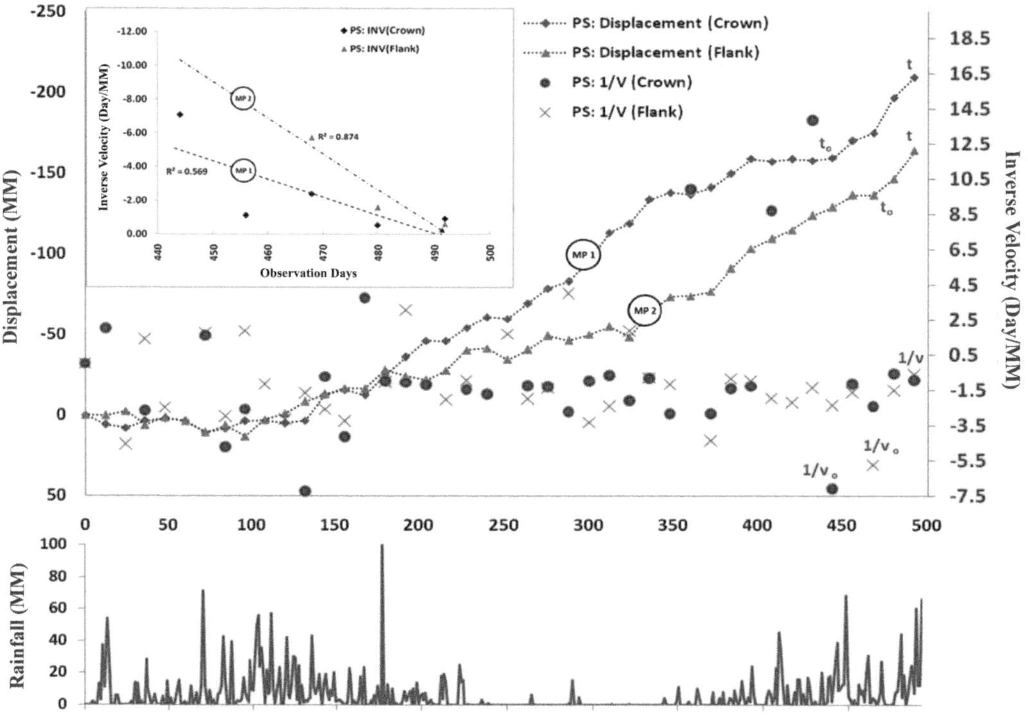

FIGURE 7.6 Landslide prediction using MT InSAR. *Top panel*: The OOA (t0) and last observation point (t) are shown in the displacement time series and inverse velocities for the Kikruma landslide. *Inset*: Estimation of failure day Using INV. *Bottom panel*: Variation in rainfall in the area throughout the observed period. *Predicted day of failure*: 2018, 26 July. *Exact day of failure*: 2018, 29 July.

Source: Jain et al., 2023

accurately representing real-world conditions and improving the reliability of the model's predictions (Kuang et al., 2022). The outputs of debris flow modeling can help with various aspects of hazard assessment and risk management, such as identifying high-risk areas, designing effective mitigation strategies, and developing early warning systems to protect communities and infrastructure from the destructive impact of debris flows.

The equations employed in debris flow modeling may vary depending on the specific methods and approach used. In this context, (Gomes et al., 2013) discuss the SHALSTAB/FLO-2D models. Their study focused on the Quitite and Papagaio basins in Brazil, where mass movements, particularly debris flows, are prevalent, especially during periods of heavy rainfall and in areas with illegal settlements on slopes. The research involved the utilization of the SHALSTAB model for identifying landslide-prone areas and the FLO-2D model for forecasting debris flow volume and travel distances. The primary objectives of this approach were to minimize casualties, provide valuable input for public policy, and support urban planning initiatives. The methodology's reliability was validated through comparison with a 1996 event, illustrating its effectiveness in predicting such disasters (Figure 7.7).

UAV data in landslide modeling: Landslide modeling incorporates UAV data, which entails the utilization of Unmanned Aerial Vehicles (UAVs), generally referred to as drones, to collect high-resolution and up-to-date data for landslide modeling and monitoring (Karantanellis et al., 2020). UAVs have become valuable tools in various geospatial applications (Preethi Latha et al., 2019), including landslide research, due to their ability to capture detailed and accurate data over difficult-to-access or hazardous terrains (Karantanellis et al., 2020) (Figure 7.8). UAV data significantly enhances landslide modeling efforts, enabling more accurate and informed assessments and contributing to better landslide risk management strategies. In this prospect, (Yang et al., 2022) conducted a valuable research study addressing the need for efficient landslide recognition, emphasizing the importance of rapid access to disaster information for effective disaster reduction and relief.

Classical satellite remote sensing can be hindered by orbital cycles and weather conditions, necessitating improvements in data acquisition. The study focuses on landslide recognition using UAV remote sensing images, particularly in the Zhangmu Port region of Tibet, affected by the Nepal earthquake in 2015. To overcome limited training data, transfer learning is applied. The evaluation demonstrates that this approach effectively identifies landslide disasters. Compared to the SSD model, the proposed method exhibits superior detection performance, offering precise data for informed disaster rescue decision-making. This research represents a significant step forward in leveraging UAV technology for disaster management and response.

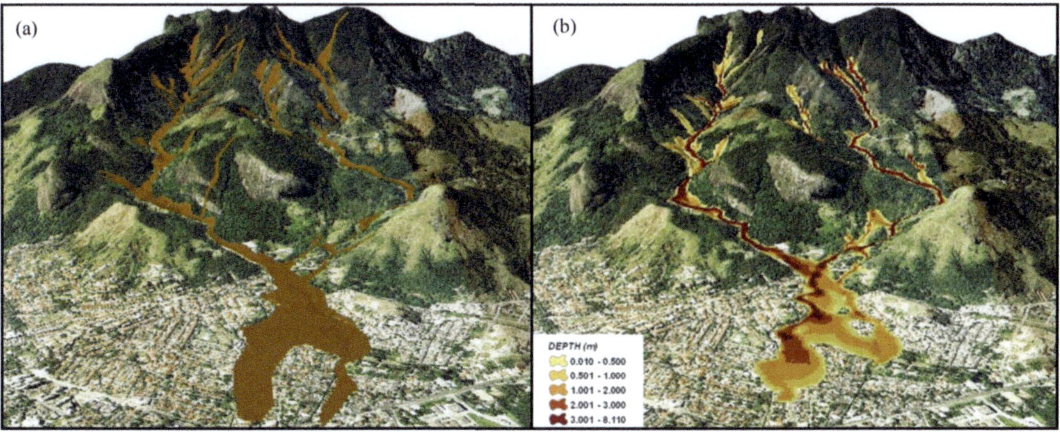

FIGURE 7.7 Perspective views: (a) Debris-flow bounded area just after the incident, (b) Simulation findings from the linked models SHALSTAB/FLO-2D.

Source: Gomes et al., 2013

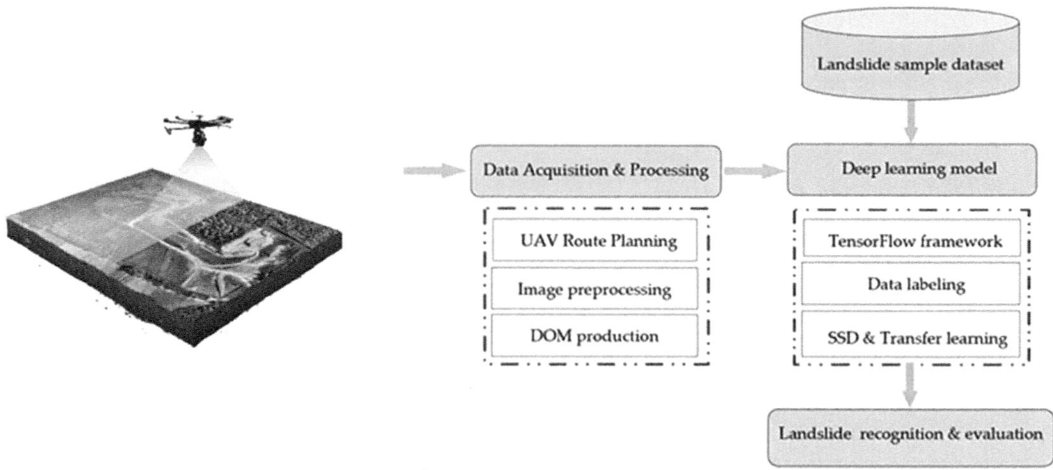

FIGURE 7.8   UAV data in landslide modeling based on SSD Model.

Source: Yang et al., 2022.

Landslide Susceptibility Zonation: LSZ is a geospatial technique used to map and assess the relative susceptibility of an area to landslides (Pradhan, 2011). It involves dividing a region into different zones based on the likelihood of landslides occurring, ranging from low susceptibility to high susceptibility. The primary goal of landslide susceptibility zonation (Leonardi et al., 2016; Shano et al., 2020) is to identify and delineate areas more prone to landslides, enabling better land-use planning, risk management, and hazard mitigation strategies. The zonation map categorizes different areas based on their relative susceptibility to landslides (Kumra et al., 2018; P. Singh et al., 2023). The map provides valuable information for land-use planners, disaster management authorities, and policymakers to prioritize areas for development, implement appropriate land-use regulations, and plan measures for landslide risk reduction and prevention (Kumari et al., 2023; Onagh et al., 2012b). The landslide susceptibility zonation does not predict specific landslide events; instead, it identifies areas more likely to experience landslides under certain conditions. It serves as a tool to guide decision-making and risk-reduction efforts, contributing to safer and more sustainable development in landslide-prone regions.

There are various methods to derive landslide susceptibility zonation LSZ, and one commonly used approach is the AHP (Analytical Hierarchy Process) for landslide susceptibility zonation. In this scenario, (Sonker et al., 2021) conducted valuable research employing remote sensing, GIS, and the AHP to generate landslide susceptibility maps for the Sikkim Himalaya region in India. Various factors such as geology, geomorphology, slope, drainage density, lineament density, rainfall, soil, and LULC were considered. Thematic layers for these factors were generated using remote sensing and ground data, and their relative importance was determined using the AHP technique. The study then integrated these layers to produce landslide susceptibility maps, which were categorized into five levels: Very High - 2.76%, High - 33.60%, Moderate - 0.06%, Low - 63.34%, and Very Low - 0.18%. The accuracy assessment, based on the landslide inventory and AHP model, showed an AUC (Area Under the Curve) of 73%. This research provides valuable information for identifying landslide susceptibility zones in the study area (Figure 7.9).

Early landslide warning system based on rainfall threshold: Rainfall threshold-based landslide early warning systems are predictive systems that use rainfall data to provide advance warning of potential landslide events. These systems are designed to monitor and analyze rainfall patterns and trigger alerts when certain critical rainfall thresholds are exceeded (Aleotti, 2004), indicating conditions that could lead to landslides. Early warning systems are valuable tools for reducing the risk associated with landslides (Perotto-Baldiviezo et al., 2004), especially in areas where rapid-onset

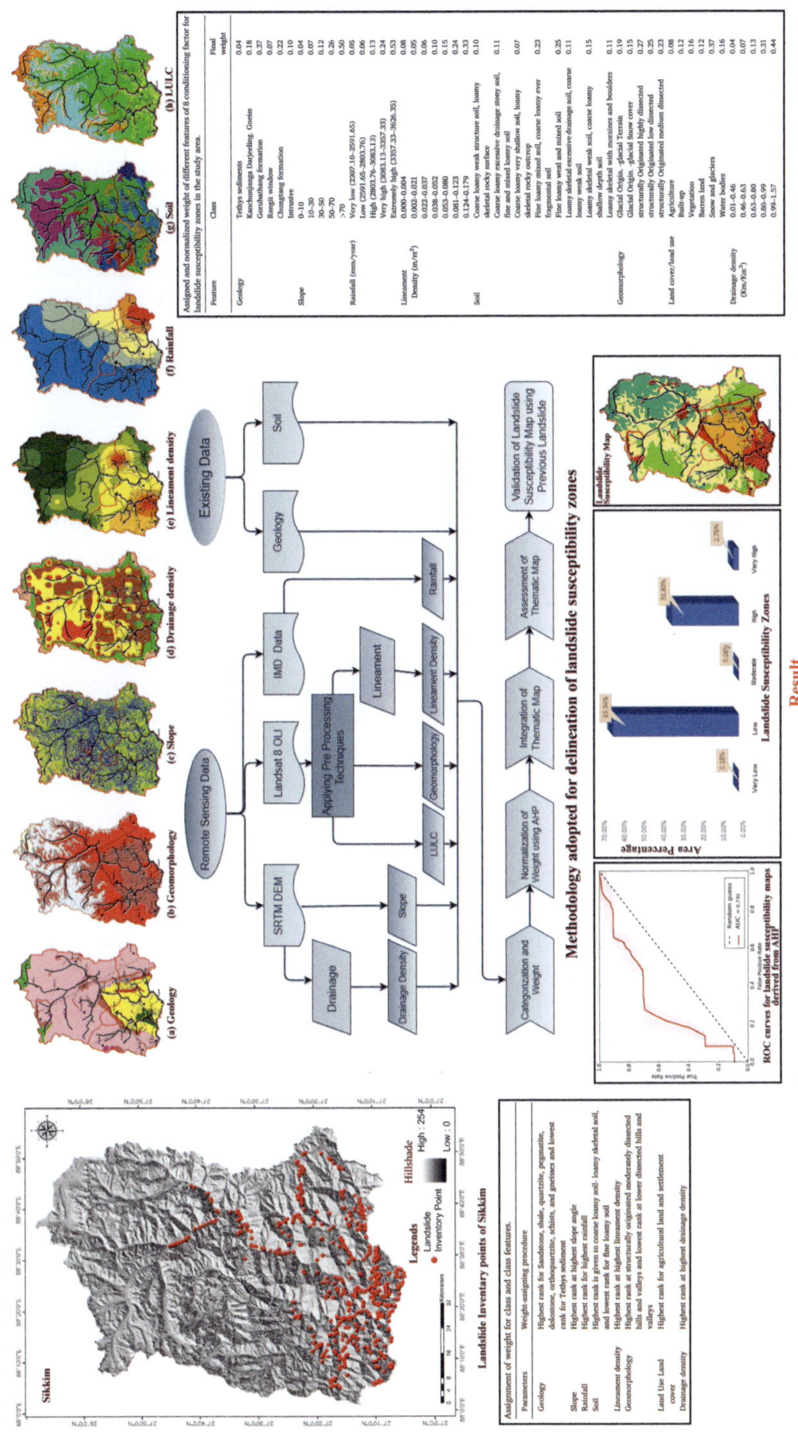

FIGURE 7.9 Landslide susceptibility zonation using Analytical Hierarchy Process of Sikkim.

Source: Sonker et al., 2021.

rainfall-induced landslides (Baum et al., 2010) are common. These systems can provide valuable lead time for emergency response and preparedness (Hong et al., 2007), potentially saving lives and mitigating damage to property and infrastructure. Early warning systems are effective in certain regions and conditions, but they may not be sufficient for all landslide types or situations. For example, slow-moving or long-term landslides may not be triggered by short-term heavy rainfall events alone, and other factors such as geological conditions, soil moisture, and land-use changes may also play significant roles.

The combination of multitemporal SAR interferometry, debris flow modeling, UAV data collection, landslide susceptibility zonation, and an early landslide warning system based on rainfall-threshold has significantly advanced our ability to forecast, monitor, and mitigate landslide events. Together, they empower decision-makers, researchers, and communities to better understand landslide dynamics, reduce risks, and enhance disaster preparedness, ultimately fostering safer environments for all.

7.11   OPPORTUNITIES AND RECOMMENDATIONS

Addressing climate change-driven landslide hazards is multifaceted. Leveraging geospatial technology for improved early warning systems, comprehensive risk assessments, and climate-resilient infrastructure presents significant opportunities. Encouraging interdisciplinary collaboration among experts from various fields, integrating climate data, and enhancing public awareness can enhance preparedness. International cooperation and standardized regulations are vital for global solutions.

Investment in advanced monitoring techniques, like SAR interferometry and UAV data, should be prioritized. Additionally, conducting socioeconomic assessments and targeting vulnerable populations are essential for effective risk reduction. By embracing these opportunities and recommendations, we can mitigate the impacts of landslides in the face of a changing climate.

7.12   DISCUSSION

This chapter highlights the increasing importance of monitoring climate change-driven landslide hazards due to the rise in heavy rainfall events caused by climate change. With slopes becoming more vulnerable and rapid urbanization occurring in landslide-prone areas, the need for effective monitoring becomes crucial to implement early warning systems and mitigate potential losses. The impact of landslides on the environment, such as soil erosion and disruption of ecosystems, underscores the significance of comprehensive monitoring and land management strategies. Geospatial technology emerges as a crucial tool in landslide monitoring, offering a wide range of benefits in understanding and managing landslide hazards.

Satellite remote sensing, aerial imagery, LiDAR, radar-based systems, and GIS software contribute valuable data for landslide monitoring. These technologies provide detailed information on topography, land cover, and surface characteristics, essential for assessing landslide risks and enabling early warning systems. Geospatial technology also aids in risk assessment, zoning, modeling, and post-disaster response, making it indispensable in landslide hazard management. It also discusses the influencing factors behind landslides, both natural phenomena (excess water, vegetation, earthquakes) and human activities (excavation, mining, construction). Understanding these factors is vital for effective landslide detection and risk mitigation strategies.

It is evident that landslides are complex events influenced by various natural and anthropogenic elements, necessitating a comprehensive and integrated approach to hazard assessment. The chapter has been a comprehensive review of previous literature on climate change-driven landslide hazards, outlining various studies that analyze the impact of climate change on landslides, landslide risk management strategies, and predictive modeling using climate scenarios. These research endeavors offer valuable perspectives on the connection between climate change and the occurrence of landslides, enhancing our comprehension and readiness to manage landslide risks effectively.

Furthermore, the study sheds light on recent developments in monitoring landslides in India, including the use of multitemporal SAR interferometry, debris flow modeling, UAV data integration, landslide susceptibility zonation, and rainfall-threshold-based early warning systems. These advancements have the potential to significantly enhance landslide monitoring efforts in the country, enabling better risk assessment and mitigation.

7.13 CONCLUSION

The aim of this chapter was to investigate how geospatial technology can promote sustainability by keeping an eye on landslide risks brought by climate change. The chapter followed a systematic approach to select and analyze the most relevant literature on landslide detection techniques, climate change and landslide hazards, and landslide monitoring in different regions and contexts. The chapter identified four main themes: conventional and modern methods of landslide detection, the influence of climate change on landslide occurrences and hazard management, recent advancements and hurdles in geospatial technology based on landslide monitoring, and opportunities and recommendations for enhancing landslide monitoring for sustainability. This chapter's findings augment the existing body of literature by furnishing a thorough assessment of the current status of knowledge and practical applications concerning geospatial technology in landslide monitoring across diverse settings and scales. This chapter also has practical implications for policymakers, practitioners, and communities in landslide-prone regions, as it provides valuable information for developing adaptation strategies, risk management plans, and early warning systems.

Lastly, this chapter shows that geospatial technology is a key tool for comprehensive landslide monitoring that can help protect lives, reduce environmental damage, and enhance resilience in the context of climate change. Future study should concentrate on determining the feasibility and effectiveness of implementing these techniques in different contexts and scales, as well as exploring how to integrate geospatial technology with other sources of data and knowledge for landslide hazard management.

REFERENCES

Abdalla, R., & Li, J. (2010). Towards effective application of geospatial technologies for disaster management. *International Journal of Applied Earth Observation and Geoinformation, 12*(6), 405–407. https://doi.org/10.1016/j.jag.2010.09.003

Adinehvand, M., Rai, P. K., & Mindrescu, M. (2018). Remote Sensing Data and GIS Application for Landslides Mapping: A Case Study from India. *Georeview: Scientific Annals of Stefan Cel Mare University of Suceava. Geography Series, 28*(1), 13.

Ahmed, B., Rahman, M. S., Sammonds, P., Islam, R., & Uddin, K. (2020). Application of geospatial technologies in developing a dynamic landslide early warning system in a humanitarian context: the Rohingya refugee crisis in Cox's Bazar, Bangladesh. *Geomatics, Natural Hazards and Risk, 11*(1), 446–468. https://doi.org/10.1080/19475705.2020.1730988

Aleotti, P. (2004). A warning system for rainfall-induced shallow failures. *Engineering Geology, 73*(3–4), 247–265. https://doi.org/10.1016/j.enggeo.2004.01.007

Alvioli, M., Melillo, M., Guzzetti, F., Rossi, M., Palazzi, E., von Hardenberg, J., Brunetti, M. T., & Peruccacci, S. (2018). Implications of climate change on landslide hazard in Central Italy. *Science of the Total Environment, 630*, 1528–1543. https://doi.org/10.1016/j.scitotenv.2018.02.315

Anbalagan, R., Kumar, R., Lakshmanan, K., Parida, S., & Neethu, S. (2015). Landslide hazard zonation mapping using frequency ratio and fuzzy logic approach, a case study of Lachung Valley, Sikkim. *Geoenvironmental Disasters, 2*(1), 6. https://doi.org/10.1186/s40677-014-0009-y

Andersson-Sköld, Y., Bergman, R., Johansson, M., Persson, E., & Nyberg, L. (2013). Landslide risk management - A brief overview and example from Sweden of current situation and climate change. *International Journal of Disaster Risk Reduction, 3*(1), 44–61. https://doi.org/10.1016/j.ijdrr.2012.11.002

Baum, R. L., Godt, J. W., & Savage, W. Z. (2010). Estimating the timing and location of shallow rainfall-induced landslides using a model for transient, unsaturated infiltration. *Journal of Geophysical Research: Earth Surface, 115*(3), F03013. https://doi.org/10.1029/2009JF001321

Bello, O. M., & Aina, Y. A. (2014). Satellite Remote Sensing as a Tool in Disaster Management and Sustainable Development: Towards a Synergistic Approach. *Procedia - Social and Behavioral Sciences, 120*, 365–373. https://doi.org/10.1016/j.sbspro.2014.02.114

Bijan, K. (2017). Monitoring of Incomati River Basin with Remote Sensing [Lund University, Sweden]. https://rammb.cira.colostate.edu/dev/hillger/environmental.htm

Bovenga, F., Argentiero, I., Refice, A., Nutricato, R., Nitti, D. O., Pasquariello, G., & Spilotro, G. (2022). Assessing the Potential of Long, Multi-Temporal SAR Interferometry Time Series for Slope Instability Monitoring: Two Case Studies in Southern Italy. *Remote Sensing, 14*(7), 1677. https://doi.org/10.3390/rs14071677

Ciabatta, L., Camici, S., Brocca, L., Ponziani, F., Stelluti, M., Berni, N., & Moramarco, T. (2016). Assessing the impact of climate-change scenarios on landslide occurrence in Umbria Region, Italy. *Journal of Hydrology, 541*, 285–295. https://doi.org/10.1016/j.jhydrol.2016.02.007

Dai, F. C., Lee, C. F., & Ngai, Y. Y. (2002). Landslide risk assessment and management: An overview. *Engineering Geology, 64*(1), 65–87. https://doi.org/10.1016/S0013-7952(01)00093-X

Dewan, A. M., Islam, M. M., Kumamoto, T., & Nishigaki, M. (2007). Evaluating flood hazard for land-use planning in greater Dhaka of Bangladesh using remote sensing and GIS techniques. *Water Resources Management, 21*(9), 1601–1612. https://doi.org/10.1007/s11269-006-9116-1

Dwivedi, R., Narayan, A. B., Tiwari, A., Dikshit, O., & Singh, A. K. (2016). Multi-temporal SAR Interferometry for landslide monitoring. *International Archives of the Photogrammetry, Remote Sensing and Spatial Information Sciences - ISPRS Archives, 41*, 55–58. https://doi.org/10.5194/isprsarchives-XLI-B8-55-2016

Easton-Gomez, S. C., Mouritz, M., & Breadsell, J. K. (2022). Enhancing Emotional Resilience in the Face of Climate Change Adversity: A Systematic Literature Review. *Sustainability, 14*(21), 13966. https://doi.org/10.3390/su142113966

Emberson, R., Kirschbaum, D., & Stanley, T. (2021). Global connections between El Nino and landslide impacts. *Nature Communications, 12*(1), 2262. https://doi.org/10.1038/s41467-021-22398-4

Fayaz, M., Meraj, G., Khader, S. A., Farooq, M., Kanga, S., Singh, S. K., Kumar, P., & Sahu, N. (2022). Management of Landslides in a Rural–Urban Transition Zone Using Machine Learning Algorithms—A Case Study of a National Highway (NH-44), India, in the Rugged Himalayan Terrains. *Land, 11*(6), 884. https://doi.org/10.3390/land11060884

Fowze, J. S. M., Bergado, D. T., Soralump, S., Voottipreux, P., & Dechasakulsom, M. (2012). Rain-triggered landslide hazards and mitigation measures in Thailand: From research to practice. *Geotextiles and Geomembranes, 30*, 50–64. https://doi.org/10.1016/j.geotexmem.2011.01.007

Fuchs, S., Heiss, K., & Hübl, J. (2007). Towards an empirical vulnerability function for use in debris flow risk assessment. *Natural Hazards and Earth System Science, 7*(5), 495–506. https://doi.org/10.5194/nhess-7-495-2007

Gariano, S. L., & Guzzetti, F. (2016). Landslides in a changing climate. *Earth-Science Reviews, 162*, 227–252. https://doi.org/10.1016/j.earscirev.2016.08.011

Gomes, R. A. T., Guimarães, R. F., de Carvalho Júnior, O. A., Fernandes, N. F., & Do Amaral Júnior, E. V. (2013). Combining spatial models for shallow landslides and debris-flows prediction. *Remote Sensing, 5*(5), 2219–2237. https://doi.org/10.3390/rs5052219

Guan, S., Shi, Z., Zheng, H., Shen, D., Hanley, K. J., Yang, J., & Xia, C. (2023). Effects of soil properties and geomorphic parameters on the breach mechanisms of landslide dams and prediction of peak discharge. *Acta Geotechnica*. https://doi.org/10.1007/s11440-023-01908-2

Hong, Y., Adler, R. F., Negri, A., & Huffman, G. J. (2007). Flood and landslide applications of near real-time satellite rainfall products. *Natural Hazards, 43*(2), 285–294. https://doi.org/10.1007/s11069-006-9106-x

Huang, B., Jiang, B., & Li, H. (2001). An integration of GIS, virtual reality and the internet for visualization, analysis and exploration of spatial data. *International Journal of Geographical Information Science, 15*(5), 439–456. https://doi.org/10.1080/13658810110046574

Jain, N., Roy, P., Martha, T. R., Jalan, P., & Nanda, A. (2023). *Landslide Atlas of India (Mapping, Monitoring and R&D studies using Remote Sensing data)* (Special Ed). National Remote Sensing Centre. https://www.nrsc.gov.in

Jiang, H., Xu, C., Adhikari, B. R., Liu, X., Tan, X., & Yuan, R. (2022). Editorial: Environmental change driven by climatic change, tectonism and landslide. *Frontiers in Earth Science, 10.* https://doi.org/10.3389/feart.2022.1076801

Johnston, E. C., Davenport, F. V., Wang, L., Caers, J. K., Muthukrishnan, S., Burke, M., & Diffenbaugh, N. S. (2021). Quantifying the Effect of Precipitation on Landslide Hazard in Urbanized and Non-Urbanized Areas. *Geophysical Research Letters, 48*(16). https://doi.org/10.1029/2021GL094038

Jurchescu, M., Micu, D., Sima, M., Bălteanu, D., Bojariu, R., Dumitrescu, A., Dragotă, C., Micu, M., & Senzaconi, F. (2018). An approach to investigate the effects of climate change on landslide hazard at a national scale (Romania). *Proceedings of the 33rd Romanian Geomorphology Symposium,* 121–124. https://doi.org/10.15551/prgs.2017.121

Kafle, J., Dangol, B. R., Tiwari, C. N., & Kattel, P. (2023). Dynamics of landslide-generated tsunamis and their dependence on the particle concentration of initial release mass. *European Journal of Mechanics, B/Fluids, 97,* 146–161. https://doi.org/10.1016/j.euromechflu.2022.10.003

Karantanellis, E., Marinos, V., Vassilakis, E., & Christaras, B. (2020). Object-based analysis using unmanned aerial vehicles (UAVs) for site-specific landslide assessment. *Remote Sensing, 12*(11), 1711. https://doi.org/10.3390/rs12111711

Kim, H. G., Lee, D. K., Park, C., Kil, S., Son, Y., & Park, J. H. (2015). Evaluating landslide hazards using RCP 4.5 and 8.5 scenarios. *Environmental Earth Sciences, 73*(3), 1385–1400. https://doi.org/10.1007/s12665-014-3775-7

Kuang, P., Li, R., Huang, Y., Wu, J., Luo, X., & Zhou, F. (2022). Landslide Displacement Prediction via Attentive Graph Neural Network. *Remote Sensing, 14*(8), 1919. https://doi.org/10.3390/rs14081919

Kumar Rai, P., Mohan, K., & Kumra, V. K. (2014). Landslide Hazard and Its Mapping Using Remote Sensing and Gis. *Journal of Scientific Research, 58*(1), 1–13.

Kumari, A., Rai, P. K., Mishra, V. N., Singh, P., & Mehra, A. (2023). Remote sensing-based study of landslide hazard zonation in Namchi and its surrounding area of Sikkim, India. *Atmospheric Remote Sensing,* 429–456. https://doi.org/10.1016/b978-0-323-99262-6.00013-4

Kumra, V. K., Shahi, A. P., & Rai, P. K. (2018). GEOSPATIAL TECHNOLOGY IN LANDSLIDE MAPPING: A CASE STUDY OF UTTRAKHAND HIMALAYA, INDIA. *Practical Geography and XXI Century Challenges,* 567–568.

Lacasse, S., & Nadim, F. (2009). Landslide risk assessment and mitigation strategy. In *Landslides - Disaster Risk Reduction* (pp. 31–62). Springer Berlin Heidelberg. https://doi.org/10.1007/978-3-540-69970-5_3

Leonardi, G., Palamara, R., & Cirianni, F. (2016). Landslide Susceptibility Mapping Using a Fuzzy Approach. *Procedia Engineering, 161,* 380–387. https://doi.org/10.1016/j.proeng.2016.08.578

Lorentz, J. F., Calijuri, M. L., Marques, E. G., & Baptista, A. C. (2016). Multicriteria analysis applied to landslide susceptibility mapping. *Natural Hazards, 83*(1), 41–52. https://doi.org/10.1007/s11069-016-2300-6

Mateos, R. M., López-Vinielles, J., Poyiadji, E., Tsagkas, D., Sheehy, M., Hadjicharalambous, K., Liscák, P., Podolski, L., Laskowicz, I., Iadanza, C., Gauert, C., Todorović, S., Auflič, M. J., Maftei, R., Hermanns, R. L., Kociu, A., Sandić, C., Mauter, R., Sarro, R., … Herrera, G. (2020). Integration of landslide hazard into urban planning across Europe. *Landscape and Urban Planning, 196,* 103740. https://doi.org/10.1016/j.landurbplan.2019.103740

Mejri, O., Menoni, S., Matias, K., & Aminoltaheri, N. (2017). Crisis information to support spatial planning in post disaster recovery. *International Journal of Disaster Risk Reduction, 22,* 46–61. https://doi.org/10.1016/j.ijdrr.2017.02.007

Mountjoy, J. J., Pecher, I., Henrys, S., Crutchley, G., Barnes, P. M., & Plaza-Faverola, A. (2014). Shallow methane hydrate system controls ongoing, downslope sediment transport in a low-velocity active submarine landslide complex, Hikurangi Margin, New Zealand. *Geochemistry, Geophysics, Geosystems, 15*(11), 4137–4156. https://doi.org/10.1002/2014GC005379

Niculiță, M. (2020). Landslide Hazard Induced by Climate Changes in North-Eastern Romania. In *Climate Change Management* (pp. 245–265). https://doi.org/10.1007/978-3-030-37425-9_13

Onagh, M., Kumra, V. K., & Rai, P. K. (2012a). Landslide Susceptibility Mapping in a Part of Uttarkashi District (India) By Multiple Linear Regression Method. *International Journal of Geology, Earth and Environmental Sciences, 2*(2), 102–120.

Onagh, M., Kumra, V., & Rai, P. (2012b). Application of Multiple Linear Regression Model in Landslide Susceptibility Zonation Mapping the Case Study Narmab Basin. *International Journal of Geology, Earth and Environmental Sciences, 2*(2), 87–101. http://www.researchgate.net/publication/259572499_Application_of_Multiple_Linear_Regression_Model_in_Landslide_Susceptibility_Zonation_Mapping_The_Case_Study_of_Narmab_Basin_(Iran)/file/e0b4952ca61ba504aa.pdf

Pánek, T., & Klimeš, J. (2016). Temporal behavior of deep-seated gravitational slope deformations: A review. *Earth-Science Reviews*, *156*, 14–38. https://doi.org/10.1016/j.earscirev.2016.02.007

Perotto-Baldiviezo, H. L., Thurow, T. L., Smith, C. T., Fisher, R. F., & Wu, X. B. (2004). GIS-based spatial analysis and modeling for landslide hazard assessment in steeplands, southern Honduras. *Agriculture, Ecosystems and Environment*, *103*(1), 165–176. https://doi.org/10.1016/j.agee.2003.10.011

Picarelli, L., Lacasse, S., & Ho, K. K. S. (2021). *The Impact of Climate Change on Landslide Hazard and Risk* (pp. 131–141). https://doi.org/10.1007/978-3-030-60196-6_6

Pourghasemi, H. R., & Rossi, M. (2017). Landslide susceptibility modeling in a landslide prone area in Mazandarn Province, north of Iran: a comparison between GLM, GAM, MARS, and M-AHP methods. *Theoretical and Applied Climatology*, *130*(1–2), 609–633. https://doi.org/10.1007/s00704-016-1919-2

Pradhan, B. (2011). Use of GIS-based fuzzy logic relations and its cross application to produce landslide susceptibility maps in three test areas in Malaysia. *Environmental Earth Sciences*, *63*(2), 329–349. https://doi.org/10.1007/s12665-010-0705-1

Preethi Latha, T., Naga Sundari, K., Cherukuri, S., & Prasad, M. V. V. S. V. (2019). Remote Sensing uav/drone technology as a tool for urban development measures in Apcrda. *The International Archives of the Photogrammetry, Remote Sensing and Spatial Information Sciences*, XLII-2/W13, 525–529. https://doi.org/10.5194/isprs-archives-XLII-2-W13-525-2019

Pun, W. K., Chung, P. W. K., Wong, T. K. C., Lam, H. W. K., & Wong, L. A. (2020). Landslide Risk Management in Hong Kong - Experience in the Past and Planning for the Future. *Landslides*, *17*(1), 243–247. https://doi.org/10.1007/s10346-019-01291-8

Reichenbach, P., Rossi, M., Malamud, B. D., Mihir, M., & Guzzetti, F. (2018). A review of statistically-based landslide susceptibility models. *Earth-Science Reviews*, *180*, 60–91. https://doi.org/10.1016/j.earscirev.2018.03.001

Sammonds, P., Shamsudduha, M., & Ahmed, B. (2021). Climate change driven disaster risks in Bangladesh and its journey towards resilience. *Journal of the British Academy*, *9s8*, 55–77. https://doi.org/10.5871/jba/009s8.055

Sawaya, K. E., Olmanson, L. G., Heinert, N. J., Brezonik, P. L., & Bauer, M. E. (2003). Extending satellite remote sensing to local scales: Land and water resource monitoring using high-resolution imagery. *Remote Sensing of Environment*, *88*(1–2), 144–156. https://doi.org/10.1016/j.rse.2003.04.006

Shano, L., Raghuvanshi, T. K., & Meten, M. (2020). Landslide susceptibility evaluation and hazard zonation techniques – a review. *Geoenvironmental Disasters*, *7*(1), 18. https://doi.org/10.1186/s40677-020-00152-0

Sharma, A., Rai, P. K., Singh, P., & Srivastava, P. K. (2020). Probabilistic Landslide Hazard Assessment using Statistical Information Value (SIV) and GIS Techniques: A Case Study of Himachal Pradesh, India. *Techniques for Disaster Risk Management and Mitigation*, 197–208. https://doi.org/10.1002/9781119359203.ch15

Sharma, M., Upadhyay, R. K., Tripathi, G., Kishore, N., Shakya, A., Meraj, G., Kanga, S., Singh, S. K., Kumar, P., Johnson, B. A., & Thakur, S. N. (2023). Assessing Landslide Susceptibility along India's National Highway 58: A Comprehensive Approach Integrating Remote Sensing, GIS, and Logistic Regression Analysis. *Conservation*, *3*(3), 444–459. https://doi.org/10.3390/conservation3030030

Singh, B. (2014). Landslide Disaster Management. In *Geospatial Technologies and Climate Change* (pp. 53–63). Springer International Publishing. https://doi.org/10.1007/978-3-319-01689-4_4

Singh, P., Sharma, A., Sur, U., & Rai, P. K. (2021). Comparative landslide susceptibility assessment using statistical information value and index of entropy model in Bhanupali-Beri region, Himachal Pradesh, India. *Environment, Development and Sustainability*, *23*(4), 5233–5250. https://doi.org/10.1007/s10668-020-00811-0

Singh, P., Sur, U., Rai, P. K., & Singh, S. K. (2023). Landslide susceptibility prediction using frequency ratio model: a case study of Uttarakhand, Himalaya (India). *Proceedings of the Indian National Science Academy*, 1–13. https://doi.org/10.1007/s43538-023-00171-z

Singh, S. K., & Pandey, A. C. (2014). Geomorphology and the controls of geohydrology on waterlogging in Gangetic Plains, North Bihar, India. *Environmental Earth Sciences*, *71*(4), 1561–1579. https://doi.org/10.1007/s12665-013-2562-1

Siriwardana, C. S. A., Jayasiri, G. P., & Hettiarachchi, S. S. L. (2018). Investigation of efficiency and effectiveness of the existing disaster management frameworks in Sri Lanka. *Procedia Engineering*, *212*, 1091–1098. https://doi.org/10.1016/j.proeng.2018.01.141

Sonker, I., Tripathi, J. N., & Singh, A. K. (2021). Landslide susceptibility zonation using geospatial technique and analytical hierarchy process in Sikkim Himalaya. *Quaternary Science Advances*, *4*, 100039. https://doi.org/10.1016/j.qsa.2021.100039

Stumpf, A., Malet, J. P., Kerle, N., Niethammer, U., & Rothmund, S. (2013). Image-based mapping of surface fissures for the investigation of landslide dynamics. *Geomorphology*, *186*, 12–27. https://doi.org/10.1016/j.geomorph.2012.12.010

Suh, J., Choi, Y., Roh, T. D., Lee, H. J., & Park, H. D. (2011). National-scale assessment of landslide susceptibility to rank the vulnerability to failure of rock-cut slopes along expressways in Korea. *Environmental Earth Sciences*, *63*(3), 619–632. https://doi.org/10.1007/s12665-010-0729-6

Sur, U., Singh, P., Rai, P. K., & Thakur, J. K. (2021). Landslide probability mapping by considering fuzzy numerical risk factor (FNRF) and landscape change for road corridor of Uttarakhand, India. *Environment, Development and Sustainability*, *23*(9), 13526–13554. https://doi.org/10.1007/s10668-021-01226-1

Tarchi, D., Casagli, N., Fanti, R., Leva, D. D., Luzi, G., Pasuto, A., Pieraccini, M., & Silvano, S. (2003). Landslide monitoring by using ground-based SAR interferometry: An example of application to the Tessina landslide in Italy. *Engineering Geology*, *68*(1–2), 15–30. https://doi.org/10.1016/S0013-7952(02)00196-5

Tiwari, P. C. (2000). Land-use changes in Himalaya and their impact on the plains ecosystem: Need for sustainable land use. *Land Use Policy*, *17*(2), 101–111. https://doi.org/10.1016/S0264-8377(00)00002-8

Tripathi, G., Shakya, A., Upadhyay, R. K., Singh, S. K., Kanga, S., & Pandey, S. K. (2023). Landslide Susceptibility Mapping of Tehri Reservoir Region Using Geospatial Approach. In *Climate Change Adaptation, Risk Management and Sustainable Practices in the Himalaya* (pp. 135–156). Springer International Publishing. https://doi.org/10.1007/978-3-031-24659-3_7

Uzielli, M., Rianna, G., Ciervo, F., Mercogliano, P., & Eidsvig, U. K. (2018). Temporal evolution of flow-like landslide hazard for a road infrastructure in the municipality of Nocera Inferiore (southern Italy) under the effect of climate change. *Natural Hazards and Earth System Sciences*, *18*(11), 3019–3035. https://doi.org/10.5194/nhess-18-3019-2018

Van Beek, L. P. H., & Van Asch, T. W. J. (2004). Regional assessment of the effects of land-use change on landslide hazard by means of physically based modelling. *Natural Hazards*, *31*(1), 289–304. https://doi.org/10.1023/B:NHAZ.0000020267.39691.39

Yang, K., Li, W., Yang, X., & Zhang, L. (2022). Improving Landslide Recognition on UAV Data through Transfer Learning. *Applied Sciences (Switzerland)*, *12*(19), 10121. https://doi.org/10.3390/app121910121

Zhao, F., Mallorqui, J. J., Iglesias, R., Gili, J. A., & Corominas, J. (2018). Landslide monitoring using multi-temporal SAR interferometry with advanced persistent scatterers identification methods and super high-spatial resolution terraSAR-X Images. *Remote Sensing*, *10*(6), 921. https://doi.org/10.3390/rs10060921

8 Detection of Tsunami-induced Ionospheric Disturbances Using GNSS Measurements

Sanjay Kumar, Arti Mishra, and Abhay Kumar Singh

8.1 INTRODUCTION

Tsunami comprises two words originating from the Japanese language, "tsu" and "nami," where "tsu" stands for harbor and "nami" means wave. Tsunami is a natural disaster in which a series of giant ocean waves progresses at a speed of more than 500 km/hour caused by geophysical phenomena such as impacts of meteorites, large vertical displacements associated with earthquakes inside water, landslides, etc. [1]. These disastrous waves generate a huge amount of deadly energy, e.g., 31 kilotons, which is almost equivalent to the Hiroshima–Nagasaki atom bombs.

Most tsunamis occur due to undersea earthquakes, which can claim millions of lives. This horrific rare event occurs naturally, and it can cause massive destruction. In 2004, one of the deadliest tsunamis on record occurred at Oceans of India, with the earthquake having the magnitude 9.1, 18 miles below the seabed between Burma and Indian plates. A Ring of Fire, located in Pacific Ocean, is known as the deadliest place where earthquakes and volcanic eruptions occur periodically. Most tsunamis are associated with earthquakes at the surface of the ocean and can be predicted by the DART system and other AI models. There are several seismic monitoring stations that monitor undersea earthquakes to predict the intensity of earthquakes and tsunamis as well. Tide gauges can measure tsunami waves over coastal regions, but measurements over the open ocean is challenging due to long wavelengths (~200 km) and small amplitude (few cm or less than vertical displacement in the sea surface) as compared to wind-induced waves.

8.1.1 DETECTION METHOD OF TSUNAMIS

Tsunamis that have little amplitude (< 1m) or long wavelength (~ 100 km) in the deep ocean cannot be detected by the human eye due to the small slope/steepness of the waves. Several measuring instruments that can detect tsunamis are listed below.

8.1.2 TIDE GAUGES

The principle of the tide gauge is to estimate the sea-surface height as well as the tide level. Tide gauges are operated by the Bureau of Meteorology at the National Tidal Centre; they are known as SEAFRAME (Sea Level Fine Resolution Acoustic Measuring Equipment) stations. An acoustic sensor emitting sound pulses is connected to a vertical tube open at the lower end in the water. The sound pulses move from the top of the tube downward to the water surface and are reflected back to the top. The travel time of the pulse is then used to estimate distance of the water level. The microscopic effects like wind-waves can be filtered out and can measure the changes in sea level (±1 mm). The location of Cocos Island's tide gauge is shown in Figure 8.1. The detection of the tsunami of 26 December 2004 as it passed by the island is shown in the top panel (Figure 8.1a).

DOI: 10.1201/9781032712444-10

8.1.3 Satellites

Satellite altimeters estimate the ocean surface height by taking advantage of the electromagnetic pulses transmitted from the satellite downward to the ocean surface. In determining the height of the ocean surface, information about speed of pulse, satellite position, and pulse travel time return to the satellite from ocean surface is important. Since this kind of satellite passage is very rare and some satellites at particular locations can only pass once in a month, tracking tsunamis can be difficult. But it is possible to track the tsunami using this satellite. Moreover, the Jason satellite altimeter was used to track the location and occurrence of the tsunami of December 2004 in the Indian Ocean (http://www.bom.gov.au/tsunami/info/index.shtml#observations).

The height of the sea surface (marked by blue color) estimated by the Jason satellite just two hours after the initial earthquake of December 26, 2004, which hit the region southeast of Sumatra (marked by red color), is shown in Figure 8.1(b). In this figure the data were used from a radar altimeter on board the satellite along a track traversing the Indian Ocean when the tsunami waves had covered the entire Bay of Bengal. The data shown here represents the differences of sea surface height from previous observations along the same track 20–30 days before the earthquake. The difference in sea surface height represents the signals for the tsunami.

8.1.4 The DART System

In 1995, the Deep-ocean Assessment and Reporting of Tsunamis (DART) system was developed by the National Oceanic and Atmospheric Administration (NOAA). Several such stations are

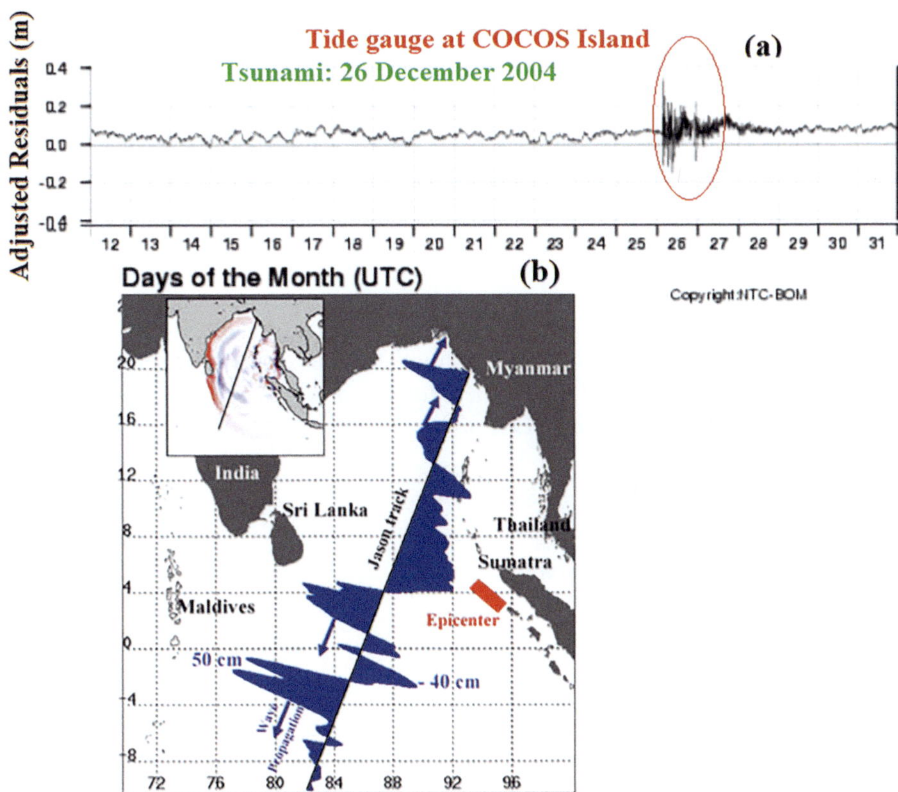

FIGURE 8.1 (a) *Recognition of tsunami signals of 26 December 2004 by using tide gauge at COCOS island,* (b) sea surface height (in blue color) measured by the Jason satellite-based data observed just after the 02 h occurrence of the earthquake of Sumatra of December 2004 (after Li [2]).

positioned in the Pacific Ocean to provide widespread data sets of tsunamis that are too far from the coast. A pressure recorder is available at the bottom of the seabed to measure and observe the passage of a tsunami at each station. The data is then transmitted to a surface buoy using the SONAR, which further wirelesses the information to the Pacific Tsunami Warning Center (PTWC) through satellite application. The surface buoy is replaced every year, whereas replacement time for bottom pressure is about two years. The forecasting/warning of tsunamis in the Pacific region now has been improved considerably, and such forecasting is useful to minimize damages [2].

8.2 GLOBAL NAVIGATION SATELLITE SYSTEMS (GNSS)

GNSSs consist of (i) collections of satellites orbiting the Earth broadcasting their positions/locations (ii) chain of ground control stations, and (iii) receivers. The receivers calculate ground positions using trilateration anywhere on the globe in all weather, all the time. Space stations, aviation, maritime, rail, road, and mass transit railways, etc., use GNSS Technology for positioning. In addition, the positioning services provided by GNSSs, along with navigation and timing information, are useful in other areas of science/technology including telecommunications, land surveying, law enforcement, emergency response, precision agriculture, mining, finance, air traffic, power grids, and scientific research, and so on. Many countries are actively involved in the development of their own GNSS systems, which include European Galileo, USA GPS, Russian GLONASS, China's COMPASS/Bei-Dou, Japan's Quasi Zenith Satellite system (QZSS), and India's IRNSS or NavIC (Navigation with Indian Constellation) [3, 4].

8.2.1 MONITORING OF THE IONOSPHERE USING GNSS/GPS

The positioning accuracies determined by GPS are affected errors in GPS measurements. These errors are mainly grouped into three categories, namely satellite-based errors, atmospheric errors, and receiver-based errors. The atmosphere-related errors, also known as propagation errors, are ionospheric error and tropospheric error. Receiver clock error, receiver inter-frequency biases, multipath error, and receiver noises are taken as receiver-related errors. The satellite-based errors include broadcast ephemeris, satellite clock error, and inter-frequency biases [5]. The ionospheric delay permits us to estimate TEC either by carrier phase measurement or code pseudorange measurements in the GPS signals [5]. The GPS receiver stores observation and navigation data files in standard FORMAT and RINEX (Receiver Independent Exchange), which is widely used [4]. The GPS observation data is applicable to calculate STEC (slant total electron content). The STEC estimated from observation data is converted into VTEC (vertical total electron content) [6]. The following equations are utilized to covert VTEC from STEC:

$$\text{VTEC} = \left(\left(\text{STEC} - \lfloor b_R + b_S \rfloor \right) \right) \big/ S(B) \tag{8.1}$$

where $S(B)$ is the mapping function given by the following relation

$$S(B) = \frac{1}{CosZ} = \left[1 - \left[\frac{R_E \times \cos E}{R_E + h_{\text{eff}}} \right]^2 \right]^{-\frac{1}{2}} \tag{8.2}$$

where R_E is the radius of the earth taken in kilometer, h_{eff} denotes the effective height of the ionosphere above the Earth's surface, Z is used to denote the zenith angle and E for the elevation angle (in degrees). The value of h_{eff} is taken as 350 km to ascertain the ionospheric pierce point (IPP) location. The b_S, and b_R are used for satellite and receiver biases, respectively. The GPS data can be downloaded from the IGS (International GNSS Services) website at cddis.gsfc.nasa.gov.

8.3 INTERACTION OF TSUNAMI WITH ATMOSPHERE: ATMOSPHERIC-IONOSPHERIC COUPLING

Peltier & Hines [7] have proposed that atmospheric gravity waves are key parameters in tracing tsunamis and the means to detect tsunamis by atmospheric coupling through gravity waves. The possible source for detection of tsunami, vertical displacement of the sea surface, can be considered and may be responsible for the generation of atmospheric gravity waves (AGWs). Tsunami waves are expected to propagate large distances when traveling through the open ocean surface. A tsunami can induce AGWs through coupling at the surface, and tsunami waves are expected to couple with AGWs [7]. The AGWs thus generated at the sea surface travel through the atmosphere up to the ionosphere.

The density decreases exponentially in the atmosphere and energy conservation requires exponential increase (amplification) in the wave amplitude. For example, at the F-region, (150–600 km), the wave's velocity is amplified by a factor of 10^4 times, as compared to that at the ground level [8]. This makes their detection easy using ground- or satellite-based measurements. During propagation of AGWs, they perturb neutral density and ambient temperature in the upper atmosphere [9–15]. As AGWs reach the ionosphere, they perturb the local plasma there and hence, may induce perturbations on radio signals, which can be detected using soundings including GPS, ionosonde, and VLF techniques. This atmospheric-ionospheric coupling is quantified and discussed in the following section.

8.3.1 CHARACTERISTICS OF TSUNAMIS AND GRAVITY WAVES

Generally, the nature of tsunamis is non-dispersive, whose speed depends on acceleration due to gravity (g) and depth (d) of water, which can be obtained from this shallow water equation:

$$V = \sqrt{gd} \tag{8.3}$$

If g = 9.8 ms-2, d = 500 m, v = 221 ms-1. The period of the wave typically lies between 10 and 30 mins. Tsunamis are high-speed waves and have fantastic energy, and the velocity of tsunamis varies with the depth of water. For example, the velocity of a tsunami in deep ocean is as much as that of a jet plane, over 500 mph (800 km/h). It means tsunamis can travel entire oceans in less than a day (https://www.noaa.gov/jetstream/tsunamis/tsunami-propagation).

A 2-D description of tsunami waves is described by Peltier & Hines [7] in which a tsunami propagates along x direction as a plane wave, and a comparative study of dispersion relations for tsunami waves and acoustics gravity waves in simple isothermal atmosphere show some basic properties of the waves. Figure 8.2 shows the cross section of the atmosphere for ideal tsunamis, which propagate at a speed of 140 ms-1.

8.3.2 ACOUSTIC GRAVITY WAVES

Considering the isothermal atmosphere, the following equations describe gravity waves and are discussed in detail by Hines [16].

$$\textbf{Continuity equation}: \frac{\partial \rho}{\partial t} + v.\nabla\rho = -\rho\nabla.v \tag{8.4}$$

$$\textbf{Momentum equation}: \frac{\partial v}{\partial t} + v.\nabla v = g - \frac{1}{\rho}\nabla p \tag{8.5}$$

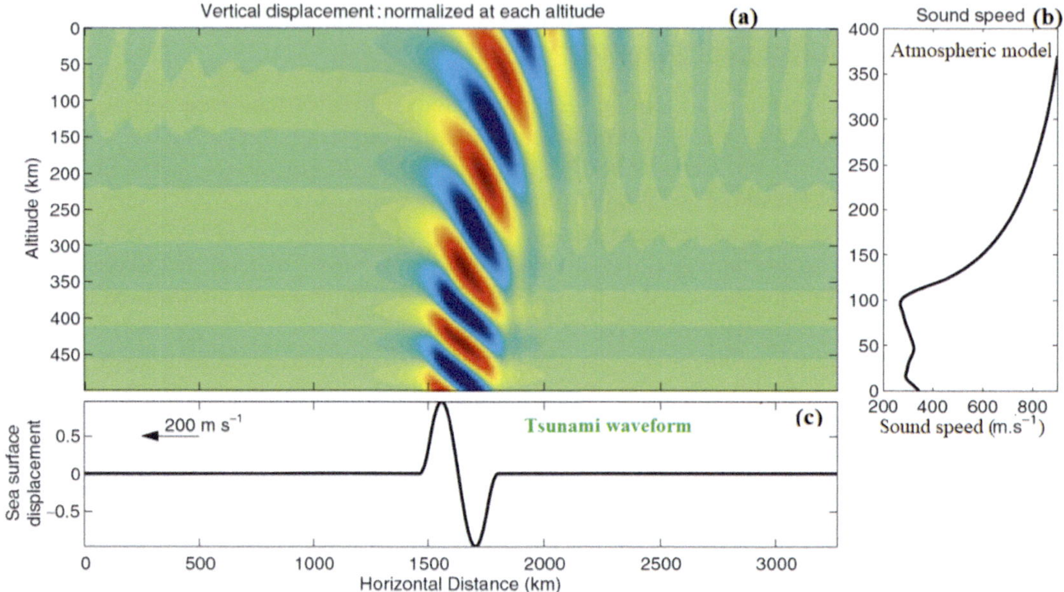

FIGURE 8.2 (a) Description of the gravity wave generated from a typical tsunami having a speed of 140 ms⁻¹, (b) The model for representation of the atmosphere, (c) The generated tsunami waveform (c) (after Artru et al. [10]).

$$\textbf{Adiabatic condition}: \frac{\partial p}{\partial t} + v.\nabla p = V_s^2\left(\frac{\partial \rho}{\partial t} + v.\nabla p\right) \tag{8.6}$$

p represents the pressure, v is the velocity of neutral gas, ρ is the density, g is the acceleration due to gravity, and V_s is the constant sound speed.

For equilibrium, the following conditions are required:

$$v_0 = 0 \tag{8.7}$$

$$\rho_{0\infty} \exp\left(\frac{z}{2H}\right) \tag{8.8}$$

$$p_{0\infty} \exp\left(\frac{z}{2H}\right) \tag{8.9}$$

where lower script "0" stands for equilibrium values. The density scale height H is given by $H = \frac{\gamma g}{V_s}$, and $\gamma = \frac{c_p}{c_v}$ is the ratio of two specific heats. Let us assume p_1, ρ_1, and v is a small perturbation along x-axis and do not have any dependency along the y-axis. The solution of the linear system of equations is required to get harmonic solutions for p_1/p_0, ρ_1/ρ_0, and v, which are proportional to exp $[i(\omega t - k_x x - k_z z)]$. Let's define $k'_z = k_z + i/2H$; we can obtain the dispersion relation given below:

$$\omega^4 - \omega^2 V_s^2\left(k_x^2 + k_z'^2\right) + \left(\gamma - 1\right)g^2 k_x^2 - \frac{\gamma^2 g^2 \omega^2}{4V_s^2} = 0 \tag{8.10}$$

Solution of the dispersion relation with k'_z as a real value, following are two modes of propagations acoustic modes and gravity modes.

8.3.2.1 Acoustic Modes ($\omega > \omega_a$)

For the acoustic mode of propagation, the wave frequency should be greater than the acoustic frequency defined by

$$\omega > \omega_a = \frac{\gamma g}{2V_s} \tag{8.11}$$

The acoustic waves are governed by compression, also called compressional waves. The usual value of acoustic frequency for lower atmosphere is 3.3 mHz.

8.3.2.2 Gravity Modes ($\omega < \omega_g$)

For the gravity mode of propagation, the wave frequency should be smaller than ω_g, which is the Brunt-V̈ais̈al̈a frequency defined by following relation:

$$\omega < \omega_g = \left(\gamma - 1\right)^{1/2} \frac{g}{V_s} \tag{8.12}$$

The ω_g is governed by buoyancy and sometime also called buoyancy frequency and is defined as the frequency with which an air parcel oscillates when the parcel is displaced vertically from the background position. This can be measured using potential temperature. The typical value of ω_g in the lower atmosphere is 2.9 mHz. Results reported in North Western China by He et al. [17] indicate that the mean value of ω_g increases between 10.5 and 13 km height in the atmosphere and for the case of troposphere (< 10 km) ω_g is ~ 0.0076 rad/s, which are equivalent to ~1 mHz. The waves oscillating at a frequency that is less or equal to the ω_g could not be gravity waves [18].

8.4 DETECTION TSUNAMI-INDUCED IONOSPHERIC DISTURBANCES

8.4.1 TSUNAMI-INDUCED PERTURBATION IN LOWER IONOSPHERE USING VLF SIGNALS

VLF and LF probing techniques are frequently utilized to measure plasma density of the D-region ionosphere. The VLF frequency ranges from 3 to30 kHz are propagated between the surface of the lower layers of the ionosphere, mainly the D-layer. The reflections of such waves occur from an altitude of 60 km during the day, whereas they occur from an altitude of 85 km at night. The reflected signals, when measured by using a VLF receiver on the ground, provide knowledge about the reflecting layer [19]. The sub-ionospheric VLF signals propagate over long distances (~1000 km) and hence, are useful to study a large portion of the upper atmosphere.

The ionospheric variations in plasma density during the tsunami are observed, which further affect the amplitude and phase of the VLF signal when received on the ground [20]. A diagram depicting the mechanism of tsunami-induced GWs perturbation of the path of VLF signal propagation is shown in Figure 8.3a. The study of the reflected VLF signal from the lower ionosphere has the potential to measure electron density variation. Using a network of ground-based VLF receivers at three selected stations (Figure 8.3), Rozhnoi et al. [20] reported perturbation in the ionosphere to tsunamis, which was induced by the earthquakes. The receivers can track the signals from transmitters positioned in Japan, Australia, Hawaii, and the USA. They have reported significant perturbations in amplitude as well as the phase of VLF signals.

For the tsunami of November 15, 2006, the variation in phase and amplitude is indicated (Figure 8.3b). The observed and averaged signals are represented by black and red lines, respectively. The residual for the NPM transmitter (in Hawaii) is computed by differentiation observed from the averaged signals and is depicted in the bottom panel of the figure. In this figure, the vertical line represents the occurrence time of the earthquake of 15 November 2006, and the blue circles indicate the amplitude and phase perturbation in the VLF signal caused by the tsunami.

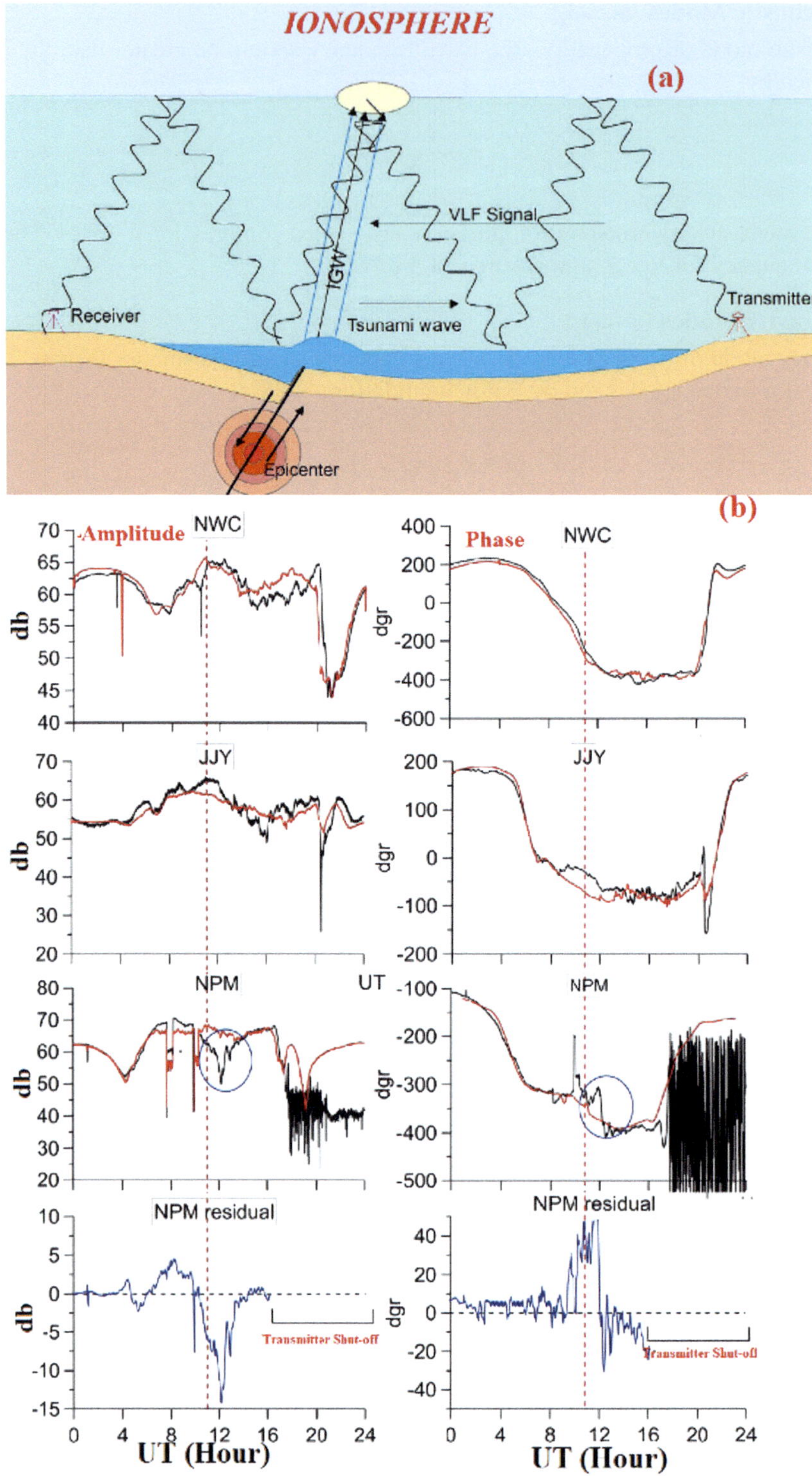

FIGURE 8.3 (a) A drawing depicting the mechanism gravity wave generation from the tsunami, which in turn perturbs the propagation path of the VLF signal, (b) Variations of amplitude and phase of the signals received from the three transmitters: NWC (19.8 kHz), JJY (40.0 kHz), and NPM (21.4 kHz). The observed and average signals are shown by the black and red colors, respectively (after Rozhnoi et al. [20]).

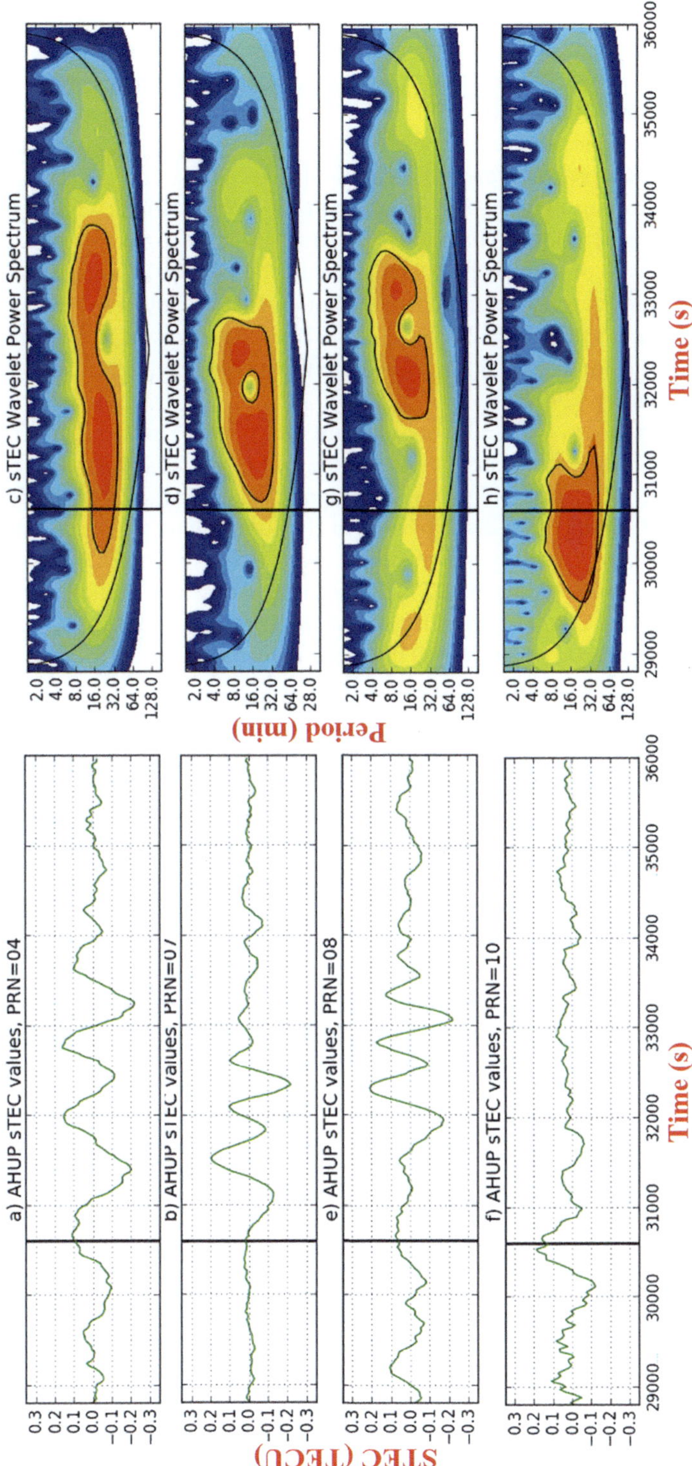

FIGURE 8.4 (a,b,e,f) wavelet analysis (Paul wavelet) at the selected station AHUP observed from the four GPS satellites designated by PRNs 4, 7, 8, 10 (c,d,g,h). The period in min from Fourier analysis is shown in vertical axis; time(s) is represented with horizontal axis. The vertical black line is the arrival time of the tsunami at the Hawaiian Islands. The color panel represents the magnitude of the power spectrum analysis (after Savastano et al. [29]).

8.4.2 Tsunami-induced Ionospheric Perturbation Using GNSS

Several studies show that vertical movement of tsunami waves cause acoustic waves to travel to the ionosphere [21], resulting in TIDs (Traveling Ionospheric Disturbances). Tsunamis also generate gravitational waves that propagate towards the ionosphere through the atmosphere and can also cause perturbation in TEC [7, 16, 22]. This tsunami-induced ionospheric perturbation can be detected by ground-based GNSS measurements, and GPS is one among them [23, 24] and other observational techniques [11, 25–27]. The ionospheric responses to tsunamis were originally proposed in 1970, which formed a basis for their propagation and detection [7, 16]. To support this hypothesis, using GPS-based TEC data measured from GEONET in Japan, the first observations were reported in 2005.

The results showed tsunamis produced ionospheric perturbation in the trans-Pacific region, which was induced by the earthquake of magnitude Mw = 8.2 on June 23, 2001, in Peru [24]. Apart from this, several similar observations have been reported for other cases of tsunamis; for example, Kuril (in 2006), Samoa (in 2009), and Chile in 2010 [23, 28]. Thus, tsunamis have proven the ability to produce GWs, which travel from the lower atmosphere to the ionosphere, producing disturbances in F- and E-layers' plasma density [24, 29].

These ionospheric disturbances induced from tsunamis have been studied using TEC measurements collected from GNSS-based measurements for the 2012 Haida Gwaii tsunami event. They have used a new approach called VARION to compute slant TEC (sTEC) over about 56 GPS receivers in Gwaii. They have observed TEC disturbances in the ionosphere with amplitude up to 0.25 TECU. The TIDs generated in this study are traveling away from the epicenter of the earthquake with the mean speed of 316 ms^{-1}. Using the STEC data in the Hawaiian Islands, detection of TEC perturbations before the actual tsunami arrival has also been reported. Further, they used Power wavelet analysis of TIDs to estimate periodicity GWs in the ionosphere, which is shown in Figure 8.4. The periods of TIDs are found between 10 mins and 30 mins.

8.5 FUTURE SCOPE AND CHALLENGES

The tsunami and its consequence in generating perturbation in atmospheric/ionosphere through the atmospheric-ionospheric coupling by means of generation/propagation of GWs has been discussed. GNSS-based measurements are able to detect such perturbations in the ionosphere with high accuracy and precision. The real-time capability VARION algorithm, which is used to detect characteristic features of TIDs from tsunami-induced GWs, has also been discussed [29]. This technique may be integrated in the future for tsunami early warning systems. Over the equatorial and low latitude region, the dynamics of the ionosphere is quite complex, which makes it challenging to distinguish the contribution of tsunami-induced perturbation in the ionosphere from other causes.

REFERENCES

1. Jia, J. (2017). Introduction to Tsunamis. In: *Modern Earthquake Engineering*. Springer, Berlin, Heidelberg. https://doi.org/10.1007/978-3-642-31854-2_12
2. Li, H. (2022). Deep-Ocean Assessment and Reporting of Tsunamis (DART) BUOY. In: Cui, W., Fu, S., Hu, Z. (eds) *Encyclopedia of Ocean Engineering*. Springer, Singapore. https://doi.org/10.1007/978-981-10-6946-8_79
3. Xu, G. (2007). "*GPS Theory Algorithms and Applications*," Springer-Verlag, Berlin, Heidelberg, New York, 2nd Edition, ISBN: 978-3-540-72714-9.
4. Kumar, S., Singh, A. K., Singh, R. P. 2021. GPS and GNSS technology in Geosciences, *Science Direct*, 119–134.
5. Grewal, M.S., Weill, L.R., Andrews, A.P. (2007). *Global Positioning System Inertial Navigation, and Integration*," A John Wiley & Sons, Inc., Publication, 2nd Edition, ISBN: 9780470041901.
6. Rama Rao, P.V.S., Gopi Krishna, S., Niranjan, K., Prasad, D.S.V.V.D. (2006). Temporal and spatial variations in TEC using simultaneous measurements from the Indian GPS network of receivers during the low solar activity period of 2004–2005. *Ann. Geophys.*, 24(12), 3279–3292.

7. Peltier, W. R. & Hines, C.O. (1976). On the possible detection of tsunamis by a monitoring of the ionosphere. *J. Geophys. Res.*, 81(12), 1995–2000.

8. Blanc, E. (1985). Observations in the upper atmosphere of infrasonic wavesfrom natural or artificial sources. A summary. *Ann. Geophys.* 3(6), 673–688.

9. Tarasick, D.W., Hines, C.O. (1990). The observable effects of gravity waves on air glow emissions. *Planet. Space Sci.* 38(9), 1105–1119.

10. Artru, J., Ducic, V., Kanamori, H., Lognonne, P., and Murakami, M. (2005). Ionospheric detection of gravity waves induced by tsunamis. *Geophys. J. Int.* 20, 840–848, https://doi.org/10.1111/j.1365-246X.2005.02552.x

11. Occhipinti, G., Lognonné, P., Kherani, E.A., and Hébert, H. (2006). Three dimensional waveform modelling of ionospheric signature induced by the 2004 Sumatra tsunami. *Geophys. Res. Lett.* 33, L20104, https://doi.org/10.1029/2006GL026865

12. Hickey, M. P., G. Schubert, and R. L. Walterscheid (2010), Atmospheric airglow fluctuations due to a tsunami-driven gravity wave disturbance. *J. Geophys. Res.* 115, A06308.

13. Garcia, R. F., E. Doornbos, S. Bruinsma, and H. Hébert (2014), Atmospheric gravity waves due to the Tohoku-Oki tsunami observed in the thermosphere by GOCE, *J. Geophys. Res. Atmos.* 119, 4498–4506.

14. Meng, X., A. Komjathy, O. P. Verkhoglyadova, Y.-M. Yang, Y. Deng, and A. J. Mannucci (2015), A new physics-based modeling approach for tsunami-ionosphere coupling, *Geophys. Res. Lett.*, 42, 4736–4744.

15. Coïsson, P., Lognonné, P., Walwer, D., and Rolland, L. M. (2015), First tsunami gravity wave detection in ionospheric radio occultation data, *Earth Space Sci.*, 2, 125–133.

16. Hines, C.O. (1972), Gravity waves in atmosphere, *Nature*, 239, 7378, https://doi.org/10.1038/239073a0

17. He, Y., Sheng, Z., and He, M. (2020), Spectral analysis of gravity waves from near space high-resolution balloon data in Northwest China, *Atmosphere*, 11, 133, https://doi.org/10.3390/atmos11020133

18. Kumar, S., Chen, W., Louis, O.P. (2023), Ionospheric and atmospheric response to extremely severe cyclonic storm Nida of 29-July-02 August 2016, *J. Geophys. Res., Space Physics (AGU)*, 128, e2023JA031422, https://doi.org/10.1029/2023JA031422

19. Barr, R., Llanwyn Jones, D., and Rodger, C. J. (2000), ELF and VLF radio waves, *J. Atmos. Sol. Terr. Phys.*, 62, 1689–1718, https://doi.org/10.1016/S1364-6826(00)00121-8

20. Rozhnoi, A., Shalimov, S., Solovieva, M., Levin, B., Hayakawa, M., and Walker, S. (2012), Tsunami-induced phase and amplitude perturbations of subionospheric VLF signals, *J. Geophys. Res. Sp. Phys.* 117, A09313.

21. Putra, M. E., Cahyadi, M.N., Muslim, B., Muafity, I.N., Wulansari, M. (2023), Analysis of tsunami traveling ionospheric disturbances due to tsunami using the GNSS-TEC method, Earth, Environ. Sci., 1127, 012003.

22. Hines, C.O. (1960). Internal atmospheric gravity waves at ionospheric heights. *Can. J. Phys.* 38, 1441–1481, https://doi.org/10.1139/p60-150

23. Rolland, L.M., Occhipinti, G., Lognonne, P., and Loevenbruck, A. (2010). Ionospheric gravity waves detected offshore Hawaii after tsunamis. *Geophys. Res. Lett.* 37, L17101, https://doi.org/10.1029/2010GL044479

24. Artru, J., Ducic, V., Kanamori, H., Lognonne, P., and Murakami, M. (2005). Ionospheric detection of gravity waves induced by tsunamis. *Geophys. J. Int.* 20, 840–848, https://doi.org/10.1111/j.1365-246X.2005.02552.x

25. Galvan, D. A. et al. (2012), Ionospheric signatures of Tohoku-Oki tsunami of march 11,2011: Model comparisons near the Epicenter. *Radio Sci.* 47, RS4003, https://doi.org/10.1029/2012RS005023

26. Mai, C.L. and Kiang, J.F. (2009). Modeling of ionospheric perturbation by 2004 Sumatra tsunami. *Radio Sci.* 44, RS3011, https://doi.org/10.1029/2008RS004060

27. Lee, M.C. et al. (2008). Did tsunami-launched gravity waves trigger ionospheric turbulence over Arecibo? *J. Geophys. Res.* 113, A01302, https://doi.org/10.1029/2007JA012615

28. Galvan, D.A., Komjathy, A., Hickey, M.P., Stephens, P., Snively, J., Tony Song, Y., Butala, M.D., and Mannucci, A.J. Ionospheric signatures of Tohoku-Oki tsunami of March 11, 2011: Model comparisons near the epicenter. *Radio Sci.* 2012:47, 1–10.

29. Savastano, G., Komjathy, A., Verkhoglyadova, O., Mazzoni, A., Crespi, M., Wei, Y., and Mannucci, A. (2017). Real-time detection of tsunami ionospheric disturbances with a stand-alone GNSS receiver: A preliminary feasibility demonstration. *Sci. Rep.* 7, 46607.

9 Monitoring of Calbuco Volcanic Eruption Using GNSS-derived TEC Perturbation

Mahesh N. Shrivastava, Ajeet Kumar Maurya,
Francisca Sánchez, and Susana Layana Guerrero

9.1 INTRODUCTION

The Calbuco volcano eruption occurred on 22 April 2015 at 21:05 UTC near the cities of Puerto Montt, Puerto Varas, and Ensenada, inhabited by a total of ~260,000 people. The eruption started suddenly with only ~3 h of precursory seismic activity (SERNAGEOMIN, 2015a). There was a huge spread of an ash plume more than 15 km above the crater. The Oficina Nacional de Emergencia del Ministerio del Interior (National Office of Emergency of the Interior Ministry, ONEMI) called for the evacuation of all people within 20 km of the volcano, and ~5,000 residents evacuated. Pre-eruptive deformation was not detected by the InSAR interferograms up to one day before the eruption (SERNAGEOMIN, 2015a; Valderrama et al., 2015; Delgado et al., 2017). Only co-eruptive deformation was detected during the eruption (Delgado et al., 2017). The first eruptive phase generated an eruptive column that reached up to ~15 km above the crater level and lasted ~1.5 h (SERNAGEOMIN, 2015a; Romero et al., 2016; Van Eaton et al., 2016; Castruccio et al., 2016; Pardini et al., 2018). After a pause of 5.5 h, a second and more energetic eruptive phase started on 23 April at 04:08 UTC and lasted for 6 h, produced an eruptive column >15 km above the crater level (SERNAGEOMIN, 2015b; Romero et al., 2016; Van Eaton et al., 2016; Castruccio et al., 2016).

Volcanoes produce a variety of materials when they erupt. The characteristics of the eruptions themselves also vary from one volcano to the other, and sometimes from one eruption to the other for the same volcano. Eruptions are classified according to how explosive they are and the height of their **eruption column**—how high they blast material into the air. There are four types of eruptions with properties determined mostly by the silica content of magma and the amount of gas it contains. In order of increasing explosiveness, these four types of volcanoes are (i) **Hawai'ian**, (ii) **Strombolian**, (iii) **Vulcanian**, and (iv) **Plinian** eruptions. The Calbuco volcano was a plinian-type eruption, i.e., the largest explosive volcanic event in terms of the mass discharge rate (intensity) and erupted magma volume (magnitude) (Walker, 1980).

The temporal recurrence of sub-Plinian eruptions (0.1–1.0 km³ ejecta volume, >10 km plume height) is about every year, while Plinian eruptions (1.0–10.0 km³ ejecta volume, >20 km plume height) occur about every decade, and together they produce significant volcanic hazards (Newhall and Self, 1982). After a volcanic eruption, there is deformation in the Earth's surface, which can be measured with InSAR geodesy technique. There are several works that estimated the volume related to this eruption such as 0.27 km³ by Romero et al. (2016), 0.38 km³ by Castruccio et al. (2016), 0.48 km³ by Pardini et al. (2018), and 0.56 km³ by Van Eaton et al. (2016). In the first phase, 38% of the volume erupted, whereas it was 62% in the second phase.

DOI: 10.1201/9781032712444-11

The ionosphere is the ionized region where the concentration of charged particles (i.e., free electrons and +ions) of the Earth's atmosphere is located, between ~60 and ~1000 km altitude. The total mass of the ionosphere is 10^{12}–10^{13} times smaller than the mass of the neutral atmosphere, and, at all altitudes, filled with charged particles. The ionosphere is a highly irregular and variable region, and its state is determined by several competing large-scale processes of solar, tropospheric, and lithospheric origin (Hargreaves, 1992). The main characteristics of the ionosphere include: (i) The concentration of charged particles varies depending on the level of solar activity; (ii) the ionization also depends on the day/night conditions, on the season, and on latitudes (solar zenith angle dependence); (iii) the electron density distribution is dependent on the altitude, defining the ionospheric layers: D (60–90 km of altitude), E (90–120 km), F (120–800 km); the ionization maximum reached within the F-layer, at the altitude of ~250–400 km; (iv) the ionosphere is tightly coupled with the neutral upper atmosphere. The thermosphere serves as a supplier of particles that can be ionized and determines the speed of ionization and regulates the recombination processes; (v) the ionosphere can be strongly perturbed by disturbances in the geomagnetic field, such as geomagnetic storms and substorms.

The ionosphere is largely impacted by solar and magnetic activities, as well as by natural hazards. Besides these large-scale and global processes influenced from above, the ionosphere can be more strongly perturbed by sources from below by geophysical phenomena such as earthquakes, tsunamis, volcanic eruptions, severe tropospheric weather events, and man-made events (Reddy et al., 2015; Reddy et al., 2017; Shrivastava et al., 2021; Shrivastava et al., 2023; Maurya et al., 2013, 2016).

The impulsive forcing from the Earth's surface occurring due to earthquakes, explosions, volcanic eruptions, tsunamis, etc., trigger atmospheric pressure waves. Depending on their frequencies, these atmospheric waves can be distinguished as acoustic and gravity waves (Blanc, 1985; Huang et al., 2021). The acoustic waves are characterized by frequencies higher than the acoustic cut-off frequency, i.e., higher than ~3.3 mHz. Acoustic waves are longitudinal waves in which a particle moves in the direction of the wave propagation. The force restoring a displaced particle toward its original position is the change of pressure associated with the compression of the medium (Hargreaves, 1992). Therefore, acoustic waves traveling through a gas disturb the equilibrium state by compressions and rarefactions (Zel'Dovich and Raizer, 2002). The acoustic waves propagate at the speed of sound equal to ~330 m/s at the surface and grow up to 800–1000 m/s at the ionospheric altitude of 250–300 km. Therefore, it takes ~8–9 minutes for the acoustic waves to reach the ionosphere.

The first ionospheric perturbations were detected after the great 1964 Alaska earthquake by ionosondes and Doppler sounders. Since then, many other observations confirmed the responsiveness of the ionosphere to natural hazards. Within the last two decades, outstanding progress has been made in this area remaining to the development of networks of ground-based dual-frequency Global Navigation Satellite System (GNSS) receivers. The use of GNSS-sounding has substantially enlarged our knowledge about the Solid Earth/ocean/atmosphere/ionosphere coupling and natural hazards of ionospheric perturbations and their main features.

We have utilized the Sentinel-1A InSAR data and analyzed GNSS data for the ionospheric Total Electron Contents (TEC) during the Calbuco volcanic eruption, with the objective to identify the exact deformation with the InSAR data related to the Calbuco volcanic eruption and its possible impact on the ionosphere and to develop the near real-time warning system.

9.2 GLOBAL NAVIGATION SATELLITE SYSTEMS (GNSSS) FOR TEC PERTURBATION

GNSSs offer a powerful and accessible tool to monitor the ionosphere. Due to the dispersive nature of the ionosphere, signals from dual-frequency GNSS receivers make it possible to measure the total electron content (TEC)—the electron density integrated along the raypath from a satellite to a receiver (Klaffenbock & Hofmann-Wellenhof, 2008):

$$TEC = \int_{raypath} N\left(d\vec{r}\right)d\vec{r} \tag{9.1}$$

As of today, operational GNSSs comprise the American Global Positioning System (GPS), Russia's GLONASS, the European Union's Galileo, China's BeiDou, Japan's Quasi-Zenith Satellite System (QZSS), and the Indian Regional Navigation Satellite System (IRNSS, or NAVIC). From GNSS signals transmitted at two distinct carrier-frequencies, f1 and f2, it is possible to calculate the differential (relative) slant TEC:

$$TEC = \frac{1}{A} \cdot \frac{f_1^2 f_2^2}{f_1^2 - f_1^2}\left(L_1\lambda_1 - L_2\lambda_2 + const + nL\right) \tag{9.2}$$

where A = 40.308 m³/s², $L_1\lambda_1$ and $L_2\lambda_2$ are additional paths of the signal caused by the phase delay in the ionosphere, const is the unknown initial phase path caused by the unknown number of total phase rotations along the raypath, and nL are errors in determining the phase path. TEC is measured in TEC units (TECU), which is equal to 10^{16} electrons/m². The accuracy of the differential TEC estimation from phase measurements is about 0.01–0.02 TECU (e.g., Coster et al., 2013).

9.3 ANALYSIS OF TEC PERTURBATION

To investigate the ionospheric perturbations associated with the Calbuco volcanic eruption region around 1,000 km, we have analyzed Vertical TEC (VTEC) data from two continuous GNSS sites (Figure 9.1). The VTEC data from satellite Pseudo-Random Numbers (PRNs) 01, 03, 04, 11, and 32 are used to identify the eruption signatures in the ionosphere. Since the VTEC is highly variable and perturbed by Space weather (solar flare and geomagnetic storm) conditions (Rastogi and Sharma, 1971; Rao et al., 1977), we have checked the Space weather conditions from https://www.spaceweatherlive.com/ and confirmed that no solar flare activity took place during April 22, 2015. We have also checked three hourly planetary K indexes (K_p) and the disturbance storm time index (Dst) providing the conditions for the geomagnetic activity (https://wdc.kugi.kyoto-u.ac.jp/) during the same time span. The daily Kp index was in the range 2.0–3.3 during this period. The Dst index showed a moderate geomagnetic storm with minimum Dst ~91nT around 23 UTC on April 22, 2015.

9.4 DEFORMATION PATTERN OF CALBUCO VOLCANO WITH INSAR DATA

The Calbuco volcano region has been tracked by different SAR missions since 2003; however, the existing record of SAR images captured during the 2000s is fragmentary and poorly robust. After 2014, with the launch of the Sentinel-1A satellite mission, a continuous record of SAR images became available. This is fortunate, as it allows detailed observation of vertical deformation patterns related to the April 2015 Calbuco eruption through InSAR analysis.

In order to observe the deformation, the GMTSAR version 6.1 software (Sandwell et al., 2011) was used. The InSAR data analyses consist of Repeated-Pass Interferometry to study the co-eruptive ground deformation with ascending and descending images (Table 9.1). During the InSAR processing, a Shuttle Radar Topography Mission1 (SRTM) digital elevation model with resolution of 30 meter was utilized.

Based on the analysis of the Repeat-Pass Sentinel-1 interferograms, a ground deflation of the Calbuco volcano during the April 2015 eruption was observed. Deformation patterns detected from the ascending interferogram indicate a maximum co-eruptive subsidence of ~12 cm in the

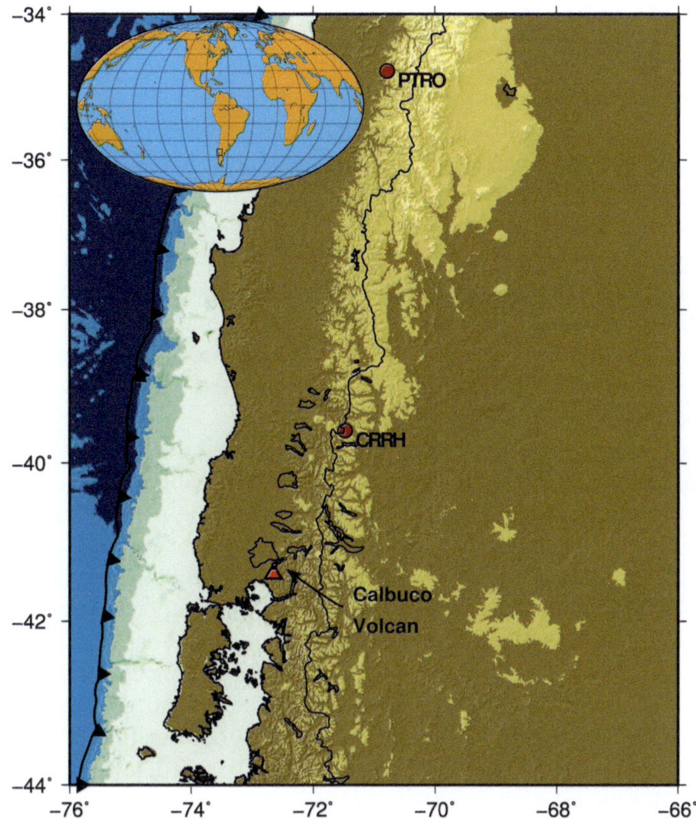

FIGURE 9.1 Geographic map of the Calbuco volcanic eruption region. The brown circles represent the locations of the GNSS sites used for the TEC perturbation study. The triangle represents the Calbuco volcano.

TABLE 9.1

Methods and Sentinel-1 Data Processed for InSAR Analysis of the Calbuco Volcano, Which the Satellite Pass Direction Can Be "A" for Ascending or "D" for Descending

Period Covered	Method	Baseline (m)	Pass	Path/frame	Dates (yyyy/mm/dd)	Beam Mode
Co-eruptive	Repeat-pass	≤ 150	A	164/1042	2015/04/14	IW
			A	164/1041	2015/04/26	IW
Co-eruptive	Repeat-pass	≤ 150	D	83/730	2015/04/21	IW
			D	83/730	2015/05/03	IW

Line-of-Sight direction (LOS) (Figure 9.3A and B). Concentric patterns observed from the ascending wrapped phase (Figure 9.3B) cover a few tens of kilometers of radius around the summit area and are possibly related to a horizontally or a deep elongated deformation source (Nikkhoo et al., 2016). The loss of coherence (< 2.5) around the summit and towards the northeast flank of the volcano caused by ash deposition and other volcano-sedimentary products during the eruption made it impossible to recognize a center of maximum subsidence from the descending interferogram, where only ranges less than 1.2 cm were observed (Figure 9.3C and D).

9.5 DISCUSSION AND CONCLUSION

On April 22, 2015, at 21:05 UTC, the Calbuco volcano erupted after more than 50 years of being dormant, producing an eruptive column that reached ~15 km in a few minutes, as per information achieved from National Service of Geology and Mineralogy of Chile (SERNAGEOMIN, 2015a). According to a special report of volcanic activity issued by SERNAGEOMIN, (2015a), this event was preceded by an increase in seismic activity associated with the volcano area, which began about an hour before the first eruption, on the eastern flank of the main vent. After this episode (20:35 UTC), the region began to experience seismic activity associated with magma flows inside the volcano. It is important to note that the seismic activity in the area has been insignificant since 2009, when seismic monitoring began in real time by the OVDAS (Southern Andes Volcano Observatory). After a big seismic event (20:54 UTC), a vigorous eruptive column reached a height greater than 15 km above the main vent. The tephra plume was mainly dispersed to the east-northeast (ENE), reaching several areas of Argentina (Romero et al., 2016).

The Sentinel images acquired on April 14, 2015 and April 26, 2015 showed that the vertical deformation of Calbuco's crater changed during the eruption. Some material seems to have been removed from the summit, potentially part of the ice cap, leaving a much deeper crater than it was previously, although there is no evidence of deformation before the eruption. The Sentinel interferogram from April 14th–26th showed that a region centered to the southwest of Calbuco subsided during the eruption. The vertical deformation has a maximum magnitude around 12 cm in the satellite LOS. The surface deformation estimated with the InSAR images is one of the most used data sets to constrain the rates of magma transport and storage in active volcanoes and the geometry and location of the plumbing systems that produce eruptions (e.g., Dvorak and Dzurisin, 1997; Dzurisin, 2003). However, not all the eruptions are related to vertical deformation, and hence, this relationship, unique to each volcano, is not always straightforward (e.g., Biggs et al., 2014; Acocella et al., 2015). The surface deformation is also related to degassing and seismicity, which provide more robust constraints on magma dynamics (Anderson and Poland, 2016) and because unrest detected by one of them might not be related to changes in the others (e.g., Heimisson et al., 2015; Delgado et al., 2016).

During the 90-minute eruption, the eruptive column collapsed, affecting the headwaters of major rivers (SERNAGEOMIN, 2015a). OVDAS reported that during the first eruptive pulse beginning April 22 at 21:04 UTC, there were pyroclastic flows up to 7 km and lahars were found up to 15 km from the vent, while there was little seismic activity. Before 10 min, there was a clear indication of the impending eruption, which produced an ash column taller than 15 km that lasted about 1:30 hours. The radar RMA0 (Lat=−41.14°, Lon=−71.15°, 830 m above Mean Sea Level) in Argentina recorded the first data of the eruption at 21:09 UTC with echoes more than 60 dBZ. According to OVDAS's reports, the first eruption ended its most energetic phase at 22:39 UTC, reaching a height of 15 km, which indicated agreement with the radar data. Atmospheric pyroclastic flow was directed towards the North-East. According to SERNAGEOMIN (2015a), Calbuco erupted explosively at 21:04 UTC on April 22, 2015, sending an ash plume to more than 15 km above the crater, where it drifted primarily ENE.

The most important peculiar feature of the present observation is the diverse VTEC variations at different stations from the vent of the Calbuco volcanic eruption (Figure 9.2). There are two GNSS sites such as CRRH 217 Km and PTRO approximately 744 Km away from the Calbuco volcanic eruption utilized for the VTEC analysis. The VTEC perturbations are estimated for five PRN's: 01, 03, 04, 11, and 32. The change in VTEC value in the case of the PTRO GNSS sites around 744 Km away detected very small TEC perturbation, can be seen in the PRN03. There are no other PRN that could sense changes in ionospheric perturbation at the PTRO GNSS site. But in contrast, the CRRH GNSS site, which is situated around 217 Km PRN 01 and PRN 03, detected 0.2 TECU and 0.5 TECU changes in the ionospheric perturbation respectively. The first signal of the VTEC of the ionosphere was observed at PRN 01 and 03 at the CRRH GNSS site approximately 21:14 UTC after

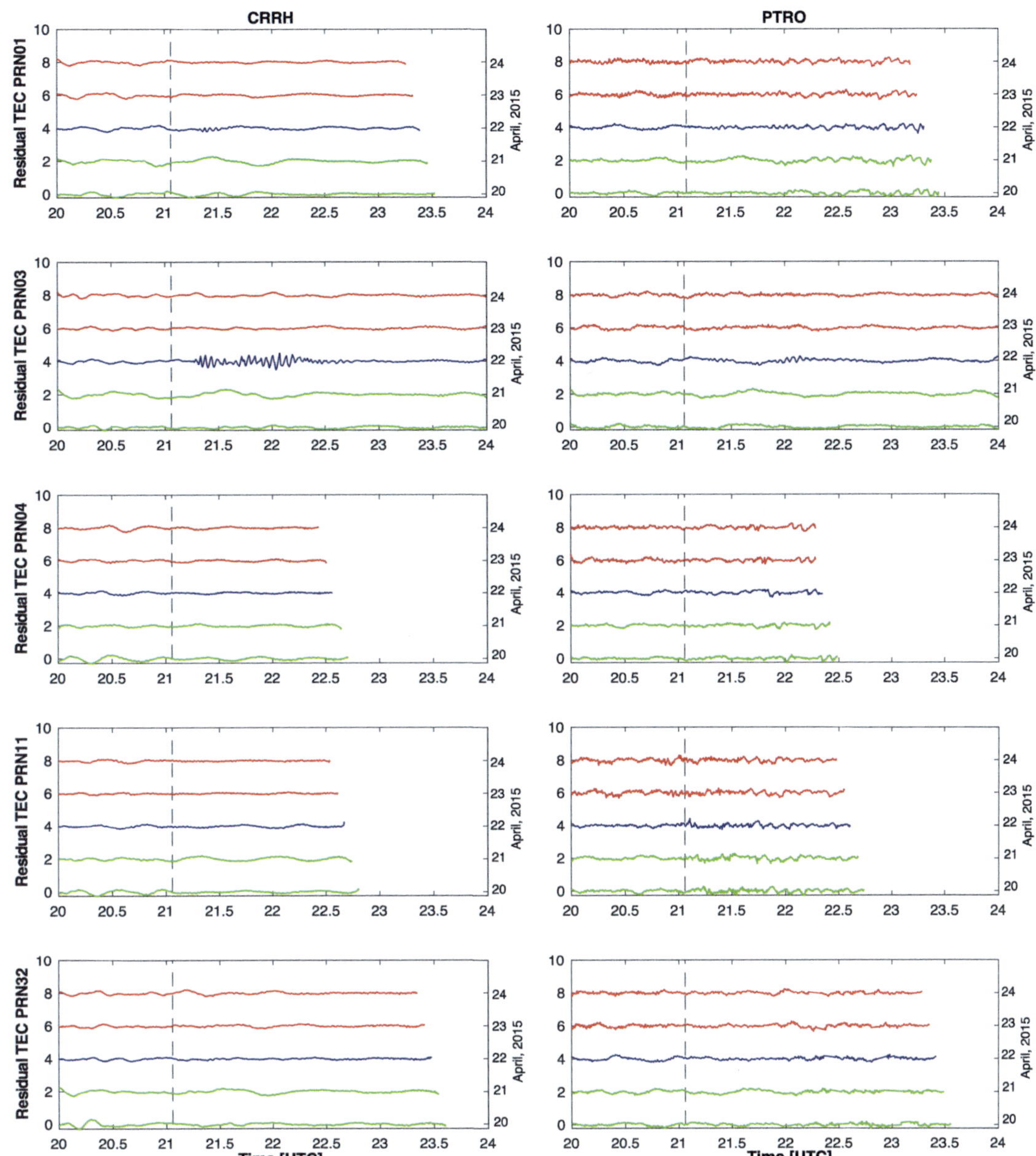

FIGURE 9.2 The VTEC time series are presented on the day of eruption (blue color), two days before (green color), and two after (red color). The residual VTEC at GNSS sites are presented for PRNs 01, 03, 04, 11, and 32 during the Calbuco volcanic eruption. The dotted grey line marks the time of the volcanic eruption.

9 mins of the volcanic eruption. In order to verify the spatial TEC changes in the ionosphere, we have provided a skyview of the GNSS satellite (Figure 9.4) path for the PRN01 and 03, which has sensed TEC perturbation. It is very clear to notice that TEC perturbation in the ionosphere is due to the Calbuco volcanic eruption more than what we observed via the GNSS satellites. None of the GNSS satellites' IPP paths were very close to the Calbuco volcanic eruption vent. This drawback of the GNSS constellation is that most of the GNSS satellites only partially cover the continents in the high latitude.

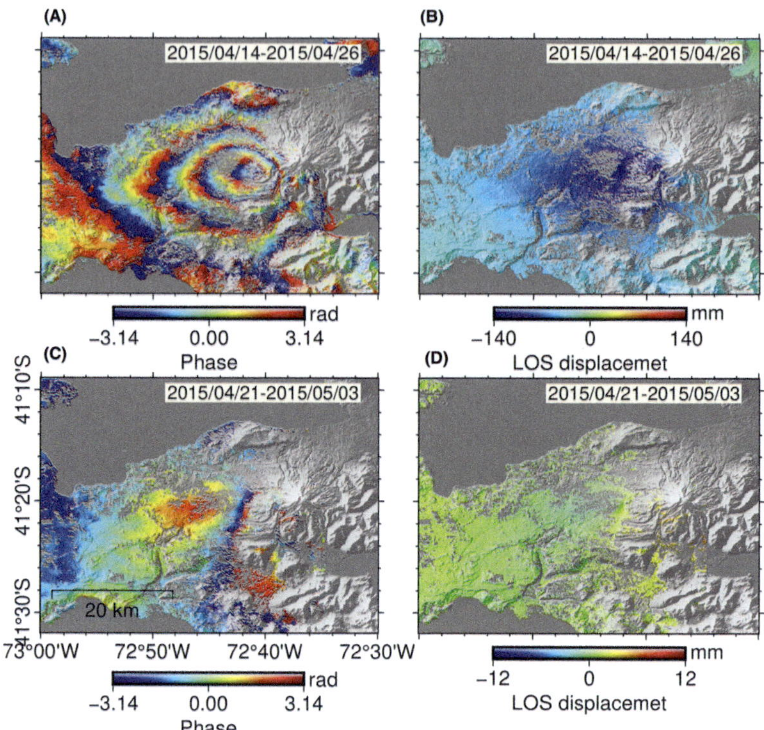

FIGURE 9.3 InSAR analysis results of the co-eruptive deformation at the Calbuco volcano area from the Sentinel-1 data. **(A)** and **(C)**. Ascending and descending wrapped phase interferograms, respectively. **(B)** and **(D)**. Ground displacements in millimeters looking from LOS direction from the unwrapped phase interferograms.

Recent technological advances have been enormous, facilitating the monitoring of earthquakes, tsunamis, and volcanic eruptions via ionospheric perturbations with Space-based advanced radio techniques. Detecting these ionospheric perturbation signals offers possible remote sensing signals (Lognonné et al., 2006), mainly for two reasons, (1) continuity of vertical displacement at the surface/water wave, where the atmosphere is then forced to move with the same vertical velocity as the surface/water, and (2) conservation of kinetic energy and the exponential decrease of air density with the height. The GNSS-derived TEC analysis of the ionospheric perturbation mainly influenced the surface/water vertical upliftment. Therefore, the range of TEC varies from different sources such as earthquakes, tsunamis, or volcanic eruptions.

In the case of subduction earthquakes like Pisagua 2014 and Illapel 2015 peak-to-peak, the TEC amplitudes were around 1.25 and 1.40 TECU (Reddy and Seemala, 2015; Reddy et al., 2017; Shrivastava et al., 2021). The deformation was not directly on the ground, but due to the tsunami waves in the ocean. The vertical deformation in these earthquakes was around 1.2 m for Pisagua and 2.05 m for the Illapel earthquake, respectively. The tsunami height is used to observe the same as vertical deformation. However, in the case of the collision zone, the Nepal earthquake in 2015, the peak-to-peak TEC amplitude was 1.2 TECU. The vertical deformation was approximately 1.6 m on the ground. Similarly, the peak-to-peak TEC amplitude for the Tonga volcanic eruption was around 3.0 TECU. In this case, the deformation was enormous, and the topography diminished ~100 m during the tsunami just after the volcanic eruption. This is the reason we have observed tsunamis in many Pacific rim countries from South America, North America, Alaska, and Japan. In our study region (Calbuco volcanic eruption), the maximum vertical deformation was approximately 12 cm.

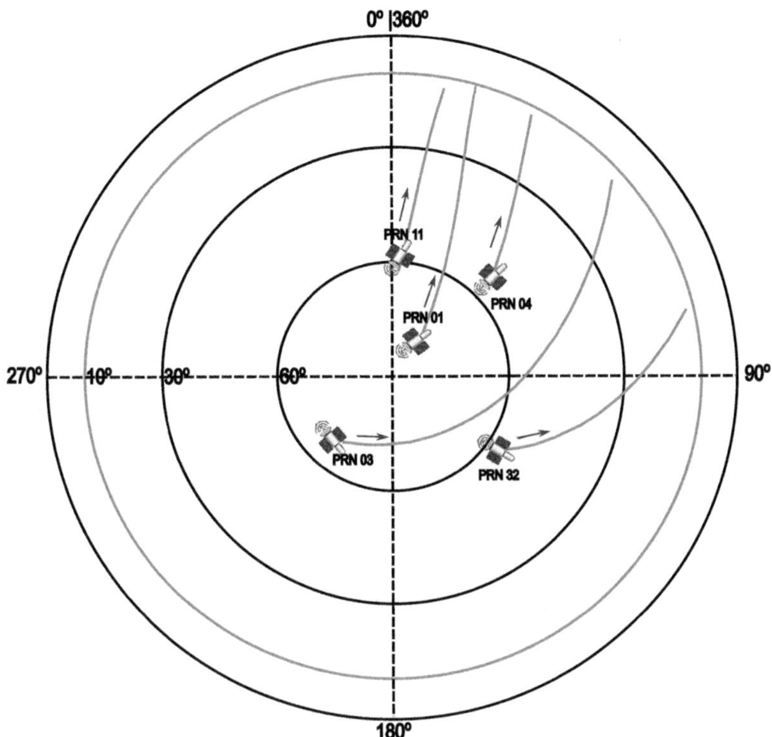

FIGURE 9.4 The skyview of the GNSS satellites from Calbuco volcanic crator during the eruption on 22nd April 2015. The green circle shows the elevation criteria for utilizing the GNSS data. The arrow shows the GNSS satellite direction. The green curved line shows the path of the satellites.

We observed the TEC changes in the ionosphere around 217 Km away from the CRRH GNSS site data but not 744 km away from the PTRO GNSS site data. The ionospheric perturbation was confined to less than 700 Km.

It is important to note that the surface deformation of the volcano is very small, approximately 2,500 m^2 (50 m x 50 m) to 10.000 m^2 (100 m x 100 m), in comparison to médium-size earthquakes (Mw 7.5–8.0), where the deformation area was ~20,000 Km^2 (200 Km x 100 Km). That is why the TEC perturbation in the case of the Calbuco volcanic eruption is significantly very small. It also impartant to note that volcanic eruption takes place in a concentric fashion, which is why TEC perturbation is very concentric. The main evidence is that PRN01 and PRN03, which lie at about 60° elevation, could be able to sense significantly less TEC perturbation. The use of dense and continuous measurement of TEC perturbation can be utilized for the detection of volcanic eruption and volcanic hazard assessment.

ACKNOWLEDGEMENTS

Mahesh N. Shrivastava, Francisca Sánchez, and Susana Layana thank Millennium Institute on Volcanic Risk Research – Ckelar Volcanes, Millennium Scientific Initiative (ANID, ICN2021_038). Mahesh N. Shrivastava thanks the National Research Center for Integrated Natural Disaster Management (CIGIDEN ANID/Fondap/1522A0005), Chilean research grant. The Sentinel-1 data were provided by the European Space Agency (ESA) and obtained from https://asf.alaska.edu, Alaska Satellite Facility, UAF, May, 2022.

REFERENCES

Acocella, V., Di Lorenzo, R., Newhall, C., & Scandone, R. (2015). An overview of recent (1988 to 2014) caldera unrest: Knowledge and perspectives. *Reviews of Geophysics, 53*(3), 896–955.

Anderson, K. R., & Poland, M. P. (2016). Bayesian estimation of magma supply, storage, and eruption rates using a multiphysical volcano model: Kīlauea Volcano, 2000–2012. *Earth and Planetary Science Letters, 447*, 161–171.

Biggs, J., Ebmeier, S. K., Aspinall, W. P., Lu, Z., Pritchard, M. E., Sparks, R. S. J., & Mather, T. A. (2014). Global link between deformation and volcanic eruption quantified by satellite imagery. *Nature Communications, 5*(1), 3471.

Blanc, E. (1985, December). Observations in the upper atmosphere of infrasonic waves from natural or artificial sources – A summary. In *Annales Geophysicae* (Vol. 3, pp. 673–687).

Castruccio, A., Clavero, J., Segura, A., Samaniego, P., Roche, O., Le Pennec, J. L., & Droguett, B. (2016). Eruptive parameters and dynamics of the April 2015 sub-Plinian eruptions of Calbuco volcano (southern Chile). *Bulletin of Volcanology, 78*, 1–19.

Coster, A., Williams, J., Weatherwax, A., Rideout, W., & Herne, D. (2013). Accuracy of GPS total electron content: GPS receiver bias temperature dependence. *Radio Science, 48*(2), 190–196.

Delgado, F., Pritchard, M. E., Basualto, D., Lazo, J., Córdova, L., & Lara, L. E. (2016). Rapid reinflation following the 2011–2012 rhyodacite eruption at Cordón Caulle volcano (Southern Andes) imaged by InSAR: Evidence for magma reservoir refill. *Geophysical Research Letters, 43*(18), 9552–9562.

Delgado, F., Pritchard, M. E., Ebmeier, S., González, P., & Lara, L. (2017). Recent unrest (2002–2015) imaged by space geodesy at the highest risk Chilean volcanoes: Villarrica, Llaima, and Calbuco (Southern Andes). *Journal of Volcanology and Geothermal Research, 344*, 270–288.

Dvorak, J. J., & Dzurisin, D. (1997). Volcano geodesy: The search for magma reservoirs and the formation of eruptive vents. *Reviews of Geophysics, 35*(3), 343–384.

Dzurisin, D. (2003). A comprehensive approach to monitoring volcano deformation as a window on the eruption cycle. *Reviews of Geophysics, 41*(1).

Hargreaves, J. K. (1992). *The solar-terrestrial environment: an introduction to geospace-the science of the terrestrial upper atmosphere, ionosphere, and magnetosphere.* Cambridge University Press.

Heimisson, E. R., Einarsson, P., Sigmundsson, F., & Brandsdóttir, B. (2015). Kilometer-scale Kaiser effect identified in Krafla volcano, Iceland. *Geophysical Research Letters, 42*(19), 7958–7965.

Huang, Y., He, T., Yan, M., Yang, L., Gong, H., Wang, W., … & Wang, J. (2021). Atmospheric transport and deposition of microplastics in a subtropical urban environment. *Journal of Hazardous Materials, 416*, 126168.

Klaffenbock, E., & Hofmann-Wellenhof, B. (2008, September). European satellite navigation-a challenge. In *2008 50th International Symposium ELMAR* (Vol. 2, pp. 561–568). IEEE.

Lognonné, P., Artru, J., Garcia, R., Crespon, F., Ducic, V., Jeansou, E., & Godet, P. E. (2006). Ground-based GPS imaging of ionospheric post-seismic signal. *Planetary and Space Science, 54*(5), 528–540.

Maurya, A. K., R. Singh, B. Veenadhari, S. Kumar and A. K. Singh (2013), Subionospheric VLF perturbations associated with the 12 May 2008 M7.9 Sichuan earthquake, *Natural Hazards and Earth System Sciences*, 13, 2331–2336, doi:10.5194/nhess-13-1-2013

Maurya, A., K Venkatesham, P Tiwari, K. V. Kumar, R. Singh, A. K. Singh, D. Ramesh (2016), 25 April 2015 Nepal Earthquake: Investigation of precursor in VLF sub-ionospheric signal, *Journal of Geophysical Research*, 121, pp 10,403–10,416, doi:10.1002/2016JA022721

Newhall, C. G., & Self, S. (1982). The volcanic explosivity index (VEI) an estimate of explosive magnitude for historical volcanism. *Journal of Geophysical Research: Oceans, 87*(C2), 1231–1238.

Nikkhoo, M., Walter, T. R., Lundgren, P. R., & Prats-Iraola, P. (2016). Compound dislocation models (CDMs) for volcano deformation analyses. *Geophysical Journal International*, ggw427.

Pardini, F., Burton, M., Arzilli, F., La Spina, G., & Polacci, M. (2018). SO2 emissions, plume heights and magmatic processes inferred from satellite data: The 2015 Calbuco eruptions. *Journal of Volcanology and Geothermal Research, 361*, 12–24.

Rao, P. V. S., Rao, M. S., & Satyam, M. (1977). Diurnal & seasonal trends in TEC values observed at Waltair, *Indian Journal of Radio & Space Physics*, 6, 233–235.

Rastogi, R. G., & Sharma, R. P. (1971). Ionospheric electron content at Ahmedabad (near the crest of equatorial anomaly) by using beacon satellite transmissions during half a solar cycle. *Planetary and Space Science, 19*(11), 1505–1517.

Reddy, C. D., & Seemala, G. K. (2015). Two-mode ionospheric response and Rayleigh wave group velocity distribution reckoned from GPS measurement following Mw 7.8 Nepal earthquake on 25 April 2015. *Journal of Geophysical Research: Space Physics, 120*(8), 7049–7059.

Reddy, C. D., Shrivastava, M. N., González, G., & Baez, J. C. (2017). Ionospheric plasma response to M w 8.3 Chile Illapel Earthquake on September 16, 2015. In *The Chile-2015 (Illapel) Earthquake and Tsunami* (pp. 145–155). Birkhäuser, Cham.

Reddy, C. D., Sunil, A. S., González, G., Shrivastava, M. N., & Moreno, M (2015). Near-field co-seismic ionospheric response due to the northern Chile Mw 8.1 Pisagua earthquake on April 1, 2014 from GPS observations. *Journal of Atmospheric and Solar-Terrestrial Physics*, 134, 1–8.

Romero, J. E., Morgavi, D., Arzilli, F., Daga, R., Caselli, A., Reckziegel, F., ... & Perugini, D. (2016). Eruption dynamics of the 22–23 April 2015 Calbuco Volcano (Southern Chile): Analyses of tephra fall deposits. *Journal of Volcanology and Geothermal Research*, 317, 15–29.

Sandwell, D., Mellors, R., Tong, X., Wei, M., & Wessel, P. (2011). GMTSAR: An InSAR processing system based on generic mapping tools.

SERNAGEOMIN, (2015a). Reporte Especial de Actividad Volcánica (REAV) Región de los Lagos. (REAV) Año 2015 Abril 22 (20:45 HL).

SERNAGEOMIN, (2015b). Reporte Especial de Actividad Volcánica (REAV) Región de los Lagos. Año 2015 Abril 22 (22:30 HL).

Shrivastava, M. N., Maurya, A. K., Gonzalez, G., Sunil, P. S., Gonzalez, J., Salazar, P., & Aranguiz, R (2021). Tsunami detection by GPS-derived ionospheric total electron content. *Scientific Reports*, 11(1), 1–13.

Shrivastava, M. N., Sunil, A. S., Maurya, A. K., Aguilera, F., Orrego, S., Sunil, P. S., ... & Moreno, M. (2023). Tracking tsunami propagation and Island's collapse after the Hunga Tonga Hunga Ha'apai 2022 volcanic eruption from multi-space observations. *Scientific Reports*, 13(1), 20109.

Valderrama, O., Franco, L., & Gil-Cruz, F. (2015). Erupción intempestiva del volcán Calbuco, Abril 2015. In *XIV Congreso Geológico Chileno*, Vol. 3, pp. 91–93.

Van Eaton, A. R., Amigo, Á., Bertin, D., Mastin, L. G., Giacosa, R. E., González, J., ... & Behnke, S. A. (2016). Volcanic lightning and plume behavior reveal evolving hazards during the April 2015 eruption of Calbuco volcano, Chile. *Geophysical Research Letters*, 43(7), 3563–3571.

Walker, G. P. (1980). The Taupo pumice: product of the most powerful known (ultraplinian) eruption? *Journal of Volcanology and Geothermal Research*, 8(1), 69–94.

Zel'Dovich, Y. B., & Raizer, Y. P. (2002). *Physics of shock waves and high-temperature hydrodynamic phenomena*. Courier Corporation.

10 Quantifying Mineral Abundance and Erosion Using Remote Sensing and Global Navigation Satellite System (GNSS) Technology

*Ramesh Kumar, Rajesh Kumar, Atar Singh, Prity Singh Pippal,
Payal Sharma, Shruti Singh, and Surjeet Singh Randhawa*

10.1 INTRODUCTION

Quantifying mineral abundance and rock erosion through remote sensing and GNSS (Global Navigation Satellite System) technology involves utilizing various data sources and techniques to gather information about the Earth's surface and subsurface. Indeed, hyperspectral infrared remote sensing has emerged as a precious and progressively accessible instrument for mineral exploration and broader geoscientific endeavors over the past four decades (Clark and Roush, 1984). Also, hyperspectral remote sensing has been the precise quantification of mineral abundance, serving the purposes of mineral exploration and geological mapping on Earth and extending to mineral mapping endeavors on other celestial bodies (Eichstaedt et al., 2020).

Quantifying key minerals in airborne hyperspectral geological exploration over expansive areas poses numerous challenges. These include the heterogeneity of solar exposure, the presence of vegetation, diverse geometries of rock materials, and the impact of moisture variations in topsoil on the accuracy of mineral quantity information, all of which can complicate the processing of data (Murphy et al., 2015; Salehi and Thaarup, 2018). The mapping of hydrothermal alterations through spatial imagery is a pivotal application of remote sensing. This approach enhances traditional mining prospecting methods and identifies potential new prospecting areas on a regional scale while significantly reducing fieldwork costs. However, the alterations in spectral characteristics, particularly in the VNIR (visible-near infrared), SWIR (shortwave infrared), and TIR (thermal infrared) spectral regions, are due to the presence of certain minerals. These alterations are typically identified by distinctive features, primarily absorption patterns, in the spectral data. This type of analysis is common in fields such as remote sensing, geology, and mineral exploration, where understanding the spectral signatures of minerals can provide valuable information about the composition and properties of geological materials. This characterization is a fast, efficient, cost-effective tool for precise mineralogical mapping.

By leveraging spectral data across multiple regions like VNIR, SWIR, and TIR, researchers and professionals can quickly identify and map mineral distributions in various geological formations (Azizi et al., 2010). However, hyperspectral imagery has recently been incorporated into digital projections (Kurz et al., 2011; Buckley et al., 2013; Kurz and Buckley, 2016; Lorenz et al., 2018; Salehi et al., 2018), but relying solely on hyperspectral data in the VNIR and SWIR regions, which lack distinctive bond-related spectral features, may pose limitations in certain cases. While VNIR

DOI: 10.1201/9781032712444-12

and SWIR spectral data can still provide valuable information about mineral compositions and surface properties, the absence of certain bond-related features might make it challenging to differentiate between minerals with similar spectral characteristics. In such scenarios, identifying and mapping minerals may be less precise than techniques incorporating additional spectral regions, such as the TIR (thermal infrared) region, which can provide complementary information about mineralogy based on temperature-dependent emission properties. However, despite these limitations, VNIR and SWIR hyperspectral data still offer advantages regarding data availability, spatial resolution, and cost-effectiveness (Hunt and Salisbury, 1970). In addition, hyperspectral long-wave infrared (LWIR) imaging serves as a valuable complement to VNIR-SWIR data in the realm of mineral mapping, as many rock-forming minerals exhibit characteristic resonant wavelengths due to molecular vibrations in the LWIR segment of the electromagnetic spectrum.

Studies have reported the potential of integrating VNIR, SWIR, and LWIR hyperspectral data for geological mapping (Kruse, 2015; McDowell and Kruse, 2015; Notesco et al., 2016). However, hyperspectral imagers have historically operated primarily from a single platform, typically airborne. Recently, the Helmholtz Institute Freiberg for Resource Technology has initiated the deployment of hyperspectral sensors on the ground (Kirsch et al., 2018; Lorenz et al., 2018) and unmanned aerial systems. This advancement signifies broader accessibility and application of hyperspectral imaging technology across various platforms, enhancing the scope and efficiency of geological mapping endeavors. This approach allows places with steep outcrops and difficult-to-access and potentially hazardous terrain since it enables higher spatial resolutions (from millimeters to tens of centimeters) and various scanning views. By combining satellite imagery and geospatial positioning data, the researchers can create highly detailed maps and models of geological features with unprecedented accuracy.

This multidisciplinary approach allows for a more comprehensive understanding of the Earth's surface dynamics and facilitates the identification of mineral deposits and erosion hotspots with great precision. By analyzing spectral signatures acquired from satellite sensors, researchers can infer the presence and abundance of various minerals across different terrains. This non-invasive method reduces the need for costly and time-consuming field surveys. It enables the study of remote or inaccessible regions where traditional geological mapping methods may be impractical. Moreover, integrating GNSS technology adds another layer of sophistication to the research methodology. By precisely geo-referencing remote sensing data with ground-based measurements, the researchers can accurately quantify changes in elevation and terrain morphology over time. By providing a more accurate and efficient means of quantifying mineral abundance and tracking rock erosion, the findings of this study have significant implications for various industries, including mining, environmental monitoring, and land management.

Therefore, this comprehensive study pushes the boundaries of traditional geological analysis by leveraging cutting-edge technologies to precisely measure mineral abundance and track rock erosion over time. One of the key strengths of this chapter lies in its integration of remote sensing data with GNSS technology. This capability is precious for studying erosion processes, as it allows monitoring of subtle topographic changes at a high temporal resolution. The implications of this research are far-reaching and extend beyond the realm of academic inquiry.

10.2 INTEGRATION OF REMOTE SENSING AND GLOBAL NAVIGATION SATELLITE SYSTEM (GNSS)

The scientific community is becoming more interested in using Global Navigation Satellite System (GNSS) signals for remote sensing for various observations. However, GNSS systems have been effectively employed to remotely detect the Earth's surface and atmosphere by utilizing their refracted, reflected, and scattered signals. It has proven capable of sensing the ocean, terrestrial surfaces (including soil moisture), the cryosphere, the atmosphere, and the ionosphere. However, lithological mapping and landform identification are all frequently accomplished using geospatial technologies (Shivkumar, 2018). Understanding the workings of the Earth's surface and subsurface processes about mineral deposits is greatly aided by this technology. Standard techniques for geological mapping using remotely sensed data are necessary. As a result, scientists and academics have created computer programs to precisely extract thematic information about geology from satellite data.

Lyon (1964) first attempted to identify rocks and minerals from their electromagnetic spectra, proposing and researching their potential in mineral discrimination on the lunar surface. This study marks the first instance where the distinct absorption patterns of various minerals across diverse wavelength ranges have been brought to light. Mainly, phyllosilicates containing Al-OH and Mg-OH exhibit precise spectral absorption characteristics within the 2.1–2.4 µm range.

In addition, the Global Navigation Satellite System (GNSS) is a typical example of military technology that has become open to the public, has been further applied, and has greatly impacted exploration technology. In common among all the airborne methods, GNSS's higher position accuracy brought about a denser and more regular fight-line pattern with a relatively more minor terrain clearance. Mapping of geological structures also holds significant importance in mineral resource studies, as many ore deposits are often situated along zones of geological features such as fractures, faults, folds, and linear or curvilinear formations. Geological structures, lithological characteristics, mineralization, alteration patterns, and biogeochemical anomalies are among the critical features in the mineral deposit assessment process using remotely sensed data. Researchers have identified various linear and circular phenomena in satellite images through remote sensing techniques. These observations have prompted extensive investigations into the relationship between minerogenetic geological backgrounds and statistical information about linear and circular structures. Consequently, heightened attention has been given toward understanding the ore-controlling function of these structures.

10.3 FUNDAMENTALS OF REMOTE SENSING FOR LITHOLOGICAL MAPPING

Remote sensing (RS) is the science and art of acquiring information about an object, area, or phenomenon without physically being in contact with it. It relies on detecting and analyzing energy reflected or emitted from the target of interest (Sabins and Ellis, 2020). The fundamental principles of remote sensing include electromagnetic radiation, interaction with the Earth's surface, sensors and detectors, and spectral signatures. In addition, remote sensing plays a crucial role in lithological mapping, which involves identifying and delineating different rock types and geological formations from a distance, typically using satellite or airborne sensors—the details of these remote sensing principles are mentioned in Table 10.1 with their concise descriptions.

Remote sensing data can be obtained from various platforms, including satellites, aerial platforms (aircraft), and unmanned aerial vehicles (UAVs) (Bhardwaj et al., 2016; Zhang and Zhu, 2023). Each type of data has its advantages and limitations. Unmanned Aerial Vehicles (UAVs), or drones, are small aircraft equipped with cameras or other sensors (Eskandari et al., 2023). They can be deployed quickly and cheaply to capture high-resolution imagery of specific areas, making them ideal for environmental monitoring, agriculture, and infrastructure inspection. Based on the major principle and different data types, the processing plays a significant role (Zin et al., 2011).

By leveraging these fundamental concepts and techniques, remote sensing is critical in lithological mapping, providing valuable insights into the Earth's surface composition and geological structures over large spatial scales. Image acquisition involves capturing raw data from remote sensing platforms, while preprocessing consists of correcting and enhancing the data for analysis (de Lange, 2023). The flowchart demonstrated the sequential order of the preprocessing steps, starting from image acquisition and proceeding through radiometric correction, geometric correction, image enhancement, and data calibration (Figure 10.1). Therefore, understanding the principles of remote sensing, types of remote sensing data, and image acquisition and preprocessing techniques is

TABLE 10.1

The Fundamental Principles of Remote Sensing and Their Respective Information

Sl. No.	Principle	Details
1.	Electromagnetic Radiation	Remote sensing relies on the interaction between electromagnetic radiation (EMR) and the Earth's surface.
2.	Interaction with Surface	When interacting with the Earth's surface materials, EMR undergoes absorption, reflection, transmission, and emission processes.
3.	Sensors and Detectors	Remote sensing instruments, such as cameras and spectrometers, detect and measure EMR reflected or emitted by the Earth's surface.
4.	Spectral Signatures	Different materials have unique spectral signatures, which are patterns of EMR reflected or emitted at specific wavelengths.

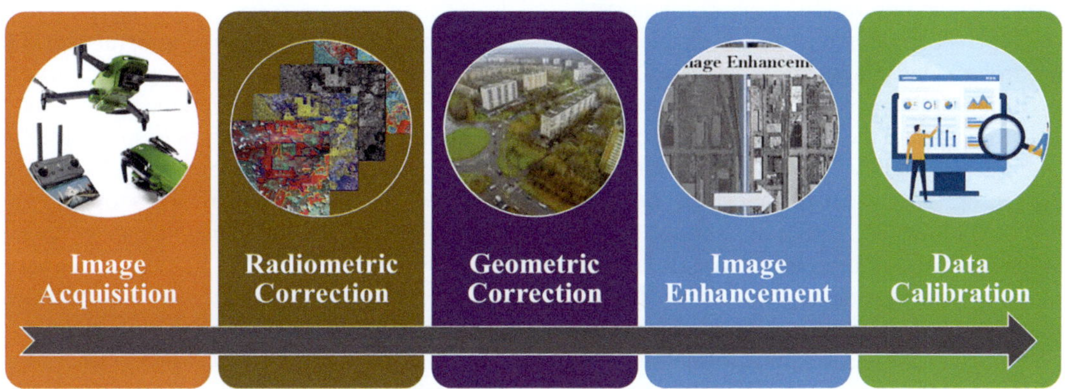

FIGURE 10.1 The flowchart demonstrates the sequential order of the preprocessing steps of a satellite image.

essential for effectively using remote sensing data for various applications, including lithological mapping and geological structures over large spatial scales.

Remote sensing techniques are used for lithological mapping, mineral detection, and many other perspectives (Gabr et al., 2015). Remote sensors can detect the energy emittance by surfaces of objects. The primary focus lies on the physical parameter called radiance, which is temperature-dependent. Spectral emissivity is a material-dependent property that is of immense help in studying the lithology and mineralogy of the Earth's crust. The Thermal Infrared Multispectral Scanner (TIMS) in a satellite was installed earlier for geological studies (Kahle and Rowan, 1980; Kahle, 1981).

Airborne and satellite remote sensing systems demonstrated their capabilities in detecting rock composition. Earlier studies through TIMS (Bihong and Xiaowei, 1998) confirmed their ability to detect silicate materials. Later, the "Advanced Spaceborne Thermal Emission Reflection Radiometer (ASTER)" sensor having three spectral bands "visible/near-infrared (VNIR)," six spectral bands of "shortwave infrared (SWIR)," and five spectral bands of "thermal infrared (TIR)" was installed on the Terra satellite. These could have resolutions of 15 m, 30 m and 90 m, respectively, and opened new exploratory capabilities for Earth surface studies (Rowan and Mars, 2003; Gad and Kusky, 2007; Gabr et al., 2010; Van der Meer et al., 2012). The spectral coverage of ASTER images is 60 km X 60 km. In general, ASTER-SWIR data are used to delineate types of rocks based on the optical spectrum of "SWIR-TIR" to interpret and measure minerals.

10.4   MINERAL MAPPING USING SWIR AND TIR INDICES

The short wavelength infrared (SWIR) and thermal infrared (TIR) spectral resolutions agree on qualitative and quantitative methods for mapping surface mineralogy. These spectral bands are available in the "Advanced Space Borne Thermal Emission Reflection Radiometer (ASTER)." Minerals are mapped through the band indices like "SWIR indices, TIR indices, and TIR emissivity." The mineralogy measurement reflects the primary presence of "quartz, feldspars, carbonates, and micas." The "SWIR and TIR indices" have also been used for quantitative estimation of mineral (Ninomiya, 2004) and creating "thematic mineral abundance maps." Short Wavelength Infrared (SWIR) mineral indices are used for wavelength-dependent absorption patterns in estimating minerals. A previous study reported (Kumar et al., 2023) for lithological mapping for layered silicate (LS), calcite (CA), hydroxyl-bearing (OH), and alunite (AL) in the Shaune Garang catchment, Western Himalayan region (Figure 10.2). The visible and near-infrared wavelength region proves valuable for mapping gossans rich in iron oxides, often associated with weathered sulphide occurrences, and for characterizing regolith. The shortwave infrared wavelength region is useful for mapping alteration haloes containing minerals like chlorites and white micas in epithermal/porphyry styles of Cu-Au mineralization and regolith characterization. The thermal infrared wavelength region provides utility in mapping various exploration targets.

Hyperspectral remote sensing enables the remote mapping of a diverse range of minerals, including iron oxides, clays, micas, chlorites, amphiboles, talc, serpentines, carbonates, quartz, garnets, pyroxenes, feldspars, and sulphates, along with their physiochemistrys. However, the varying indices are related to the varying absorption properties, which helps measure the types of minerals. For example, LS and CA, SWIR indices bands measure hydroxyl (2.2 µm) and carbonate (2.35 µm). At the sensor, "radiance band ratios" can reduce the influence of the atmosphere, a region's topography, and the illuminance variation (Abrams et al., 1983). This also signifies the nonsignificant evidence in "single-band or three-band true or false color composite images." It is also helpful in quantitatively estimating mineral abundances. Therefore, remote sensing plays a crucial role in quantifying mineral abundance and rock erosion by providing spatially and temporally explicit data on Earth's surface characteristics. This information enables geoscientists to study geological processes, monitor landscape changes, and assess the impact of natural and anthropogenic activities on the environment.

10.5   GNSS TECHNOLOGY IN ENVIRONMENTAL MONITORING AND MANAGEMENT

GNSSs are a constellation of orbiting satellites that transmits radio signals. Receivers on Earth can pick up these signals and calculate their position based on the signal arrival time from multiple satellites (Petropoulos and Srivastava, 2021). GNSS technology is crucial in geoscience and environmental science by providing accurate positioning, navigation, and timing information for various applications. GNSS systems, such as the Global Positioning System (GPS), GLONASS, Galileo, and BeiDou, utilize a constellation of satellites orbiting the Earth to transmit signals that GNSS receivers can receive on the ground or in vehicles, aircraft, and ships. GNSSs offer many applications in geoscience research and practice (Li et al., 2015; Kumar et al., 2021). It has revolutionized various fields, including geoscience, by providing a precise and efficient method for determining location, measuring Earth's movements, and studying multiple geophysical phenomena. GNSS technology is a versatile tool in geoscience, providing accurate positioning, navigation, timing, and synchronization capabilities for various applications, from basic research to practical environmental monitoring and management (Kumar et al., 2021). A detailed explanation of GNSS technology in geoscience is given in Table 10.2.

However, GNSS-enabled remote sensing techniques allow for monitoring changes in mineral composition and rock morphology, essential for geological exploration, resource management, and

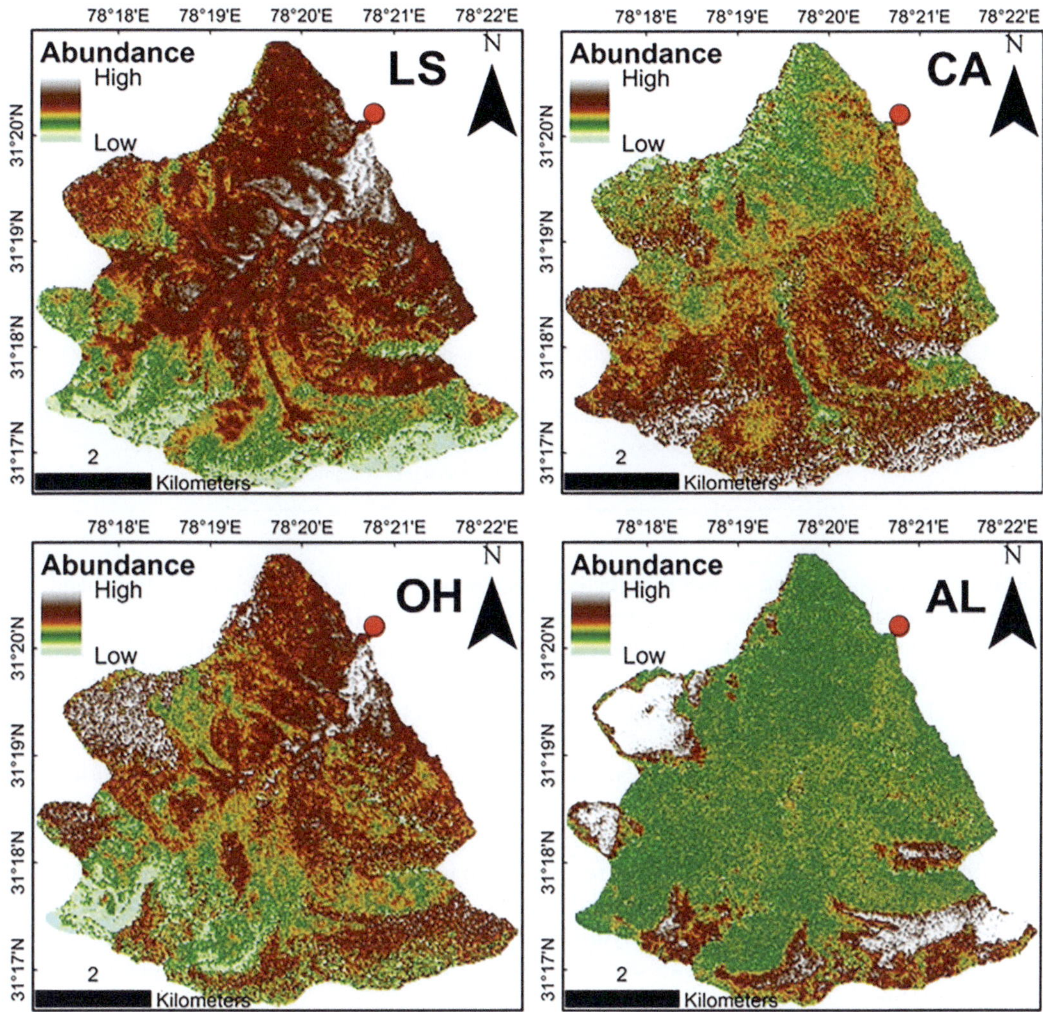

FIGURE 10.2 Mineralogical maps of Shaune Garang catchment represented through LS, CA, AL, and OH (Kumar et al., 2023).

environmental assessment. By combining GNSS-derived positional data with remote sensing imagery, geoscientists can detect subtle changes in terrain and geological features, helping to identify potential mineral deposits and understand erosion patterns. Therefore, GNSS technology enhances the effectiveness and accuracy of remote sensing applications in geoscience, particularly in quantifying mineral abundance and rock erosion. By providing precise positioning information, GNSS enables geoscientists to conduct detailed spatial analysis and monitoring, leading to a better understanding of Earth's surface dynamics and geological processes.

10.6 QUANTIFYING MINERAL ABUNDANCE

Minerals exhibit distinctive characteristics across the electromagnetic spectrum, particularly in the visible and infrared regions. The reflectance spectra of geological materials, such as rocks and soils, can be observed by specific spectral signatures or patterns akin to unique "fingerprints," which arise from the minerals and other components (Laukamp et al., 2021). This distinctive pattern emerges because of the interactions between the mineral's electronic and vibrational processes encompassing

TABLE 10.2

The Detailed Explanation of GNSS Technology in Geoscience

Sl. No.	Method	Application	Description
1.	Static Positioning (Long obs.)	Geodetic Surveys, Control Points	Highly accurate positioning for creating precise maps and establishing reference points.
2.	Real-time Kinematic (RTK)	Monitoring Tectonics	Continuously monitors position changes to study plate movements and potential earthquake precursors.
3.	Differential GNSS	Improves Accuracy	Corrects errors in satellite signals for high-precision applications across geoscience.
4.	Kinematic Positioning	Mapping Topography, Landslides	Collects position data for creating topographic maps and monitoring landscape changes.
5.	GNSS-aided InSAR	Studying Subsidence, Volcanoes	Combines GNSS data with InSAR for improved accuracy in studying ground movements like subsidence and volcanic activity.

crystal field effects, charge transfer, color centers, transitions towards the conduction band, and overtone and combination tone vibrational transitions (Clark et al., 1990).

The recorded information of spectral reflectance is a function of the chemical and physical properties of the target material, which cause different reactions to the incoming light on a molecular and atomic level (Hunt, 1980; Koerting et al., 2021). These spectral signatures of minerals can be used to identify and quantify the presence of specific minerals, and it is a critical tool in remote sensing and geological applications (Pontual et al., 2008). For example, the spectral signature of iron-rich minerals is characterized by low reflectance below 0.7 μm, while the spectral features of hydroxyls and carbonates are prominent in the Short-Wave InfraRed (SWIR) region (Clark et al., 1990; Santos et al., 1990). The Thermal InfraRed (TIR) region is also important, as it exhibits fundamental vibration features of many rock-forming mineral groups, such as silicates, carbonates, oxides, phosphates, sulphates, nitrates, nitrites, and hydroxyls (Clark et al., 1990; Gupta & Gupta, 1991).

The spectral signature of minerals can be used to determine mineral compositions for specific areas and can function as a useful aid for remote sensing image interpretation (Sabins, 1999). These spectral fingerprints serve as indicators of various attributes that enable the extraction of valuable details such as mineral types, quantities, chemical composition, and crystalline structure. This includes identifying the specific mineral species present, determining their relative abundance within the sample, assessing their chemical composition, and even evaluating their crystalline structure. Through careful analysis and interpretation of these spectral fingerprints, scientists and researchers can unlock insights into the geological history, environmental conditions, and processes that have shaped the formation and composition of the sample.

Thus, the Quantification of Mineral abundance is a crucial aspect of geoscientific research, particularly in mineral exploration and environmental monitoring. However, minerals are naturally formed inorganic substances that combine cations and anions. They may be chemically simple or highly complex. Some ions may occur as major, minor, or trace constituents. The spectrum of a mineral is governed by the effect of various factors like the Spectra of dominant anions, the Spectra of dominant cations, the Spectra of ions occurring as trace constituents, and the Crystal field effect. For instance, limonite (iron oxide) displays a broad absorption band in the UV-blue region due to the Fe-O charge-transfer effect, with additional absorption features from ferric ions in the near-IR region. When hydrated, limonite exhibits water molecule bands at specific wavelengths. Quartz, composed of simple silicate tetrahedra, primarily shows absorption bands in the thermal-IR. Still, impurities can introduce absorption bands in the visible region, leading to coloration, such as brown and green hues from iron impurities.

Pyroxenes exhibit absorption bands from ferrous ions in the VNIR region and silicate ions in the thermal-IR, with the crystal field effect influencing the position and intensity of these bands. Similarly, amphiboles, micas, and clays display distinct absorption bands from iron, hydroxyl ions, and silicates in different spectral regions, highlighting their unique spectral signatures. Carbonate minerals exhibit characteristic absorption bands in the SWIR and thermal-IR regions, with specific features from iron in siderite and manganese in rhodochrosite visible in the VNIR region. Conversely, clays show SWIR bands for Al-OH/Mg-OH and water molecule bands, with distinctive features in the thermal-IR and near-IR regions due to their composition and impurities.

In summary, the spectrum of a mineral is a complex interplay of its constituent elements and crystal field effects, resulting in unique spectral signatures that can be used for mineral identification and characterization (Gupta & Gupta, 1991). Various methods are available for mineral exploration studies, such as the Band ratio method in remote sensing. The property of having high reflectance values in one spectral range and lower in another spectral region of the objects may be exploited to emphasize or exaggerate the anomaly of the target object (Inzana et al., 2001; Fatima et al., 2017). Band ratio images enhance spectral characteristics by dividing the digital number values of one spectral band by another. This compensates for variations in scene illumination due to diverse topography, emphasizing the color content and compositional differences in the image data.

Band ratios are particularly effective for mineral mapping, as they minimize the influence of environmental factors and highlight the spectral signatures of different minerals in Landsat-ETM+ and ASTER data (Fatima et al., 2017). Hyperspectral remote sensing, or imaging spectroscopy, has revolutionized mineral exploration by enabling high-resolution spectral data acquisition over large areas (Sahu et al., 2023). This technology allows for identifying minerals based on their unique spectral signatures, often distinct across different wavelengths.

The Spectral Angle Mapper algorithm is another popular approach for identifying minerals using hyperspectral data, as it compares the spectral signatures of pixels with those of known minerals (Boardman & Kruse, 1994). Raman spectroscopy is another powerful mineral abundance quantification technique, particularly in planetary science and astrobiology (Breitenfeld et al., 2018). This method involves the analysis of the inelastic scattering of light by molecules, which provides a unique fingerprint of the mineral composition. Machine learning multivariate unmixing models have been developed to quantify mineral abundances using Raman spectroscopy, offering promising results for analyzing complex mineral mixtures (Breitenfeld et al., 2018). Several challenges remain despite the advances in spectral signatures and mineral abundance quantification. Non-linear mixing effects in Raman spectroscopy and the complexity of mineral mixtures continue to pose challenges for accurate quantification.

However, using band ratio indices derived from ASTER data effectively facilitated the mapping of diverse minerals in previous studies (Oliver & van der Wielen, 2006; Fatima et al., 2017; Abrams & Yamaguchi, 2019). The remote sensing techniques that have been explored hold promising applications in remote and inaccessible areas to help delineate lithologies and mineral deposits at the surface. Employing such methods, it becomes feasible to outline a target area's geological composition and mineralogy, thus refining the scope of surveys and directing focus towards specific mineral resources. New generation satellites like Sentinel may be investigated for their potential for geological and mineral mapping.

10.7   CONCLUSION

Hyperspectral sensors are one of the most significant technological advancements in remote sensing for scanning Earth's surface materials. Spatial technology is precious for collecting, processing, and analyzing spatial data and providing access to spatial information in various development sectors, including mining, disaster management, and environmental management. Hyperspectral remote sensing has long been aimed at quantifying mineral abundance for terrestrial mineral exploration, geologic mapping, and mineral mapping in broader contexts. Hence, this chapter delves into an extensive exploration of remote sensing and Global Navigation Satellite System (GNSS) technology's application in quantifying mineral abundance and rock erosion. The chapter showcases a robust methodology for accurately evaluating geological features and landscape dynamics by combining satellite imagery analysis with GNSS data processing. The findings underscore the efficacy of remote sensing techniques in discerning mineral distribution patterns and tracking erosion rates across varied terrains.

Furthermore, incorporating GNSS technology elevates spatial precision and temporal resolution, facilitating precise measurements of topographic alterations and erosion processes over time. This research substantially contributes to geology, environmental science, and remote sensing by offering valuable insights and comprehensive knowledge. Particularly, satellite processing techniques emerge as a straightforward and efficient means of lithological mapping. Also, the assessment reveals the high potential of all three sensors in mineralogical mapping, with the sensor's performance attributed to its rich spectral capabilities, particularly in the SWIR region. This technique capitalizes on the property of objects to exhibit high reflectance values in one spectral range and lower values in another, which can be utilized to accentuate target anomalies. Band ratio images enhance spectral characteristics by dividing the digital number values of one spectral band by another. This process helps mitigate variations in scene illumination stemming from diverse topography, emphasizing color content and compositional disparities in the image data.

10.8 FUTURE RESEARCH AND RECOMMENDATIONS

Quantifying mineral abundance and rock erosion using remote sensing and GNSS technology represents a significant advancement in geology. The chapter's innovative approach and interdisciplinary methodology shed new light on the dynamic processes shaping the Earth's surface and offer valuable insights for scientific research and practical applications. As remote sensing and GNSS technologies continue to evolve, the potential for further advancements in geological analysis is immense, promising a future where our understanding of the Earth's geology is more precise and comprehensive than ever. Indeed, given the uncertain long-term supply of mineral resources, collaboration between exploration engineers and researchers becomes paramount. By pooling their collective experience, knowledge, and wisdom, they can drive breakthroughs in mineral exploration technology. This collaboration is essential for revitalizing exploration efforts and steering them towards success, thereby securing mineral resources for the future. This requires leveraging technological advancements—such as remote sensing, machine learning, and data analytics—to enhance exploration efficiency and effectiveness.

REFERENCES

Abrams, M. J., Brown, D., Lepley, L., & Sadowski, R. (1983). Remote sensing for porphyry copper deposits in southern Arizona. *Economic Geology*, *78*(4), 591–604.

Abrams, M., & Yamaguchi, Y. (2019). Twenty years of ASTER contributions to lithologic mapping and mineral exploration. *Remote Sensing*, 11(11), 1394.

Azizi, H., Moore, F., & Modabberi, S. (2010). Extraction of hydrothermal alterations from ASTER SWIR data from east Zanjan, northern Iran. *Advances in Space Research*, *46*(1), 99–109.

Bhardwaj, A., Sam, L., Martín-Torres, F.J., Kumar, R. (2016). UAVs as remote sensing platform in glaciology: Present applications and prospects, *Remote Sensing of Environment*, 175, 196–204.

Bihong, F., & Xiaowei, C. (1998). Thermal infrared spectra and TIMS imagery features of sedimentary rocks in the Kalpin Uplift, Tarim Basin, China. *Geocarto Int.*, 13, 69–73.

Boardman, J. W., and Kruse, F. A., (1994). Automated spectral analysis: A geologic example using AVIRIS data, north Grapevine Mountains, Nevada: in *Proceedings, Tenth Thematic Conference on Geologic Remote Sensing, Environmental Research Institute of Michigan*, Ann Arbor, MI, p. I-407–I-418.

Breitenfeld, L. B., Dyar, M. D., Carey, C. J., Tague Jr, T. J., Wang, P., Mullen, T., & Parente, M. (2018). Predicting olivine composition using Raman spectroscopy through band shift and multivariate analyses. *American Mineralogist*, 103(11), 1827–1836.

Buckley, S. J., Kurz, T. H., Howell, J. A., & Schneider, D. (2013). Terrestrial lidar and hyperspectral data fusion products for geological outcrop analysis. *Computers & Geosciences*, 54, 249–258.

Clark, R. N., King, T. V., Klejwa, M., Swayze, G. A., & Vergo, N. (1990). High spectral resolution reflectance spectroscopy of minerals. *Journal of Geophysical Research: Solid Earth*, 95(B8), 12653–12680.

Clark, R. N., & Roush, T. L. (1984). Reflectance spectroscopy: Quantitative analysis techniques for remote sensing applications. *Journal of Geophysical Research: Solid Earth*, 89(B7), 6329–6340.

de Lange, N. (2023). Remote Sensing and Digital Image Processing. In *Geoinformatics in Theory and Practice: An Integrated Approach to Geoinformation Systems, Remote Sensing and Digital Image Processing* (pp. 435–510). Berlin, Heidelberg: Springer Berlin Heidelberg.

Eichstaedt, H., Tsedenbaljir, T., Kahnt, R., Denk, M., Ogen, Y., Glaesser, C., ... & Michalski, J. (2020). Quantitative estimation of clay minerals in airborne hyperspectral data using a calibration field. *Journal of Applied Remote Sensing*, 14(3), 034524.

Eskandari, A., Hosseini, M., & Nicotra, E. (2023). Application of Satellite Remote Sensing, UAV-Geological Mapping, and Machine Learning Methods in the Exploration of Podiform Chromite Deposits. *Minerals*, 13(2), 251.

Fatima, Khunsa, Muhammad Umar Khan Khattak, Allah Bakhsh Kausar, Muhammad Toqeer, Naghma Haider, and Asid Ur Rehman. "Minerals identification and mapping using ASTER satellite image." *Journal of Applied Remote Sensing* 11, no. 4 (2017): 046006.

Gabr, S., Ghulam, A., & Kusky, T. (2010). Detecting areas of high-potential gold mineralization using ASTER data. *Ore Geology Reviews*, 38(1), 59–69.

Gabr, S., Ghulam, A., & Kusky, T. (2015). Detecting areas of high-potential mineralization using remote sensing data: A review of methods and applications. *International Journal of Remote Sensing*, 36(23), 5971–6001.

Gad, S., & Kusky, T. (2007). ASTER spectral ratioing for lithological mapping in the Arabian-Nubian Shield, Egypt. *Journal of African Earth Sciences*, 44(3), 196–202.

Gupta, R. P., & Gupta, R. P. (1991). Spectra of minerals and rocks. *Remote Sensing Geology*, 19–34.

Hunt, G.R. (1980). Electromagnetic radiation: the communication link in remote sensing. *Remote Sensing in Geology* (ed. B. Siegal and A. Gillespie), pp. 5–44. New York: John Wiley.

Hunt, G.R. and Salisbury, J.W. (1970) Visible and near-infrared spectra of minerals and rocks: I silicate minerals. *Modern Geology*, 1, 283–300.

J. Inzana, T. Kusky, and R. Tucker, "Comparison of TM band ratio images, supervised classifications, and merged TM and radar imagery for structural geology interpretations of the central Madagascar highlands," in *Abstracts with Programs of American Society of Photogrammetry and Remote Sensing Annual Meeting*, Vol. 34 (2001).

Kahle, A. B. (1981). Remote sensing of the Earth using multispectral middle infrared scanner data. In Geological Applications of Thermal Infrared Remote Sensing Techniques (p. 72).

Kahle, A. B., & Rowan, L. C. (1980). The Thermal Infrared Multispectral Scanner (TIMS): A New Tool for Geological Studies. *Journal of Geophysical Research*, 85(B11), 7111–7117.

Kirsch, M., Lorenz, S., Jakob, S., & Gloaguen, R. (2018). Integration of Terrestrial and Drone-Borne Hyperspectral and Photogrammetric Sensing Methods for Exploration Mapping and Mining Monitoring. *Remote Sensing*, 10(9), 1366. https://doi.org/10.3390/rs10091366

Koerting, F., Koellner, N., Kuras, A., Boesche, N. K., Rogass, C., Mielke, C., ... & Altenberger, U. (2021). A solar optical hyperspectral library of rare-earth-bearing minerals, rare-earth oxide powders, copper-bearing minerals and Apliki mine surface samples. *Earth System Science Data*, 13(3), 923–942.

Kruse, F. A. (2015). Comparative analysis of Airborne Visible/Infrared Imaging Spectrometer (AVIRIS) and Hyperspectral Thermal Emission Spectrometer (HyTES) longwave infrared (LWIR) hyperspectral data for geologic mapping. In *Proceedings of the 8th JPL Airborne Earth Science Workshop* (Vol. 99-17, pp. 247–258).

Kumar, R., Kumar, R., Bhardwaj, A., Singh, A., Singh, S., Kumari, A., & Sinha, R. K. (2023). Multivariate statistical analysis and Geospatial approach for evaluating Hydro-geochemical characteristics of meltwater from Shaune Garang Glacier, Himachal Pradesh, India. *Acta Geophysica*, 71(1), 323–339.

Kumar, P., Srivastava, P. K., Tiwari, P., & Mall, R. K. (2021). Application of GPS and GNSS technology in geosciences. In *GPS and GNSS Technology in Geosciences* (pp. 415–427). Elsevier.

Kurz, T. H., & Buckley, S. J. (2016). A review of hyperspectral imaging in close range applications. *The International Archives of the Photogrammetry, Remote Sensing and Spatial Information Sciences*, 41, 865–870.

Kurz, T. H., Buckley, S. J., Howell, J. A., & Schneider, D. (2011). Integration of panoramic hyperspectral imaging with terrestrial lidar data. *The Photogrammetric Record*, 26(134), 212–228.

Laukamp, C., Rodger, A., LeGras, M., Lampinen, H., Lau, I. C., Pejcic, B., … & Ramanaidou, E. (2021). Mineral physicochemistry underlying feature-based extraction of mineral abundance and composition from shortwave, mid and thermal infrared reflectance spectra. *Minerals*, 11(4), 347.

Li, X., Zhang, X., Ren, X., Fritsche, M., Wickert, J., & Schuh, H. (2015). Precise positioning with current multi-constellation global navigation satellite systems: GPS, GLONASS, Galileo and BeiDou. *Scientific Reports*, 5(1), 8328.

Lorenz, S., Salehi, S., Kirsch, M., Zimmermann, R., Unger, G., Vest Sørensen, E., & Gloaguen, R. (2018). Radiometric correction and 3D integration of long-range ground-based hyperspectral imagery for mineral exploration of vertical outcrops. *Remote Sensing*, 10(2), 176.

Lyon, R. J. P. (1964). Evaluation of infrared spectrophotometry for compositional analysis of lunar and planetary soils. part ii-rough and powdered surfaces (No. NASA-CR-100).

McDowell, R. J., & Kruse, F. A. (2015). Integrated visible to near infrared, short wave infrared, and long wave infrared spectral analysis for surface composition mapping near Mountain Pass, California. In *Proceedings of the 8th JPL Airborne Earth Science Workshop* (Vol. 99-17, pp. 259–270).

Murphy, R. J., Taylor, Z., Schneider, S., & Nieto, J. (2015). Mapping clay minerals in an open-pit mine using hyperspectral and LiDAR data. *European Journal of Remote Sensing*, 48(1), 511–526.

Ninomiya, Y. (2004). Lithologic mapping with multispectral ASTER TIR and SWIR data. In *Sensors, Systems, and Next-Generation Satellites VII* (Vol. 5234, pp. 180–190). SPIE.

Notesco, G., Ben-Dor, E., & Levin, N. (2016). Comparison of lithological mapping results from airborne hyperspectral VNIR-SWIR-LWIR and combined data. In *Proceedings of the 8th JPL Airborne Earth Science Workshop* (Vol. 99-17, pp. 271–282).

Oliver, S., & van der Wielen, S. (2006). Mineral mapping with ASTER. *AUSGEO NEWS (Geoscience Australia)*, (28).

Petropoulos, G. P., & Srivastava, P. K. (Eds.). (2021). *GPS and GNSS Technology in Geosciences*. Elsevier.

Pontual, S., Merry, N., Gamson, P., & Birch, I. (2008). *The Spectral Geologist (TSG) – A Spectral Analysis Tool for Exploration and Mining*. 7th International Congress for Applied Mineralogy (ICAM), 153–160.

Rowan, L.C., & Mars, J.C. (2003). Lithologic mapping in the Mountain Pass, California area using Advanced Spaceborne Thermal Emission and Reflection Radiometer (ASTER) data. *Remote Sensing of Environment*, 84(3), 350–366.

Sabins, F. F. (1999). Remote sensing for mineral exploration. *Ore Geology Reviews*, 14(3–4), 157–183.

Sabins Jr, F. F., & Ellis, J. M. (2020). *Remote sensing: Principles, interpretation, and applications*. Waveland Press.

Sahu, D. K., Homeshvari, V. R., Singh, V., & Upadhyay, S. K. (2023). *Use of hyperspectral remote sensing as advanced tools for study in soil characteristics: A review*.

Salehi, S., Lorenz, S., Vest Sørensen, E., Zimmermann, R., Fensholt, R., Henning Heincke, B., … & Gloaguen, R. (2018). Integration of vessel-based hyperspectral scanning and 3D-photogrammetry for mobile mapping of steep coastal cliffs in the arctic. *Remote Sensing*, 10(2), 175.

Salehi, S., & Thaarup, S. M. (2018). Mineral mapping by hyperspectral remote sensing in West Greenland using airborne, ship-based and terrestrial platforms. *GEUS Bulletin*, 41, 47–50.

Santos, E., Castañeda, C., & Fuentes, J. (1990). Analysis of spectral features of minerals and their applications to remote sensing. *International Journal of Remote Sensing*, 11(7), 1147–1159.

Shivkumar, V (2018). Advances and Applications of Geospatial Technology in Earth Science. *Journal of Emerging Technologies and Innovative Research (JETIR)* 5(12), 217–221.

Van der Meer, F., van der Werff, H., van Ruitenbeek, F., Hecker, C., Bakker, W., Noomen, M., … & Woldai, T. (2012). Multi- and hyperspectral geologic remote sensing: A review. *International Journal of Applied Earth Observation and Geoinformation*, 14(1), 112–128.

Zin, T. T., Takahashi, H., Toriu, T., & Hama, H. (2011). Fusion of infrared and visible images for robust person detection. Image Fusion, 239–264.

Zhang, Z., & Zhu, L. (2023). A review on unmanned aerial vehicle remote sensing: Platforms, sensors, data processing methods, and applications. *Drones*, 7(6), 398.

11 Forest Fire Risk (FFR) Mapping Using Analytical Hierarchy Process (AHP) and Geospatial Technique and Advantage of GNNS to Monitor the Forest Fire

A Case Study in the Southern Districts of Uttarakhand

Jayant Nath Tripathi, Swarnim, Irjesh Sonker, Roopa Maurya, and Digvijay Dubey

11.1 INTRODUCTION

In order to preserve the various types of wildlife and plant life essential to maintaining the equilibrium of the Earth's ecosystem, forests play a significant role. The density of forest in a region reveals its ecological state. Forest fires have occurred all across the world at varying times and locations since forests have existed. Terrain and geomorphology, weather, and forest vegetation all play a role in the frequency and severity of its incidence (Somashekar et al., 2008). Forest fires are dangerous events that destroy biomasses and kill local species by burning them or making them starve because they decimate fruit trees (Mina et al., 2023; Robinne et al., 2018). India has an overall forest cover area of 76.4 million ha, which accounts for 22 percent of the country's total area and 2% of the world's forested area (including 1% of the world's main forests) (FSI 2019).

In India, forest degradation is a persistent concern. The major problem of degradation of forests is the natural and anthropogenic activity, which has caused frequent forest fire events in the Himalayan region (Saklani, 2008). Half of India's forestland is at high risk of being destroyed by fire. It is estimated that 33% of forested area is at risk, with that number rising to 90% in some states, resulting in a loss of around 440 crores in economic activity (Gubbi S., 2003). As forest fires pose such a serious threat to the ecosystem, they provide a problem for the Indian environment and policy-makers (Kushla & Ripple, 1997; Rawat, 2003). There is no way to eliminate forest fires completely, but making a forest fire risk map can help reduce the damage they cause.

Geospatial technology is most important when it comes to studying the quality and quantity of plant cover and other parts of an ecosystem. For fast and accurate implementation of complex fire models, geospatial technologies offer the spatial and time data required (Burrows et al., 2011). The exact site of forest fires can be detected and monitored using geospatial technologies. The most significant method for modeling in relation to environmental and natural hazard prediction, such as

DOI: 10.1201/9781032712444-13

135

forest fire risk mapping, is the combination of Multi-Criteria Decision Analysis (MCDA) methodologies and Geospatial Information System (GIS) (Vadrevu and Badarinath, 2009). Many researchers use methods like the Analytical Hierarchy Process (AHP), fuzzy logic, frequency ratio (FR), and artificial neural networks (ANN) to map the risk of forest fires (Ardakani, 2010; Chhetri and Kayastha, 2015; Pourghasemi et al., 2016; Eugenio et al., 2016; Bui et al., 2017; Sonker et al., 2021).

Employing Remote Sensing (RS) and Geographic Information Systems (GIS), several researchers have carried out work on forest fire mapping utilizing factors such as land use land cover (LULC), Wind Speed, Road map, Normalized Differential Vegetation Index (NDVI), Elevation, etc. (Maeda et al., 2009; Shamra et al., 2012; Motazeh et al., 2013; Chhetri and Kayastha, 2015; Feizizadeh et al., 2015; Goldarag et al., 2016; Pourghasemi, 2016; Pourtaghi et al., 2016; Hofmann, 2017; Triapathi et al., 2017).

In India, forest fire incidents happen frequently in Uttarakhand state. Utilizing Global Navigation Satellite Systems (GNSSs) in forest fire monitoring is crucial due to its ability to offer improved information, consistent mapping, and real-time monitoring. Firefighters and authorities can expand their capability to coordinate reaction operations, allocate resources effectively, and monitor the progress of fires thanks to GNSS technology (Belenguer-Plomer et al., 2019a, 2019b; Eroglu et al., 2019; Dente et al., 2020). Since 2002 the Uttarakhand State Forest Department has recorded how many forest fires occurred in each district throughout the months of April, May, and June. The forest fire season peaks towards the end of April and the beginning of June. In both Himalayan and non-Himalayan nations, this phenomenon persists longer than the forest fire season (Singh 2014).

In the southern districts (Dehradun, Tehri Garhwal, Pauri Garhwal, Haridwar, Almora, Nainital, Champawat, and Udham Singh Nagar) of Uttarakhand, a map of the zone of FFR was generated by integrating parameters such as elevation, slope, Land Surface Temperature (LST), wind speed, rainfall, aspect, and land use land cover (LULC). These districts of Uttarakhand are the most sensitive to forest fires.

11.2   STUDY AREA

The districts of Uttarakhand (an Indian state) studied for fire risk mapping include Dehradun, Tehri Garhwal, Pauli Garhwal, Haridwar, Almora, Nainital, Champawat, and Udham Singh Nagar. The study area covers approximately 12,559.13 km^2 and lies between 28°40'0" N to 31°10'0" N and 77°38'0" E to 80°05'0" E, as shown in Figure 11.1. The area covered by forest is 64% of the total study area. This region has many different types of forest biomes, such as tropical moist deciduous, subtropical pine, tropical dry deciduous, Himalayan dry temperate, Himalayan moist temperate, subalpine, and alpine (Suresh Babu et al., 2016). These forest types develop at varying elevations and are influenced by a range of climatic conditions. Both higher and lower elevations have areas covered with tropical woods. From a physical point of view, the study area is split into two zones: the Terai region and the Siwalik region. The plain area has a tropical climate, whilst the hilly region area has a temperate environment. Temperatures in the study region vary from 7 to 38 degrees centigrade (FSI 2019).

11.3   DATA USED AND METHODOLOGY

The factors used in the study for the forest fire risk mapping are elevation, Land Surface Temperature (LST), aspect, rainfall, wind speed, slope, land use land cover (LULC), Normalized Difference Vegetation Index (NDVI), geology, road buffer, and Topographic Wetness Index (TWI). The elevation, slope, and aspect map was generated using ALOS PALSAR Dem (12.5-m resolution) data with the help of Arc GIS 10.4 software, and the Landsat 8 (OLI) (30-m resolution) satellite data was used to generate the LULC and NDVI map. The LST map is prepared by the MODIS/Terra Surface Reflectance 8-Day L3 Global 500m data. The Lithology and Road map was downloaded by the BHUKOSH portal. Rainfall data was taken from Indian metrological department (IMD), Pune. The

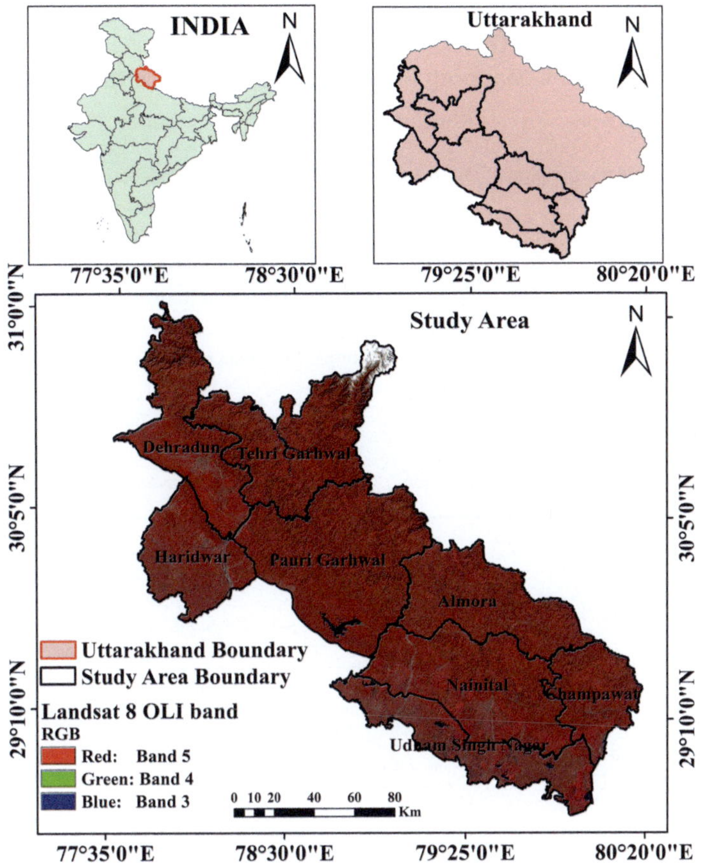

FIGURE 11.1 Study area of Southern District of Uttarakhand (Dehradun, Tehri Garhwal, Haridwar, Pauri Garhwal, Almora, Nainital, Champawat, and Udham Singh Nagar).

Wind speed map was downloaded from Global wind atlas site. All maps are generated with the help of Arc GIS 10.4 software. The resources of data are given below in the Table 11.1.

11.3.1 Factors that Contribute to Forest Fires

Forest fires pose a significant threat to biomass, driven by various factors including elevation, slope, aspect, Land Surface Temperature (LST), rainfall, wind speed, land use land cover (LULC), Normalized Difference Vegetation Index (NDVI), geology, road buffer, and Topographic Wetness Index (TWI). Fuel availability, mainly represented by vegetation, plays a crucial role in fire occurrence (Gupta & Nair, 2013). This chapter employs these factors to map forest fire zones using the AHP method, ensuring accuracy through validation against past fire occurrences. The resulting ROC curve, illustrated in Figure 11.2, confirms the reliability of the mapped zones.

11.3.1.1 Land Use Land Cover Factor

Lamat et al.'s (2021) analysis of the effects of human expansion on ecosystems emphasizes how crucial it is to comprehend the dynamics of land use land cover (LULC). The importance of LULC classes in determining a forest fire's vulnerability was highlighted by Baker et al. (1975). Eight different LULC categories are identified by this study: Bare ground, built-up area, forest, flooded vegetation, water body, snow/ice, agriculture plantation, and agriculture crop land (Figure 11.3). A framework for assessing ecological shifts and possible dangers, particularly with regard to forest fire

TABLE 11.1

Data Used for FFR Mapping in Southern District of Uttarakhand

Data Type	Resolution	Source
Landsat 8–9, OLI (operational land imager)	30 meters Raster	USGS (https://earthexplorer.usgs.gov/)
ALOS PALSAR	12.5 m Raster	Earth data https://www.earthdata.nasa.gov/
Wind Data	30 m Raster	Global wind Atlas (https://globalwindatlas.info)
MODIS	500 m Raster	https://earthdata.nasa.gov/
Rainfall	Gridded Point	https://dsp.imdpune.gov.in/
Geology, road map	Polygon	https://bhukosh.gsi.gov.in

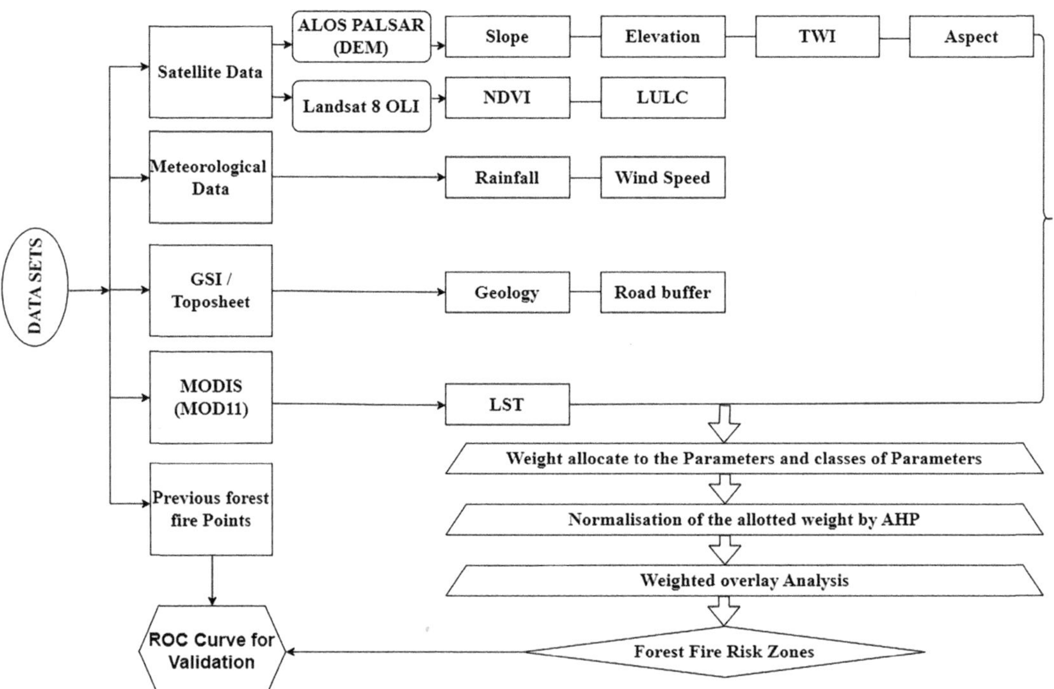

FIGURE 11.2   Flowchart of methodology for FFR mapping in the southern district of Uttarakhand.

hazards, is provided by this classification. The trends of habitat fragmentation, ecosystem resilience, and human influence by drawing boundaries between these categories is examined.

11.3.1.2 Elevation Factor

The impact of elevation on forest fires is well documented, with higher elevations correlating to reduced fire severity due to favorable climate conditions like higher humidity and lower temperatures, as noted by Rothermel (1983). This relationship between elevation and climate plays a

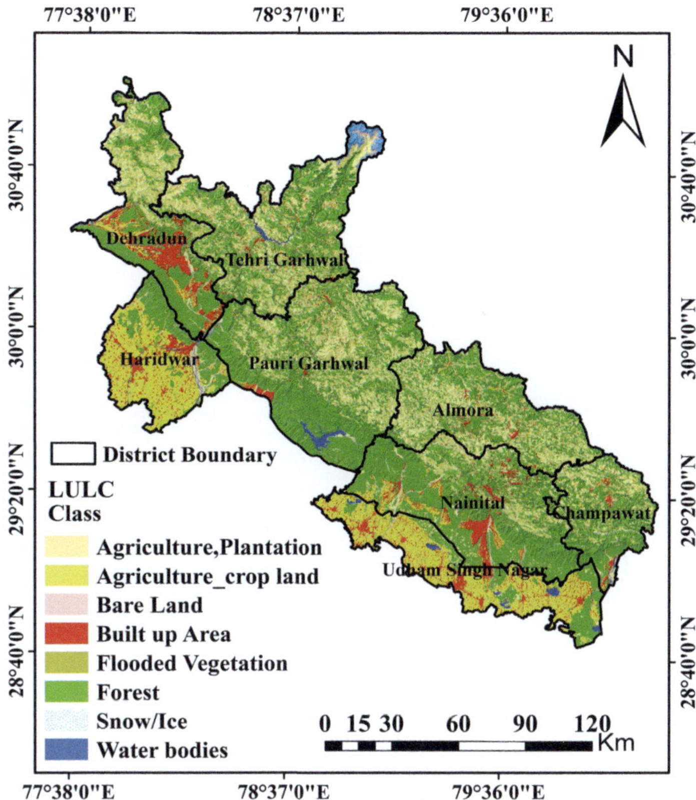

FIGURE 11.3 LULC map of southern district of Uttarakhand.

significant role in fire behavior. Figure 11.4 illustrates the elevation map of the research area, spanning 176 meters to 6609 meters. The elevation is categorized into five distinct classes: Very high (3413–6609 m), High (1943–3413 m), Moderate (1305–1943 m), low (685–1305 m), and Very low (176–685 m). This categorization provides a clear understanding of the elevation distribution across the study region and its potential influence on forest fire dynamics.

11.3.1.3 Land Surface Temperature (LST) Factor

The Land Surface Temperature (LST) emerges as an essential determinant of FFR, as affirmed by Bonora et al. (2013). Elevated temperatures accelerate evapotranspiration rates, leading to the desiccation of forest fuels such as dry leaves and dead trees, thus predisposing them to ignition. Within the study area, monthly average temperature fluctuations range from −7° to 38°C. Figure 11.5 delineates a spatial distribution wherein higher temperatures are observed in the southern region, while the northernmost part experiences comparatively lower temperatures. This disparity underscores the varying thermal conditions across the study area, accentuating the potential fire risk in zones exhibiting elevated temperatures.

11.3.1.4 Slope Angle Factor

The slope in the study area is one of the main factors for forest fires because generally, the fire increases upward rather than the downward direction of a hill. If the slope is high, it is very prone to forest fire because the moisture content is very low in the fuel on the steeper slope rather than the gentle slope (García-Ortega et al., 2011). In this chapter, this area has been classified into five classes such as Very low slope (<15°), Low slope (15–30°), Moderate slope (30–45°), High slope

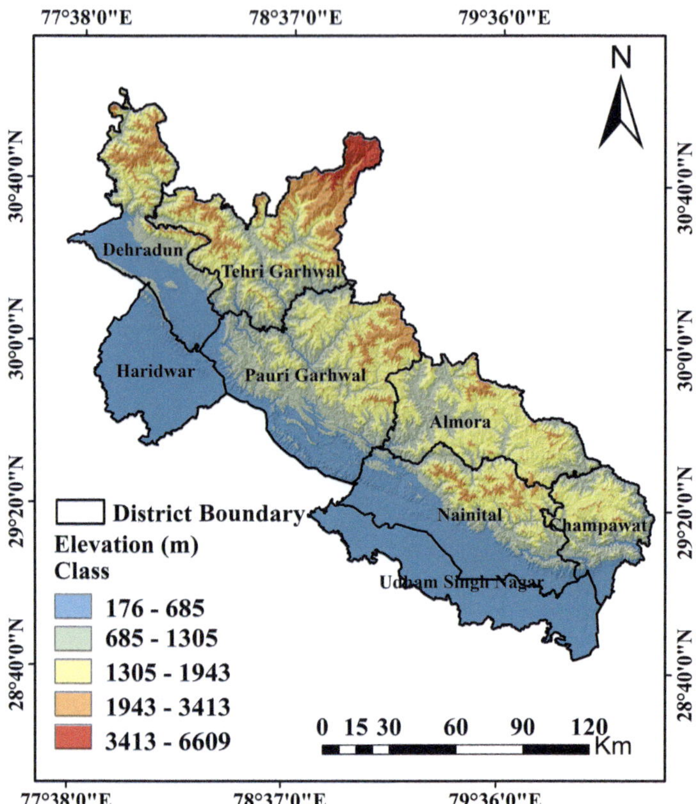

FIGURE 11.4   Elevation map of southern district of Uttarakhand.

(45–60°), and Very high slope (>60°), as shown in Figure 11.6. In the study area, forest fires occurred continuously on slopes that are 15–45°, which indicates a low to moderate slope of the area because most of the forest cover is at that slope angle.

11.3.1.5 Aspect

The Aspect map delineates the directional trend of slope angles, a pivotal factor influencing forest fire susceptibility. Particularly in the northern hemisphere, southern and eastern slopes receive amplified sunlight exposure, intensifying fuel desiccation—a key precursor to forest fires, as highlighted by Mukherjee & Raj (2014). In the study area, the Aspect map is classified into ten distinct classes, including Flat, North, Northeast, East, Southeast, South, Southwest, West, Northwest, and North. This classification scheme offers a nuanced understanding of slope orientations across the landscape, emphasizing regions prone to heightened solar radiation and consequent fuel drying. Figure 11.7 visually elucidates these aspect variations, aiding in the identification of high-risk areas vulnerable to forest fire ignition and propagation.

11.3.1.6 Wind Speed

Wind velocity stands out as the most critical weather factor influencing forest fires, as it dictates the rate and direction of fire spread. The combined impact of heightened temperatures and warm wind intensities exacerbates the risk of forest fires, as noted by Rathaur (2005). In the study area, wind speed (depicted in Figure 11.8) is classified into five distinct classes: Very low (0.11–1.54 knots), Low (1.54– 2.27 knots), Moderate (2.27–3.06 knots), High (3.06–4.19 knots), and Very high

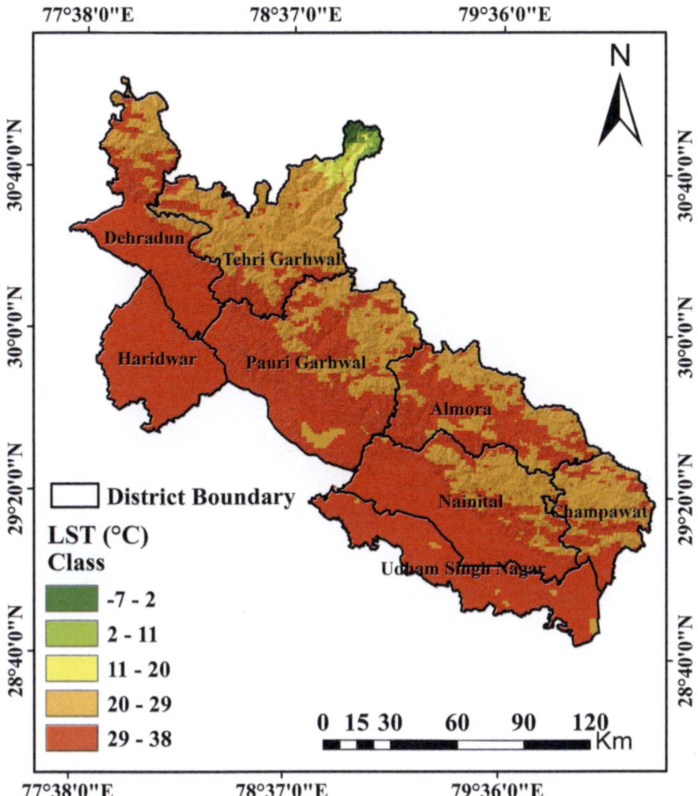

FIGURE 11.5　LST map of southern district of Uttarakhand.

(4.19–12.66 knots). Notably, the prevailing wind velocities within the study area predominantly range from low to moderate, suggesting a substantial influence on fire behavior and propagation dynamics. This classification offers valuable insights into wind patterns, aiding in the identification of regions at heightened risk of forest fire occurrence.

11.3.1.7　Rainfall

Rainfall plays a pivotal role in mitigating forest fire incidents by augmenting moisture levels in the land surface. Conversely, reduced rainfall contributes to drier conditions, heightening the risk of fire occurrence. The relationship between rainfall and forest fires is underscored by García-Ortega et al. (2011), who elucidated how increased vegetation growth resulting from higher rainfall can reduce fuel fire propagation. The rainfall map presented in Figure 11.9 has been divided into five separate groups: Very low (607–707 mm), Low (707–806 mm), Moderate (806–905 mm), High (905–1005 mm), and Very high (1005–1104 mm). These categories are based on the amount of rainfall in the research area. This classification offers insights into regional precipitation patterns, facilitating the identification of areas vulnerable to forest fires due to varying moisture levels. By comprehensively assessing rainfall dynamics, this classification scheme aids in the development of targeted fire prevention and management strategies.

11.3.1.8　TWI Factor

This factor is one of the significant controlling factors for forest fires that indicate water content of the area with the differences of height. It is a hydrological parameter, which indicates the soil wetness condition on the basis of areal topography and soil features (Kanga et al., 2017).

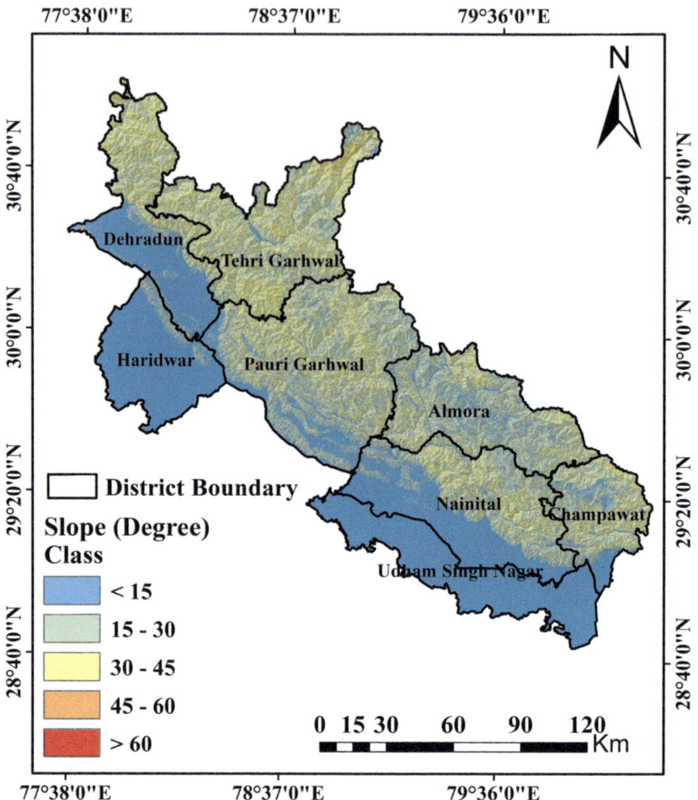

FIGURE 11.6 Slope map of southern district of Uttarakhand.

It is calculated using formula

$$TWI = \ln \frac{C}{\tan d}$$

where C represents the area of the upward slope in square meters and d represents the angle of the slope in radians. The TWI map is categorized into five classes: Very low wetness index (2.1–6.04), Low wetness index (6.04–9.97), Moderate wetness index (9.97–13.9), High wetness index (13.9–17.83), and Very high wetness index (17.83–21.76), as shown in Figure 11.10.

11.3.1.9 Normalized Differential Vegetation Index (NDVI) Factor

The factor of NDVI indicates the density of the vegetation based on the reflectance of the biomass present. This factor is significant for forest fire because certain vegetation types are more combustible for forest fire. (Hoseinpoor and Esmaeili, 2012). NDVI value varies from −1 to +1 in which the NDVI map has been categorized into five classes: Very low vegetation density (−0.2–0.08), Low vegetation density (0.08–0.17), Moderate vegetation density (0.17–0.23), High vegetation density (0.23–0.29), and Very high vegetation density (0.29–0.74), as shown in Figure 11.11.

11.3.1.10 Road Buffer Factor

Anthropogenic activities, such as road construction, significantly contribute to forest fires. Roads serve as potential ignition sources, increasing the likelihood of fires occurring nearby. To account for

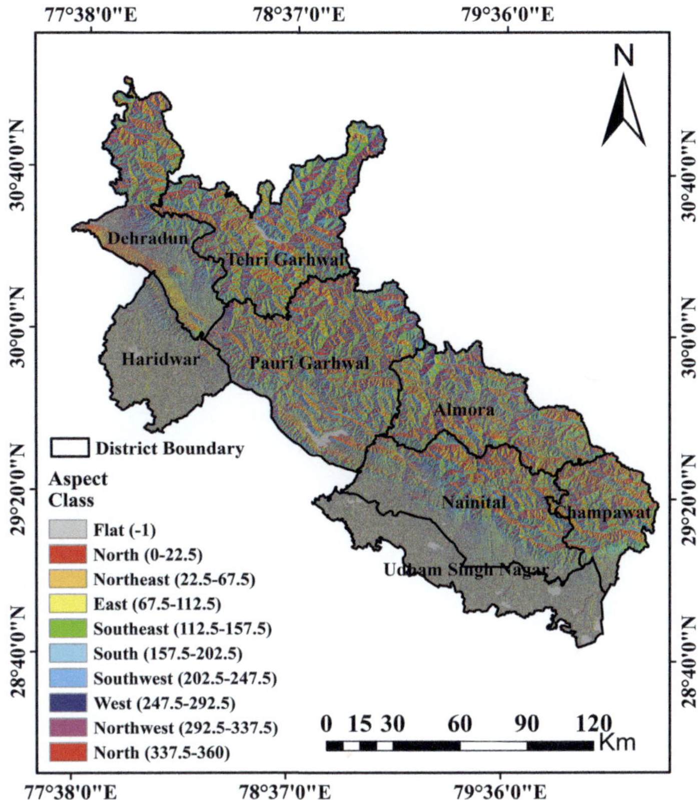

FIGURE 11.7   Aspect map of the southern district of Uttarakhand.

this risk, a road buffer map is generated with 50-meter intervals, as suggested by Akay & Şahin (2019), to identify areas with heightened susceptibility to fire in proximity to roads. The buffer map, depicted in Figure 11.12, is classified into five distinct categories based on distance to roads: Very low distance (< 100 m), Low distance (100–200 m), Moderate distance (200–300 m), High distance (300–400 m), and Very high distance (> 400 m). This classification enables the identification of areas where the danger of fires in forests is enhanced by their proximity to roads, thereby facilitating targeted fire prevention and management efforts.

11.3.1.11   Geology

The study area is found in the Siwalik and lower Himalayan ranges. The succession in the area is mainly shale, sandstone, limestone, and schist. The southern districts of Uttarakhand, situated within the Indo-Gangetic alluvial zone, is part of the Upper Gangetic Plain region. Mostly sedimentary rocks fall under the Siwalik ranges. Sandstone rock indicates good moisture condition because it has sufficient pore spaces, but schist rock does not indicate a good moisture condition. Thus, a forest situated on Schist rock is more susceptible to forest fire, as compared to sandstone. The geology map is shown in Figure 11.13.

11.3.2   ANALYTIC HIERARCHY PROCESS (AHP)

The Analytic Hierarchy Process (AHP) is a Multi-Criteria Decision Analysis (MCDA) technique. This methodology, created by Thomas L. Saaty in 1990, is based on cognitive awareness and

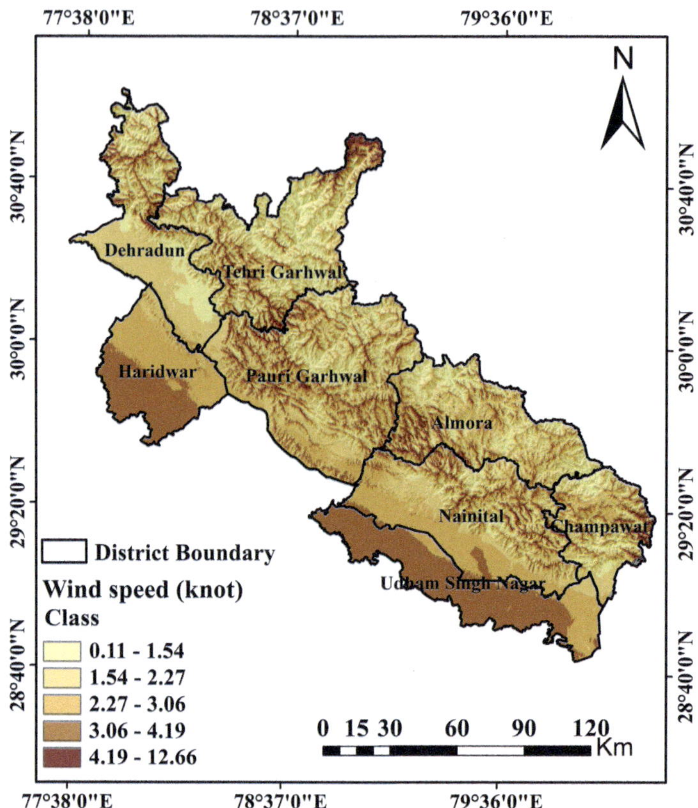

FIGURE 11.8   Wind map of the southern district of Uttarakhand.

mathematics, enabling individuals to compare a variety of alternatives to improve their decision-making. (Saaty, 1990, Saaty, 1987). Below are some steps followed for this method:

Step 1: The objective comes first in the hierarchy, followed by the criteria, and finally the options.

Step 2: Choose the factors based on how important they are to the goal using the Saaty scale (1990) to make a pair-wise comparison matrix (Table 11.2).

TABLE 11.2
Saaty Scale Explanation

Criteria weight	Explanation
1	Equal importance of factors
3	Moderate importance of factors
5	Strong importance of factors
7	Very strong importance of factors
9	Extreme importance of factors
(2, 4, 6, 8)	Intermediate importance of factors
(1/3, 1/5, 1/7, 1/9)	Values for inverse comparing factors

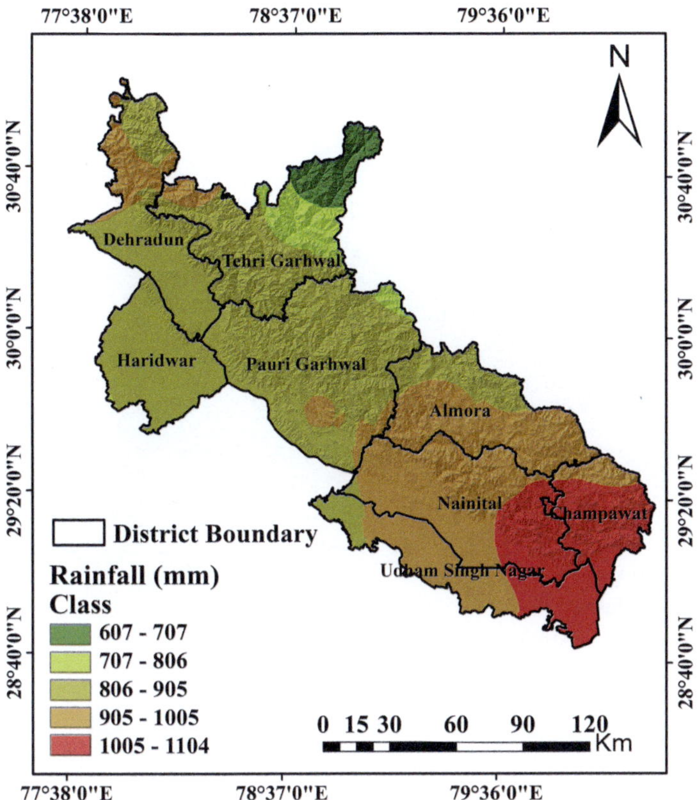

FIGURE 11.9 Rainfall map of the southern district of Uttarakhand.

1. Using the below equation (11.1), we can add up all the values in the columns of the pairwise comparison matrix in Table 11.4.

$$A_{mn} = \sum_{n=1}^{n} B_{mn} \tag{11.1}$$

where A_{mn} implies the summation of column value of Matrix, and B_{mn} implies the criteria weight.

2. To create the normalized pair-wise comparison matrix (Table 11.5), all given weight is divided by its addition of the row with the help of equation (11.2)

$$I_{mn} = \frac{B_{mn}}{A_{mn}} \tag{11.2}$$

where I_{mn} = pair-wise comparison matrix for normalization.

3. The addition of normalized row (Table 11.5) is divided by the number of factors to create the standard weight with the help of equation (11.3)

$$X_{mn} = \frac{\sum_{j=1}^{n} I_{mn}}{N} \tag{11.3}$$

where X_{mn} = standard weight.

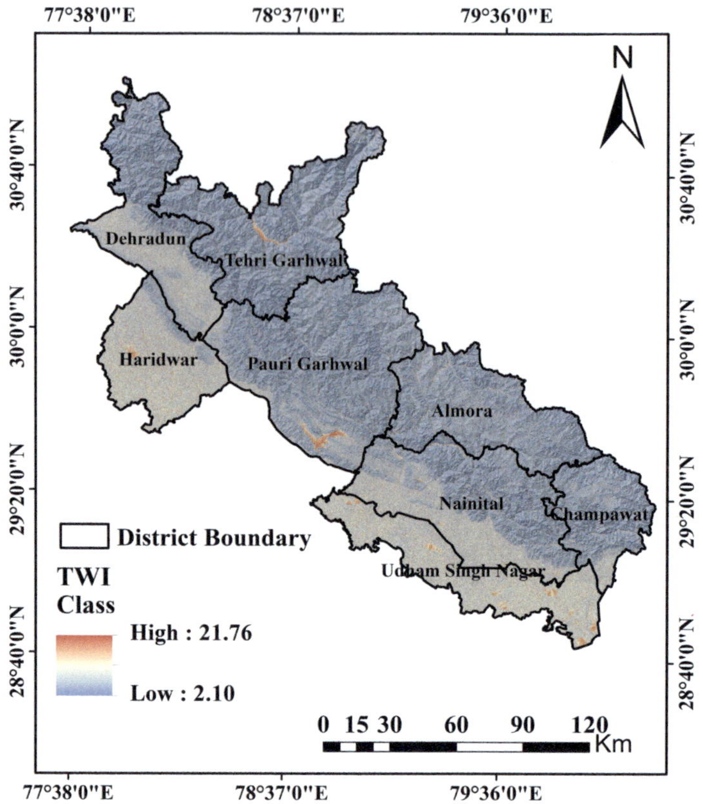

FIGURE 11.10 TWI map of the southern district of Uttarakhand.

4. λ_{max} The maximum value is required for the consistency ratio (CR).

$$\lambda_{max} = \left(\frac{\sum_{i=1}^{N} W_i * X_{mn_i}}{X_{mn_i}} \right) / N \tag{11.4}$$

Here, the λ_{max} is the Maximum eigen value.

W_i = Column of weight

X_{mn_i} = Corresponding weight of Column of weight

5. Consistency index (*CI*) was calculated using equation (11.5)

$$CI = \frac{\lambda_{max} - n}{n - 1} \tag{11.5}$$

where *CI* = Consistency Index, n = Number of Parameters.

6. Consistency ratio (Cr) is estimated with the help of equation (11.6)

$$Cr = \frac{CI}{RI} \tag{11.6}$$

where RI denote random inconsistency if $Cr <= 0.10$ means the suitable degree of consistency.

Random inconsistency is based on the number of criteria (Table 11.3) (Saaty, 1987; Saaty, 2008).

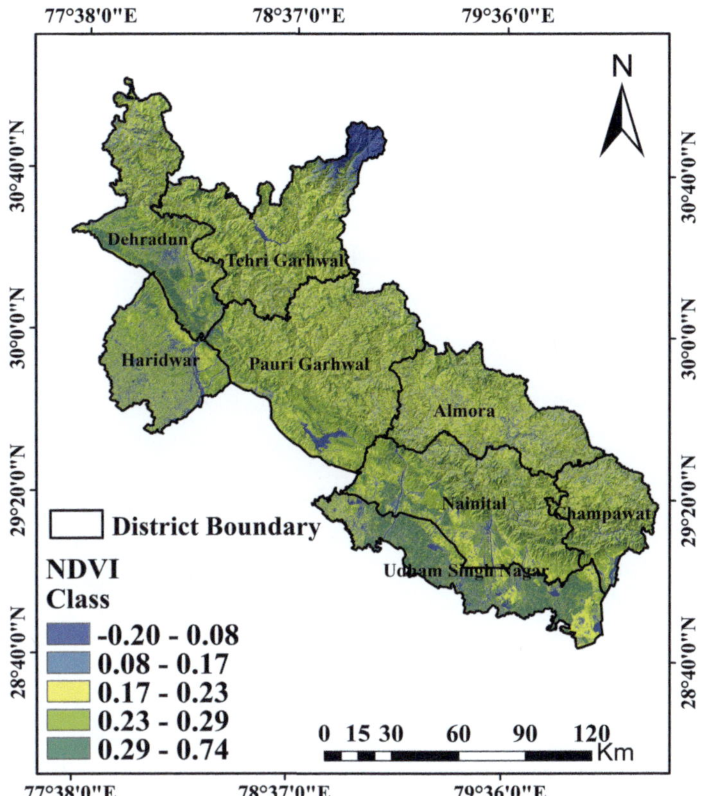

FIGURE 11.11 NDVI map of the southern district of Uttarakhand.

11.3.3 Forest Fire Risk Zone Assessment Using AHP Modal

The weighted overlay method is applied to detect the forest fire risk zone using the equation (11.7).

$$\text{FFI} = \Sigma W \times R \tag{11.7}$$

where FFI = forest fire index, W represents the designated weight for the thematic layers, while R represents the designated weight for the class within the thematic layer.

11.4 RESULTS AND DISCUSSION

11.4.1 AHP Calculation for Forest Fire Risk (FFR) Mapping

In order to conduct the analysis of the FFR zone in the Uttarakhand districts, 11 characteristics factors were utilized. These variables contain the following: LULC, NDVI, elevation (EL), (LST), slope (SL), aspect (AS), wind speed (WS), TWI, geology (GG), rainfall (RF), and road buffer (RB). All these factors are integrated for the AHP model.

Calculations are shown in the Tables 11.4, 11.5, and 11.6.

11.4.2 Forest Fire Risk Mapping

With the Arc GIS 10.4 software, the Forest fire danger zone is developed with the assistance of an integration of the Analytical Hierarchy Process (AHP) and the Geospatial approach. Every factor's

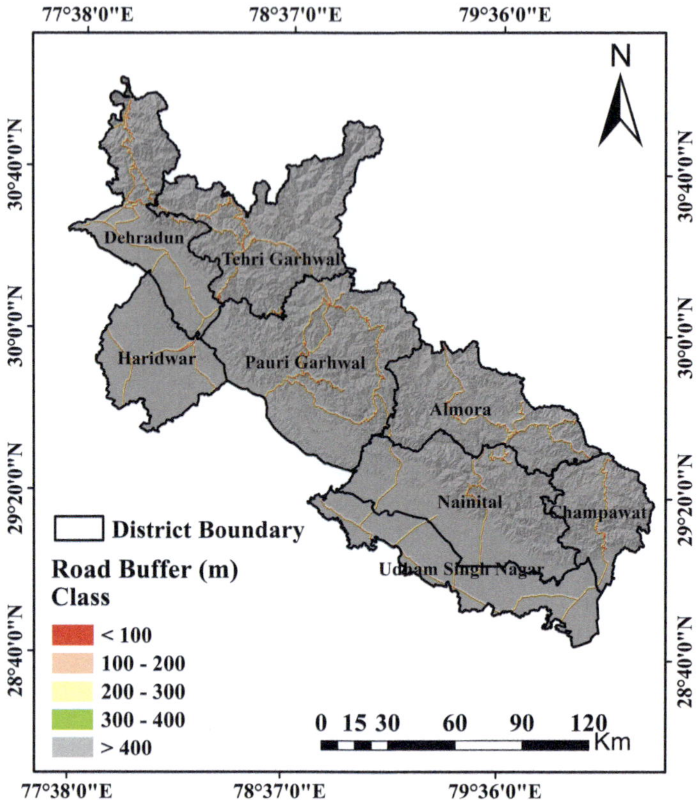

FIGURE 11.12   Road buffer map of the southern district of Uttarakhand.

thematic layer is assigned by weightage according to the literature review and estimation of the final weight of all factors using the weighted overlay tool in the Arc GIS software. Among the many methods are natural breaks, standard deviation, equal interval, geometrical interval, etc., in which the natural breaks method has been used to classify the FFR map. So, this map (Figure 11.14) is categorized into five classes (Table 11.7): No risk (60.9 km^2; 0.24%), Low risk (3035.7 km^2; 11.76%), Moderate risk (19512.5 km^2; 75.5%), High risk (3195.26 km^2; 12.38%), and Very high risk (6.89 km^2; 0.03%).

The most likely areas for fire lie in the districts of Pauri Garhwal, Nainital, and Haridwar, which have lower to moderate elevation, low to moderate slope, low to moderate rainfall, higher wind speed, high surface temperature, low to moderate moisture conditions, and contain high vegetation that faces northward.

11.4.3 VALIDATION OF FFR MAP

The FFR zone mapping was verified by 225 previous fire sites that occurred in 2020 and 2022, which can be downloaded from NRSC (ISRO) (https://bhuvan-app1.nrsc.gov.in/disaster/disaster. php?id=fire#). Using the data from the fire risk map, a ROC (Receiver Operating Characteristics) curve can be created. The AUC model defines the relationship between the false-positive and true-positive rates as a curve, with values ranging from 0.5 to 1. If the value is closer to 1, this indicates that the model being applied for the research region is performing exceptionally well. For this

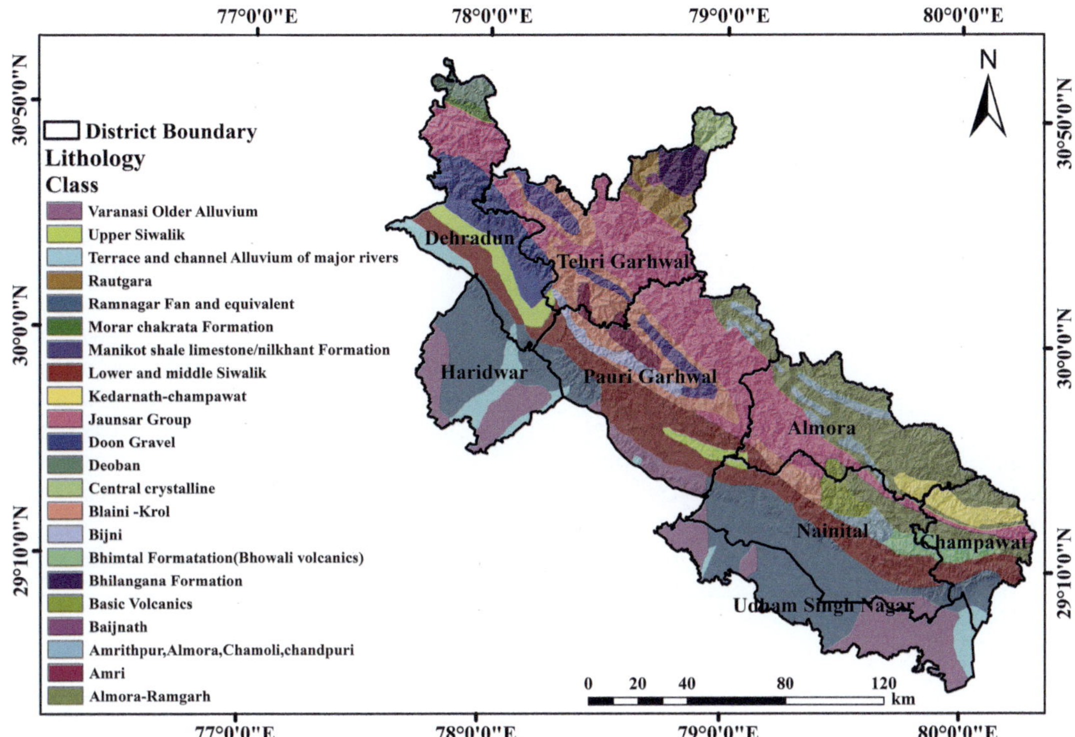

FIGURE 11.13 Geology map of the southern district of Uttarakhand.

TABLE 11.3

Random Inconsistency for the Criteria

n	2	3	4	5	6	7	8	9	10
RI	0.00	0.58	0.90	1.12	1.24	1.32	1.41	1.45	1.49

chapter, the AUC value is 82%, which implies good performance of the model, as shown in Figure 11.15a, b.

11.5 GNSS AND REMOTE SENSING

Combining GNSS with remote sensing technologies such as satellite photography and unmanned aerial vehicles (UAVs) can produce precise maps of impacted areas and fire boundaries, aiding resource distribution and decision-making. GNSSs are a complex network of satellites, ground stations, and user equipment that provides certain location, navigation, and timing services (Ferrazzoli et al., 2011; Belenguer-Plomer et al., 2019a, 2019b; Santi et al., 2022). These components operate collectively to deliver critical services in a variety of industries, including transportation, agriculture, telecommunications, and disaster management. The satellite constellation, which encompasses multiple satellites orbiting the Earth, sends signals containing timing and positional data (Santi et al., 2022; Jiang et al., 2017). Ground stations monitor and supervise satellite operations, ensuring the dependability and accuracy of transmitted signals. GPS receivers, GNSS modules, and antennas receive and interpret these signals to establish their precise location, velocity, and time.

TABLE 11.4

Pair-Wise Comparison Matrix

Thematic Layers	Weight	LULC	Road Buffer	NDVI	LST	Elevation	Slope	Wind Speed	Geology	Rainfall	Aspect	TWI
LULC	8	1.00	1.14	1.14	1.14	1.33	1.33	1.33	1.60	1.60	2.00	2.00
Road Buffer	7	0.88	1.00	1.00	1.00	1.17	1.17	1.17	1.40	1.40	1.75	1.75
NDVI	7	0.88	1.00	1.00	1.00	1.17	1.17	1.17	1.40	1.40	1.75	1.75
LST	7	0.88	1.00	1.00	1.00	1.17	1.17	1.17	1.40	1.40	1.75	1.75
EL	6	0.75	0.86	0.86	0.86	1.00	1.00	1.00	1.20	1.20	1.50	1.50
Slope	6	0.75	0.86	0.86	0.86	1.00	1.00	1.00	1.20	1.20	1.50	1.50
Wind Speed	6	0.75	0.86	0.86	0.86	1.00	1.00	1.00	1.20	1.20	1.50	1.50
Geology	5	0.63	0.71	0.71	0.71	0.83	0.83	0.83	1.00	1.00	1.25	1.25
Rainfall	5	0.63	0.71	0.71	0.71	0.83	0.83	0.83	1.00	1.00	1.25	1.25
Aspect	4	0.50	0.57	0.57	0.57	0.67	0.67	0.67	0.80	0.80	1.00	1.00
TWI	4	0.50	0.57	0.57	0.57	0.67	0.67	0.67	0.80	0.80	1.00	1.00
		8.13	9.29	9.29	9.29	10.83	10.83	10.83	13.00	13.00	16.25	16.25

TABLE 11.5

Normalized Pair-Wise Comparison Matrix

	LULC	Road Buffer	NDVI	LST	Elevation	Slope	Wind Speed	Geology	Rainfall	Aspect	TWI	Total Weight	Norm. Weight
LULC	0.12	0.12	0.12	0.12	0.12	0.12	0.12	0.12	0.12	0.12	0.12	1.35	0.12
Road Buffer	0.11	0.11	0.11	0.11	0.11	0.11	0.11	0.11	0.11	0.11	0.11	1.18	0.11
NDVI	0.11	0.11	0.11	0.11	0.11	0.11	0.11	0.11	0.11	0.11	0.11	1.18	0.11
LST	0.11	0.11	0.11	0.11	0.11	0.11	0.11	0.11	0.11	0.11	0.11	1.18	0.11
EL	0.09	0.09	0.09	0.09	0.09	0.09	0.09	0.09	0.09	0.09	0.09	1.02	0.09
Slope	0.09	0.09	0.09	0.09	0.09	0.09	0.09	0.09	0.09	0.09	0.09	1.02	0.09
Wind Speed	0.09	0.09	0.09	0.09	0.09	0.09	0.09	0.09	0.09	0.09	0.09	1.02	0.09
Geology	0.08	0.08	0.08	0.08	0.08	0.08	0.08	0.08	0.08	0.08	0.08	0.85	0.08
Rainfall	0.08	0.08	0.08	0.08	0.08	0.08	0.08	0.08	0.08	0.08	0.08	0.85	0.08
Aspect	0.06	0.06	0.06	0.06	0.06	0.06	0.06	0.06	0.06	0.06	0.06	0.68	0.06
TWI	0.06	0.06	0.06	0.06	0.06	0.06	0.06	0.06	0.06	0.06	0.06	0.68	0.06

Overall CR < 0.10 occur, which is (−0.18).

TABLE 11.6

Final Weights and Normalized Weight of Different Factors and Their Classes for Evaluation of Forest Fire Zone

Factors	Normalized Weight (%)	Classes	Final Weight
LULC	12	Agriculture, Plantation	4
		Built-up Area	5
		Water Bodies	1
		Forest	7
		Agriculture, Crop Land	3
		Bare Land	1
		Snow/Ice	1
		Flooded Vegetation	1
RD Buffer (m)	11	<100	8
		200	6
		300	5
		400	4
		>400	1
NDVI	11	−0.2–0.08	2
		0.08–0.17	2
		0.17–0.23	4
		0.23–0.29	7
		0.29–0.74	8
LST (Degree Celsius)	11	−7–2	1
		2–11	3
		11–20	5
		20–29	6
		29–38	8
EL (m)	9	176–685	8
		685–1305	6
		1305–1943	5
		1943–3413	3
		3413–6609	1
Slope (degrees)	9	<15	1
		15–30	3
		30–45	5
		45–60	6
		>60	7
Wind Speed (knots)	9	0.11–1.54	1
		1.54–2.27	3
		2.27–3.06	5
		3.0–4.19	6
		4.19–12.66	8

(Continued)

TABLE 11.6 (Continued)

Factors	Normalized Weight (%)	Classes	Final Weight
Geology	8	Jaunsar Group	3
		Deoban	2
		Doon Gravel	2
		Varanasi Older Alluvium	4
		Ramnagar Fan and Equivalent	3
		Lower and Middle Siwalik	3
		Terrace and Channel Alluvium of Major Rivers	3
		Blaini-Krol	2
		Almora-Ramgarh	1
		Basic Volcanics	1
		Amrithpur, Almora, Chamoli, Chandpuri	3
		Central Crystalline	6
		Bhilangana Formation	3
		Baijnath	3
		Manikot Shale Limestone/Nilkhant Formation	3
		Bhimtal Formation (Bhowali Volcanics)	4
		Kedarnath-Champawat	2
		Bijni	1
		Amri	1
		Rautgara	1
		Morar Chakrata Formation	1
		Upper Siwalik	4
Rainfall (mm)	8	607–707	7
		707–806	5
		806–905	4
		905–1005	3
		1005–1104	2
Aspect	6	Flat	1
		North	1
		Northeast	1
		East	3
		Southeast	5
		South	8
		Southwest	6
		West	2
		Northwest	2
		North	1
TWI	6	2.1–6.04	7
		6.04–9.97	6
		9.97–13.9	5
		13.9–17.83	2
		17.83–21.76	1

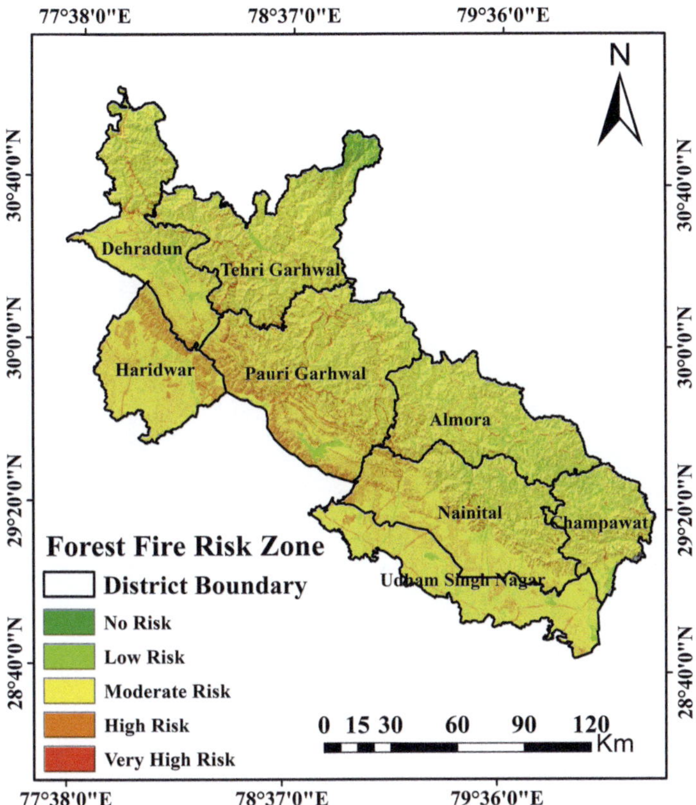

FIGURE 11.14   Fire risk zone map of the southern district of Uttarakhand.

TABLE 11.7

Distribution of the Forest Fire Risk Zone

	Area km²	Area %
No Risk	60.867	0.235816
Low Risk	3035.73	11.76127
Moderate Risk	19512.5	75.59689
High Risk	3195.26	12.37933
Very High Risk	6.8886	0.026688

11.6   CONCLUSION

This chapter investigates the impact of population growth on forest fires in tropical regions, focusing on the southern district of Uttarakhand, India. Employing an Analytic Hierarchy Process (AHP) model combined with Geographic Information System (GIS) techniques, the research aims to delineate fire risk zones. Eleven distinct factors including elevation, slope, aspect, Land Surface Temperature (LST), rainfall, wind speed, land use land cover (LULC), Normalized Difference Vegetation Index (NDVI), geology, road buffer, and Topographic Wetness Index (TWI) were combined for FFR assessment in this region. The resultant FFR map was classified into five zones: Very

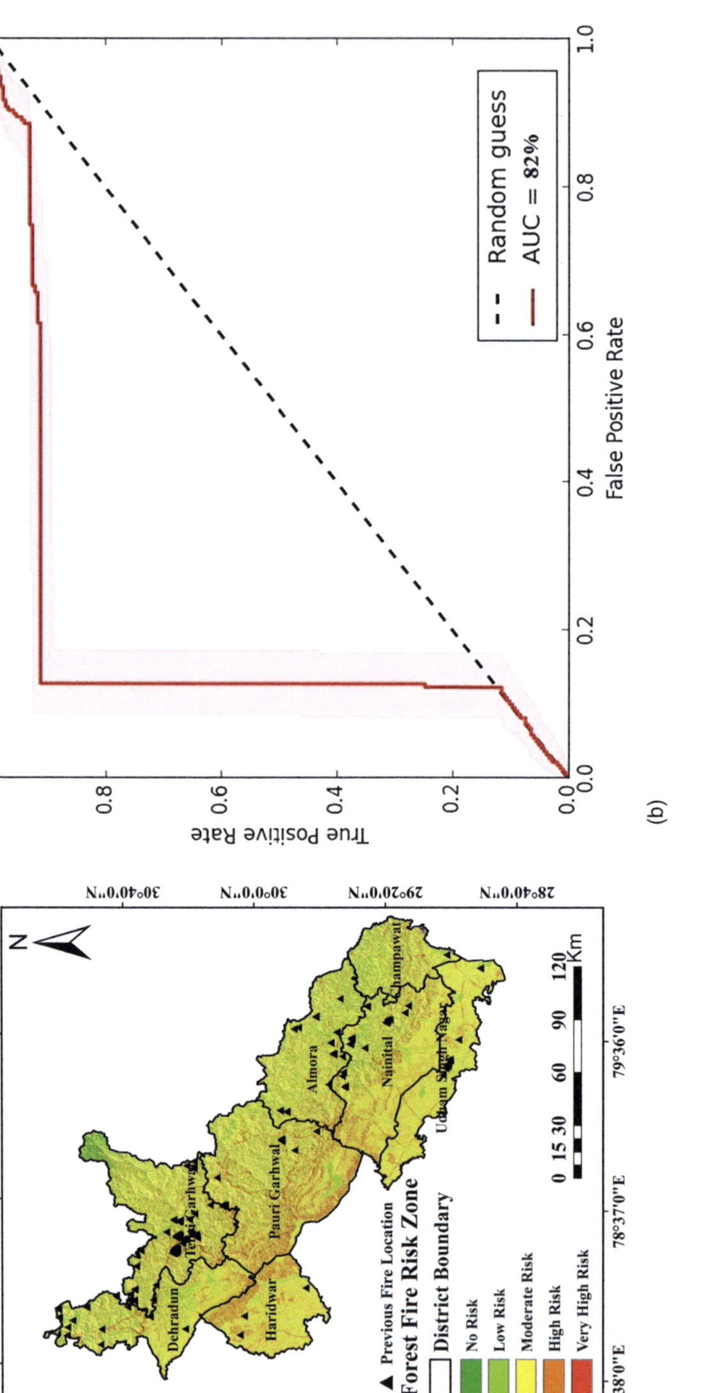

FIGURE 11.15　(a) Validation map of fire risk zone map of the southern district of Uttarakhand, and (b) ROC curve for validation of fire risk zone map.

high risk, High risk, Moderate risk, Low risk, and No risk zones. Notably, the Very high risk zone constituted approximately 0.02% of the total study area. To validate the accuracy of the final map, the Receiver Operating Characteristic (ROC) curve analysis was employed, yielding an accuracy of approximately 82%.

Moreover, the unique characteristics of GNSS technology enable the tracking of moisture levels, the calculation of fuel quantities, and the identification of alterations in surface reflectance. These capabilities greatly aid in the detection and control of forest fires.

The findings indicate that forest fire risk predominantly occurs in Pauri Garhwal. This comprehensive assessment provides valuable insights for mitigating losses attributed to forest fires, thereby contributing to effective forest management strategies.

REFERENCES

Ardakani, A., M. J. Valadanzooj, & A. Mansourian (2010). Spatial Analysis of Fire Potential in Iran Using RS and GIS. *Journal of Environmental Studies* 35, 7–9.

Akay, A. E., & Şahin, H. (2019). Forest Fire Risk Mapping by using GIS Techniques and AHP Method: A Case Study in Bodrum (Turkey). *European Journal of Forest Engineering*, 25–35. https://doi.org/10.33904/ejfe.579075

Baker, R. R., Dowdall, M. J., & Whittaker, V. P. (1975). The Involvement of Lysophosphoglycerides in Neurotransmitter Release; The Composition and Turnover of Phospholipids of Synaptic Vesicles of Guinea-pig Cerebral Cortex and Torpedo Electric Organ and the Effect of Stimulation. *Brain Research*, 100(3), 629–644. https://doi.org/10.1016/0006-8993(75)90162-6

Belenguer-Plomer, M. A., Chuvieco, E., & Tanase, M. A. (2019a). Temporal Decorrelation of C-Band Backscatter Coefficient in Mediterranean Burned Areas. *Remote Sensing*, *11*(22), 2661. https://doi.org/10.3390/rs11222661

Belenguer-Plomer, M. A., Tanase, M. A., Fernandez-Carrillo, A., & Chuvieco, E. (2019b). Burned Area Detection and Mapping Using Sentinel-1 Backscatter Coefficient and Thermal Anomalies. *Remote Sensing of Environment*, *233*, 111345. https://doi.org/10.1016/j.rse.2019.111345

Bonora, L., Claudio Conese, C., Marchi, E., Tesi, E., & Montorselli, N. B. (2013). Wildfire Occurrence: Integrated Model for Risk Analysis and Operative Suppression Aspects Management. *American Journal of Plant Sciences*, 04(03), 705–710. https://doi.org/10.4236/ajps.2013.43a089

Bui, D. T., Q. T. Bui, Q. P. Nguyen, B. Pradhan, H. Nampak, & P. T. Trinh (2017). A Hybrid Artificial Intelligence Approach Using GIS-Based Neural-Fuzzy Inference System and Particle Swarm Optimization for Forest Fire Susceptibility Modeling at a Tropical Area. *Agricultural and Forest Meteorology* 233, 32–44. https://doi.org/10.1016/j.agrformet.2016.11.002

Burrows, J. P., Platt, U., & Borrell, P. (2011). *The remote sensing of tropospheric composition from space.* Springer, New York.

Chhetri, S. K., & P. Kayastha (2015). Manifestation of an Analytic Hierarchy Process (AHP) Model on Fire Potential Zonation Mapping in Kathmandu Metropolitan City, Nepal. *ISPRS International Journal of Geo-Information* 4 (1), 400–417. https://doi.org/10.3390/ijgi4010400

Dente, L., Guerriero, L., Comite, D., & Pierdicca, N. (2020). Space-Borne GNSS-R Signal Over a Complex Topography: Modeling and Validation. *IEEE Journal of Selected Topics in Applied Earth Observations and Remote Sensing*, *13*, 1218–1233. https://doi.org/10.1109/JSTARS.2020.2975187

Eroglu, O., Kurum, M., Boyd, D., & Gurbuz, A. C. (2019). High Spatio-Temporal Resolution CYGNSS Soil Moisture Estimates Using Artificial Neural Networks. *Remote Sensing*, 11 (19), 2272. https://doi.org/10.3390/rs11192272

Eugenio, F. C., A. R. Dos Santos, N. C. Fiedler, G. A. Ribeiro, A. G. Da Silva, A. B. Dos Santos, G. G. Paneto, & V. R. Schettino (2016). Applying GIS to Develop a Model for Forest Fire Risk: A Case Study in Esp_irito Santo, Brazil. *Journal of Environmental Management* 173, 65–71. https://doi.org/10.1016/j.jenvman.2016.02.021

Feizizadeh, B., K. Omrani, & F. B. Aghdam. (2015). Fuzzy Analytical Hierarchical Process and Spatially Explicit Uncertainty Analysis Approach for Multiple Forest Fire Risk Mapping. *Journal for Geographic Information Science* 3, 72–80. https://doi.org/10.1553/giscience2015s72

Ferrazzoli, P., Guerriero, L., Pierdicca, N., & Rahmoune, R. (2011). Forest Biomass Monitoring with GNSS-R: Theoretical Simulations. *Advances in Space Research*, 47 (10), 1823–1832. https://doi.org/10.1016/j.asr.2010.04.025

FSI [Forest Survey of India]. (2019). India State of Forest Report 2019. Dehradun: Forest Survey of India, Ministry of Environment and Forests. Available online at https://fsi.nic.in/isfr2019/isfr-fsi-vol2.pdf

García-Ortega, E., Trobajo, M. T., López, L., & Sánchez, J. L. (2011). Synoptic Patterns Associated with Wildfires Caused by Lightning in Castile and Leon, Spain. *Natural Hazards and Earth System Sciences*, 11(3), 851–863. https://doi.org/10.5194/nhess-11-851-2011

Goldarag, Y. J., A. Mohammadzadeh, & A. S. Ardakani (2016). Fire Risk Assessment Using Neural Network and Logistic Regression. *Journal of the Indian Society of Remote Sensing*, 44 (6), 885–894. https://doi.org/10.1007/s12524-016-0557-6

Gubbi S (2003). Fire, Fire Burning Bright! Deccan Herald. Bangalore.India. http://www.wildfirefirst.info/images/wordfiles/fire.doc. Accessed Dec 2018.

Gupta, A. K., & Nair, S. S. (2013). *The role of ecosystems in disaster risk reduction*. United Nations University Press.

Hofmann, P. (2017). A Fuzzy Belief-Desire-Intention Model for Agent-Based Image Analysis. *Chapter 14 in Modern Fuzzy Control Systems and Its Applications*, edited by S. Ramakrishnan. London: IntechOpen. https://doi.org/10.5772/67899

Hoseinpoor, A., & Esmaeili, A. (2012). An AHP method for forest fire hazard modeling using remote abstract: Forest fires are one of the major natural risks in the north and west of Iran. In. July 2015.

Jiang, W., Yuan, P., Chen, H., & Zhang, K. (2017). Annual Variations of Monsoon and Drought Detected by GPS: A Case Study in Yunnan, China. *Scientific Reports*, 7(1), 5874. https://doi.org/10.1038/s41598-017-06095-1

Kanga, S., Tripathi, G., & Singh, S. K. (2017). Forest Fire Hazards Vulnerability and Risk Assessment in Bhajji Forest Range of Himachal Pradesh (India): A Geospatial Approach. *Journal of Remote Sensing & GIS*, 8(1), 25–40. https://www.researchgate.net/publication/325848317

Kushla, J. D., & Ripple, W. J. (1997). Forest Ecology and Management ELSEVIER The Role of Terrain in a Fire Mosaic of a Temperate Coniferous Forest. *Forest Ecology and Management*, 95, 97–107.

Lamat, R., Kumar, M., Kundu, A., & Lal, D. (2021). Forest Fire Risk Mapping Using Analytical Hierarchy Process (AHP) and Earth Observation Datasets: A Case Study in the Mountainous Terrain of Northeast India. *SN Applied Sciences*, 3(4), 1–15. https://doi.org/10.1007/s42452-021-04391-0

Maeda, E. E., A. R. Formaggio, Y. E. Shimabukuro, G. F. B. Arcoverde, & M. C. Hansen. (2009). Predicting Forest Fire in the Brazilian Amazon Using MODIS Imagery and Artificial Neural Networks. *International Journal of Applied Earth Observation and Geoinformation* 11 (4), 265–272. https://doi.org/10.1016/j.jag.2009.03.003

Mina, U., Dimri, A. P., & Farswan, S. (2023). Forest Fires and Climate Attributes Interact in Central Himalayas: An Overview and Assessment. *Fire Ecology*, 19(1). https://doi.org/10.1186/s42408-023-00177-4

Motazeh, A. G., E. F. Ashtiani, R. Baniasadi, & F. M. Choobar. (2013). Rating and Mapping Fire Hazard in the Hardwood Hyrcanian Forests Using GIS and Expert Choice Software. *Forestry Ideas* 19 (46), 141–150. https://forestry-ideas.info/issues/issues_Download.php?download=195

Mukherjee, S., & Raj, K. (2014) Analysis of forest fire of Meghalaya using geospatial tools. In: *15th ESRI India User Conference 2014*. https://www.esri.in/~/media/esri.india/files/pdfs/events/uc2014/proceedings/papers/UCP0053.pdf

Pourghasemi, H. R., M. Beheshtirad, & B. Pradhan. (2016). A Comparative Assessment of Prediction Capabilities of Modified Analytical Hierarchy Process (M-AHP) and Mamdani Fuzzy Logic Models Using Netcad-GIS for Forest Fire Susceptibility Mapping. *Geomatics, Natural Hazards and Risk* 7 (2), 861–885. https://doi.org/10.1080/19475705.2014.984247

Pourtaghi, Z. S., H. R. Pourghasemi, R. Aretano, & T. Semeraro. (2016). Investigation of General Indicators Influencing on Forest Fire and Its Susceptibility Modeling Using Different Data Mining Techniques. *Ecological Indicators*, 64, 72–84. https://doi.org/10.1016/j.ecolind.2015.12.030

Rathaur S. (2005) Fire risk assessment for tiger preybase in Chilla Range and vicinity, Rajaji National Park using remote sensing and GIS. M.Sc. thesis, Indian Institute of Remote Sensing, ISRO, Dept. of Space, Govt. of India, 4 Kalidas Road, Dehradun-248001, Uttarakhand, India.

Rawat, G. S. (2003). Fire risk asessment for fire control management in Chilla Forest Range of Rajaji National Park Uttaranchal (India). Thesis, International Institute for Geo-information Science and Earth Observation, Enschede, The Netherlands. https://www.iirs.gov.in/iirs/sites/default/files/StudentThesis/gsr_thesis.pdf

Robinne, F. N., Bladon, K. D., Miller, C., Parisien, M. A., Mathieu, J., & Flannigan, M. D. (2018). A Spatial Evaluation of Global Wildfire-water Risks to Human and Natural Systems. *Science of the Total Environment*, 610–611, 1193–1206. https://doi.org/10.1016/j.scitotenv.2017.08.112

Rothermel, R. C. (1983). How to Predict the Spread and Intensity of Forest and Range Fires. US Department of Agriculture, Forest Service, General Technical Report, INT-143.

Saaty, T. L., 1990. How to Make a Decision: The Analytic Hierarchy Process. *European Journal of Operational Research* 48 (1), 9–26. https://doi.org/10.1016/0377-2217(90)90057-I

Saaty, R. W. (1987). *The Analytic Hierarchy Process-What It Is and how it is used* (Vol. 9, Issue 5).

Saaty, T. L. (2008). Decision Making with the Analytic Hierarchy Process. In International Journal of Services Sciences (Vol. 1, Issue 1).

Saklani P. (2008). Forest fire risk zonation, a case study Pauri Garhwal, Uttarakhand, India [dissertation]. Enschede: *International Institute for Geo-information Science and Earth Observation.*

Santi, E., Clarizia, M. P., Comite, D., Dente, L., Guerriero, L., & Pierdicca, N. (2022). Detecting Fire Disturbances in Forests by Using GNSS Reflectometry and Machine Learning: A Case Study in Angola. *Remote Sensing of Environment*, 270, 112878. https://doi.org/10.1016/j.rse.2021.112878

Sharma, R. K., Sharma, N., Shrestha, D. G., Luitel, K. K., Arrawatia, M. L., & Pradhan, S. (2012). Study of Forest Fires in Sikkim Himalayas, India using Remote Sensing and GIS Techniques. *Climate Change in Sikkim–Patterns, Impacts and Initiatives*, May 2009, 233–244.

Singh, S. P. 2014. Attributes of Himalayan forest ecosystems: they are not temperate forests. *Proceedings of the Indian National Science Academy* 80: 221–233. https://doi.org/10.16943/ptinsa/2014/v80i2/55103

Somashekar RK, Nagaraja BC, Urs K (2008) Monitoring of Forestfires in Bhadra Wildlife Sanctuary. *Journal of the Indian Society of Remote Sensing* 36(1): 99–104.

Sonker, I., Tripathi, J. N., & Singh, A. K. (2021). Landslide Susceptibility Zonation using Geospatial Technique and Analytical Hierarchy Process in Sikkim Himalaya. *Quaternary Science Advances*, 4, 100039. https://doi.org/10.1016/j.qsa.2021.100039

Suresh Babu, K. V., Roy, A., & Prasad, P. R. (2016). Forest Fire Risk Modeling in Uttarakhand Himalaya using TERRA Satellite Satasets. *European Journal of Remote Sensing*, 49, 381–395. https://doi.org/10.5721/EuJRS20164921

Tripathi, D. K., Sahdev, S., & Kumar, M. (2017). Forest Fire Risk Zone Mapping using GIS-A Case Study. *International Journal of Computer Science*, 5(7), 9–15.

Vadrevu, K. P., Badarinath, K. V. S. (2009). Spatial Pattern Analysis of Fire Events in Central India—A Case Study. *Geocarto International* 24(2), 115–131.

Section 3

GNSS Applications for Climate Studies

12 Monitoring of Sea Level Variations Using GNSS Technology

Ashutosh Kumar Singh and Abhay Kumar Singh

12.1 INTRODUCTION

Global navigation satellite systems (GNSSs) are critical technologies that influence various aspects of everyday life. GNSSs extend beyond satellites orbiting Earth, comprising multiple constellations that send signals to master control stations and users worldwide (Gao and Shen, 2002). The systems include three primary segments: the space segment (satellites), the control segment (master control stations), and the user segment (devices and users) (Hofmann-Wellenhof, Lichtenegger, and Wasle, 2008). These satellites broadcast signals detailing their identity, time, orbit, and status.

GNSSs encompass several systems, such as the U.S. Global Positioning System (GPS), China's BeiDou Navigation Satellite System (BDS), Russia's GLONASS, and Europe's Galileo, alongside regional augmentation systems like Japan's Quasi-Zenith Satellite System (QZSS) and India's Regional Navigation Satellite System (IRNSS) (Parkinson and Spilker, 1996). These systems are renowned for their capabilities under all weather conditions, real-time availability, global reach, and high precision. They continuously transmit L-band signals and are extensively used for Positioning, Navigation, and Timing (PNT).

With advancements in navigation satellite systems and the increasing number of GNSS observation stations, the range of applications is expanding. Beyond PNT, GNSSs can also employ surface reflection signals for remote sensing (Garrison, Komjathy, and Zavorotny, 2007). Satellites send radio signals to the Earth, some of which are reflected off from the surfaces. These reflected signals, experiencing delays, provide valuable information about the paths of both direct and reflected signals, including variations in waveform, amplitude, phase, and frequency, with polarization changes linked to the properties of the reflecting surface (Larson and Nievinski, 2013). By combining delay measurements with receiver antenna position and medium information, surface roughness and properties can be determined using a technique known as GNSS-Reflectometry (GNSS-R).

In 1988, Hall and Cordey first proposed using GPS-reflected signals for scatterometry. ESA scientist Manuel Martin-Neira (Martin-Neira, 1993) suggested using GPS reflection measurements for altimetry, termed the Passive Reflectometry and Interferometry System (PARIS). In 1994, French scientists discovered that receivers could inadvertently lock onto sea surface-reflected signals during flight tests, though such multipath signals are typically rejected due to their impact on positioning accuracy (Zavorotny and Voronovich, 2000). In 1996, NASA's Langley Research Center scientists utilized dual-frequency GPS signals for sea forward scatter, compensating for ionospheric delays and improving traditional satellite altimetry (Komjathy and Born, 1996). Ground-based experiments revealed that traditional receivers struggled to track GNSS-reflected signals over long periods, leading to the development of new receiver types (Gleason and Gebre-Egziabher, 2009).

In 1997, ESA conducted the GNSS-R sea surface altimetry experiment, known as the PARIS altimeter Zeeland Bridge-I experiment in the Netherlands (Martin-Neira et al., 2001). In October 2000, NOAA's "Hurricane Hunter" aircraft, equipped with GNSS-R technology, flew into

Hurricane "Michael" to analyze reflections from the tropical cyclone sea surface, successfully obtaining wind speed data from the returned GPS signals. In 2003, the UK-Disaster Monitoring Constellation (UK-DMC) satellite used GNSS-R equipment to measure physical properties of the Earth's surface, including sea surface roughness (Hyatt et al., 2010). GNSS reflection signals in calm sea areas also provided high-precision altimetry results (Garrison and Katzberg, 2000; Cardellach et al., 2011).

The launch of the first space-borne GNSS-R Technology Demonstration Satellite-1 (TDS-1) in 2014 was a significant milestone, offering Delay-Doppler Map (DDM) data products and enabling spaceborne GNSS reflection measurements. Numerous scientific institutions have since conducted theoretical and experimental studies on GNSS-reflected signals, developed new GNSS-R receivers, and tested applications across various platforms, including ground, coastal, bridge, and aircraft environments (Jin et al., 2010). These studies have explored using GNSS-R for estimating sea surface properties such as height and wind speed and land surface characteristics, leading to significant progress and numerous findings (Lowe et al., 2002).

Monitoring sea level variations is crucial for understanding climate change, coastal erosion, and the health of the planet's oceans. Traditionally, this monitoring has relied on tide gauges and satellite altimetry. However, advancements in GNSS technology have provided a new, precise, continuous, and cost-effective method for measuring sea level changes (Lowe et al., 2002). This chapter explores the principles, methodologies, applications, and challenges of using GNSS technology to monitor sea level variations.

12.2 PRINCIPLES OF GNSS TECHNOLOGY

GNSS technology utilizes a network of satellites that send signals to receivers on Earth. By calculating the travel time of these signals, the receivers can determine their exact position in terms of latitude, longitude, and altitude. Figure 12.1 illustrates the GNSS signal reflection process, which includes the following components:

1. A GNSS satellite transmitting signals.
2. A GNSS receiver located in a coastal area or on a buoy, capturing the satellite signals.
3. Direct Signal Paths: (i) from the satellite to the sea surface, and (ii) from the satellite to the GNSS receiver.
4. Reflected Signal: from the sea surface to the GNSS receiver.
5. A traditional tide gauge near the GNSS receiver for comparison.
6. Sea Level: depicted as a horizontal line with waves.

The main GNSS systems currently in use include:

- GPS (United States)
- GLONASS (Russia)
- Galileo (European Union)
- BeiDou (China)

These systems provide global coverage and high precision, making them useful for geodesy, navigation, and environmental monitoring.

Various GNSS positioning techniques offer different levels of precision:

- **Standard Positioning Service (SPS)**: Provides accuracy within a few meters.
- **Precise Point Positioning (PPP)**: Achieves centimeter-level accuracy by using dual-frequency receivers and correcting for satellite and atmospheric errors (Gao and Shen, 2002).

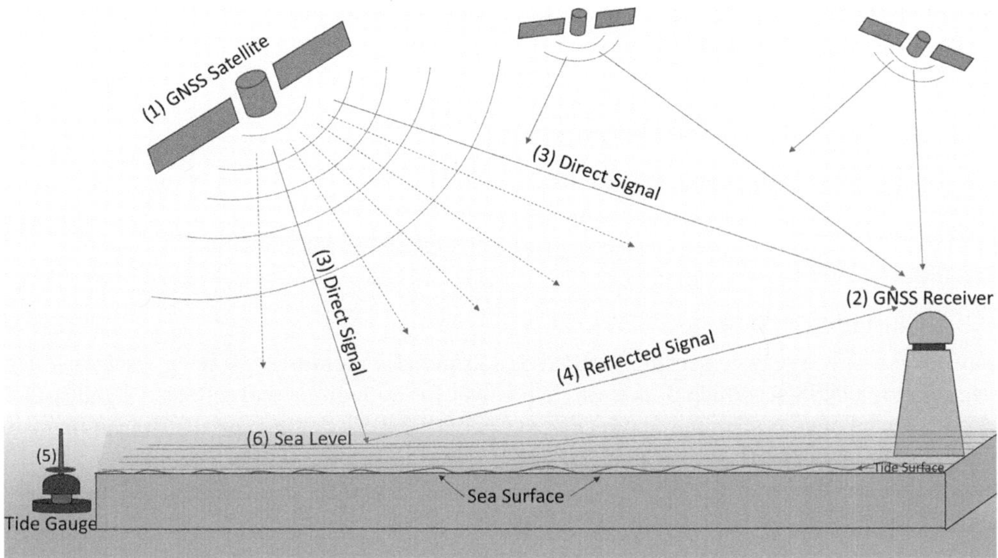

FIGURE 12.1 illustrates the GNSS signal reflection process, including the following elements: (1) A GNSS satellite in the sky, transmitting signals. (2) A GNSS receiver located on a coastal area or buoy, capturing the satellite signals. (3) Direct Signal paths: (i) dotted lines from the satellite to the sea surface, and (ii) dotted lines from the satellite to the GNSS receiver on the ground. (4) Reflected Signal: dotted lines from the sea surface to the GNSS receiver. (5) A traditional tide gauge positioned near the GNSS receiver for comparison. (6) Sea Level: represented as a horizontal line with waves indicating the ocean.

- **Differential GNSS (DGNSS)**: Enhances accuracy to sub-meter levels using ground-based reference stations that provide correction signals to GNSS receivers.
- **Real-Time Kinematic (RTK)**: Provides centimeter-level accuracy in real time using a base station and a rover receiver (Gleason and Gebre-Egziabher, 2009).

12.3 METHODOLOGIES FOR SEA LEVEL MONITORING USING GNSS

12.3.1 GNSS TIDE GAUGES

GNSS tide gauges integrate traditional tide gauge technology with GNSS receivers to measure the vertical position of the tide gauge relative to the Earth's surface. This allows for correcting land movements and accurately measuring sea level changes. Key components include:

- **GNSS Receiver**: Captures signals from GNSS satellites.
- **Antenna**: Mounted at a known height above the sea surface.
- **Tide Gauge Sensor**: Measures sea level relative to a fixed point on the tide gauge structure.
- **Data Logger**: Records GNSS and tide gauge data for analysis.

12.3.2 GNSS BUOYS

GNSS-equipped buoys provide a mobile solution for sea level monitoring. Floating on the sea surface, they directly measure sea level changes. They are especially useful in areas where installing fixed tide gauges is impractical. Key features include:

- **Floating Platform**: Stabilized to minimize wave-induced movement.
- **GNSS Receiver and Antenna**: Mounted on the buoy.

- **Telemetry System**: Transmits data to shore-based stations.
- **Power Supply**: Often solar-powered for continuous operation.

12.3.3 GNSS-R FOR SEA LEVEL MONITORING

GNSS-R (GNSS Reflectometry) is a promising technique for high-resolution sea level monitoring over large areas. It involves receiving both direct and reflected GNSS signals and analyzing the time delay and phase difference to determine sea surface height and characteristics.

12.3.3.1 Bistatic GNSS-R

Bistatic GNSS-R is a remote sensing method that accurately measures sea level variations. Unlike Interferometric GNSS-R, which uses a single receiver for both direct and reflected signals, Bistatic GNSS-R uses separate transmitter and receiver systems. GNSS satellites serve as transmitters, emitting signals that reflect off the sea surface. These reflected signals are then captured by dedicated receivers positioned on various platforms such as buoys, ships, coastal stations, or low Earth orbit satellites.

The core principle of Bistatic GNSS-R involves measuring the travel time and phase shift between direct GNSS signals and those reflected from the sea surface. By analyzing the delay and phase changes of the reflected signals, scientists can determine the height of the sea surface with high precision. This method provides continuous and real-time data, which is essential for monitoring dynamic sea level changes. GNSS-R can be implemented using various platforms:

- **Ground-Based GNSS-R**: Uses a GNSS receiver near the coast to capture reflected signals from the sea surface.
- **Airborne GNSS-R**: Utilizes GNSS receivers on aircraft to measure sea surface height over larger areas.
- **Spaceborne GNSS-R**: Employs satellites equipped with GNSS receivers to monitor sea level on a global scale.

By deploying GNSS-R on various platforms, extensive spatial coverage and continuous monitoring are achieved, significantly advancing our understanding and management of coastal and oceanic environments. GNSS-R complements traditional sea level monitoring methods such as tide gauges and satellite altimetry. While tide gauges provide precise local measurements and satellite altimetry offers global coverage, Bistatic GNSS-R delivers high-resolution, real-time data over extensive areas. This integration enhances the overall accuracy and reliability of sea level monitoring systems.

12.3.3.2 Interferometric GNSS-R (GNSS-IR)

Interferometric GNSS Reflectometry (GNSS-IR) is a technique that precisely measures sea level variations with high temporal resolution. Unlike conventional GNSS Reflectometry, which requires separate transmitters and receivers, GNSS-IR uses a single receiver to detect both direct and reflected GNSS signals. This method allows for the analysis of interference patterns created by the interaction of these signals, enabling accurate measurements of sea surface height.

The GNSS-IR technique relies on measuring the phase difference between direct signals from GNSS satellites and the signals reflected off the sea surface. These phase differences are directly related to the height of the sea surface, allowing for precise calculation of sea level changes. GNSS-IR systems are typically installed on coastal stations or floating buoys, providing continuous, real-time data essential for understanding dynamic sea level variations.

12.4 APPLICATIONS IN SEA SURFACE MONITORING

GNSS technology can be deployed on various platforms, including satellites, aircraft, ships, and buoys, to monitor sea surface variations. Some specific applications include:

Sea Level Monitoring: GNSS-R and GNSS-IR can measure sea surface height with high precision, enabling the detection of absolute sea level changes over time.

Wave Height Measurement: By analyzing the signal's amplitude and phase, GNSS can be used to determine wave heights and periods, providing valuable data for oceanographic studies.

Surface Current Mapping: GNSS-based systems can track the movement of floating buoys equipped with GNSS receivers to map surface currents, aiding in understanding ocean circulation patterns.

12.4.1 LONG-TERM SEA LEVEL TRENDS

GNSS-R utilizes the reflection of GNSS signals off the Earth's surface. Receivers, located on satellites, aircraft, or coastal stations, capture these reflected signals. The time delay and the change in signal characteristics (such as amplitude and phase) are analyzed to derive precise sea level measurements. GNSS-R provides high spatial and temporal resolution. The technique can offer continuous monitoring, essential for capturing short-term variations and long-term trends (Gao & Shen, 2002).

GNSS-R has been successfully employed in various scientific missions, including the UK's TechDemoSat-1 and the NASA CYGNSS mission, which focus on ocean and hurricane monitoring (Camps et al., 2016). As technology advances, GNSS-R is expected to play a crucial role in comprehensive climate monitoring systems, providing critical data for policymakers and researchers (Hyatt, Zuffada, & Cardellach, 2010).

12.4.2 MEASURING ABSOLUTE SEA LEVEL VARIATIONS

12.4.2.1 Tsunamis

Tsunamis, caused by underwater seismic activity, volcanic eruptions, or landslides, produce significant and rapid changes in sea level. Early detection and accurate measurement of these changes are critical for issuing timely warnings and reducing the impact on coastal communities. Detecting a tsunami involves real-time analysis of sea level measurements. The detection model must be relatively simple to be integrated into low-cost hardware with limited computational power, which is typically sufficient for the necessary calculations.

A tsunami manifests as a significant deviation from the expected tide, particularly in terms of its temporal characteristics. Understanding local tide patterns is crucial for accurate tsunami alerts. However, precise tide data is often unavailable, especially at new installations without historical records. Additionally, the algorithm must distinguish between tsunami-induced anomalies and those caused by meteorological events like storm surges, as well as other disturbances such as harbor activities or the presence of small boats, especially for radar sensors.

Extreme storm surges can also exceed sensor measurement ranges, causing false alarms. Therefore, an effective detection algorithm must isolate tsunami-specific events from other anomalies, with various models proposed in literature addressing this challenge (Trnkoczy, 2012; Bressan et al., 2013; Chierici et al., 2017; Wang and Satake, 2021). One of the first real-time implementations of such an algorithm was performed by NOAA in the DART buoys (Bernard & Meinig, 2011, Bernard & Titov, 2015), which used an interpolation of the previous three hours to determine the low-frequency tide. This approach enabled higher frequency data acquisition (one point every 15 seconds) to optimize satellite transmission time.

Numerous studies have explored real-time event analysis. For instance, Chierici et al. (2017) proposed a method for estimating local tides by removing tidal effects using a harmonics analysis of the least squares method. Bressan et al. (2013) introduced TEDA, which defined an alerting function based on the signal's instantaneous slope and the difference between two windows of varying lengths. Wang et al. (2020) developed a technique that adaptively decomposed time series data into intrinsic mode functions, allowing for automatic separation of tsunami signals from tidal signals, seismic activity, and background noise. In a subsequent paper, Wang and Satake (2021) demonstrated a tsunami detection algorithm using ensemble empirical mode decomposition to extract tsunami signals, simulating real-time operations for early warning systems.

GNSS technology can enhance tsunami monitoring and warning systems in the following ways:

Real-Time Data: GNSS-R and GNSS-IR provide real-time measurements of sea level variations, enabling early detection of tsunami waves.

High-Resolution Measurements: GNSS can capture small, rapid changes in sea level, providing detailed information about the tsunami's amplitude and propagation speed.

Integration with Other Systems: GNSS data can complement traditional tide gauges and deepocean pressure sensors, improving the overall accuracy and reliability of tsunami detection systems.

12.5 SIGNIFICANT STORM SURGES

Understanding the dynamics of sea level variation during storm surges is crucial for improving coastal resilience and preparedness. Storm surges occur when high winds from a storm system push water towards the coast, combined with the low pressure at the storm's center, which causes the water to rise. The magnitude of a storm surge depends on various factors, including storm intensity, forward speed, angle of approach, coastal topography, and the depth of the continental shelf. For instance, a slow-moving storm approaching perpendicularly to a shallow, gently sloping coast can generate a higher surge compared to a fast-moving storm hitting a steep coastline at an oblique angle (Jelesnianski, 1972).

The impact of storm surges on sea level can be profound, with surges raising sea levels by several meters. During events like Hurricane Katrina in 2005, storm surges reached up to 8.5 meters in some areas, leading to catastrophic flooding (Fritz et al., 2007). These extreme sea level variations are not only due to the surge itself but also due to wave setup, which is the increase in mean water level due to wave breaking, and astronomical tides. The interaction between these factors can lead to even higher sea levels, exacerbating the flooding potential (Pugh, 1987).

GNSS applications in storm surge monitoring include: *Continuous Monitoring*: GNSS-equipped buoys and coastal stations can provide continuous, real-time data on sea level changes during storms. *Early Warning Systems*: By integrating GNSS data with meteorological models, it is possible to enhance the accuracy of storm surge predictions and issue timely warnings to affected regions. *Post-Event Analysis*: GNSS data collected during storm events can be used for post-event analysis to improve future forecasting models and response strategies.

Thus, accurate monitoring and prediction of storm surges are essential for effective early warning systems. Tide gauges, radar altimeters, and satellite remote sensing are commonly used to measure sea level changes. Numerical models, such as the Sea, Lake, and Overland Surges from Hurricanes (SLOSH) model, are used to simulate storm surge dynamics and predict their impact (Jelesnianski et al., 1992). These models incorporate atmospheric data, sea surface temperatures, and topographic features to forecast surge heights and potential flooding areas. Recent advancements in technology have improved the accuracy of storm surge predictions. For example, ensemble forecasting, which uses multiple model runs with slightly varying initial conditions, provides a range of possible outcomes and improves the reliability of surge forecasts (Brecht et al., 2012). Additionally, integrating real-time data from tide gauges and satellites into predictive models enhances their precision and timeliness. Thus, monitoring these surges is essential for coastal hazard mitigation and emergency response.

12.6 ADVANTAGES OF GNSS-R OVER TRADITIONAL METHODS

Global Navigation Satellite System Reflectometry (GNSS-R) has emerged as a groundbreaking technique in the field of remote sensing and environmental monitoring. Unlike traditional methods that often rely on direct measurements or active sensing, GNSS-R utilizes reflected signals from GNSS satellites to infer various environmental parameters. This innovative approach offers several advantages, making it a superior alternative to traditional remote sensing methods in many applications.

One of the most significant advantages of GNSS-R is its cost-effectiveness. Traditional remote sensing systems, such as radar and optical sensors, typically require expensive dedicated satellites or aircraft equipped with specialized instruments. In contrast, GNSS-R leverages the existing infrastructure of GNSS constellations, including GPS, GLONASS, Galileo, and BeiDou. This utilization of already deployed satellites eliminates the need for additional costly infrastructure, leading to substantial savings in both initial setup and ongoing operational expenses.

Katzberg et al. (2006) highlight that using existing GNSS signals can significantly reduce the financial burden associated with environmental monitoring projects. GNSS-R also offers enhanced spatial and temporal coverage compared to traditional methods. The numerous satellites in GNSS constellations continuously transmit signals, providing near-global coverage and frequent revisit times. This continuous data stream enables real-time monitoring and more comprehensive spatial analysis, which is particularly beneficial for applications requiring high temporal resolution. For instance, GNSS-R has been shown to improve the accuracy of sea surface height measurements, which are crucial for oceanographic models and climate studies (Zuffada et al., 2005).

Another advantage of GNSS-R is its all-weather capability. Traditional optical remote sensing methods can be significantly affected by atmospheric conditions such as clouds, rain, and fog, which obstruct the view of the sensors. GNSS-R operates in the L-band frequency range, which is less affected by these atmospheric conditions, ensuring consistent data acquisition regardless of weather. This capability is particularly valuable in regions with frequent cloud cover or adverse weather, where traditional methods might fail to provide reliable data (Lowe et al., 2002).

The high precision and accuracy of GNSS-R in measuring various environmental parameters is another key advantage. Reflected GNSS signals can be used to derive information about surface roughness, soil moisture, vegetation biomass, and sea ice thickness with remarkable accuracy. Larson et al. (2008) demonstrated that GNSS-R could achieve soil moisture measurement accuracies comparable to those obtained by traditional methods such as microwave radiometry. The precise positioning capabilities of GNSS also contribute to the high accuracy of GNSS-R measurements, enhancing the reliability of the data collected.

The versatility of GNSS-R allows it to be used in a wide range of applications across different scientific disciplines. It is employed in hydrology for monitoring soil moisture and flooding, in oceanography for measuring sea surface height and wind speed, and in cryospheric studies for assessing sea ice and snow cover. Additionally, GNSS-R is used in agriculture for precision farming and crop monitoring, as well as in forestry for estimating biomass and forest health (Garrison et al., 1998). This multi-disciplinary applicability makes GNSS-R a valuable tool for integrated environmental monitoring and resource management.

Traditional remote sensing methods, particularly those involving active sensors like radar, can have environmental impacts due to their energy consumption and electromagnetic emissions. GNSS-R, being a passive remote sensing technique, relies on existing GNSS signals and does not require additional energy expenditure or emissions. This passive nature reduces the environmental footprint of GNSS-R, aligning with sustainable and eco-friendly monitoring practices (Chew et al., 2014).

The ability of GNSS-R to penetrate clouds and operate in harsh weather conditions makes it particularly valuable for polar regions, where traditional methods face significant challenges. For example, in cryospheric studies, GNSS-R is used to monitor sea ice thickness, snow cover, and glacier dynamics. The high precision of GNSS-R measurements allows for accurate monitoring of cryospheric changes, contributing to an improved understanding of polar processes and their impact on global climate (Garrison et al., 1998). Thus, GNSS-R has an advantage over traditional methods in high temporal resolution, wide coverage, its all-weather capability, and cost effectiveness.

12.7 CASE STUDIES

12.7.1 Tsunami Monitoring in the Pacific Ocean

Tsunamis are among the most devastating natural disasters, and the Pacific Ocean is particularly vulnerable due to its seismic activity, primarily along the Pacific Ring of Fire. Effective tsunami monitoring and early warning systems are essential for minimizing the impact of these catastrophic events on coastal communities. The Pacific Tsunami Warning Center (PTWC), established by the National Oceanic and Atmospheric Administration (NOAA), is a cornerstone of tsunami monitoring efforts in the Pacific region. It uses a sophisticated network of seismic stations, Deep-ocean Assessment and Reporting of Tsunamis (DART) buoys, and coastal tide gauges to detect and analyze potential tsunami events. These systems work in tandem to provide early warnings and mitigate damage caused by tsunamis (Bernard & Robinson, 2009 Bernard & Titov, 2015).

The effectiveness of this monitoring system was demonstrated during the 2011 Tōhoku earthquake and tsunami. The earthquake, with a magnitude of 9.1, struck off the coast of Japan, generating a massive tsunami that caused widespread devastation. Within minutes of the earthquake, seismic data indicated the potential for a significant tsunami, prompting the PTWC to issue warnings. DART buoys detected the tsunami as it traveled across the ocean, confirming its magnitude and helping refine warnings for other Pacific nations. This timely information was crucial for initiating evacuations and saving lives, although the disaster still resulted in significant loss (Titov & Synolakis, 1998).

Another significant event highlighting the importance of tsunami monitoring was the 2004 Indian Ocean tsunami, which, while not in the Pacific, prompted advancements in monitoring systems globally, including the Pacific. This tsunami was caused by a massive undersea earthquake off the coast of Sumatra, Indonesia. The lack of an effective early warning system at the time led to over 230,000 deaths across 14 countries. In response, international efforts, including those by the PTWC, were intensified to enhance tsunami monitoring and warning capabilities worldwide, leading to the deployment of more DART buoys and improved seismic networks (Bernard & Titov, 2015).

Despite advancements, challenges remain in tsunami monitoring. The vast expanse of the Pacific Ocean means some areas are sparsely covered by monitoring instruments, potentially delaying detection and warnings. Additionally, the unpredictable nature of tsunamis, influenced by underwater topography and wave interactions, complicates accurate forecasting. Efforts to address these challenges include expanding the network of DART buoys and improving data integration and analysis techniques. Deploying more buoys and tide gauges, particularly in under-monitored regions, aims to enhance detection capabilities and reduce response times (González et al., 1998).

International collaboration is essential for effective tsunami monitoring. The Intergovernmental Oceanographic Commission (IOC) of UNESCO coordinates the Pacific Tsunami Warning and Mitigation System (PTWS), facilitating information sharing and cooperation among Pacific Rim countries. This collaborative approach ensures all nations, regardless of technological capabilities, benefit from timely and accurate tsunami warnings (Weinstein & Lundgren, 2008). Public education and preparedness are also critical components of tsunami risk mitigation. Coastal communities must be aware of the risks and understand how to respond to warnings. Regular drills and educational programs reinforce the importance of swift evacuation and adherence to safety protocols (Titov & Synolakis, 1998).

12.7.2　STORM SURGE MONITORING DURING HURRICANE KATRINA

Hurricane Katrina, which struck the Gulf Coast of the United States in August 2005, remains one of the most catastrophic natural disasters in American history. The storm surge associated with Katrina caused unprecedented flooding and extensive damage, particularly in New Orleans. One of the most significant insights from the monitoring data was the extreme height and extent of the storm surge. In some areas, the surge exceeded 25 feet, far surpassing previous records. This immense surge was driven by Katrina's large size, slow movement, and the shallow bathymetry of the Gulf Coast, which allowed the surge to build up and reach inland areas with devastating force (Fritz et al., 2007).

The integration of satellite data with ground-based observations enhanced the accuracy of surge modeling and predictions. Satellite altimetry provided a broad overview of sea surface height anomalies, which, when combined with tide gauge data, allowed for more comprehensive surge maps. These maps were crucial for emergency managers and responders, enabling them to identify the most vulnerable areas and prioritize evacuations (Wang et al., 2020, Wang and Satake, 2021). Moreover, the data collected during Hurricane Katrina highlighted the importance of continuous monitoring and the need for robust infrastructure. Many tide gauges and monitoring stations were damaged or destroyed by the surge, underscoring the need for resilient and redundant systems that can withstand such extreme events (Irish et al., 2008). This realization has since driven improvements in monitoring technologies and the deployment of more resilient equipment.

12.8　CONCLUSION

The use of GNSS for remote sensing of sea surface variations offers a powerful tool for measuring absolute sea level changes caused by tsunamis, significant storm surges, and other coastal hazards. The high precision, real-time capabilities, and versatility of GNSS technology make it invaluable for enhancing coastal hazard mitigation efforts. By leveraging the global network of GNSS satellites, this approach provides continuous and real-time data, which is crucial for understanding the dynamics of sea level variations and their impacts on coastal regions.

One of the primary benefits of GNSS-based monitoring is its ability to deliver accurate measurements under various environmental conditions. Unlike traditional tide gauges and satellite altimetry, GNSS reflectometry (GNSS-R) can operate effectively regardless of weather conditions, ensuring consistent data acquisition. This capability is particularly valuable for monitoring in remote or harsh environments, where other methods may fail or provide less reliable data. Thus, the application of GNSS technology in monitoring sea level variations offers unparalleled accuracy, reliability, and comprehensive coverage. Its integration with existing observation networks and its resilience to environmental challenges make it an indispensable tool for modern climate science and coastal management. By integrating GNSS data with traditional monitoring systems and advanced modeling techniques, it is possible to improve early warning systems, reduce the impact of coastal disasters, and protect vulnerable communities.

REFERENCES

Bernard, E. N., & Robinson, A. R. (2009). *The Sea, Volume 15: Tsunamis.* Harvard University Press, Cambridge, MA, and London, England.

Bernard, E., & Meinig, C. (2011). History and future of deep-ocean tsunami measurements. *Proceedings of Oceans '11 MTS/IEEE Kona.* Waikoloa, HI, USA, September 19–22, 1–7, https://doi.org/10.23919/ OCEANS.2011.6106894

Bernard, E. N., & Titov, V. V. (2015). Evolution of tsunami warning systems and products. *Philosophical Transactions of the Royal Society A: Mathematical, Physical and Engineering Sciences*, 373(2053), 20140371.

Brecht, H. H., Jelesnianski, C. P., Shaffer, W. A., & Anderson, B. (2012). Advances in storm surge forecasting: The role of ensemble forecasting. *Journal of Coastal Research*, 28(3), 486–497.

Bressan, L., Zaniboni, F., & Tinti, S. (2013). Calibration of a real-time tsunami detection algorithm for sites with no instrumental tsunami records: Application to stations in Eastern Sicily, Italy. *Natural Hazards and Earth System Sciences Discussions*, 1, 2455–2493.

Camps, A., Park, H., Mateu, J., & Castellvi, J. (2016). Advances in GNSS-R Earth Observation and Experimentation. *IEEE Journal of Selected Topics in Applied Earth Observations and Remote Sensing*, 9(10), 4610–4622. https://doi.org/10.1109/JSTARS.2016.2575501

Cardellach, E., Ao, C. O., de la Torre Juarez, M., & Hajj, G. A. (2011). Carrier phase delay altimetry with GPS-reflected signals. *Geophysical Research Letters*, 38(21), L21102. https://doi.org/10.1029/2011GL049243

Chew, C. C., Small, E. E., & Larson, K. M. (2014). An algorithm for soil moisture estimation using GPS-interferometric reflectometry for bare and vegetated soil. *IEEE Transactions on Geoscience and Remote Sensing*, 52(1), 537–543.

Chierici, F., Embriaco, D., & Pignagnoli, L. (2017). A new real-time tsunami detection algorithm. *Journal of Geophysical Research: Oceans*, 122, 636–652.

Fritz, H. M., et al. (2007). Hurricane Katrina storm surge distribution and field observations on the Mississippi Barrier Islands. *Estuarine, Coastal and Shelf Science*, 74(1–2), 12–20.

Gao, G. X., & Shen, H. (2002). Global navigation satellite systems. *IEEE Transactions on Aerospace and Electronic Systems*, 38(1), 54–58.

Garrison, J. L., & Katzberg, S. J. (2000). Detection of ocean reflected GPS signals: theory and experiment. *IEEE Transactions on Geoscience and Remote Sensing*, 38(3), 1560–1573. https://doi.org/10.1109/36.843034

Garrison, J. L., Katzberg, S. J., & Hill, M. I. (1998). Effect of sea roughness on bistatically scattered range coded signals from the Global Positioning System. *Geophysical Research Letters*, 25(13), 2257–2260.

Garrison, J. L., Komjathy, A., & Zavorotny, V. U. (2007). Wind speed measurement using forward scattered GPS signals. *IEEE Transactions on Geoscience and Remote Sensing*, 45(2), 342–353. https://doi.org/10.1109/TGRS.2006.886199

Gleason, S., & Gebre-Egziabher, D. (2009). *GNSS Applications and Methods*. Artech House.

González, F. I., Milburn, H. B., Bernard, E. N., & Newman, J. C. (1998). Deep-ocean assessment and reporting of tsunamis (DART): Brief overview and status report. *Proceedings of the International Workshop on Tsunami Disaster Mitigation*.

Hofmann-Wellenhof, B., Lichtenegger, H., & Wasle, E. (2008). *GNSS – Global Navigation Satellite Systems: GPS, GLONASS, Galileo, and More*. Springer.

Hyatt, S. L., Zuffada, C., & Cardellach, E. (2010). Analysis of GNSS-Reflected Signals for the Remote Sensing of Ocean Roughness. *IEEE Transactions on Geoscience and Remote Sensing*, 48(5), 2545–2550.

Irish, J. L., Resio, D. T., & Ratcliff, J. J. (2008) The Influence of Storm Size on Hurricane Surge. *Journal of Physical Oceanography*, 38(9), 2003–2013. https://doi.org/10.1175/2008JPO3727.1

Jelesnianski, C. P. (1972). SPLASH (Special Program to List Amplitudes of Surges from Hurricanes): Vol. I. *NOAA Technical Report NWS 34*.

Jelesnianski, C. P., Chen, J., & Shaffer, W. A. (1992). SLOSH: Sea, lake, and overland surges from hurricanes. *NOAA Technical Report NWS 48*.

Jin, S. G., Park, J. G., Cho, J. H., & Park, P. H. (2010). Satellite-observed sea level change in the Korean Peninsula from GPS and GRACE. *Earth, Planets and Space*, 62(11), 943–951. https://doi.org/10.5047/eps.2010.11.005

Katzberg, S. J., Garrison, J. L., & Lin, B. (2006). Development of a low-cost GNSS instrument for soil moisture estimation. *IEEE Transactions on Geoscience and Remote Sensing*, 44(1), 104–115.

Komjathy, A., & Born, G. H. (1996). Sea surface height measurement using GPS signals. *Journal of Geophysical Research*, 101(C7), 17629–17640. https://doi.org/10.1029/96JC01648

Larson, K. M., & Nievinski, F. G. (2013). GPS snow sensing: results from the earthscope plate boundary observatory. *GPS Solutions*, 17(1), 41–52. https://doi.org/10.1007/s10291-012-0259-7

Larson, K. M., Gutmann, E. D., Zavorotny, V. U., Braun, J. J., Williams, M. W., & Nievinski, F. G. (2008). Can we measure snow depth with GPS receivers? *Geophysical Research Letters*, 36(17), L17502.

Lowe, S. T., LaBrecque, J. L., & Zuffada, C. (2002). First spaceborne observation of an Earth-reflected GPS signal. *Radio Science*, 37(1), 1–28.

Martin-Neira, M. (1993). A Passive Reflectometry and Interferometry System (PARIS): Application to ocean altimetry. *ESA Journal*, 17, 331–355.

Martin-Neira M., et al. (2001). The PARIS concept: an experimental demonstration of sea surface altimetry using GPS reflected signals, *IEEE Transactions on Geoscience and Remote Sensing*, 39(1), 142–150.

Parkinson, B. W. & Spilker Jr., J. J. (1996). *Global Positioning System: Theory and Applications, Vol. I, Progress in Astronautics and Aeronautics*, American Institute of Aeronautics and Astronautics, Washington, D.C.

Pugh, D. T. (1987). *Tides, Surges and Mean Sea-Level: A Handbook for Engineers and Scientists*. Wiley.

Titov, V. V., & Synolakis, C. E. (1998). Numerical modeling of tidal wave runup. *Journal of Waterway, Port, Coastal, and Ocean Engineering*, 124(4), 157–171.

Trnkoczy, A. (2012). Understanding and parameter setting of STA/LTA trigger algorithm. In Bormann, P. (Ed.), *New Manual of Seismological Observatory Practice 2 (NMSOP-2)* (pp. 1–20). Deutsches GeoForschungsZentrum GFZ: Potsdam, Germany.

Wang, Y., & Satake, K. (2021). Real-time tsunami data assimilation of S-Net pressure gauge records during the 2016 Fukushima earthquake. *Seismological Research Letters*, 92, 2145–2155.

Wang, Y., Satake, K., Maeda, T., Shinohara, M., & Sakai, S. (2020). A method of real-time tsunami detection using ensemble empirical mode decomposition. *Seismological Research Letters*, 91, 2851–2861.

Weinstein, S. A., & Lundgren, P. R. (2008). Finite fault modeling in a tsunami warning center context. *Pure and Applied Geophysics* 165, 451–474, https://doi.org/10.1007/s00024-008-0316-x

Zavorotny, V. U., & Voronovich, A. G. (2000) Scattering of GPS signals from the ocean with wind remote sensing application. *IEEE Transactions on Geoscience and Remote Sensing*, 38(2), 951–964. https://doi.org/10.1109/36.841977

Zuffada, C., Lowe, S. T., & Kroger, P. M. (2005). Sea surface height determination using GNSS reflections. *IEEE Transactions on Geoscience and Remote Sensing*, 43(2), 340–347.

13 Monitoring Climate Impacts on Himalayan Glacier Dynamics Using GNSS Radio Occultation

Atar Singh, Rajesh Kumar, Ramesh Kumar, Varsha Tanu, Shruti Singh, and Prity Singh Pippal

13.1 INTRODUCTION

The Himalayan River systems that drain from glaciated catchments are affected by changes in temperature, precipitation, and glacier response (Kaushik et al., 2023; Shrestha and Aryal, 2011). Three major river basins (Indus, Ganga, and Brahmaputra) are found in the Indian Himalayan Region, with many tributaries originating from the Indian and Nepalese regions of the Himalayas. These river basins mainly rely on the Himalayan cryosphere (Tayal & Sarkar, 2019). These river systems receive significant contributions from snow and ice melt, particularly during the lean period of the year. Any disruptions in the natural hydrological cycle due to climate change can profoundly affect these river systems' water availability and sustainability, impacting downstream communities and ecosystems. In this context, understanding the complex interactions between climate change, temperature, precipitation patterns, and the cryosphere in the Himalayas is of utmost importance.

Scientific research and monitoring efforts are essential to assess the region's current state, predict future changes, and develop effective adaptation and mitigation strategies. Climatic parameters can be monitored or measured at stations and on a space-based basis. Particularly in the Himalayan region, there are very few stations that need to be increased for the scientific research outcomes for future scenarios of modeling validation. Space-based monitoring helps model and validate the future aspects of the climate scenario. Space-based monitoring is quite popular, and remote sensing applications for monitoring the meteorological parameters from space are one of them.

Space-based monitoring is a refraction of GNSS radio signals as they travel through the atmosphere on nearly horizontal pathways. A collection of space-based GNSS occultation sensors for worldwide long-term observation of atmospheric change in temperature and other variables with high vertical resolution and precision represents a promising potential contribution to the Global Climate Observing System (GCOS) (Steiner et al., 2001), offering excellent vertical resolution and long-term stability that spans from the surface to the upper stratosphere (Steiner et al., 2013, 2020; Gleisner et al., 2022). To improve numerical weather forecasts, it is essential to have access to high-quality data, which is of utmost importance in light of climate change.

A technique based on positioning, navigation, and timing (PNT) called GNSS Radio Occultation (RO) emerged recently after the launch of the GPS (Kursinski et al., 1997). GNSS technology is about global positioning systems (GPS), the Global Navigation Satellite System (GLONASS), Galileo, and BeiDou, which have revolutionized our observations. By employing GNSS receivers and precise positioning techniques, we can track the movements of glaciers with unprecedented accuracy and examine their response to climate variables. Integrating GNSS with traditional

DOI: 10.1201/9781032712444-16

glaciological methods, such as field observations, remote sensing, and numerical modeling, offers a comprehensive approach to studying glacier dynamics. GNSS provides high-temporal-resolution data on glacier surface motion, allowing researchers to quantify ice flow velocities, detect changes in glacier volume, and estimate mass balance.

These measurements enable us to assess glaciers' overall health and behavior, helping us unravel the complex interplay between climate forcing and glacier response. In this chapter, we explore the applications of the Global Navigation Satellite System (GNSS) for monitoring and analyzing the impact of climate change on glacier dynamics. The main objective of this chapter is to bridge the gap between the pressing concern of climate change impacts on Himalayan glaciers and the innovative GNSS RO technology, ultimately contributing to the sustainable management of water resources and the safeguarding of downstream communities.

13.2 GLOBAL NAVIGATION SATELLITE SYSTEM (GNSS)

A network of satellites used for navigation is called the Global Navigation Satellite System. An independent geo-positioning system that utilizes satellite technology is a satellite navigation system. Four navigation systems available worldwide—GPS of the United States, GLONASS of Russia, BeiDou Navigation Satellite System of China, and Galileo of the European Union—are operational. While the Japanese Quasi-Zenith Satellite System (QZSS) improves GPS accuracy, the Indian Regional Navigation Satellite System (IRNSS) or Navigation with Indian Constellation (NavIC) aims to eventually become a regional navigation system. Using time signals of satellite navigation enables exact location determination (longitude, latitude, and elevation) with an accuracy of a few centimeters to several meters. It can be used for satellite-based location, navigation, and tracking of receiver-equipped objects. These signals also help with time synchronization and allow electronic receivers to detect local time precisely. Positioning, Navigation, and Timing (PNT), or precise location, accurate navigation, and timing, are the three main uses for satellite navigation. Although they can improve the precision and accuracy of location data in satnav systems, cellular and internet connections are unnecessary for the system to work.

13.3 GNSS RADIO OCCULTATION (RO): PRINCIPLES AND TECHNOLOGY

The GNSS-RO method measures the refraction of GNSS radio signals as they travel through the atmosphere on nearly horizontal routes. Figure 13.1 shows the GNSS radio signals observed by RO equipment on satellites in low-Earth orbit (LEO) as they rise or set at the Earth's limb. To obtain a vertical profile of refraction angle (α), the delay time and amplitude of the GNSS are recorded through an occultation event. Suppose we assume that the region around the occultation point is spherically symmetric. In that case, we may apply the Abel transform to determine the refractive index (n) profile vertically from the bending angles (Gleisner et al., 2022), which can be calculated using the given equation (13.1).

$$\ln n(x) = \frac{1}{\pi} \int_{x}^{\infty} \frac{\alpha(\alpha)}{\sqrt{\alpha^2 - x^2}} \tag{13.1}$$

Where α = influence parameter, and $x = nr$, with r as the radius of a point on the radio signal ray path, the microwave refractive index is a function of pressure (p), temperature (T), and water vapor partial pressure (q), and the equation (Gleisner et al., 2022) can be written as given below:

$$N = (n-1) \times 10^6 = k_1 \frac{p}{T} + k_2 \frac{q}{T} + k_3 \frac{q}{T^2} \tag{13.2}$$

where N is devoted to refractivity.

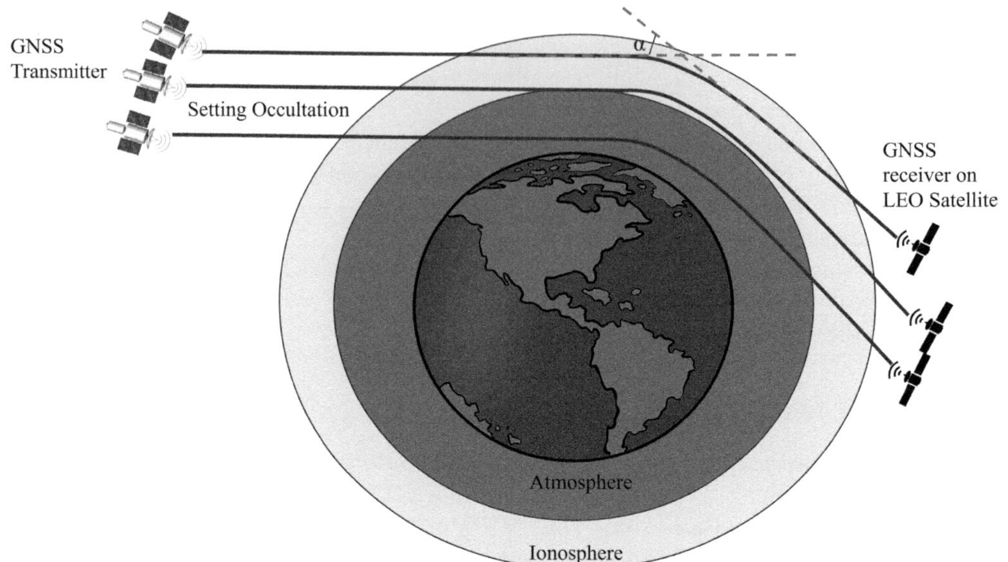

FIGURE 13.1   Geometry of the radio occultation observation (figure by L. Sánchez, GGOS).

Every day, a single instrument observes hundreds of occultation events unaffected by clouds or the Earth below and distributed relatively randomly throughout the planet. These observations span the upper stratosphere and the near-surface, with high vertical resolution and long-term stability (Steiner et al., 2013, 2020). As the time series of RO measurements lengthens, they become increasingly helpful in studying the climate and validating climate models. The vertical profile of the GNSS radio signal's refraction, or bending, angles is measured at the occultation event. The bending angle, which is nearly bias-free up to the upper stratosphere, is the parameter adjacent to the raw measurement. This is an exceptional feature of observations of the atmosphere and is used as the advantage of numerical weather prediction (NWP) (Bauer et al., 2014).

Instead of using recovered geophysical variables, most NWP centers now utilize bending-angle profiles because they can be used without bias correction (Hersbach et al., 2020; Long et al., 2017; Ho et al., 2020). Bending-angle data are unquestionably compatible with climate observation when combined with RO's all-weather capability and long-term stability across subsequent satellite missions. We emphasize that the RO bending angle is a more fundamental number than the RO geophysical retrievals, needing fewer presumptions, and is thus appropriate once minimizing tiny experimental biases over the stratosphere.

13.4 APPLICATION OF GNSS

GNSS applications are extensively active in rapidly acquiring information referring to specific fields. Various commercial uses involve consumers, transportation, GIS, precision agriculture, construction, marine mining, unmanned vehicle surveying, defense, climate-related aspects, aerial photogrammetry, etc. This technology integration into the consumer market has expanded, finding its way into an increasing array of products. GNSS receivers are now seamlessly integrated into smartphones to facilitate applications that display maps and highlight store and restaurant locations and optimal routes. Rail transportation employs GNSSs to track locomotives, rail cars, maintenance vehicles, and wayside equipment, and their positions are displayed at central monitoring stations. GNSSs are leveraged for aircraft navigation across take-off, flight, and landing in aviation.

In precision agriculture, GNSS-based applications support farm planning, soil sampling, tractor guidance, and crop assessment. Precise dispensation of fertilizers, pesticides, and herbicides curtails costs and environmental impacts. Mining management and waste movement are made more efficient with GNSS-equipped shovels and haul trucks, directed optimally by a computer-controlled dispatch system. GNSSs also prove valuable in climate research, measuring parameters like atmospheric water vapor (AWV), tropospheric delays, ozone levels, and ice cover. It significantly aids in climate change understanding by offering precise measurements of climate-linked factors within the atmosphere, oceans, and ice sheets. GNSSs also contribute to tracking the rise of sea levels and monitoring changes in sensitive areas such as the Arctic. The protection of the ozone layer against harmful UV radiation is also observed with GNSS technology. As GNSS technology advances, we can anticipate further innovative applications in climate monitoring.

13.5 DATA USED

The RO climate data sets are time series of zonal monthly means represented in a 2D latitude-height grid. The resolution of the data set is 200 m vertical profiling, and we have used monthly mean and zonal average data of EUMETSAT RO Meteorology Satellite Application Facility (ROM SAF) (https://rom-saf.eumetsat.int/index.php), and extending from 2017 to the present, to study the atmospheric changes over the last five years. Data on radio occultation is usually gathered by satellite-based equipment. A vertical profile of bending angle, refractivity, temperature, and humidity has been taken for the present study. These profiles are frequently offered throughout the atmosphere at different elevations.

The most well-known collection, which consists of spacecraft outfitted with GNSS receivers and radio transmitters, is the Global Navigation Satellite System (GNSS) Radio Occultation (RO) group. The data were obtained from the Radio Occultation Meteorology Satellite Application Facility (ROM SAF), provided by the Danish Meteorological Institute (DMI) in collaboration with several organizations like the European Centre for Medium-Range Weather Forecasts (ECMWF), the Institute of Space Studies of Catalonia (IEEC) in Barcelona, Spain, and the UK Met Office. ROM SAF utilizes atmospheric profile data from the Radio Occultation Meteorology Satellite Application Facilities for its operations. Sam and others (2016) conducted a case study of the glacier flow velocity for ground-based observation data. They used a DGPS (GNSS receiver) survey for the glacier velocity monthly as well as yearly. The field survey was carried out from September 2013 to September 2014 for yearly velocity and September to October 2013 for monthly velocity.

13.6 ATMOSPHERIC PROFILE

The bending angle of the refractive index profile gives the vertical profile of the atmosphere. The parameter adjoining the raw measurement is the bending angle, which is bias-free up to the upper stratosphere (Bauer et al., 2014; Gleisner et al., 2022). The RO bending angle can be a climate variable (Scherllin-Pirscher et al., 2015; Ringer and Healy, 2008). It primarily depends on air density, pressure, and temperature in the upper troposphere and above, where moisture is negligible. This indicates that the bending angle alteration results from the atmosphere's changing density.

Offline v1 Metop data RO ERA difference from 2017 to 2023 represented (Figure 13.2A) global bending angle. A sudden upliftment from mid-2019 has been recorded in the lower atmosphere (0–4 km), while refractivity in the upper atmosphere (40–50 km) showed a dramatic decrease (Figure 13.2B). These changes showed a slope of bending angle of +3.68%/decade; for the upper stratosphere (40–50 km), the slope change has been recorded as -1.73%/decade. This change in bending angle may be due to increased temperature (Figure 13.2C), which influences the atmosphere's physical properties. The bending angle depends upon the change in the density of the medium.

Temperature plots have been shown in Figure 13.2C, which indicate a sharp change in temperature difference for high altitude for RO ERA data sets. The humidity difference plot showed a minute decrease in the lower atmosphere from July 2019 onward (Figure 13.2D). This might be due to the effect of climate change and an evolving El Niño in 2019. The year 2019 started with a

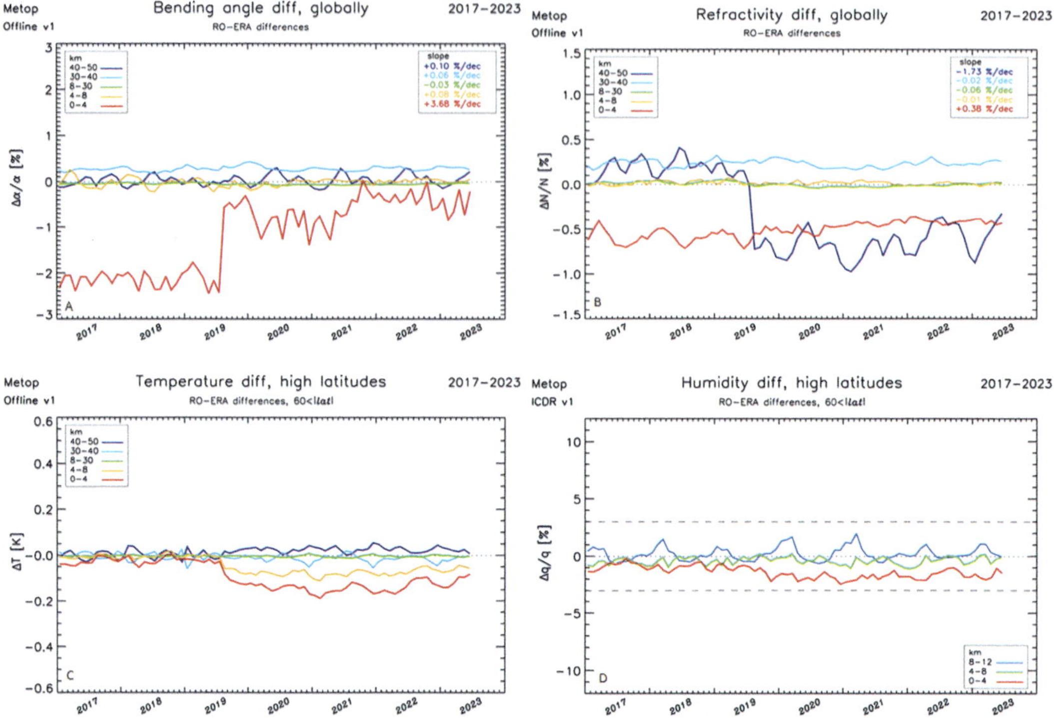

FIGURE 13.2   Radio Occultation deviation profile for different elevations, (A) RO bending angle at different levels represents the different color, (B) RO ERA refractivity of different elevations in a different color, (C) Temperature deviation profile of atmosphere from different color, (D) Humidity difference for elevation up to troposphere. The red line represents the average monthly grid value for the lower troposphere from 0–4 km, orange for 4–8 km, parrot green for 8–30 km, green for 30–40 km, and blue for 40–50 km.

moderate La Niña phenomenon, which generally has a cooling effect on global climate. Still, 2019 ended up being the fourth warmest year after the 2015, 2016, and 2017 records, which clearly shows the warming trend and climate change (Rao et al., 2019). According to Rao et al. (2019), sudden stratospheric warming has been recorded. Wang and Cai (2020) also showed that the Indian Ocean Dipole and Pacific El Niño are favorable for two consecutive years, 2019 and 2020, which might lead to a drastic change in the bending angle, refractivity, and temperature from July 2019 onward.

According to the previous study and the GNSS RO climate data analysis, climate change impacts the lower atmosphere more from 0–4 km. The high mountain region's changing climate impacts can be seen as melting glaciers.

13.7   MONITORING THE CHANGES IN THE CRYOSPHERE USING GNSS GROUND-BASED RECEIVER

GNSS receivers are used for annual change assessment for glaciers and ice sheet mass loss (Yang, 2016). Using global positioning system (GPS) observations for coastal uplift, regional and temporal fluctuations in coastal ice mass loss were observed, with results suggesting that warm atmospheric and oceanic conditions were the driving factors. Global warming, which is causing glaciers, sea ice, and snow cover to melt, leads to increased warming due to a reduced albedo caused by the increased range and period of the dark surface of the snow cover area (Slaymaker and Kelly, 2007). This is possible with the help of satellite altimetry and remote sensing instruments like NASA's ICESat-2 (ice, cloud, and land elevation satellite). Satellite altimetry is one of the methods used to monitor ice sheet movement over time. By applying GNSS to glaciology, scientists can better estimate the gross flow velocities, snowfall rate, and the isostatic adjustment related to changes in glacier mass. For example, researchers may now gather and investigate glacier flow data with ocean and atmospheric parameters with the help of real-time GPS (Hammond et al., 2010).

13.8 A CASE STUDY: SHAUNE GARANG GLACIER DYNAMIC AND FLOW VELOCITY

The glacier dynamics behave like the plastic deformation of the ice due to its basal movement. This is followed by sliding over the glacier bed, causing distortion in glacial mass. Direct measurement of basal slide is difficult and measured through the movement of subglacial crevasses and boreholes (Cuffey and Paterson, 2010). Glacial deformation can be directly measured with the help of borehole angle (Gudmundsson and Bauder, 1999). Glacier surface flow velocity can be calculated to measure the ground control points (GCPs). These GCPs are the natural or installed points over the glacier surface.

The current study used DGPS points as the GCPs to calculate the Shaune Garang glacier flow velocity. We have used the DGPS data from Sam and others for the Shaune Garang glacier (Sam et al., 2016). Data were collected in the glacier expedition in September and October of 2013. The monthly glacier velocity was calculated from 0.06 to 1.59 m per month. The glacier's velocity is represented in Figure 13.3, and DGPS points or GCPs are also represented on the glacial surface. The glacial field expedition was normally conducted at the end of the melting season, from mid-September to mid-October. In this context, data were collected for monthly velocity from September to October 2013 for the yearly velocity of Shaune Garang glacier GCPs through the stakes installed in September 2013 using GNSS-based DGPS. The exact stake location was recorded in the next year, September 2014, which gives the yearly displacement of the glacier surface. In this regard, Shaune Garang Glacier's surface velocity has been calculated to range from 0.49 to 19.18 m/year (Figure 13.4) for the different parts of the glacier. The glacier's velocity depends on the slope, topography, temperature, size, geometry, and basal condition. The higher velocity is found at the maximum slope area and over the accumulation area of the glacier.

In contrast, the minimum velocity is found around the flat and debris cover part of the glacier. Shaune Garang glacier's maximum velocity is observed at the high-altitude area or over the

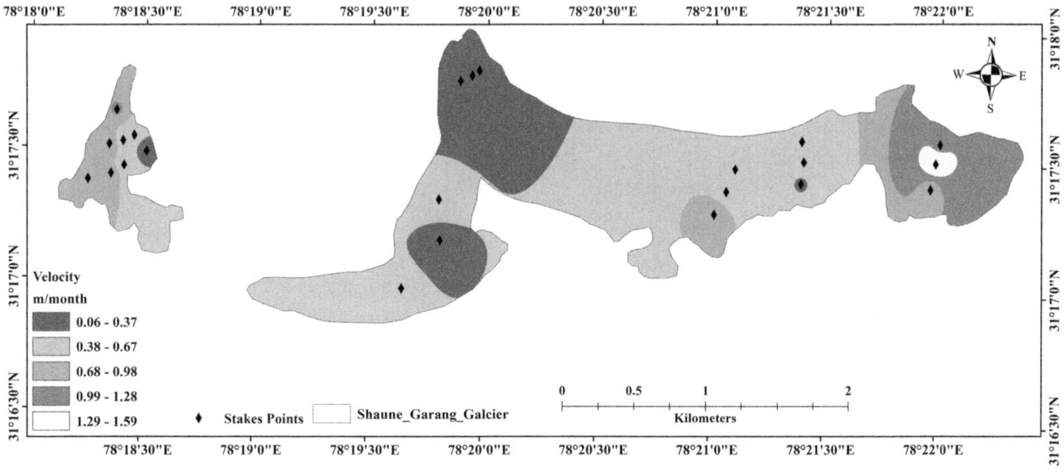

FIGURE 13.3 Shaune Garang glacier flow velocity map using the GNSS-based DGPS point monthly stake displacement.

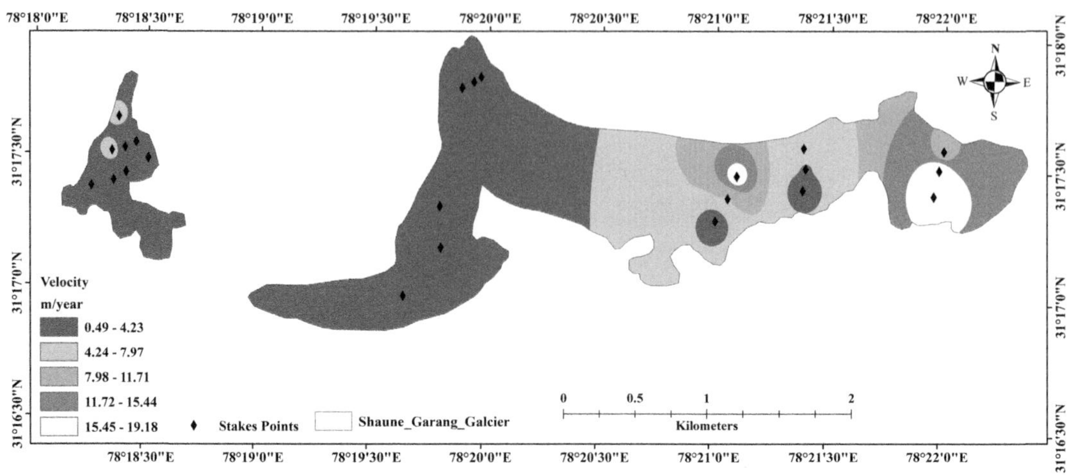

FIGURE 13.4 Shaune Garang glacier flow velocity map using the GNSS-based DGPS point yearly stake displacement.

accumulation zone of the glacier, and the minimum velocity is found over the debris cover zone of the glacier. Sam and others compared the velocity of the Shaune Garang glacier observed using a GNSS receiver and calculated it using pixel-based Landsat 8 data using COSI-Cor software.

13.9 LIMITATIONS OF THE STUDY

The GNSS occultation observation to represent the climate parameters measurement is at a primitive stage due to a lack of availability in remote and inaccessible locations. This is so that GNSS satellites can adequately communicate with receivers, which landscape features like mountains and trees can obscure. The ionospheric scintillation and tropospheric delays can impact the GNSS RO data. This data is sensitive to the glacier's mass distribution. As a result, adjustments in the glacier's surface elevation may not be attributable to climate change but may also result from calving or buildup. It is only sometimes possible to resolve minute changes in glacier dynamics using GNSS RO data due to the data's broad area of averaging, which can obscure changes at a smaller scale. There is substantial ambiguity regarding data accuracy, being a relatively new technology. This is especially true for glaciers tucked away in difficult-to-reach places. It is expensive to gather and process GNSS RO data, which might hamper the accessibility of the data to academics and decision-makers. Despite these limitations, GNSS RO data are helpful in researching glacier dynamics. It can be used with other data sources, such as altimetry and geodetic surveys, to provide a more comprehensive picture of glacier evolution.

13.10 CONCLUSION

The utilization of GNSS RO technology in monitoring climate impacts on Himalayan glacier dynamics represents a significant leap forward in understanding and responding to the changes occurring in this fragile ecosystem. This technology's precision and real-time data offer invaluable insights into the effects of climate change on glaciers, aiding scientists and policymakers in making decisions to mitigate its consequences. To understand the changes, this study used the EUMETSAT RO Meteorology Satellite Application Facility and ground satellite data for the Shune Garang glacier. Significant changes have been observed in flow velocity at Shaune Garang, as well as global vertical temperature difference, refractory difference, bending angle, and humidity difference.

Despite the limitations, GNSS RO data is essential for studying glacier movements and the consequences of climate change. It can be used with other data sources, such as altimetry and geodetic surveys, for a more complete view of glacier evolution. This approach has several advantages over traditional methods of glacier monitoring. First, it is non-invasive and requires no physical contact with the glacier. Second, it can monitor glaciers in remote and inaccessible areas. Third, it provides high-resolution data that can be used to study glacier dynamics in detail.

The use of GNSS RO to monitor Himalayan glacier dynamics is still in its early stages of development. However, it can potentially revolutionize how glaciers are studied and managed. By providing accurate and timely data on glacier dynamics, GNSS RO can help researchers better understand the effects of climate change on glaciers and develop effective strategies for glacier conservation. Monitoring climate impacts on Himalayan glacier dynamics using GNSS RO is a promising new approach to intensify the study of the changing climatic effects on glaciers of the Himalaya.

REFERENCES

Bauer, P., Radnóti, G., Healy, S., & Cardinali, C. (2014). GNSS radio occultation constellation observing system experiments. *Monthly Weather Review*, 142(2), 555–572.

Cuffey, K. M., & Paterson, W. S. B. (2010). *The physics of glaciers*. Academic Press.

Gleisner, H., Ringer, M.A., & Healy, S.B. Monitoring global climate change using GNSS radio occultation. *npj Climate and Atmospheric Science* 5, 6 (2022). https://doi.org/10.1038/s41612-022-00229-7

Gudmundsson, G. H., & Bauder, A. (1999). Towards an indirect determination of the mass-balance distribution of glaciers using the kinematic boundary condition. *Geografiska Annaler: Series A, Physical Geography*, 81(4), 575–583. https://doi.org/10.1111/1468-0459.00085

Hammond WC, Brooks BA, Bürgmann R, Heaton T, Jackson M, Lowry AR, Anandakrishnan S (2010). The scientific value of high-rate, low-latency GPS data, a white paper. http://www.unavco.org/community_science/science_highlights/2010/realtimeGPSWhitePaper2010.pdf

Hersbach, H., Bell, B., Berrisford, P., Hirahara, S., Horányi, A., Muñoz-Sabater, J., … & Thépaut, J. N. (2020). The ERA5 global reanalysis. *Quarterly Journal of the Royal Meteorological Society*, 146(730), 1999–2049.

Ho, S. P., Anthes, R. A., Ao, C. O., Healy, S., Horanyi, A., Hunt, D., … & Zeng, Z. (2020). The COSMIC/FORMOSAT-3 radio occultation mission after 12 years: Accomplishments, remaining challenges, and potential impacts of COSMIC-2. *Bulletin of the American Meteorological Society*, 101(7), E1107–E1136.

Kaushik, H., Soheb, M., Biswal, K., Ramanathan, A. L., Kumar, O., & Patel, A. K. (2023). Understanding the hydrochemical functioning of glacierized catchments of the Upper Indus Basin in Ladakh, Indian Himalayas. *Environmental Science and Pollution Research*, 30(8), 20631–20649. https://doi.org/10.1007/s11356-022-23477-9

Kursinski, E. R., Hajj, G. A., Schofield, J. T., Linfield, R. P., & Hardy, K. R. (1997). Observing Earth's atmosphere with radio occultation measurements using the Global Positioning System. *Journal of Geophysical Research: Atmospheres*, 102(D19), 23429–23465. https://doi.org/10.1029/97JD01569

Long, C. S., Fujiwara, M., Davis, S., Mitchell, D. M., & Wright, C. J. (2017). Climatology and interannual variability of dynamic variables in multiple reanalyses evaluated by the SPARC Reanalysis Intercomparison Project (S-RIP). *Atmospheric Chemistry and Physics*, 17(23), 14593–14629.

Rao, J., Garfinkel, C. I., Chen, H., & White, I. P. (2019). The 2019 new year stratospheric sudden warming and its real-time predictions in multiple S2S models. *Journal of Geophysical Research: Atmospheres*, 124(21), 11155–11174.

Ringer, M. A., & Healy, S. B. (2008). Monitoring twenty-first-century climate using GPS radio occultation bending angles. *Geophysical Research Letters*, 35(5). https://doi.org/10.1029/2007GL032462

Sam, L., Bhardwaj, A., Singh, S., & Kumar, R. (2016). Remote sensing flow velocity of debris-covered glaciers using Landsat 8 data. *Progress in Physical Geography*, 40(2), 305–321.

Scherllin-Pirscher, B., Syndergaard, S., Foelsche, U., & Lauritsen, K. B. (2015). Generation of a bending angle radio occultation climatology (BAROCLIM) and its use in radio occultation retrievals. *Atmospheric Measurement Techniques*, 8(1), 109–124.

Shrestha, A. B., Aryal, R. (2011). Climate change in Nepal and its impact on Himalayan glaciers. *Reg Environ Change* 11 (Suppl 1), 65–77. https://doi.org/10.1007/s10113-010-0174-9

Slaymaker, O., & Kelly, R. E. J. (2007). *The cryosphere and global environmental change (environmental systems and global change series)*, 1st ed. Wiley-Blackwell, Victoria.

Steiner, A. K., Hunt, D., Ho, S. P., Kirchengast, G., Mannucci, A. J., Scherllin-Pirscher, B., … & Wickert, J. (2013). Quantification of structural uncertainty in climate data records from GPS radio occultation. *Atmospheric Chemistry and Physics*, 13(3), 1469–1484.

Steiner, A. K., Kirchengast, G., Foelsche, U., Kornblueh, L., Manzini, E., & Bengtsson, L. (2001). GNSS occultation sounding for climate monitoring. *Physics and Chemistry of the Earth, Part A: Solid Earth and Geodesy*, 26(3), 113–124.

Steiner, A. K., Ladstädter, F., Ao, C. O., Gleisner, H., Ho, S. P., Hunt, D., … & Wickert, J. (2020). Consistency and structural uncertainty of multi-mission GPS radio occultation records. *Atmospheric Measurement Techniques*, 13(5), 2547–2575.

Tayal, S., & Sarkar, S. K. (2019). Climate change impacts on Himalayan glaciers and implications on energy security of the country. *TERI Discussion Paper*. The Energy and Resources Institute, New Delhi, India.

Wang, G., & Cai, W. (2020). Two-year consecutive concurrences of positive Indian Ocean Dipole and Central Pacific El Niño preconditioned the 2019/2020 Australian "black summer" bushfires. *Geoscience Letters*, 7(1), 1–9.

Yang, Q. (2016). Applications of Satellite Geodesy in Environmental and Climate Change. Graduate Theses and Dissertations. http://scholarcommons.usf.edu/etd/6440

14 Global Navigation Satellite Systems (GNSS) and Their Applications in Climate Monitoring and Modeling

Neelam Dahiya, Gurwinder Singh, Vishakha Sood, Apoorva Sharma, and Sartajvir Singh

14.1 INTRODUCTION

The Global Navigation Satellite System (GNSS) represents a network of satellites with worldwide coverage that helps users locate their position, speed, and time on the Earth's surface (Zhu et al., 2018). GNSS systems may be used for civil navigation and offer improved accuracy and integrity. The first-generation GNSS-1 combines ground-based or satellite-based augmentation systems (SBAS) with the two current systems, i.e., (a) Global Positioning System (GPS) (Zabalegui *et al.*, 2020); and (b) Global Navigation Satellite System (GLONASS) (Yan and Huang, 2019). Some of the major GNSS systems include (a) GPS with global coverage provided by the United States Department of Defense (Pipitone et al., 2018), (b) GLONASS with global coverage provided by Russia (Edokossi et al., 2020), (c) Galileo with global coverage provided by the European Union (Qu et al., 2022), (d) BeiDou Navigation Satellite System (BDS) with regional (Asia-pacific) provided by China (Qiu et al., 2023), (e) Navigation with Indian Constellation (NavIC) with regional (Indian subcontinent) provided by India (Xuerui et al., 2023), and (f) Quasi-Zenith Satellite System (QZSS) with regional Asia-pacific) provided by Japan. With the potential for future extensions, these technologies will improve the accuracy and dependability of GNSS services for users in their specific locations (Defourny et al., 2018).

GNSS systems serve numerous applications including precision of agriculture, marine navigation, and mapping of lands and waterways. Furthermore, GNSS is essential for applications like telecommunications network synchronization and geodynamics research. It has provided precise location, navigation, and timing data for more than 30 years. But using an inversion process, GNSSs may also function as an atmospheric sounding sensor (Kelly, Hehlen, and McGee, 2023). This number has been widely utilized in meteorology by being integrated by several meteorological services organizations into their numerical weather prediction (NWP) models. With the advent of continuous GNSS observations spanning over three decades, there has never been a better opportunity to leverage geodetic data analytics to optimize the value of these invaluable measurements for climate research (Rajabi et al., 2023). The Earth's climate is largely dependent on the water vapor trend (WVT), a fundamental climatic variable. It is the primary natural greenhouse gas and is the most significant known feedback mechanism that contributes to climate change (Kossieris, Asgarimehr, and Wickert, 2023). As with meteorological research (such as nowcasting apps), there's an increasing interest in determining the value of GNSS readings for climate studies and optimizing their advantages. This involves feeding and verifying climate models, as well as evaluating precipitable water vapor (PWV) trends and variability (Kim et al., 2023).

DOI: 10.1201/9781032712444-17

Previous literature covers a wide range of topics under GNSS including (a) radio occultation (RO), (b) ionospheric modeling, (c) PWV analysis, and (d) NWP assessment (Ghiasi et al., 2023). The calculation of WVT or NWP models has been explored in various past studies (Gallotti et al., 2023), analyzing NWP models using GNSS-estimated tropospheric parameters, which are then used to construct ZTD climatological or precipitation models. This chapter provides an overview of different types of GNSS and its emerging applications in various scientific domains, especially in climate change analysis. Some examples included reflectometry, ionosphere and Space weather, GNSS meteorology, positioning and orbiting, earthquake analysis, and land monitoring. This paper presents an overview of GNSS for Climate monitoring and the development of emerging applications. The focus of this chapter is to categorize various GNSS sensors. Section 14.2 presents numerous applications of GNSS in climate change studies. Section 14.3 provides a detailed assessment of GNSS's upcoming applications or future scope.

14.2 GNSS AND ITS TYPES

The GNSS is categorized into various forms, as shown in Figure 14.1. The brief description of each category is mentioned below.

14.2.1 GPS

The United States Space Force operated the GPS as the first constellation to be formed in orbit, having launched in 1978 and completed its first series of satellites in 1993 (Cahyadi et al., 2023). The GPS measures the distance from the satellites to enable exact position determination on Earth. The GPS enables you to navigate to and from locations on Earth by recording or creating locations from those locations. The System was not made accessible for public usage until the 1980s; initially, it was primarily intended for military use. The GPS segment is further divided into three segments, i.e., space, control, and user segment. The 24 satellites that make up the space segment orbit the Earth at a height of 12,000 miles. The signals can now reach a wider region because of their high altitude. A GPS receiver on Earth can always receive a signal from at least four satellites at any given moment, according to the satellites' orbital arrangement (Xing and Jin, 2023).

Every satellite sends out low-frequency radio signals on several frequencies that are uniquely coded, which the GPS receiver can detect. These coded signals serve the primary function of enabling the computation of the transit time between the GPS receiver and the satellite. The travel

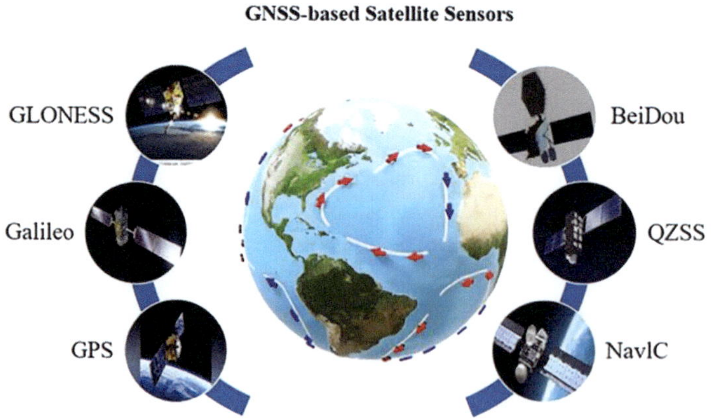

FIGURE 14.1 An overview of GNSS satellite sensors.

time multiplied by the speed of light determines the distance between the GPS receiver and the satellite. Given that these signals have minimal strength and cannot pass through solid objects, the control segment is used to provide correct orbit information (Kyuroson et al., 2023). It consists of four unmanned control stations and a single master control station. After receiving data from the satellites, the four unmanned stations relay it to the master control station for correction before sending it back to the GPS satellites. Users and their GPS receivers make up the user sector. There can never be an infinite number of users at once (Gyagenda et al., 2022).

14.2.2 GLONASS

GLObalnaya NAvigatsionnaya Sputnikovaya Sistema is known as GLONASS in Russian. It functions similarly to the US GPS and China's BDS. The Soviet Union was the first to establish the system, and Russia carried on the initiative. GLONASS development got underway in 1976 (Qin et al., 2022). Many satellites were added to the system between October 12, 1982, and October 12, 1995, thanks to various rocket launches; sadly, the nation's economy plummeted during this period, which also caused the system's health to deteriorate (White et al., 2022). GLONASS is presently run by Roscosmos State Corporation for Space Activities. It was initially created in the Soviet Union in the 1970s to rival GPS. There are now 24 GLONASS satellites in orbit, the first one being launched in 1982. The constellation was launched and finished in its entirety in 1995, and in 2010 it was entirely covered by satellites throughout Russia (Gyagenda et al., 2022).

14.2.3 GALILEO

Galileo was originally unveiled in 2011 and is run by the European Global Navigation Satellite Systems Agency. As of September 2024, it consists of 32 satellites in orbit. These satellites are designated: (a) E1 at 1575.42 MHz, (b) E5 at 1191.795 MHz, (c) E5a at 1176.45 MHz, (d) E5b at 1207.14 MHz, and (e) E6 at 1278.75 MHz when they broadcast along the L-Band band (Siemuri et al., 2022). The Galileo spacecraft made two flybys of Earth and collected close to 6000 multispectral photos of the planet. Galileo offers a different representation of our planet since none of the Earth's orbital remote sensing systems currently in use features four NIR filters. It effectively maps spectral changes caused by minerals, vegetative cover, and condensed water. These photos serve as a helpful foundation for comparative planetary geology by vividly illustrating global tectonic and volcanic processes. Because the Galileo photos include narrowband infrared filters that have never been used before, they have the potential to significantly advance Earth research in the field of geoscience and remote sensing. The Galileo data set's enormous scope and almost worldwide coverage enhance the maximum resolution of data through Earth-orbiting systems and might be a useful point of reference for further research on global change (Zhu et al., 2018).

14.2.4 BeiDou

The China National Space Administration (CNSA) is responsible for the operation of BDS, which was first launched in 2000. More than 20 years later, BeiDou now has 48 satellites in orbit. The BDS has yielded significant social and economic benefits. However, there is only one monitoring station per 1,000 square meters, and they are dispersed so far, installing Beidou subscriber machines at each monitoring site will undoubtedly raise construction and operating costs (Zabalegui et al., 2020; Bali and Singla, 2022). Consequently, a foundation platform for a local wireless sensor network has been created using Zigbee wireless sensor network technology that allows the network to integrate wireless sensor nodes into remote monitoring and warning systems and send focused data using Beidou satellite terminal devices, greatly extending the transmission distance of the monitoring data (Yan and Huang, 2019; Kaur et al., 2022).

14.2.5 QZSS

The Japan Aerospace Exploration Agency (JAXA) oversees the Quasi-Zenith Satellite System (QZSS) in Japan and was initially launched in 2010. QZSS continues to give coverage over Asia-Oceana between Japan and Australia, whereas the other constellations have provided worldwide coverage. Only four of QZSS's regional constellation's satellites are already in orbit; however, three more are scheduled to launch in the next few years. Due to significant differences in the needs of the Japanese system with regard to service, service area, and, most crucially, the national space development strategy that underpins them, the design idea of QZSS is quite different from the implementations of the GPS, GLONASS, and Galileo systems (Edokossi et al., 2020; Dahiya, Gupta and Singh, 2023). The QZSS aims to offer a more secure and better navigational reference system that may be utilized across the country for a wide range of applications in every aspect of contemporary life; the Japanese Diet adopted a measure to that effect in August 2007. Once fully operational, the QZSS constellation will function more like autonomous GPS or Galileo navigation satellites than the limited, ground-controlled mode often associated with SBAS GEO augmentation satellites (Xuerui et al., 2023).

14.2.6 NavIC

Another important regional-based Indian Regional Navigation Satellite System (IRNSS), also known as Navigation with Indian Constellation (eight satellites in orbit), or NavIC is being operated by the Indian Space Research Organization (ISRO). The coverage area is centered on India, extending to the west to encompass Saudi Arabia, the north and east to encompass all of China, and as far south as to encompass Western Australia and Mozambique. Both the GPS L5 frequency (1176.45 MHz) and the S-Band (2492.028 MHz) are used by NavIC signals for transmission. For customers in the Indian area, NavIC provides a short messaging service (Rajabi et al., 2023). Messaging services using NavIC satellites are beneficial to end users situated in remote areas where cellular or internet-based communication is difficult to reach (such as open oceans, isolated terrain, etc. Web-based Interface for Messaging Service (WIMS) portals are given to message broadcasters so they may send messages via the Internet. User-specific interfaces are also offered for the messaging service's security and prioritization (Kossieris, Asgarimehr, and Wickert, 2023). Table 14.1 specifies the technical specifications, along with the merits and demerits of some of the main GNSS categories.

14.3 GNSS APPLICATIONS IN CLIMATE MONITORING AND MODELING

GNSSs serve various applications in climate monitoring and modeling, as shown in Figure 14.2.

14.3.1 SEA LEVEL DETECTION

Global warming-related sea level rise poses a danger to coastal areas. Monitoring and measuring coastal sea level increase is crucial for assessing the possible effects of climate change on coasts, such as sea level rise. But most human interactions with the ocean happen in coastal zones, where there is still a lack of observation in our understanding of sea level. Tide gauges located on land and satellite altimetry are the two most widely used methods for monitoring sea levels directly (Singh et al., 2022a, 2022b; Kim et al., 2023).

Since the installation of the first automated tidal gauge in 1831, tide gauges have proven to be an invaluable resource for learning about variations in sea level along the coast. However, since the first satellite mission devoted to studying sea levels was launched in 1992, satellite altimetry has been continually measuring sea surface heights in open oceans. During past decades, a geodetic

TABLE 14.1

GNSS-based Technical Specifications, Along with Merits and Drawbacks

GNSS	Band Name	Center Frequency	Full Signal Bandwidth	Merits of GNSS	Drawbacks of GNSS
GPS	L2	1227.600 MHz	24.0 MHz	1. System is updated regularly by the United States government and hence is extremely advanced. 2. Its signal is out there worldwide; the users will not be bereft of it anywhere.	1. It does not penetrate solid walls. 2. It has more possibilities for failure.
	L5	1176.450 MHz	24.0 MHz		
GLONASS	L2	1246.000 MHz	6.125 MHz	1. It indicates the constellation visibility analysis. 2. It indicates the statistical analysis of coordinate estimates.	1. Errors with the satellite signal due to clock and orbital uncertainties. 2. Perturbation with the satellite signals due to the local environment of the user.
	L3	1202.250 MHz	24.0 MHz		
	L5	1176.450 MHz	16.38 MHz		
Galileo	E5a	1176.450 MHz	25.575 MHz	1. It provides more accurate and reliable positioning for the receiving device. 2. Galileo provides a significant contribution to the dual-frequency service landscape.	1. It's highly susceptible to RFI, which is difficult to mitigate. 2. The ionospheric parameters are not covered by any issue of data.
	E5b	1202.025 MHz	25.575 MHz		
	E6	1278.75 MHz	40.92 MHz		
Compass	B2	1207.140 MHz	20.46 MHz	1. To determine the true heading with high accuracy. 2. To deliver real-time position accuracy of 10 mm.	
	B2-altBO C(15,10)	1191.795 MHz	30.69 MHz		
	B3-BOC (15, 2.5)	1268.520 MHz	35.805 MHz		
	L5	1176.450 MHz	24.0 MHz		

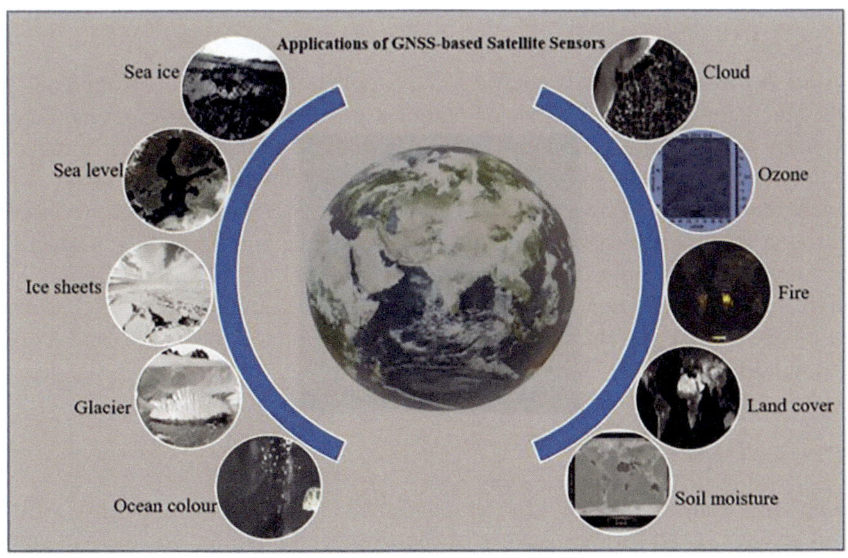

FIGURE 14.2 GNSS applications in climate monitoring and modeling.

GNSS station situated near the shore became a viable substitute for coastal sea level monitoring systems, offering several significant benefits such as obtaining both measuring changes (Ghiasi et al., 2023).

14.3.2 Sea Ice Detection

Over the past few decades, there has been a significant decrease in Arctic Sea ice, a phenomenon linked to global warming. Maintaining a high surface albedo, reducing air-sea interaction, and adjusting the distribution of freshwater and saltwater are all made possible by sea ice (Gallotti et al., 2023). In addition, the management and security of human endeavors like offshore oil and gas extraction and the world's shipping sectors are directly impacted by sea ice conditions. Understanding sea ice is essential, as it has a substantial impact on both human activity and the climate globally. On the other hand, in-situ sea ice measurement is time-consuming and has a small geographic coverage. Remote sensing methods offer a more economical and effective way to gather information on sea ice. The GNSS-IR can assess changes in sea level along the coast. It is one of the primary technologies utilized for sea ice remote sensing detection. Nowadays, it is also used to track winds and ocean surface roughness. Additionally, efforts are being made to develop land and cryosphere observation (Cahyadi et al., 2023).

14.3.3 Sea Surface Temperature Detection

With satellite altimetry observations, sea level variations at various scales may be fully comprehended. Accurate altimetry data collection in these regions is still difficult, as terrain can pollute satellite radiometer and altimeter signals (Xing and Jin, 2023). A thorough understanding of the fine-scale spatial fluctuations of the geoid and the adjustments for local hydrodynamics is necessary to connect in-situ data to altimetric observations when they are not situated underneath satellite tracks. With the advancement of GNSS, certain recognized limitations of tide gauges may now be overcome by accurately determining sea surface height above a selected ellipsoid near the satellite ground track. Similar to tidal gauges, they only offer pointwise observations; however, they are frequently utilized in altimetry satellite calibration sites (Kyuroson et al., 2023).

14.3.4 Ocean Color Detection

Users may view ocean color through satellites in a way that is not possible from land or on a ship. The way incident light interacts with water particles or other things determines the color of the ocean. Depending on what makes up the water and how these elements alter light reflections, these light interactions give the ocean its varied hues. Ocean color fluctuations can result in a shift from the typical clear blue color to a range of hues of green. These fluctuations can be produced by several components, such as the biomass of phytoplankton or zooplankton. Satellites measure things that can be used to compute the color and material concentrations of the ocean. GNSSs help provide information related to ocean color, identification of river plumes, the impact of coastal floods, and the location of toxic algal blooms that can taint shellfish and kill other fish and marine animals (Gyagenda et al., 2022).

14.3.5 Glacier Detection

Understanding glacier morphological evolution and tracking the effects of climate require ongoing monitoring of glaciers. Therefore, changes in glaciers may have an immediate effect on human livelihoods. They are also excellent indicators of climate change since variations in temperature and precipitation have a significant impact on them. To anticipate water supply, track climatic change, and get a better understanding of glacier morphological evolution over time, glaciers must be continuously monitored.

Mass balance is one of the most crucial criteria used for glacier monitoring. The mass balance may be calculated using the total volume loss, which is often calculated using the multitemporal difference of the glacier Dense Point Clouds (DPCs) or Digital Surface Models (DSMs), which measure the height change of a glacier body that happens between two successive survey epochs. Accurately retrieving changes in glacier morphology is a challenging endeavor that can be resolved with the help of GNSSs (Qin et al., 2022).

14.3.6 SOIL MOISTURE DETECTION

A crucial factor in meteorology, geology, agriculture, and ecology is soil moisture (SM). To evaluate agricultural development, climate and weather forecasts, and natural hazards, it is crucial to precisely and promptly monitor changes in soil moisture (Singh et al., 2022a, 2022b). Now, the primary methods used for SM monitoring are model simulation or data assimilation, satellite radar remote sensing monitoring, and conventional in situ observations. High temporal resolution is possessed by conventional in-situ measuring techniques, including drying-weighing, tensiometer, and time-domain reflectometry. Nevertheless, they are limited to local point measurements and pose challenges for broader monitoring. Regional SM can be obtained from Soil Moisture and Ocean Salinity (SMOS). A novel technique for remote sensing detection called GNSS interferometric reflectometry (GNSS-IR) may extract surface SM (White et al., 2022).

14.3.7 CLOUD DETECTION

Massive volcanic eruptions have the potential to produce gas and ash clouds that rise to the stratosphere and spread over the whole planet. Numerous dangers stemming from these volcanic characteristics include acid rain, ash fallout, aircraft engine damage, short-term climatic changes, and health risks. One of the biggest risks associated with volcanic eruptions is the constantly increasing number of airplanes in the sky (Singh, Sethi and Singh, 2021a; Xuerui et al., 2023). Recent research has demonstrated how volcanic ash affects engines and the necessity of rerouting aircraft to avoid potentially hazardous regions. Therefore, to avoid these risks, it is essential to continuously monitor the altitude and dispersion of volcanic clouds.

Significant technical advancements over the last several decades have made it possible to monitor volcanic characteristics more quickly and accurately. Examples of these advancements include satellite photography with varying spectral ranges, aerial aerosol measurements, and ground-based high-speed and high-resolution cameras. When it comes to studying massive volcanic clouds, satellite-based methods are the most effective, as they provide the most comprehensive coverage of the planet. Yet, depending on the method employed, there are still notable differences in the characterizations of volcanic clouds, for example, variations of up to 4 km in cloud altitude measurements (Singh, Sethi and Singh, 2021b; Gyagenda et al., 2022).

14.4 CHALLENGES FACED BY GNSS

GNSS technology has completely changed the way we navigate and is now a necessary component of our everyday existence. However, there are several difficulties with GNSS technology. Some of the challenges are mentioned below.

14.4.1 ACCESSIBLE

It is imperative that positioning data is constantly accessible, irrespective of the time of day, climate, or other variables (Zhu et al., 2018).

14.4.2 Coverage

Satellites must be seen from anywhere on Earth, and their orbits must be closely monitored for services to be provided everywhere in the world (Zabalegui et al., 2020).

14.4.3 Reliability

To recognize small land features, find specific structures, and enable ships, vehicles, and airplanes to avoid risks, precise locations must be ascertained (Pipitone et al., 2018).

14.4.4 Multipath Signals

The GPS receiver may become confused if signals from mobile towers or GPS satellites bounce off buildings because of the additional time the signal requires to reach it. You could notice abrupt positional mistakes in certain situations. In these situations, there is little that can be done to lessen the impact of multipath errors. Simply said, GPS is less precise in these circumstances (Edokossi et al., 2020).

14.4.5 GPS Distortion

The GPS trace veers off course from the path. Although it does so with far less accuracy, you can see that the route largely follows the contour of the road (Qu et al., 2022).

14.5 META-ANALYSIS

This study also presents a meta-analysis to understand the research contribution made by different authors in the field of GNSS. To compute the meta-analysis, the SCOPUS database has been utilized, and the term "GNSS" was searched on 12-Dec-2023 at 1:35 PM (Indian Standard Time – IST). The search has been performed within the "Article title, Abstract, and Keywords." The term, i.e., "GNSS," has been cross-verified concerning its relevance and eliminated redundant publications. Moreover, the Scopus-indexed publications from conferences, book chapters, and journals have been considered, as shown in Figure 14.3 (a), (b), and (c).

14.6 SUMMARY

In this chapter, GNSS applications in climate monitoring and modeling have been explored. We have summarized the GNSS categories, applications in climate monitoring and challenges. It has been also concluded that GNSSs have proven to be a powerful and sophisticated tool for climate monitoring. The review of some of the GNSSs, their applications, and challenges has been summarized. This chapter presents a thorough overview of multi-GNSS for Earth observation and the development of its new applications. One of the primary conclusions drawn from this research is that, in addition to its widespread usage in classic locating applications, one of the most popular uses of the GNSS method these days is in remote sensing applications.

As extensively covered in the chapter, multi-GNSS has the potential to be a game-changer for future applications with ongoing advancements and improvements in performance, availability, modernization, and hybridization. GNSS atmospheric modeling is the subject of another study. For many applications, including weather forecasting and natural hazard monitoring, atmospheric impacts on GNSS signals can yield useful information, while being unpleasant factors for locating and navigation applications.

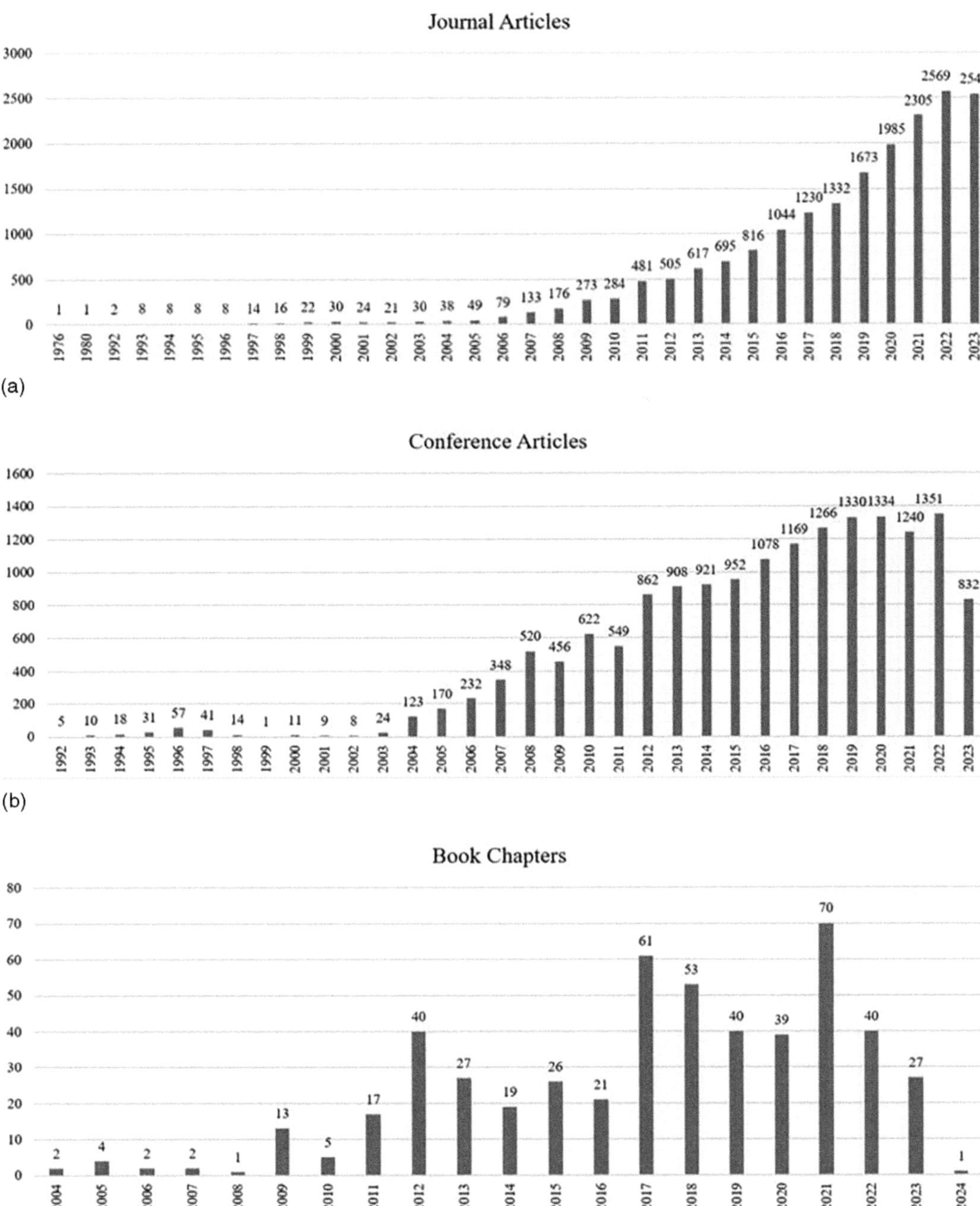

FIGURE 14.3 (a) A stacked bar graph of Journal Articles related to GNSS, (b) A stacked bar graph of Conference Articles related to GNSS. (c) A stacked bar graph of articles book chapters related to GNSS.

REFERENCES

Bali, N. and Singla, A. (2022) 'Emerging trends in machine learning to predict crop yield and study its influential factors: A survey', *Archives of computational methods in engineering*, pp. 1–18.

Cahyadi, M. N. et al. (2023) 'Analysis of the effect of the 2021 Semeru eruption on water vapor content and atmospheric particles using GNSS and remote sensing', *Geodesy and Geodynamics*. doi: https://doi.org/10.1016/j.geog.2023.04.005

Dahiya, N., Gupta, S. and Singh, S. (2023) 'Qualitative and quantitative analysis of artificial neural network-based post-classification comparison to detect the earth surface variations using hyperspectral and multispectral datasets,' *Journal of Applied Remote Sensing*, 17(3), p. 32403.

Defourny, P. et al. (2019) 'Near real-time agriculture monitoring at national scale at parcel resolution: Performance assessment of the Sen2-Agri automated system in various cropping systems around the world', *Remote Sensing of Environment*, 221(March 2018), pp. 551–568. doi: 10.1016/j.rse.2018.11.007

Edokossi, K. et al. (2020) 'GNSS-reflectometry and remote sensing of soil moisture: A review of measurement techniques, methods, and applications', *Remote Sensing*, 12(4). doi: 10.3390/rs12040614

Gallotti, G. et al. (2023) 'The tsunamigenic potential of landslide-generated tsunamis on the Vavilov seamount,' *Journal of Volcanology and Geothermal Research*, 434, p. 107745.

Ghiasi, Y. et al. (2023) 'Potential of GNSS-R for the Monitoring of Lake Ice Phenology,' *IEEE Journal of Selected Topics in Applied Earth Observations and Remote Sensing*, pp. 1–14. doi: 10.1109/jstars.2023.3330745

Gyagenda, N. et al. (2022) 'A review of GNSS-independent UAV navigation techniques,' *Robotics and Autonomous Systems*, 152, p. 104069. doi: https://doi.org/10.1016/j.robot.2022.104069

Kaur, S. et al. (2022) 'Transfer learning-based automatic hurricane damage detection using satellite images,' *Electronics*, 11(9), p. 1448.

Kelly, J. T., Hehlen, M., and McGee, S. (2023) 'Uncertainty of Satellite-Derived Glacier Flow Velocities in a Temperate Alpine Setting (Juneau Icefield, Alaska),' *Remote Sensing*, 15(15). doi: 10.3390/rs15153828

Kim, S. B. et al. (2023) 'Joint inversion of ocean-bottom pressure and GNSS data from the 2003 Tokachi-oki earthquake,' *Earth, Planets and Space*, 75(1). doi: 10.1186/s40623-023-01864-x

Kossieris, S., Asgarimehr, M., and Wickert, J. (2023) 'Unsupervised Machine Learning for GNSS Reflectometry Inland Water Body Detection,' *Remote Sensing*, 15(12). doi: 10.3390/rs15123206

Kyuroson, A. et al. (2023) 'Multimodal Dataset from Harsh Sub-Terranean Environment with Aerosol Particles for Frontier Exploration,' *2023 31st Mediterranean Conference on Control and Automation, MED 2023*, pp. 716–721. doi: 10.1109/MED59994.2023.10185906

Pipitone, C. et al. (2018) 'Monitoring water surface and level of a reservoir using different remote sensing approaches and comparison with dam displacements evaluated via GNSS,' *Remote Sensing*, 10(1), pp. 1–24. doi: 10.3390/rs10010071

Qin, Z. et al. (2022) 'Review of GNSS landslide monitoring and early warning,' *Acta Geodaetica et Cartographica Sinica*, 51(10), p. 1985.

Qiu, T. et al. (2023) 'An Innovative Signal Processing Scheme for Spaceborne Integrated GNSS Remote Sensors,' *Remote Sensing*, 15(3). doi: 10.3390/rs15030745

Qu, X. et al. (2022) 'Experimental Study of Accuracy of High-Rate GNSS in Context of Structural Health Monitoring,' *Remote Sensing*, 14(19), pp. 1–17. doi: 10.3390/rs14194989

Rajabi, M. et al. (2023) 'Tidal harmonics retrieval using GNSS-R dual-frequency complex observations,' *Journal of Geodesy*, 97(10), p. 94.

Siemuri, A. et al. (2022) 'A Systematic Review of Machine Learning Techniques for GNSS Use Cases,' *IEEE Transactions on Aerospace and Electronic Systems*, 58(6), pp. 5043–5077. doi: 10.1109/TAES.2022.3219366

Singh, G., Singh, S., Sethi, G., et al. (2022a) 'Deep Learning in the Mapping of Agricultural Land Use Using Sentinel-2 Satellite Data,' *Geographies*, 2(4), pp. 691–700. doi: 10.3390/geographies2040042

Singh, G., Singh, S., Sethi, G. K., et al. (2022b) 'Detection and Mapping of Agriculture Seasonal Variations with Deep Learning-based Change Detection using Sentinel-2 Data,' *Arabian Journal of Geosciences*, 15(9), pp. 1–19. doi: 10.1007/s12517-022-10105-6

Singh, G., Sethi, G. K. and Singh, S. (2021a) *Performance Analysis of Deep Learning Classification for Agriculture Applications Using Sentinel-2 Data*. Springer Singapore. doi: 10.1007/978-981-16-3660-8_19

Singh, G., Sethi, G. K. and Singh, S. (2021b) 'Survey on machine learning and deep learning techniques for agriculture Land,' *SN Computer Science*, 2(6). doi: 10.1007/s42979-021-00929-6

White, A. M. et al. (2022) 'A Review of GNSS/GPS in Hydrogeodesy: Hydrologic Loading Applications and Their Implications for Water Resource Research,' *Water Resources Research*, 58(7). doi: 10.1029/2022WR032078

Xing, H. and Jin, S. (2023) 'Structure and variations of global planetary boundary layer top from 2008–2022 multiple GNSS RO observations,' *Journal of Atmospheric and Solar-Terrestrial Physics*, 252, p. 106155. doi: https://doi.org/10.1016/j.jastp.2023.106155

Xuerui, W. et al. (2023) 'A review of GNSS-R/SoOP-R for essential hydrological climate variables detection,' *Geomatics and Information Science of Wuhan University*.

Yan, Q. and Huang, W. (2019) 'Sea ice remote sensing using GNSS-R: A review,' *Remote Sensing*, 11(21). doi: 10.3390/rs11212565

Zabalegui, P. et al. (2020) 'A Review of the Evolution of the Integrity Methods Applied in GNSS,' *IEEE Access*, 8, pp. 45813–45824. doi: 10.1109/ACCESS.2020.2977455

Zhu, N. et al. (2018) 'GNSS Position Integrity in Urban Environments: A Review of Literature,' *IEEE Transactions on Intelligent Transportation Systems*, 19(9), pp. 2762–2778. doi: 10.1109/TITS.2017.2766768

15 Real-time Computation of Precipitable Water Vapor from GPS Data

Reshma Raskar-Phule and S. Bala Subramaniyam

15.1 INTRODUCTION

In recent years, the accurate prediction of rainfall has become increasingly crucial for various sectors such as agriculture, hydrology, and disaster management. Traditional methods of rainfall prediction often rely on meteorological models that incorporate historical data and atmospheric variables. However, these methods can sometimes be limited in their accuracy and timeliness. In this chapter, an innovative approach is explored that utilizes near real-time PWV derived from GPS data to enhance rainfall prediction. This method has shown promising results in improving the accuracy and lead time of rainfall forecasts, offering valuable insights for various applications.

15.1.1 UNDERSTANDING PWV AND ITS SIGNIFICANCE

Before delving into the methodology, it is essential to grasp the concept of PWV and its importance in rainfall prediction. PWV refers to the amount of water vapor present in a vertical column of the atmosphere. It plays a vital role in the development of clouds and subsequent precipitation. By measuring and analyzing PWV, we can gain insights into the moisture content of the atmosphere, which is a critical factor in determining the likelihood and intensity of rainfall events.

15.1.2 GPS AND NEAR REAL-TIME PWV ESTIMATION

GPS is a satellite-based navigation system widely used for positioning and timing purposes. However, GPS signals are also affected by the Earth's atmosphere, specifically by the delay caused by water vapor content. By precisely measuring this delay, it is possible to estimate PWV values in near real time. This approach, known as GPS meteorology, has gained traction in recent years due to its ability to provide high-resolution and continuous PWV estimates over large spatial areas. GPS Meteorology refers to remote sensing of the troposphere and the stratosphere by gauging the refraction (slowing and bending) of GPS signals that propagate through the atmosphere (Bevis et al., 1992). The fundamental physical basis for GPS meteorology is the pressure, temperature, and humidity dependence of microwave refractivity in the neutral atmosphere.

Various studies conducted in the last decade have shown that the amount of PWV contained in the neutral atmosphere can, in fact, be retrieved using ground-based GPS receivers, with the accuracy comparable to radiosondes and microwave radiometers (Bevis et al., 1992; Melbourne, 1994; Ware et al., 2000; Rocken et al., 1997; Kursinski et al., 1995; Kuo et al., 1993; Tregoning and Dam, 1998).

The first and most mature use of GPS for this purpose is in the estimation of PWV above a fixed site (Duan and Bevis, 1996). The techniques currently used by the National Oceanic and Atmospheric Administration's Forecast Systems Laboratory (NOAA/FSL) to collect, process, and distribute GPS

DOI: 10.1201/9781032712444-18"

water vapor observations are mature and almost ready for transition to operational use. NOAA/FSL has shown that GPS Integrated PWV (IPWV) data can be used effectively in objective and subjective weather forecasting.

Monitoring atmospheric water vapor with greater accuracy in real or near real time is the need in the random variation of weather conditions. GPS observable is contaminated by various errors. In this chapter, the focus is on the computation of PWV using GPS data with greater accuracy and comparing it with the standard data from radiosonde.

15.2 METHODOLOGY

Before delving into the methodology, it is essential to grasp the concept of PWV and its importance in rainfall prediction. PWV refers to the amount of water vapor present in a vertical column of the atmosphere. It plays a vital role in the development of clouds and subsequent precipitation. By measuring and analyzing PWV, we can gain insights into the moisture content of the atmosphere, which is a critical factor in determining the likelihood and intensity of rainfall events.

Radiosonde (RS) observations date remain the most reliable basis to date for determining PWV, although they are unevenly distributed over the globe and taken twice a day. Radiosonde measurements of relative humidity (RH) are the main source of uncertainty in precipitable water vapor (PWV) calculation from pressure, temperature, and RH/dewpoint (PTU) data. In this context, in the last two decades GPS has proven to be particularly useful for PWV monitoring regardless of weather conditions.

The passage of GPS radio waves through the troposphere results in delays caused by the presence of natural gases and water vapor. This introduces errors in determining the relative position of the station. However, these delays serve as valuable signals for atmospheric studies, contributing to the estimation of water vapor, short-term weather forecasting, and long-term weather forecasting. The integrated total water vapor content can be computed from the zenith equivalent total path delay affecting the radio signal propagation in the troposphere. The zenith total delay (ZTD) will be decomposed into the zenith wet delay (ZWD) and zenith hydrostatic delay (ZHD).

In this research work, an attempt has been made to evaluate the differences and study the reliability of use of GPS data for the estimation of the zenith delays and PWVs at IIT Bombay GPS station with the OPT and GPT data.

Several software options are available for calculating total zenith delay, and from these, PWV can be extracted. The choice of software is a critical aspect of the research methodology. While initially contemplating the development of custom software, the focus shifted on the established GPS processing software packages to meet the research objectives effectively.

In this context, the options considered for the present chapter include the commercial software BERNESE, developed by the University of Bern, and GAMIT, a GPS analysis package jointly developed by the Massachusetts Institute of Technology and Scripps Institute of Oceanography. Ultimately, GAMIT 10.3 is the chosen software for processing the data in the present research, aligning with its established reputation in the field of GPS data analysis and processing.

The research involves processing raw data for 2006 from the Indian Institute of Technology Bombay (IITB) as a local permanent GPS reference station [Lat. 19.128 deg., Lon. 72.928 deg., Ht. −3.981 m], combined with data from four other International GPS Service (IGS) stations: BAHRAIN (BAHR) [Lat. 26.209 deg., Lon. 50.608 deg., Ht. 17.074 m], Indian Institute of Science Bangalore (IISC) [Lat. 13.021 deg., Lon. 77.570 deg., Ht. 843.726 m], LHASA (LHAS) [Lat. 29.657 deg., Lon. 91.103 deg., Ht. 362655 m], and KITAB (KIT3) [Lat. 38.946 deg., Lon. 66.885 deg., Ht. 622.649 m]. The purpose of selecting the IGS stations was to eliminate the common errors from the one-time observations. The accuracy in the estimate of the parameters can be improved using the IGS network stations.

Various models use meteorological data, i.e., surface values of temperature and pressure, as input to calculate hydrostatic delay. However, these models are designed for specific geographical locations. Therefore, there is a necessity to create a site-specific model for precise determination of hydrostatic delay, enhancing the overall accuracy in estimating PWV.

The meteorological data collected for the present study for 2006 is in the Receiver INdependent EXchange [RINEX] format, the industry standard for GNSS data. The data file contains the observation file, navigation file, almanac file, and meteorological file. The meteorological data is collected in the meteorological file. These met files can be downloaded from Scripps Orbit and Permanent Array Center (SOPAC) for various IGS stations and for IITB through the MET 3A package. This meteorological data can be used for various meteorological applications. The data format of the instrument is given in Figure 15.1.

To process GPS satellite observation data, it is essential to incorporate corresponding satellite ephemeris data, which provides the necessary information to calculate each GPS satellite's Earth-Centered Earth-Fixed (ECEF) position at any given time within the validity window. Various types of ephemerides exist with differing accuracy and availability. The key characteristics for each of the four standard ephemeris types are given in Table 15.1.

Ultra-rapid orbits offer a real-time option at the expense of some accessibility and are produced by the IGS. Despite being a predicted orbit product, ultra-rapid orbits exhibit an order of magnitude improvement in accuracy compared to broadcast ephemerides.

Processing the data involves deriving Zenith delay estimates through IGS precise orbits. The archived estimated Zenith delays (GAMIT outputs) for the reference station were retrieved from the day directories. To calculate PWV, the output files and z-files (model) or RINEX met data files were utilized.

Zenith delays and PWV estimates were obtained using both model and surface meteorological parameters. However, the accuracy of Zenith delay and PWV estimates is significantly influenced

2.10	METEOROLOGICAL DATA (GPS)				RINEX VERSION / TYPE
DAT2RINW 3.10 001 MK			09NOV07 14:28:15		PGM / RUN BY / DATE
					COMMENT
TEST					MARKER NAME
3		P	T	R	#/TYPES OF OBSERVATION
					END OF HEADER
07 11 9 6 5 50	1006.8	30.9	33.4		
07 11 9 6 6 30	1006.8	31.0	33.3		
07 11 9 6 7 30	1006.8	31.0	33.4		
07 11 9 6 8 30	1006.8	31.0	32.8		
07 11 9 6 9 30	1006.8	31.0	32.3		

P- Pressure in mb or hPa
T-Temperature in Degree Celsius
R-Relative humidity in percentage

FIGURE 15.1   Rinex Meteorological file.

TABLE 15.1
Table Satellite Ephemeris Types

GPS Satellite Ephemeris	Accuracy	Latency
Broadcast	~260 cm/~7 ns	Real time
Ultra-Rapid (predicted)	~25 cm/~5 ns	Real time
Rapid	5 cm / 0.2 ns	17 hours
Precise (final)	< 5 cm / 0.1 ns	~13 days

by the station coordinates. GAMIT RINEX observation files and GAMIT 'L' files or 'itrfyy.apr' files take the station coordinates as input. Service providers such as SOPAC regularly update IGS station coordinates weekly, ensuring precise station coordinates from the IGS network GPS observation data. However, local/regional station coordinates must be adjusted or considered for plate movements, which induce a continuous drift in coordinate components. To address this, average coordinates (weekly or monthly) for local stations were determined by manually running the Global Kalman filter (GLOBK) 'sh_glred' script for the IITB station, referencing highly accurate IGS station coordinates. The resulting average coordinates were then updated in the 'itrf00.apr' file, containing station coordinates for the fixed reference epoch. Subsequently, the data was reprocessed using GAMIT 10.3, and estimates were obtained using the 'sh_metutil' script.

15.3 RESULTS AND DISCUSSIONS

At first, to assure the accuracy of the ZTD estimations from the GAMIT 10.3, the GPS observation data of IISC Bangalore station for 2006 was processed using GAMIT 10.3. The ZTD estimates are shown in Figure 15.2.

Next, these ZTDs were compared with the ZTDs given by IISC Bangalore IGS station for 2006, which were downloaded from the SOPAC archive. The difference between ZTDs estimated using GAMIT 10.3 and ZTDs obtained from the SOPAC archive is shown in Figure 15.3.

The minimum, maximum, and average difference between the ZTDs computed from GAMIT 10.3 and SOPAC is 5.12 mm, 8.30 mm, and 6.27 mm, respectively. The coefficient of correlation is 0.993418, respectively. The comparison thus shows that the ZTD values derived from GAMIT 10.3 are in good agreement with those obtained from SOPAC. Hence, the ZTDs were then estimated for the IITB GPS permanent station using GAMIT 10.3 / GLOBK.

Next, the GPS data for the IITB station was processed using GAMIT 10.3 for the period from January 2007 to June 2007, which in Julian days is 001 to 181, and in Julian weeks is 1408 to 1433. The Zenith Delay estimates were extracted from the output files generated after processing the data

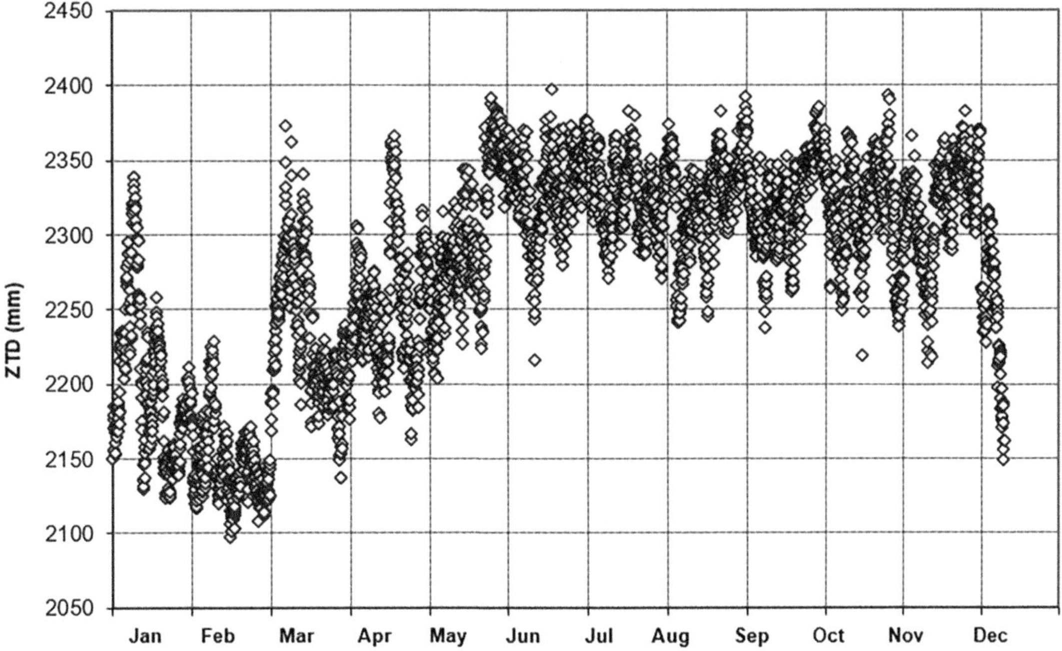

FIGURE 15.2 Seasonal Variation of ZTD at IISC Bangalore for 2006 computed by using GAMIT 10.3.

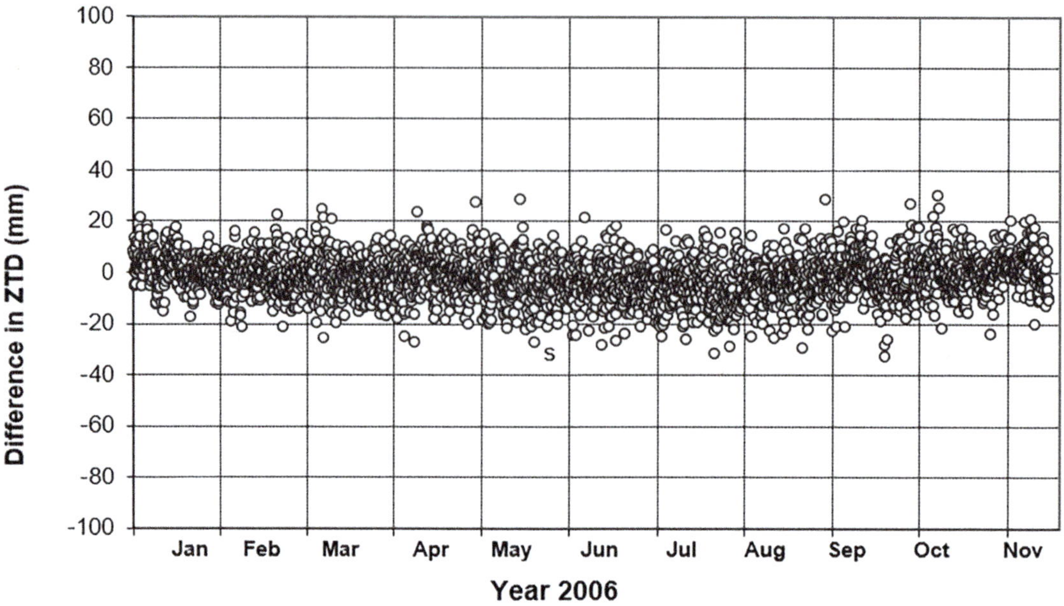

FIGURE 15.3 Difference of ZTDs—GAMIT and SOPAC, at IISC Bangalore for 2006.

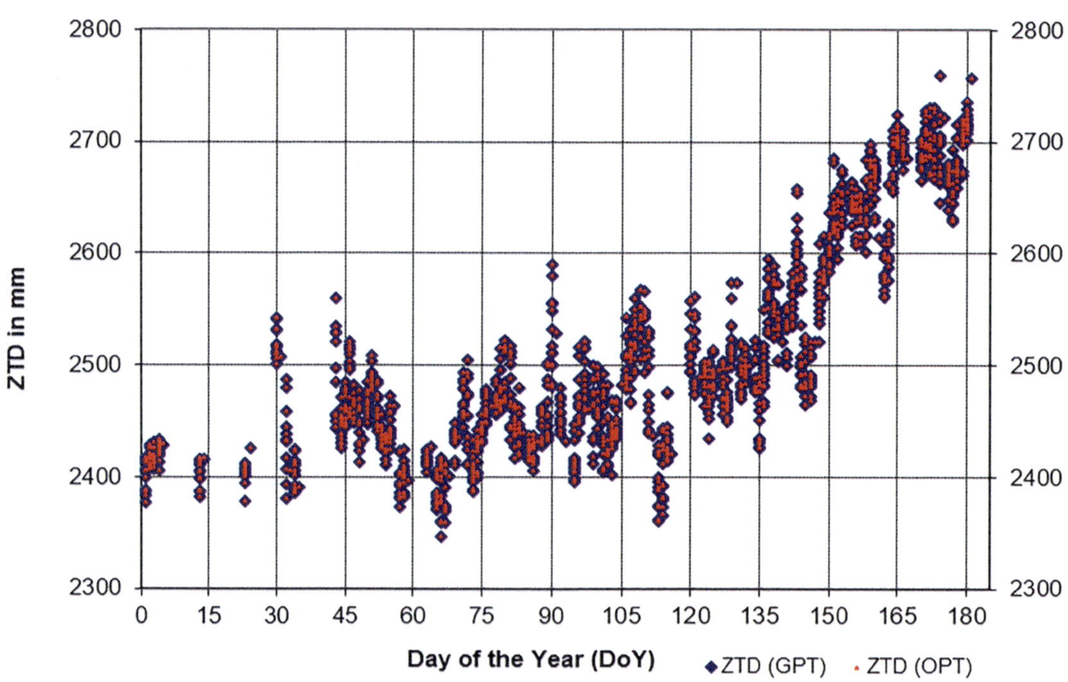

FIGURE 15.4 Comparison of the estimated ZTDs—GPT and OPT, at IIT Bombay for 2007.

by running the GAMIT (sh_metutil) script using the model (z-files obtained in day directories after processing) or RINEX met (.yym) file. The ZTDs were estimated at a 2-hr interval using Observed Pressure and Temperature (OPT). These ZTDs were compared with the ZTDs estimated with GPT of 1013.25 hPa and 20 deg. C respectively, as shown in Figure 15.4. In both estimates, from OPT and GPT, the ZTDs ranged from approximately 2347 mm to 2760 mm. The average difference

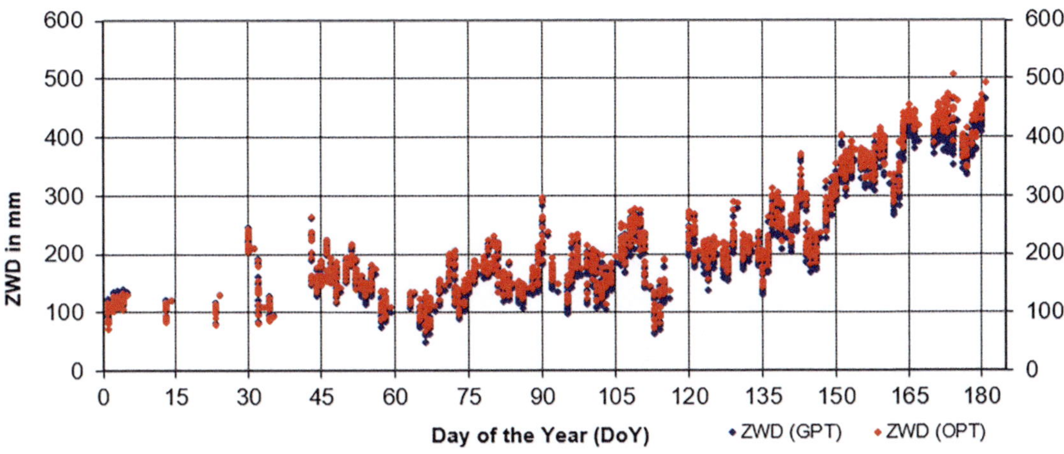

FIGURE 15.5   Comparison of the estimated ZWDs—GPT and OPT, at IIT Bombay for 2007.

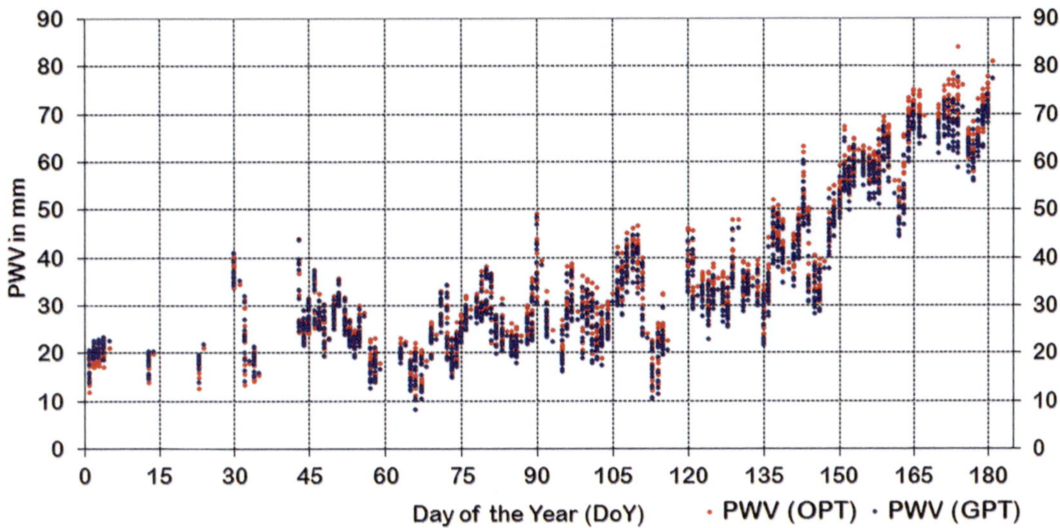

FIGURE 15.6   Comparison of the estimated PWVsGPT and OPT, at IIT Bombay for 2007.

between the ZTDs estimated from OPT and GPT was ± 0.19 mm, and the correlation coefficient between the estimates was 0.9994444.

Similarly, the ZWDs were estimated using OPT and GPT in GAMIT 10.3/GLOBK, as shown in Figure 15.5. It was found that the average difference between the estimates, ZWDs from GPT, and ZWDs from OPT, was ± 11.80 mm and the correlation coefficient was 0.997841.

The PWVs were then estimated for the same GPS data of IIT Bombay for 2007 in GAMIT 10.3/GLOBK using OPT and GPT. The comparison of the estimates is shown in Figure 15.6. The average difference between the PWV estimates, OPT and GPT, was ± 2.03 mm, and the correlation coefficient of PWVs estimated using GPT and OPT 0.997390, respectively.

Finally, the PWVs estimated using the OPT and GPT in GAMIT 10.3/GLOBK and the PWVs obtained from RS for days 1 to 177 (January 2007 to June 2007) at 12 Greenwich Mean Time (GMT) were compared, as shown in Figure 15.7.

An average error of ± 6.25 mm, RMS error of 8.04854 mm and correlation coefficient of 0.854234 between PWV (epoch & GPT) and PWV (RS); an average error of ± 5.84 mm, RMS error of

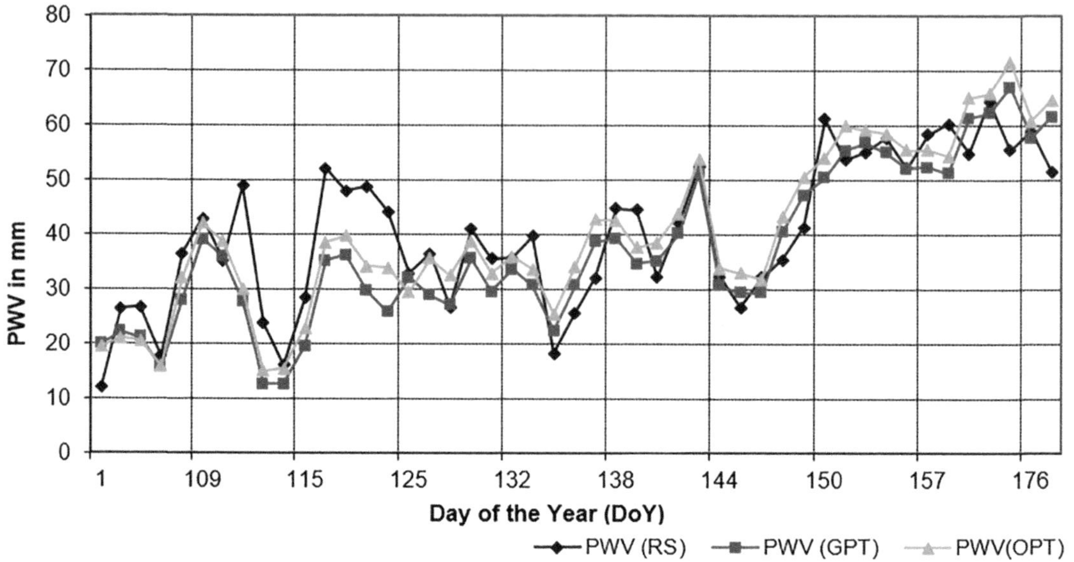

FIGURE 15.7 Comparison of the estimated PWVs—GPT and OPT and RS, at IIT Bombay for 2007 at 12 GMT.

7.30294 mm and correlation coefficient of 0.86569 between PWV (globk & GPT) and PWV (RS); and an average error of ± 5.05 mm, RMS error of 6.50850 mm and correlation coefficient of 0.892119 between PWV (globk & OPT) and PWV (RS) have been evaluated at 12 GMT.

The comparison between the PWV estimates from GPT and PWV estimates from RS at 12 GMT shows a minimum difference of -21.14 mm, a maximum difference of 11.42 mm, and an average error of ±5.84 mm. The root mean square (RMS) error of 7.30294 mm and correlation coefficient of 0.86569 have been evaluated for the estimates. On the other hand, the comparison between the PWV estimates from OPT and PWV estimates from RS at 12 GMT shows a minimum difference of −18.84 mm, a maximum difference of 16.01 mm, and an average difference of ± 5.05 mm. The root mean square (RMS) error of 6.50850 mm and correlation coefficient of 0.892119 have been evaluated for the estimates.

The above results clearly depict that the PWVs estimated from the GPS data are reliable and accurate, as compared to the PWVs estimated from RS data, specifically on the day 111, days between 115 and 125, day 139, and day 150. On the other hand, PWVs estimated from the GPS data with OPT shows improved results.

15.4 CONCLUSION

The GPS observation data for the IIT Bombay station was routinely processed using GAMIT/GLOBK software in RINEX format. The input data for processing using GAMIT/GLOBK may be affected by the meteorological parameters, pressure, and temperature, and hence the impact of GPT and OPT on the zenith delays and PWVs was studied in the present work.

The following conclusions are drawn from the present study:

The results for the ZTDs estimated by GAMIT 10.3 for IISC Bangalore for 2006 have been compared with the ZTD at the IGS site hosted on the SOPAC archive. It shows an average difference of ± 6.27 mm and correlation coefficient of 0.993418 mm. The ZTD values derived from GAMIT 10.3 are thus in good agreement with those obtained by IGS.

Next, using the OPT and GPT measurements, the variations in ZTD, ZWD, and PWV were studied. It can be seen from the results that an average difference in ZTD is only ± 0.19 mm, whereas for

ZWD, the average difference is ± 11.80 mm. An average difference in the PWV estimates is 2.03 mm. The correlation coefficient of ZTDs, ZWDs, and PWVs between OPT and GPT are 0.9994444, 0.997841, and 0.997390, respectively. These results were obtained for a small period of 180 days, January 2007 to June 2007. However, it has to be confirmed with analyses over longer time periods. Although the observed surface measurements do not have a significant effect on ZTD estimates, the ZWD vary significantly, and in turn have an effect on PWV estimates.

This shows that the use of GPT data for modeling the ZHD will underestimate the ZWD and PWV. Using OPT data, ZHD can be modeled more accurately, thereby improving the ZWD and PWV estimates.

Finally, the estimates of PWVs using GPT and OPT were compared with the PWVs obtained from radiosonde, i.e., PWV (RS) from IMD at 12 GMT. The comparison between the PWV estimates from GPT and PWV estimates from RS at 12 GMT shows a minimum difference of −21.14 mm, a maximum difference of 11.42 mm, and an average error of ±5.84 mm. The root mean square (RMS) error of 7.30294 mm and correlation coefficient of 0.86569 have been evaluated for the estimates. On the other hand, the comparison between the PWV estimates from OPT and PWV estimates from RS at 12 GMT shows a minimum difference of -18.84 mm, a maximum difference of 16.01 mm, and an average difference of ± 5.05 mm. The root mean square (RMS) error of 6.50850 mm and correlation coefficient of 0.892119 have been evaluated for the estimates.

This clearly shows that PWV estimates from GPS with OPT data and the PWV (RS) values show good correlation with each other. It is recommended that the OPT data should be used in monthly intervals to improve the accuracy of the estimates.

15.5 LIMITATIONS

The data of the IITB permanent station has been loosely constrained while processing for the estimation of the zenith delays and PWV. But for better solutions, the coordinates have to be tightly constrained. Secondly, the interval in estimation of ZTD used in GAMIT software is two hours. This is done to compare the results with the RS data, which is available at 2-hour intervals. However, for short-term weather forecasts, this interval has to be reduced.

REFERENCES

Bevis, Michael, Steven Businger, Thomas A. Herring, Chris Rocken, Richard A. Anthes, and Richard H. Ware. "GPS Meteorology: Remote Sensing of Atmospheric Water Vapor Using the Global Positioning System." *Journal of Geophysical Research* 97, no. D14 (1992): 15787–15801.

Bevis, Michael, Steven Businger, Stephen R. Chiswell, Thomas A. Herring, Richard Anthes, Chris Rocken, and Richard Ware. "GPS Meteorology: Mapping Zenith Wet Delays onto Precipitable Water." *Journal of Applied Meteorology* 33, no. 9 (1996): 379–386.

Borbas, E. "Derivation of Precipitable Water from GPS Data: An Application to Meteorology." *Physics and Chemistry of the Earth* 23, no. 1 (1998): 87–90.

David, D. T., and Barry M. "An Analysis of Satellite, Radiosonde, and Lidar Observations of Upper Tropospheric Water Vapor from the Atmospheric Radiation Measurement Program." *Journal of Geophysical Research* 109, no. D04105 (2000): 1029–1049.

Davis, J., Thomas A. Herring, I. Shapiro, A. Rogers, and G. Elgered. "Geodesy by Radio Interferometry: Effects of Atmospheric Modeling Errors on Estimates of Baseline Lengths." *Radio Science* 20, no. 6 (1985): 1593–1607.

Dixon, T. "An Introduction to the Global Positioning System and Some Tectonic Applications." *Journal of Geophysical Research* 29, no. 2 (1991): 249–276.

Duan, J., and Michael Bevis. "GPS Meteorology: Direct Estimation of the Absolute Value of Precipitable Water." *Journal of Applied Meteorology* 35 (1996): 830–838.

Emardson, T. R., G. Elgered, and J. M. Johansson. "Three Months of Continuous Monitoring of Atmospheric Water Vapor with a Network of Global Positioning System Receivers." *Journal of Geophysical Research* 103, no. D2 (1998): 1807–1820.

Fang, P., G. Gendt, T. Springer, and T. Mannucci. "IGS Near Real-Time Products and Their Applications." *GPS Solutions* 4, no. 4 (2001): 2–8.

GAMIT Manual. "Documentation for the GAMIT GPS Analysis Software, Release 10.30." 2007. Available at http://chandler.mit.edu/~simon/gtgk/GAMIT.pdf

GLOBK Documentation. "Global Kalman Filter VLBI and GPS Analysis Program, Version 10.0." 2007. Available at http://chandler.mit.edu/~simon/gtgk/GLOBK.pdf

Hofmann, W., B. Lichtenegger, and J. Collins. *GPS: Theory and Practice*. Springer Wien, New York, 2001.

Hopfield, H. S. "Improvements in the Tropospheric Refraction Correction for Range Measurement." *Phil. Trans., Royal Society, London* (1969): 341–352.

Huang, C., F. Zhang, and H. Yan. "Controls of the Sounding Points in Space-Based GPS/LEO Meteorology." *Journal of Atmospheric and Solar-Terrestrial Physics* 63 (2001): 1601–1607.

Iwabuchi, T., I. Naito, and N. Mannoji. "A Comparison of Global Positioning System Retrieved Precipitable Water Vapor with the Numerical Weather Prediction Data Over the Japanese Islands." *Journal of Geophysical Research* 105, no. D4 (2000): 4573–4585.

Jacobson, M. Z. *Fundamentals of Atmospheric Modeling*. Department of Civil and Environmental Engineering, Stanford University, Stanford, 1999.

Johansson, J. M., T. R. Emardson, and G. Elgered. "Three Months of Continuous Monitoring of Atmospheric Water Vapor with a Network of Global Positioning System Receivers." *Journal of Geophysical Research* 103, no. D2 (1998): 1807–1820.

Kursinski, E. R., G. J. Hajj, K. R. Hardy, L. J. Romans, and J. T. Schofield. "Observing Tropospheric Water Vapor by Radio Occultation Using the Global Positioning System." *Geophysical Research Letters* 22, no. 17 (1995): 2365–2368.

Kuo, Y. H., X. Zou, and W. Huang. "The Impact of Global Positioning System Data on the Prediction of an Extratropical Cyclone: An Observing System Experiment." *Dynamics of Atmospheres and Oceans* 27, no. 1–4 (1998): 439–470.

Kuo, Y. H., X. Zou, and Y. R. Guo. "Variational Assimilation of Precipitable Water Using Nonhydrostatic Mesoscale Adjoint Model." *Monthly Weather Review* 124 (1993): 122–147.

Leick, A. *GPS Satellite Surveying*. John Wiley and Sons, New York, 1990.

Melbourne, W. G. "The Case for Ranging in GPS Based Geodetic Systems." In *Proceedings of 1st International Symposium on Precise Positioning with the Global Positioning System*, U.S. Dept. of Commerce, Rockville, 1994.

McCorkle, D. "The Earth's Atmosphere." Course Lecture Notes, Department of Physics and Astronomy, University of Tennessee, 1998. Available at http://csep10.phys.utk.edu/astr161/lect/earth/atmosphere.html

MacDonald, A., Y. Xie, and R. H. Ware. "Use of a Network of Surface GPS Receivers to Recover Three Dimensional Temperature and Moisture." In *IAMAS 2001*, Innsbruck, Austria, 2000.

Niell, A. E. "Global Mapping Functions for the Atmospheric Delay at Radio Wavelengths." *Journal of Geophysical Research* 101, no. B2 (1996): 3227–3246.

Parkinson, C. L., D. J. Cavalieri, P. Gloersen, H. J. Zwally, and J. C. Comiso. "Arctic Sea Ice Extents, Areas, and Trends." *Journal of Geophysical Research* 104, no. C9 (1996): 20837–20856.

Rabbany, Ahmed EI. *A Textbook of Introduction to Global Positioning System*. Artech House, Norwood, 1997.

Rocken, C., J. M. Johnson, R. E. Neilan, M. Cerezo, J. R. Jordan, M. J. Falls, L. D. Nelson, R. H. Ware, and M. Hayes. "The Measurement of Atmospheric Water Vapor: Radiometer Comparison and Spatial Variations." *IEEE Transactions on Geoscience and Remote Sensing* 29 (1991): 3–8.

Rocken, C., R. H. Ware, T. Van Hove, F. Solheim, C. Alber, J. Johnson, and M. G. Bevis. "Sensing Atmospheric Water Vapor with the Global Positioning System." *Geophysical Research Letters* 20, no. 23 (1993): 2631–2634.

Rocken, C., T. Van Hove, J. Johnson, F. Solheim, and R. H. Ware. "GPS/Storm - GPS Sensing of Atmospheric Water Vapor for Meteorology." *Journal of Atmospheric and Oceanic Technology* 12 (1995): 468–478.

Rocken, C. et al. "Analysis and Validation of GPS/MET Data in the Neutral Atmosphere." *Journal of Geophysical Research* 102, no. D25 (1997): 29849–29866.

Saastamoinen, J. "Atmospheric Correction for the Troposphere and Stratosphere in Radio Ranging of Satellites." In *The Use of Artificial Satellites for Geodesy*, Geophysics Monogram Series, 15:247–251. Edited by S. W. Henriksen, A. Mancini, and B. H. Chovitz. AGU, Washington, D.C., Science 9 (1972): 803–807.

Schueler, T., A. Posfay, G. W. Hein, and R. Biberger. "A Global Analysis of the Mean Atmospheric Temperature for GPS Water Vapor Estimation." In *IONGPS2001—14th International Technical Meeting of Satellite Division of the Institute of Navigation*, Salt Lake City, Utah, 2001.

Smith, E. K., and S. Weintraub. "The Constants in the Equation for Atmospheric Refractive Index at Radio Frequencies." *Proceedings of the Institute of Radio Engineers (I.R.E.)* 41 (1953): 1035–1037.

Solheim, F., J. R. Godwin, E. R. Westwater, Y. Hang, S. Keihm, K. Marsh, and R. H. Ware. "Radiometric Profiling of Temperature, Water Vapor and Cloud Liquid Water Using Various Inversion Methods." *Radio Science* 33, no. 2 (1998): 393–404.

Thayer, G. D. "An Improved Equation for the Radio Refractive Index of Air." *Radio Science* 9, no. 10 (1974): 803–807.

Tregoning, P., and T. Dam. "Atmospheric Pressure Loading Corrections Applied to GPS Data at the Observation Level." *Geophysical Research Letters* 32, no. 3 (1998): 393–396.

Tregoning, P., R. Boers, and D. O'Brien. "Accuracy of Absolute Precipitable Water Vapor Estimates from GPS Observations." *Journal of Geophysical Research D: Atmospheres* 103, no. D22 (1999): 28701–28710.

Ware, H., W. Fulker, A. Stein, N. Anderson, K. Avery, D. Clark, K. Droegemeier, P. Kuettner, J. Minster, and S. Sorooshian. "Real-Time National GPS Networks for Atmospheric Sensing." *Journal of Atmospheric and Solar-Terrestrial Physics* 63, no. 12 (2000): 1315–1330.

Yang, X., B. Sass, G. Elgered, J. M. Johansson, and T. R. Emardson. "A Comparison of Precipitable Water Vapor Estimates by an NWP Simulation and GPS Observations." *Journal of Applied Meteorology* 38, no. 7 (1999): 941–956.

Zheng, Y. "Interpolating Residual Zenith Tropospheric Delays for Improved Wide Area Differential GPS Positioning." In *ION GNSS* 2004. September 21–24, Long Beach Convention Center, Long Beach, California.

LIST OF WEBSITES

Official website of The Ohio State of University, http://geodesy.eng.ohio-state.edu/GPSlab/wpafb.pdf (last accessed on 10th June '08)

Official website of National Oceanic and Atmospheric Administration, www.gpsmet.noaa.gov/jsp/index.jsp (last accessed on 10th June '08)

Official website of UNAVCO funded by the National Science Foundation (NSF) and National Aeronautics and Space Administration, www.unavco.org/facility/software/preprocessing/preprocessing.html (last accessed on 12th Feb '08)

Official website of Crustal Dynamics Data Information System, http://cddis.nasa.gov/gnss_datasum.html (last accessed on 6th June '08)

Official website of National Geodetic Survey, www.ngs.noaa.gov/CORS (last accessed on 6th June '08)

Official website of National Geodetic Survey, http://www.ngs.noaa.gov/GPS/SP3_format.html (last accessed on 15th May '08)

Official website of Scripps Orbit and Permanent Array Center, http://sopac.ucsd.edu/search/index.html (last accessed on 6th June '08)

Official website of the University of Leipzig, http://www.uni-leipzig.de/~meteo/GPSTomo/SPPposter-fin.pdf (NASA), (last accessed on 15th Dec '08)

Official website of the Geodetic Institute Norwegian Mapping Authority, http://www.gdiv.statkart.no/igsworkshop/book/gueldner.pdf (last accessed on 24th Nov '08)

Official website of MIT intended for the educational purpose to GAMIT/ GLOBK users, http://chandler.mit.edu/~simon/gtgk/GAMIT.pdf (last accessed on 10th June '08)

16 Water Vapor Retrieval and Its Climatic Implications Using GNSS Signals

Sanjay Kumar

16.1 INTRODUCTION

The greenhouse gases present in the atmosphere play a significant role in climate and weather, and hence, affect human life on Earth. Its large abundance affects the atmosphere significantly by changing the biosphere and climate. The water vapor is one among them and found as the majority greenhouse gas in Earth's atmosphere, which accounts for about half of greenhouse effects. Due to its high concentration, the atmosphere becomes more humid, causing breathing problems for human beings and is therefore a matter of risk for cardio patients [1]. As a greenhouse gas, water vapor plays a significant role in climate change. The radiation trapped in water vapor factors into the rising of the Earth's surface temperature. The tomography of water vapor is useful for climate study because it is an essential factor for the cloud formation and its growth. Water vapor is also considered an important component for the survival of life because it affects the development of crops/vegetables.

Global Navigation Satellite Systems (GNSSs) have three main segments: the control segment, the user segment, and the space segment. The user segment includes receivers that compute user positions using the trilateration method [2]. Today, GNSSs have become very popular and are used in several areas—space stations, aviation, maritime, rail, road and mass transit railways, etc. In addition, the positioning services provided by GNSSs along with navigation and timing information are useful in other areas of science/technology, including telecommunications, land surveying, law enforcement, emergency response, precision agriculture, mining, finance, air traffic, power grids, and scientific research. Several countries have developed the different kind of GNSSs, which include Europe's Galileo, USA's GPS, Russia's GLONASS, China's COMPASS/Bei-Dou, Japan's Quasi Zenith Satellite system (QZSS), and India's IRNSS [2–4].

16.2 INTERACTION OF ATMOSPHERE TO GNSS SIGNALS

When GNSS signals pass through the atmosphere, multiple errors from different sources are introduced, which include ephemeris error, satellite clock error, receiver noise, and atmospheric errors. GNSS signals are delayed within the atmosphere due to the existence of plasma density in the ionosphere and dry/wet components in the lower atmosphere/troposphere. Thus, atmospheric delay in GNSS/GPS signals are caused by ionospheric delay and tropospheric delay. Details about atmospheric delays in GNSS signals are described below.

16.3 IONOSPHERIC ERROR IN GNSS/GPS SIGNAL

In the ionosphere, the carrier phase and codes of GNSS signals are significantly affected. A phase advance in the carrier and a group delay in the code of GNSS signals occurs, resulting in two kinds

DOI: 10.1201/9781032712444-19

of refractive index of the ionosphere: phase refractive index and code refractive index. The code delay in the GNSS signal affects the pseudorange measurements in comparison to the geometric distance between satellite and receiver [2], and the magnitude of delay is directly proportional to the TEC present in the ionosphere along the signal path. TEC is defined as the integration of electron density from satellite to receiver, or, in other words, it is defined as the number of free electrons in a cylindrical column of 1 m^2 cross section area. The ionospheric delay is dispersive in nature, i.e., it depends on the frequency of the signal, which can be eliminated using a dual frequency GPS receiver. Users with a single frequency GPS receiver can use the Klobuchar model to model the effects originated from the ionosphere [2].

During passage of the GPS signal in the ionosphere, the excess phase delay Δt_p (in seconds) in the signal is given as [5]

$$\Delta t_p = -\frac{40.3 \times TEC}{f^2} \tag{16.1}$$

where c being velocity of light c and f is the frequency of GPS signals. The negative phase delay indicates that the phase is advanced. The negative sign in equation (16.1) is an indicator for phase advance in the signal. The phase delay measured in meters is denoted by I_Φ and given by the following equation:

$$I_\Phi = c.\Delta t_p = -\frac{40.3 \times TEC}{f^2} \tag{16.2}$$

Similarly, one can write the group delay measured in meters as follows:

$$I_\rho = c.\Delta t_g = \frac{40.3 \times TEC}{f^2} \tag{16.3}$$

From equation (16.2) and (16.3), it can be noted that both the group delay and phase delay have the opposite sign, but they are equal in magnitude.

For known location, date, and time, if ionosphere having TEC value of 50 TECU, then time delay in GPS L1 signal is 27 ns (or 8.1 mrange), whereas for GPS L2 signal and a time delay of 44.5 ns (or 13.35 m range). Moreover, TEC variations are a function of local time of the day, days of the month (day-to-day variations), month/season of the year, years of the solar cycle, and hence the ionospheric errors in the GPS signals also vary with these parameters. Usually, ionospheric errors in GNSS/GPS signals can be eliminated using the model developed by Klobuchar, named the Klobuchar model after him [2, 6].

16.4　ATMOSPHERIC ERROR IN GNSS/GPS SIGNAL

Aerosols, dry air (O_2, N_2), water vapor, and other gases are components of the atmosphere. In the troposphere, O_2 and N_2 are the main components that affect GPS signals, and these components are known as dry gas components and water vapor as wet components. There are also other components, which include aerosols, tiny particles, and gases (few μm - nanometer), that are unable to scatter the GNSS signals because the wavelength of GNSS signals are not comparable with the size of these atmospheric components. With GPS as an example, the two signals L1 and L2 have the wavelengths $\lambda 1 = 19$ cm and $\lambda 2 = 24$ cm, respectively. Even though O_2 and N_2 do not possess any dipole moment, a small dipole moment is induced when a GPS signal (electromagnetic wave) passes through them. The consequences of this raise the dielectric permittivity of the medium (or refractive index) because induced dipoles in the sphere of influence of radio waves increases.

Moreover, water vapor in the atmosphere has the permanent dipole moment and hence significantly affects the tropospheric radio refractive index (RRI). The dipoles are distributed randomly in normal situations, but partial alignment of the dipole occurs in response to the magnetic field of radio waves (GPS signals) when it passes through the troposphere. Consequently, polarization of the medium takes place, which contributes to the tropospheric radio refractive index. The phase and group velocity changes due to refractive index variation results in a change in the total phase, and path length is used to estimate water vapor present in the troposphere [4]. Computation of Zenith tropospheric delay (ZTD) using GPS and other data sets is useful in water vapor retrieval following two data processing approaches.

16.5 NETWORK GPS DATA PROCESSING

Processing the GPS raw data requires the GPS measurement high-quality, geodetic seek GPS products better demands of accuracy all obtained on the base of ground-based GPS-based and GPS. To use GPS-based products in a very rapid time demands products-automated system. The network of stations using GPS data needs high-demand simultaneous processing approach using the so-called least square adjustment. Least squares adjustment is used to estimate some parameters, like station coordinates, clock parameters, phase ambiguities, etc. The software available for such processing approach are GAMIT and BERNESE. GMAIT software is free and available at http://www-gpsg. mit.edu/gg/. This software may be used on LINUX platform. To get BERNESE software, payment is required.

16.6 PRECISE POINT POSITIONING (PPP) GPS DATA PROCESSING

In order to eliminate/model errors in the GNSS/GPS signals for achieving a high-level position, the igs.orgaccuracy PPP positioning technique is used. In this technique, one by one, data processing for each GPS stations is applied. The satellite orbits and clock parameters that are already estimated as priory data, only station coordinates, clock parameters, ZTD, and phase ambiguities can be estimated. Processing of single station data in this approach is simple but time-consuming. Nowadays, numerous software are available for PPP processing, and GIPSY software is popular among them (https://gipsy-oasis.jpl.nasa.gov/gipsy/software.html). Results using GAMIT software with a network approach has been discussed here [4].

16.7 GPS DATA SETS USED FOR PWV ESTIMATION

In addition to GPS observations, various supporting data are needed to get water vapor using the GPS data. The different data sets that are required to derive water vapor are described below.

16.7.1 OBSERVATION DATA

The hourly GPS observation data is required in near real-time data processing method and daily observation data for offline processing. The GPS observation data at different stations across the entire world are supported by the international organization known as the International GNSS Service (IGS). The spatial distribution of GPS stations, along with their geographic coordinates, data availability, receiver types, etc., are available under the IGS network at this website: https:// network.igs.org/.

16.7.2 NAVIGATION DATA

The navigation data include three types of data, namely ultra rabit, rabit, and final orbit data, which are available here: https://cddis.nasa.gov/. The navigation data are essential to give information about the satellite position in orbit, and more information about data availability and its access through the internet, file nomenclature, etc., are discussed by previous authors [4].

16.7.2.1 BRDC Data

BRDC is the acronym for the broadcast ephemeris data recorded in the RINEX format, and this data is required by GAMIT software in DATA processing for water vapor retrieval. The information about Keplerian elements-parameters is included in BRDC data, which are further used to calculate the location of satellite position. Furthermore, the BRDC files are a combination of the individual location navigation files into one, non-redundant file. In addition, due to the availability of the BRDC data files, it can be used in both the NRT (Near Real Time) as well as in offline processing approaches [4].

16.7.2.2 Meteorological Parameters

The meteorological sensors co-located with the GPS antenna are useful to provide surface met parameters such as pressure, temperature, and humidity. These met data sets with definite sampling can be input into a GPS receiver, which makes it suitable to use the data in the NRT approach, and met data are available at https://cddis.nasa.gov/. In the absence of met data, surface measurements or interpolated value at the GPS antenna location can be used for the same purpose. The method of interpolation of MET data is discussed in the next section.

16.7.3 Computation of PWV from ZTD

Zenith tropospheric delay (ZTD) is expressed as the sum of the zenith hydrostatic delay (ZHD) and zenith wet delay (ZWD). Since ZHD is pressure-dependent, it is therefore termed as hydrostatic delay and ZHD can be modeled as [7]

(16.4)where Ps stands for the pressure measured in mb at the Earth's surface, H is the height above the ellipsoid (in km), and λ is the latitude. As soon as ZHD is calculated, the value of ZWD can be obtained by taking the difference of ZHD from ZTD.

Perceptible water vapor is often expressed as PWV, which is the amount of water vapor present in a column of unit cross-sectional area up to the atmosphere from the surface of the Earth. Sometimes PWV is expressed in the height that water can withstand and measured in mm. Figure 16.1 is a visualized description of PWV.

Knowing the ZWD, PWV can be estimated using the expression given below:

$$PWV = K \times ZWD \tag{16.5}$$

K being constant, which depends on the weighted mean temperature (T_m) of the atmosphere. Value of K is calculated by

$$K = \left[10^{-6} \left(\frac{K3}{T_m} + K2 \right) R_v \rho \right]^{-1} \tag{16.6}$$

R_v being gas constant for the water vapor, ρ is density of water, K2 = 17.4±10 K mbar^{-1} K3 = (3.776 ± 0.0012) ×10^5 K^2 mbar^{-1}. T_m is given by expression as noted below [5]:

$$T_m = \frac{\int \left(\frac{P_v}{T} \right) dz}{\int \left(\frac{P_v}{T^2} \right) dz} \tag{16.7}$$

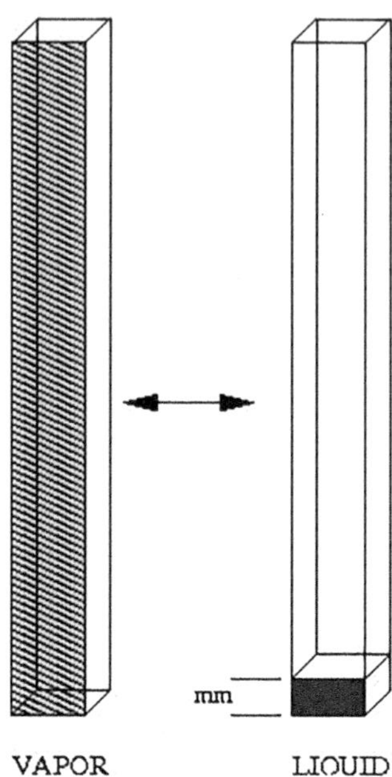

FIGURE 16.1 Description of PWV.

In equation (16.7), P_V stands for partial pressure of water vapor (mb), and T is the temperature (K). Various relationships have been derived to determine T(m) using the formula adjusted for a specific location [5]. The detail study of empirical relationships at different stations in the Indian region found that all the models for T_m agree with deviation of ±1 mm [8]. Equation (16.5–16.7) can be used to estimate the PWV of the air by considering the weighted average of the wind specified at the height of the surface and the GPS antenna. These measurements are possible from meteorological sensors connected to GPS receivers or met data taken at the antenna height. When there is no weather data, either NCEP or ECMWF weather forecast can be used.

16.7.4 National Center for Environmental Prediction (NCEP)

NCEP provides average data set over global reanalysis. These reanalysis datasets can be taken from the NOAA website: https://psl.noaa.gov/data/gridded/data.ncep.reanalysis.html. This website provides multiple weather parameters, which have been discussed elsewhere [9]. In this analysis global data sets are available with spatial resolution $2.5° \times 2.5°$ at the surface as well as at the 17 different altitudes/pressure levels from 1000 to 10 mbar at each grid point. For the data sets, the available pressure levels (in mb) are: 1000, 925, 850, 700, 600, 500, 400, 300, 250, 200, 150, 100, 70, 50, 30, 20, 10. The data sets are available from 1948 onwards with four times daily and monthly mean values. For PWV estimation, MET parameters at GPS antenna location (ϕ, λ, H) are obtained by applying the interpolation to a specific point of the GPS antenna location [4]. The GPS antenna location includes latitude (λ), longitude (ϕ), and height (H).

16.7.5 European Center for Medium Range Weather Forecasts (ECMWF)

ERA-Interim is another weather data source, which provides global reanalysis MET data sets. These data sets in real time are available from 1979 onwards. ERA-Interim reanalysis systems take apart huge data sets and use them in weather prediction models. It provides MET data sets along with global climatology of PWV at 6-hour intervals. ECMWF provides data with resolution from

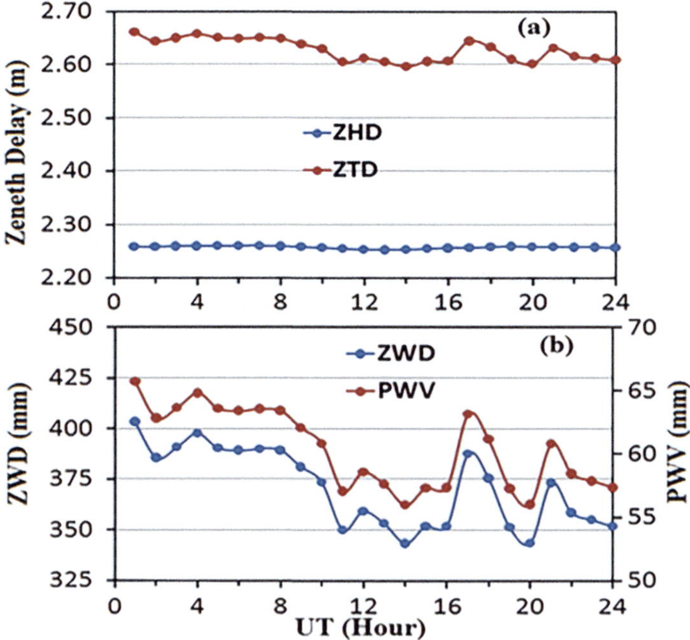

FIGURE 16.2 (a) The diurnal variation of ZTD, ZHD (m), and (b) ZWD and PWV (mm) computed at Varanasi and the analysis of the data are made on specific date 02-07-2010.

$0.125° \times 0.125°$ to $3° \times 3°$. The data sets are available here: https://www.ecmwf.int/en/forecasts. To estimate PWV, interpolation of temperature and pressure from ECMWF to specific antenna point (ϕ, λ, H) is required. The equations 10–12 can be used to obtain PWV from above interpolated data sets [4].

Figure 16.2 shows diurnal variations of PWV along with ZTD, ZHD, and ZWD against UT (Hour) on 2 July 2010 at Varanasi, India. ZWD (or PWV) shows a wide range of variability as compared to that of ZHD because the former depends nonlinearly on MET parameters: pressure and temperature.

16.8 ROLE OF WATER VAPOR IN CLIMATE AND WEATHER

Since in previous studies it was reported that PWV computed from GPS has very high accuracy, therefore, to confirm its accuracy, a comparative study between PWV data with MODIS satellite and ECMWF retrieved is required. The comparative study of GPS-PWV with those from the MODIS satellite and ECMWF data during the period 2007–2010 for Varanasi is shown in Figure 16.3.

From the comparative study, GPS-PWV is well correlated ($R^2 \geq 0.95$) with those from MODIS. Comparing GPS-PWV with those from ECMWF showed good agreement with correlation ~0.86. The comparative study of GPS-PWV with the radiosonde data at Algeria indicates well correlation between the datasets. Moreover, they reported a mean difference between two data sets < 2 mm [9]. In addition, the sharp increase in water vapor (Figure 16.3) is an indication of the monsoon beginning, and this has been used to predict a monsoon onset in the Indian region [10, 11]. The sharp increase in PWV in California on 27 July 2006 during the pre-monsoon season was used for the onset of the monsoon. Therefore, near real-time monitoring of water vapor is reasonably beneficial in weather diagnosis and forecasting applications. It has been found that prior to most precipitation cases, the PWV values increase rapidly by 4–5 mm during an interval of 1–3 hours [12].

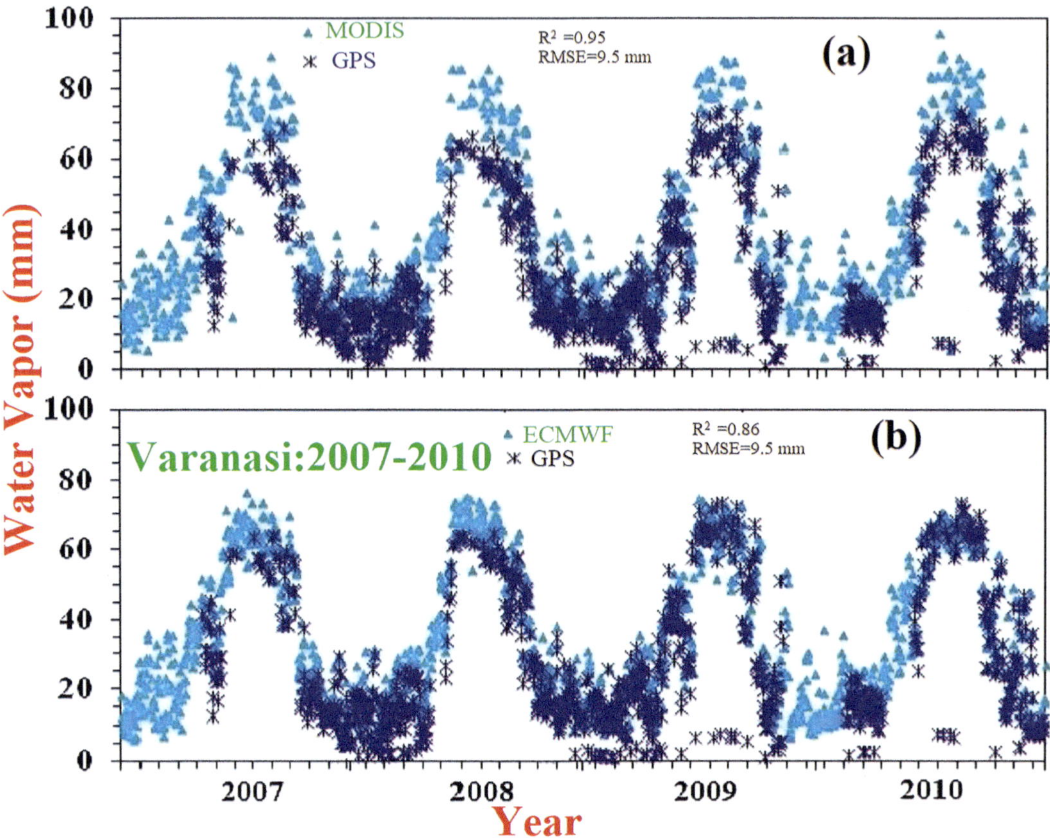

FIGURE 16.3 Comparison water vapor from GPS with those from (a) MODIS satellite (b) ECMWF at Varanasi, India, during 2007–2010.

16.9 FUTURE SCOPE AND CHALLENGES

The greenhouse effect, monsoon change, numeric weather prediction (NWP), atmospheric-thermodynamic, cloud formation, hydrological cycle, and rainfall distribution can be studied by having knowledge of the temporal/spatial distribution of atmospheric water vapor. Diagnosis of NWP modeling and the hydrological cycle requires information about geographical-distribution of PWV, taken as an input parameters while studying the global/regional climatology and circulation pattern. For this, several tomography methods of the troposphere have been suggested for a 3D representation of water vapor [12–14]. The accuracy of the NWP models or tomography techniques is governed by the mathematical procedures and experimental setup to construct tomography of PWV. ZTD are key parameters that decide accuracy of the NWP model [15]. The dependencies of PWV on the humidity make it appropriate for the building of tomography. The tomographic modeling for the real atmosphere is a challenge for the scientific community and should be addressed in future research because the presence of humidity in the atmosphere makes it difficult to model in real time.

Different mesoscale processes including tropical hurricanes/cyclones/typhoons/monsoon systems ranging between 100–2000 km are the key factors that affect the circulation of atmospheric water vapor and hence the climate/monsoon. These mesoscale transport processes have become hot topics that have attracted the scientific community in recent years [16, 17]. The study of water vapor transport along with the moisture and rainfall distribution pattern correlated with the monsoon system is one of the major areas of atmospheric research and monsoon prediction [16]. This type of study could be useful in studying the circulation pattern of monsoon systems and monsoon forecasting around the world using very high accurate and precise PWV measurements such as GPS. Temporal and spatial distribution of PWV using the dense GNSS network available from IGS website along with locally developed measurements are essential to study mesoscale transport and monsoon pattern/monsoon forecasting across the world.

16.10 SUMMARY

GNSSs are a standard tool for positioning and navigation services, but the presence of atmosphere (ionosphere) in their path is the biggest error source that degrades the positioning accuracy of GNSSs with a maximum error of ~10 m. The source of error arises from the ionosphere in L1 signals known as group/phase delay. The Process-level requirements can therefore be translated into accurate measurement signals. The user's location and actual speed are converted into estimates for pseudo-distance in the GPS signal. All this information, including the satellite location and the transmission time, must be required. In addition to measuring ionospheric and tropospheric delays, dual-frequency GPS receivers can measure so-called range, speed and range changes, Doppler frequency measurements, and time changes, which provide information on subtle changes in the atmosphere. The PWV present in atmosphere is sensitive to monsoons, weather conditions, weather events, and global warming, and GPS has demonstrated its ability to measure this. Moreover, a high concentration of water vapor in the atmosphere causes difficulties to persons suffering from bronchitis and heart disease.

REFERENCES

[1] M.C. Freitas, A.M.G. Pacheco, T.G. Verburg, and H.T. Wolterbeek, *Environ. Monit. Assess.* 162, 113 (2010).

[2] B. Hofmann-Wellenhof, H. Lichtenegger, and J. Collins, *Global Positioning System* (Springer Vienna, Vienna, 1992).

[3] G. Xu and Y. Xu, GPS Theory, *Algorithrms Appl.* Third Ed. 1 (2016).

[4] S. Kumar, R.P. Singh, and A.K. Singh, *GPS GNSS Technol. Geosci.* 119 (2021).

[5] J.L. Davis, T.A. Herring, I.I. Shapiro, A.E.E. Rogers, and G. Elgered, *Radio Sci.* 20, 1593 (1985).

[6] P. De T. Setti Júnior, D.B.M. Alves, and C.M. Da Silva, *Bol. Ciencias Geod.* 25, (2019).

[7] G. Elgered, J.L. Davis, T.A. Herring, and I.I. Shapiro, *J. Geophys. Res. Solid Earth.* 96, 6541 (1991).

[8] S. Jade, M.S.M. Vijayan, V.K. Gaur, T.P. Prabhu, and S.C. Sahu, *J. Atmos. Solar-Terrestrial Phys.* 67, 623 (2005).

[9] H. Namaoui, S. Kahlouche, A.H. Belbachir, R. Van Malderen, H. Brenot, and E. Pottiaux, *Adv. Atmos. Sci.* 34, 623 (2017).

[10] A.K. Prasad, R.P. Singh, and A. Singh, *J. Indian Soc. Remote Sens.* 32, 313 (2004).

[11] S. Kumar, A.K. Singh, A.K. Prasad, and R.P. Singh, *Phys. Chem. Earth, Parts A/B/C.* 55–57, 11 (2013).

[12] J. Braun, C. Rocken, and R. Ware, *Radio Sci.* 36, 459 (2001).

[13] A. Flores, L.P. Gradinarsky, P. Elósegui, G. Elgered, J.L. Davis, and A. Rius, *Earth, Planets Sp.* 52, 941 (2000).

[14] T. Nilsson, B. Soja, K. Balidakis, M. Karbon, R. Heinkelmann, Z. Deng, and H. Schuh, *J. Geod.* 91, 857 (2017).

[15] M. Bevis, S. Businger, T.A. Herring, C. Rocken, R.A. Anthes, and R.H. Ware, *J. Geophys. Res.* 97, 15787 (1992).

[16] N. Vigaud, Y. Richard, M. Rouault, and N. Fauchereau, *Clim. Dyn.* 32, 113 (2009).

[17] S.S. Dugam, *J. Earth Sci. India.* 3, 69 (2010).

17 GNSS-enabled Precision Agriculture

Ayushi Gupta, Damanti Murmu, Dileep Kumar Gupta, and Prasad S. Thenkabail

17.1 INTRODUCTION

Satellite navigation (SATNAV) systems are currently in use worldwide. Some operate worldwide, while others are limited to a certain area. The collection of all the SATNAV systems and their augmentation is known as the Global Navigation Satellite System (GNSS).

The Global Positioning System (GPS) was the first global navigation system. Since then, further independent systems of this type have been developed in case an event causes service to be restricted or even shut down. With the use of many GNSS systems, modern ground navigation devices track their location, increasing precision. Global navigation satellite systems (GNSSs) are utilized by scientists and engineers for various tasks, such as autonomous navigation, development, GNSS ionosphere monitoring, GNSS reflectometry, and more, using data.

Researchers in mathematics have improved navigation solutions, scientists have found ways to enhance GNSS receivers (and satellites), and several devices merge many sensors and GNSS receivers to boost the precision of GNSSs. Acquiring exact position and distant sensing has also been simpler with the combination of many navigation systems (e.g., GPS, GLONASS, Galileo, BeiDou) (Yasyukevich et al., 2024). To minimize dependability, Japan manufactured the QZSS constellation (Quasi-Zenith Satellite System), and India developed the IRNSS (Indian Regional Navigation Satellite System) (Barna et al., 2020), which was later renamed NavIC (Navigation with Indian Constellation).

GNSS, which was created for military purposes, has now become a regular element of life. It matters for the productivity of agriculture now as well. For instance, precision farming for livestock and precision crop production relies on navigation systems to give a large amount of geographical information (stationary or dynamic). Advanced farming methods are presently required to fulfill the expanding worldwide demand for agricultural food. The usage and integration of agricultural technology have revolutionized agricultural production processes. It has been demonstrated that integrating technology into agricultural production improves farming practices, increases yield and productivity, and lowers environmental externalities. PA, as discovered by Kitchen et al. (Kitchen et al., 2002), is one of many technologies that favorably affect agricultural productivity in both developed and developing nations.

Agriculture remote sensing is a broad topic extensively studied from various angles. These viewpoints may be derived from applications (such as precision farming, yield prediction, irrigation, and weed detection), from particular remote sensing platforms (such as satellites, unmanned aerial vehicles [UAVs], and unmanned ground vehicles [UGVs]), from sensors (such as active or passive sensing, spatial sampling, and wavelength domain), or specific geographic and climatic contexts (such as wetlands, drylands, countries, or continents). Precision agriculture (PA) is a collection of immediate and remote sensing techniques that employ Internet of Things (IoT) devices to monitor crop status at various growth stages. It is also referred to as "digital agriculture," which describes the use of extensive data sources and highly developed crop and environmental statistical methods to

DOI: 10.1201/9781032712444-20

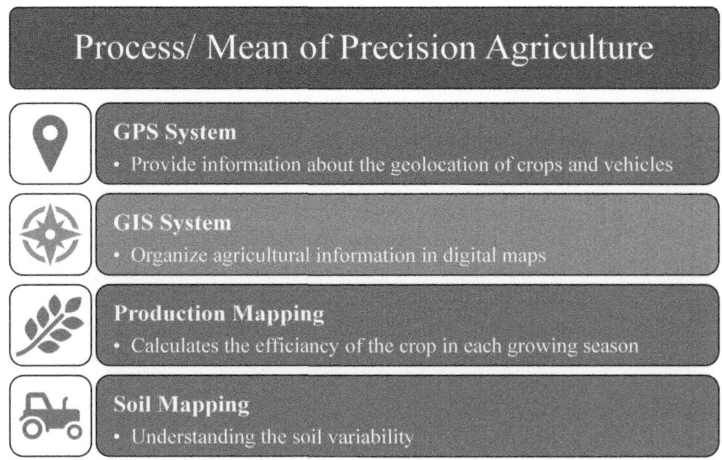

FIGURE 17.1   Illustration for a brief description of processes and means of PA.

help farmers adopt the right management practices at the right rates, times, and locations to meet economic and environmental goals.

Spatial variability plays a role in determining which measurement components to track changes in crop characteristics, such as vegetation, water availability, humidity, soil composition, terrain, and the presence or absence of pests and diseases that affect plants. Some characteristics of soil, such as its organic matter content and composition, do not change significantly or at all over time. With the use of this data, a producer may split a crop into distinct management subareas where different input quantities can be used to optimize overall performance by raising output and lowering the need for agrochemical inputs. PA may accomplish this by integrating a range of technological instruments and analysis techniques about each stage of production, from sowing to harvesting, which is briefly illustrated in Figure 17.1.

17.2   GLOBAL NAVIGATION SATELLITE SYSTEMS (GNSSS)

Precision agricultural technologies now prefer to employ global positioning systems, or GNSSs, which are a class of geolocation systems that use satellites to identify a user receiver's position in an Earth-centered global positioning system (GPS). GPS technology, along with other GNSS technologies, has developed into a potent atmospheric remote sensing tool that can measure atmospheric properties accurately. "A worldwide position and time determination system that includes one or more satellite constellations, aircraft receivers, and system integrity monitoring, augmented as necessary to support the required navigation performance for the intended operation," is how the International Civil Aviation Organization (ICAO) describes the Global Navigation Satellite System (GNSS).

The popularity of GPS has encouraged further advancements in GNSS technology. However, it is possible to characterize these two systems as complementary. The fact that this type of technology functions in all-weather situations adds to its worth. Not only does this GNSS capability offer useful information, but it also performs admirably on cloudy and rainy days—days that pose unforeseen challenges to low-orbiting satellites and radar systems. The processing power and advanced technologies that were previously integrated into GPS receivers must be taken into account when considering potential technical improvements.

The GNSS technology these instruments employ allows for the possibility of land mapping. With the former, a geo-referenced dataset assessment makes it possible to observe the characteristics of the field and outlines the subsequent planning steps (such as its size and perimeter) to be cultivated.

It is possible to conduct an information flow between the management of the farm and the rural machinery. These gadgets, such as GPS, allow for the optimization of automation in conjunction with automobiles, which could repeat (or reject) previously completed paths.

Although most of these approaches are being adopted in flat regions, mountainous contexts are paying more attention to location-specific administration in PA. Because of the work involved in the evaluation, PA is still quite restricted in the latter locations. Hills may be a significant factor in observations that are lacking. One of their primary limitations is that, in comparison to flatter regions, they have the largest mistake probability, even in minor hilly areas. A vehicle in operation can have its correct location and in-field positions continually recorded by the positioning system (e.g., GPS). Figure 17.2 illustrates the GNSS's constellation, services, and timelines, along with positioning and application in PA.

17.2.1 GNSS Constellations

The collection of satellites that comprise a specific GNSS system is known as a GNSS constellation. All the satellites in the constellation cooperate to give users on the ground position and time data. Each satellite is in a specific orbit at a certain altitude. Figure 17.3 illustrates the key parameters of the constellation.

17.2.2 Principle of GNSS

Understanding how satellites orbit the Earth to determine a location and provide directions to any specific place can be challenging; but the fundamental ideas that underpin GNSS systems and their operation can be understood. The following are three main components of every GNSS and represent these segment's (Barna et al., 2020) augmentations and are illustrated in Figure 17.4.

- **The satellites or space segment or Satellite-based augmentation system (SBAS):**
  The actual satellites that orbit the Earth are meant to be understood by this. Six planes with equal spacing comprise the GPS satellite constellation. At least four satellites are carried by each plane, and this configuration guarantees that any receiver like a phone, wristwatch, smart car, etc., always has access to at least four satellites.

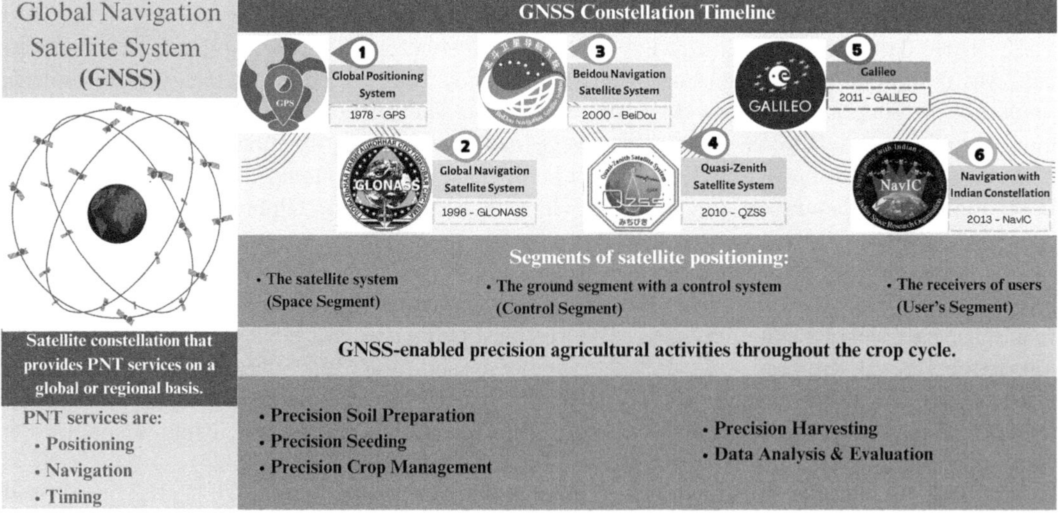

FIGURE 17.2 The image illustrates GNSS briefly with constellation, positioning, and application in PA.

Detailed Global Navigation Satellite System (GNSS) Constellations

SYSTEM	Satellites in Operation	Regime(s)	Orbital Height	Civil Signals
GPS (US)	31	MEO	20,180 km	L1 C/A, L2C, L5, L1C
GLONASS (RS)	24	MEO	19,130 km	L1OF, L2OF, L3OC
Galileo (EU)	30	MEO	23,222 km	E1, E5a, E5b
BeiDou (CH)	44	GEO IGSO MEO	35,786 km 35,786 km 21,528 km	B1I, B2I, B3I, B1C, B2a
NavCI* (IN)	7	GEO IGSO	35,786 km	L5, S-band, BOC
QZSS* (JP)	4	GSO	36000 km	L1C/A, L1C, L2C, L5

FIGURE 17.3 Briefed GNSS constellation. US - United States, RS - Russia, EU - Europe, CH - China, IN - India, JP - Japan. GEO: Geosynchronous Earth Orbit, MEO: Medium Earth Orbit, IGSO: Inclined Geosynchronous Orbit. *Satellites operational for regional level only.

- **The control segments or Ground-Based Augmentation System (GBAS):**
  Tracking the satellites to determine their orbit and timing, uploading data to them, and synchronizing time across the constellation is all done by the ground, or control, segment.
- **The user segment:**
  The user segment comprises the receivers and associated antennas used to receive and decode the signal to provide positioning, navigation, and timing (PNT) information. This includes any devices that have a receiver, including vehicles, airplanes, police enforcement equipment, and cell phones.

17.2.3 VULNERABILITIES OF GNSS IN PA

The integration of digital technology in agriculture is revolutionizing the industry, enhancing efficiency, and optimizing farm management. This transformation boosts productivity, effectiveness, and profits. Moreover, it offers opportunities for more sustainable, eco-friendly, and environmentally conscious farming practices. However, this digital shift also poses potential risks and unintended consequences, such as increased vulnerability in agricultural production, which have not received adequate scientific and public attention. Despite tremendous efforts to evaluate likely failure mechanisms and their consequences on services, as well as to create techniques for identifying and correcting mistakes, all GNSSs are susceptible to failure, disruption, and interference.

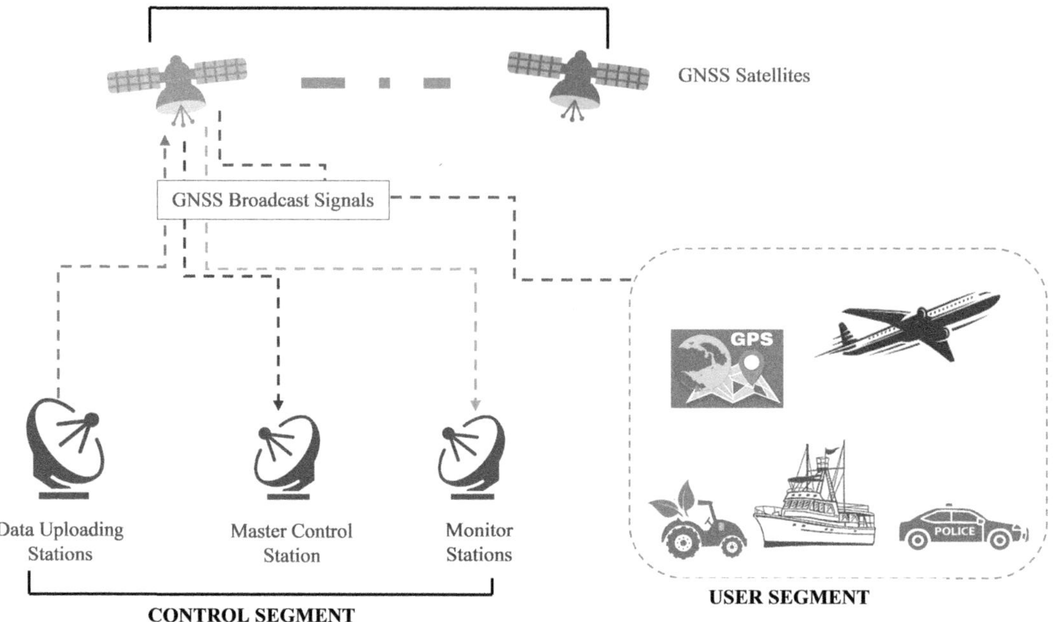

FIGURE 17.4   Working of GNSS in three major segments.

GNSS vulnerabilities are roughly grouped into three categories:

1. System-related (Chen & Zhan, 2021) (signals and receivers).
2. Propagation channels (Zidan et al., 2021) (atmospheric and multipath).
3. Interference-related (Ortega et al., 2022) (either unintentional or deliberate).

GNSS receivers can process valid signals incorrectly, leading to unanticipated results. Occasionally, GPS satellites have broadcast dangerously inaccurate signals. There may be insufficiently visible satellites to allow for the availability of a location fix.

GNSS transmissions are often weak, with power levels of less than 100 watts broadcast across 20,000 to 25,000 kilometers (Hibbert, 2011; Tegedor, 2015). With the spectrum successfully stretched out below the noise level in the receivers, the signal strength at the Earth's surface can be reduced as much as -160 dBW ($1 \times 10-16$) watts. Intentional or unintentional interference with this signal has the potential to quickly disrupt signal recovery or overload the receiver's electronics.

17.2.4 Economic and Environmental Benefits

The agricultural business has transformed thanks to the integration of technologies like Robotic Systems or Smart Machines (RSSM), Controlled Traffic Farming (CTF), Reactive or Variable Rate Technologies (VRT), Recording and Mapping Technologies (RMT), and Farm Management Information Systems (FMIS). Increased yields, reduced use of pesticides and fertilizers, water conservation, and labor, fuel, and expense savings are just a few of the substantial economic advantages brought about by these technologies. These technologies' economic benefits have led to notable improvements in crop yields, ranging from 5% to 70%, and considerable reductions in fertilizer consumption and expenditures, with savings ranging from 1% to 82%. Additionally, these technologies have led to savings in Plant Protection Products (PPP) through precision application technologies, water conservation in agriculture, with reductions in water usage reported between 16% and 90%, and significant efficiencies in fuel consumption, cost savings, and labor efficiency improvements.

The environmental benefits of these technologies have also been significant, with reported reductions in CO_2 emissions between 5% to 56% (Papadopoulos et al., 2024), notable improvements in soil health metrics, including increased organic matter levels and reduced soil erosion rates, and biodiversity improvements, showcasing a notable increase in local biodiversity hotspots by up to 18%. Furthermore, these technologies have promoted efficient water use, with reductions in water runoff and improved water-use efficiency, contributing to environmental sustainability and reduced environmental impact. Overall, these technologies have the potential to transform the agricultural industry, promoting sustainable practices, reducing environmental impact, and improving economic efficiency.

17.3 PAST GNSS-BASED STUDIES FOR PA

In most operating scenarios, PA refers to applications that need precise locations that are consistently and highly accessible down to the centimeter. Controlled traffic farming, automated steering, and machine guidance enable machines to follow predictable tracks across the field, minimizing pass-to-pass errors and overlaps. Satellite-based navigation has recently enabled the development of semi-autonomous machines for certain farming scenarios and tasks. The farming sector is increasingly aiming to deploy tiny robots to increase farm efficiency and advantages, as they can do complicated jobs that traditional large-scale agricultural gear cannot.

While the requirements of PA applications are frequently met by cutting-edge Real Time Kinematic (RTK) GNSS receivers in vast areas, propagation effects diminish performance near obstacles or beneath vegetation. To assess the actual performance of RTK GNSS-based devices in real-world settings, Pini et al. (Pini et al., 2020) describe an experimental testbed and methodology (such as functional block diagram, measurement reference system selection, key performance indicators, projection of the calculated coordinates to the referring system trajectory, and hypothetical system calibration). A testbed and methodology like this helped contrast various devices that performed similarly in open skies but differently in other types of environments.

Gawande et al. (Gawande et al., 2023) examined the potential benefits of precision farming technology for attaining ecologically responsible and sustainable agricultural practices. Precision farming employs targeted uses of resources such as water, fertilizer, and herbicides to minimize waste and lessen its environmental effects. Modern technologies like GPS, GIS, remote sensing, and data analytics are combined with precision farming to enable farmers to make well-informed decisions based on current data. Higher production and improved agricultural management are the outcomes of this.

Positional precision utilizing GNSS signals has improved, leading to a rise in the use of GPS-based devices in PA. The opportunity to include GPS into the PA, following user demands and specifications, has been greatly expanded by it. Additionally, as time has gone on, GPS devices have become more sophisticated. Pandey et al. (Pandey et al., 2021) saw the availability of GPS signals in a range of instruments and equipment as a growing trend and market opportunity in Pennsylvania. This led to the development of creative techniques for route optimization for the most effective use of fertilizer spray and associated agricultural practices, which have evolved into smart farming.

A broad definition of the Precision Agriculture Management System (PAMS) is described by Kahveci et al., (Kahveci, 2017) as a system built on cutting-edge information, communication, and navigation technology. This system divides fields into small areas rather than managing them as a single unit. For each area, various data and information about the characteristics of the soil and vegetation (such as yield, moisture, texture, structure, nutrient status, landscape position, etc.) are gathered and stored in a GIS database. In industrialized nations, PA is progressively using more current technology (e.g., Information & Communication Technology [ICT], etc.). Due to its advantages in lowering labor and mineral fuel costs and saving time by limiting the number of hours that machinery must operate in the field, the use of these technologies is unavoidable in Turkey, where there are many dispersed farming fields. Thus, it is essential to start building the required infrastructure and encourage farming's usage of GIS and GNSS. As a result, it is feasible to merge ICT, GNSS, and GIS technologies to create an efficient PAMS.

According to Shafi et al. (Shafi et al., 2019), the agriculture sector may become more dynamic and intelligent while utilizing fewer human resources by automating agricultural events using the IoT. This will replace the current static and manual sector. Wireless Sensor Networks (WSN) and PA are the primary forces for automation in the agricultural sector. Crops may be seen in high definition by satellite or unmanned aerial vehicle imagery, which is then analyzed to extract data for use in making decisions in the future.

The article reviews both near- and far-field sensor networks in the agricultural area and highlights various issues and obstacles. The platforms for obtaining spectrum images of crops, the standard vegetation indices for analyzing spectral images, and the uses of WSN in agriculture are all covered in the study, along with sensors, wireless communication technologies, and nodes for evaluating environmental behavior.

17.4 APPLICATIONS OF GNSS IN PA

GNSSs have already altered our everyday operations in a great number of ways, notwithstanding the challenges. Numerous contexts, such as homes, workplaces, and educational institutions, have been recognized as potential applications for GNSS.

GNSSs are also used in agriculture and other industries. This includes anything from monitoring approaching trucks to making farm equipment run more smoothly. It includes tracking cattle, managing forests, and keeping an eye on the soil. PA is a crucial part of the sustainable agriculture systems of the twenty-first century. Although the term PA has been defined in many ways, the fundamental idea has not changed. With the use of advanced communication, knowledge sharing, and data analysis techniques, PA is a management strategy that increases crop productivity while reducing water and nutrient losses as well as adverse environmental effects. Decisions about fertilizers, pesticides, water, seed, fuel, labor, etc., are made using this strategy. According to Misara et al. (Misara et al., 2022), multiple aims can be achieved; furthermore, when applied to prudent soil practices, precision agricultural concepts can gain the advantages shown in the illustration in Figure 17.5 below. Each of the applications discussed is further explained in the subsection.

17.4.1 Soil Sampling and Mapping

Understanding soil characteristics helps improve farming decisions and soil management practices. In PA, monitoring soil status is crucial for adjusting tillage, fertilization, and irrigation. PA leverages

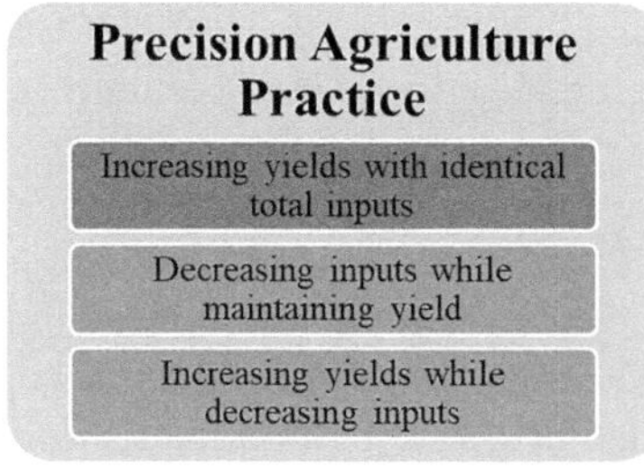

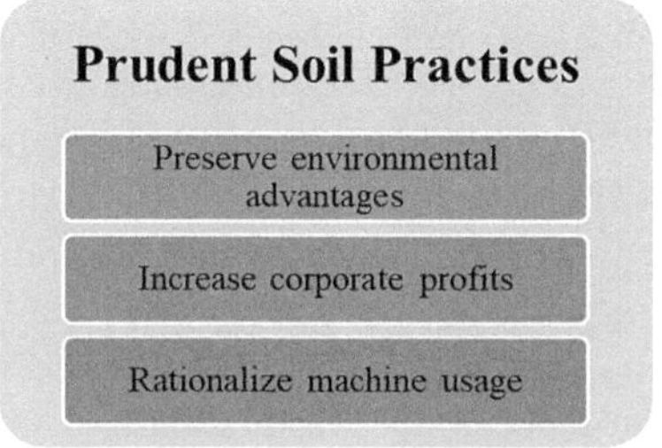

FIGURE 17.5 Chart illustrating multiple aims of PA practices and their advantages for prudent soil practices.

advanced technology to maximize crop yields and minimize waste. Soil sampling and mapping are vital components of PA, enabling farmers to identify soil type, texture, and nutrient content; detect variations in soil properties within fields; create detailed soil maps for targeted management; optimize fertilizer and irrigation applications; monitor soil health; and make data-driven decisions.

17.4.2 Variable Rate Technology (VRT)

In PA, variable rate technology, or VRT, is applied precisely to different types and amounts of inputs on different land parcels under the needs and circumstances of the moment. Regarding VRT, a multi-year analysis of one or more variables influencing crop production and the yield values of the target crop is carried out. With the help of this study, management zones with precisely specified features and management suggestions are successfully developed. This approach may be used to assess the factors impacting crop output in both weighted and non-weighted studies.

17.4.3 Precision Planting and Seeding

Currently, a range of mechanically driven devices, usually powered by hydraulic motors or wheels in contact with the ground, are used to place seeds in row planting equipment. Using these techniques to supply mechanical power for seed metering device operation is an affordable and dependable alternative. The metering devices are driven by a ground-driven wheel through a common drive shaft that varies the metering rates according to the ground speed of the implement.

Another option for planters is to utilize a hydraulic motor, which adjusts the flow rate supplied to it to manage metering rates (Iacomi & Popescu, 2015). A set of row units is usually composed of sections that are eight rows wide, which means that each of the eight-row units must be either on or off (e.g., pressurized drum-type precision seed meters), which is diagrammatically represented in Figure 17.6. Furthermore, it must plant at the same rate of population growth throughout the section. Each row-unit can be turned on or off, but doing so calls for a clutching mechanism at each row. This makes it possible to turn off any row as needed, but the mechanism to do so is pricy and complex technically.

For this reason, most of the planting implements rarely have this system. An electronically actuated seed meter's development and design should be able to measure seed population at a level comparable to that of currently on-the-market machinery or devices. Farmers will find the new seed-measuring gadget more appealing due to its increased precision.

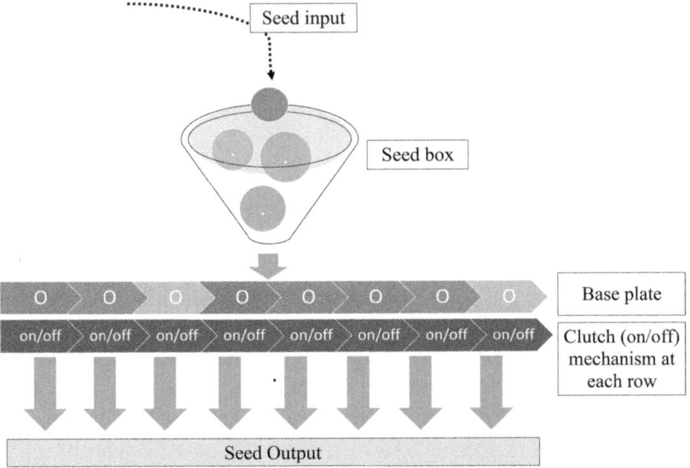

FIGURE 17.6 Diagrammatically representation of pressurized drum-type precision seed meters.

17.4.4 Harvesting Operations and Yield Mapping

For analysis and assessment in PA, spatial yield information is a must. With sensors fitted to the harvesting equipment, automated crops make yield mapping simple to perform. In addition to being used to determine input rates, yield maps offer feedback on how crops react to certain soil and crop management techniques. A significant way to integrate several crop and soil factors, such as moisture, nutrients, and insect issues, is through crop yield. This lays the foundation for managing farmlands according to the characteristics of the crops and soil. Since hand harvesting is extremely time-consuming yet yields higher sampling quality, yield mapping from such data is difficult.

However, several factors that affect data quality, such as the need for continuous calibration and inspection of sensor systems before and during harvesting, limit the yield mapping that can be done from yield monitors that collect high-density data with fewer personnel. When problems with the sensor system, including incorrect calibration, impact the yield monitor data, RS might be a useful alternative for assessing yield maps. Big data is any data produced by RS in sufficient amounts to be considered meaningful. RS techniques give non-destructive, instantaneous, quantitative information on agricultural conditions and yield estimation. To predict annual agricultural productivity, RS imagery is commonly utilized instead of yield monitoring.

17.4.5 Irrigation Management

Availability of water is a key production factor that influences crop productivity. The amount and effectiveness of fertilizer applied also depend on the moisture content of the soil. Crops experiencing water stress result from less soil water being available to fulfill evapotranspiration demands. Canopy temperature may give constant information on the condition of water, water usage, and metabolic processes in a plant, and it has been demonstrated to be a useful indication of plant stress. To plan irrigation and monitor crop water status in real time, farmers and businesses have implemented a variety of crop monitoring and irrigation techniques, such as drip irrigation, furrow irrigation, and sheet irrigation.

17.5 INTEGRATING GNSS WITH OTHER TECHNOLOGIES

17.5.1 Geographic Information Systems (GIS)

GIS and GNSS are technologies employed by modern agriculture to run their fields and manage their farms. When collecting data from field operations including soil sampling, planting, dispensing supplies, crop sampling, harvesting, yield evaluation, and irrigation, GNSS position data are utilized to georeference the GIS database. Users of GNSS positioning for PA are concerned about position alterations when they return to the same spot in the field later from a practical standpoint, where integrating GNSS with GIS enables the advantages.

17.5.2 Remote Sensing and Unmanned Aerial Vehicles (UAVs)

Aspects of RS such as deployment length and range, sensor distances from the target, time and periodicity of image acquisition, location, and coverage area are all influenced by the kind of platform. Unmanned Aerial Vehicles (UAVs) have the potential to completely transform traditional remote sensing (RS) platforms about real-time crop monitoring, weeds identification, foliage classification, water stress evaluation, disease identification, yield prediction, and various pest and nutrient management tactics. UAVs deliver high-quality data and photographs to agronomists, growers, and industry service providers. UAVs may perform a wide range of agricultural tasks to assist PA, including soil health screening, planning irrigation schedules, planting seeds, applying fertilizer, and weather analysis. Compared to other RS platforms, UAVs provide several benefits.

17.5.3 Internet of Things (IoT) and Smart Farming

The Internet of Things is one of the most important technological advancements in precision and smart agriculture. IoT architecture for agriculture, which combines agricultural sensors with Information and Communication Technology (ICT) and UAV, is used to gather data on PA (Saranya et al., 2023).

Mobile data and the growing Internet of Things are at the core of the fourth industrial revolution. Meanwhile, advancements in wireless networking and technology for communication (5G, LoRaWAN, NB-IoT, Sigfox, ZigBee, and Wi-Fi) (Liu et al., 2021) have expanded the application of IoT in multiple domains, such as high-throughput phenotyping and real-time remote control, while improving coverage, bandwidth, connection density, and end-to-end latency.

17.5.4 Sensor Technologies and Data Analytics

A foundation of integrated sensor technology and data efficiency drives modern PA, resulting in a synergy that improves resource management and agricultural output. Different protocols and data formats are employed by different sensor systems in smart agriculture. As systems get more complicated, interoperability becomes increasingly crucial. Improving interoperability allows for the consolidation and cogent analysis of data to maximize yield and resource usage.

Massive volumes of sensor data require rigorous administration and analysis, which is made easier by artificial intelligence and clever algorithms. Large-scale sensor-generated data analysis necessitates complex analytics, ranging from traditional statistical methods to modern machine learning. Table 17.1 gives a summary of various analytics and management techniques for agricultural sensor data.

17.6 CASE STUDIES

GNSSs have been used in precision farming in agriculture to increase crop yields and productivity. With further developments, smartphone technology and the Internet of Things will enable agricultural activities such as monitoring, regulating, safeguarding, and detecting to be expanded. For instance, the IoT has helped farmers monitor the health and well-being of animals by reducing the tedious process of monitoring cow conditions. Additionally, deep learning (DL) and image processing are the fundamental components of weed identification via machine vision (MV) (Javaid et al., 2022).

Implementing GNSS-guided irrigation systems for rice and wheat crop irrigation is used in India (Pandey et al., 2021). Monitoring and carrying out agricultural activities over a vast geographical coverage frequently needs advanced and precise positioning information to optimize operating costs and minimize expected completion times. Large-scale agricultural techniques are challenging to remotely manage until accurate positioning data is obtained. This requires using a GPS to obtain sample locations via the GNSS. As a result, the increased use of GPS-based sensors in PA is a result of the improved positional accuracy achieved by using GNSS signals.

TABLE 17.1

Various Digital Advanced Technologies with Applications in PA

Technique/Methodology	Application in PA	Reference
Machine Learning	Weather predictive modeling, infestation forecast	(Amani & Marinello, 2022)
Statistical Analysis	Soil analysis, yield prediction	(Alwis et al., 2022)
Edge Computing	Real-time soil monitoring	(Iftikhar et al., 2023)
Neural Networks	Early detection and image recognition	(Ramachandran et al., 2022)
Decentralized Data Storage	Multiple source data storage	(Saba et al., 2023)

Senapaty et al. (Senapaty et al., 2023) presented the IoTSNA-CR technique from their study, which employed IoT technology to classify soil nutrients and generate crop recommendations to optimize fertilizer use and boost farmers' productivity. The deployment of AI harvesters in Israel demonstrates the potential of artificial intelligence, machine learning, cloud services, sensors, and automation to give farmers real-time information and assistance. The proposed IoTSNA-CR model integrated machine learning methods, cloud storage, IoT sensors, and an enhanced algorithm (MSVM-DAG-FFO) to attain high accuracy in soil analysis. By storing soil data on the cloud, the method allowed farmers to reduce costs and boost productivity.

The model's efficacy for crop prediction and soil health management was validated by experimental validation, highlighting the significance of collecting data in real time, growing data sets, and applying frequent applications to increase soil quality and make well-informed decisions. To inspect insect traps in olive groves, Berger et al., (Berger et al., 2023) proposed the use of unmanned aerial systems (UASs), along with unmanned ground vehicles (UGVs) in PA.

Two research projects from the University Tenaga Nasional in Malaysia describe robotic and autonomous equipment for handling pesticide and fertilizer applications. An affordable agricultural robot for crop monitoring, insect identification, and fertilizer and pesticide spraying was presented by the authors, Gafar et al., (Ghafar et al., 2021). Although labor expenses were reduced by the prototype system's autonomous operation, productivity was marginally lower than that of human workers. An IoT-based automated organic fertilizer mixer was developed to reduce labor costs and increase output (Hadi Ishak et al., 2021). The updated mixer allowed for remote monitoring, updates, and alerts, which significantly streamlined the process of mixing organic fertilizer.

For greenhouses, the Beijing Research Center of Intelligent Equipment for Agriculture has developed a robot system for selecting cherry tomatoes. This novel cherry tomato harvesting robot system was designed using a manipulator, picking end-effectors, a visual servo unit, and a railed-type vehicle. According to field studies, picking tomatoes takes an average of 12 seconds per bunch with an 83% success rate. Jin et al. (Jin et al., 2019) created a compact electric harvester for vegetable seeds that features an optical fiber sensing system and a power drive, and it offers excellent accuracy and efficiency by continuously monitoring the sowing parameters for different seed sizes.

One barrier to the adoption of innovative technology in agricultural fields is inadequate power infrastructure. In the United Arab Emirates, researchers successfully created an Internet of Things (IoT)-based solar-energy-powered smart farm irrigation system that gathered renewable energy for smart farm irrigation. The results of this study opened the door to the creation of three operating modes that farmers might utilize.

Applications of PA in agroforestry were reported by other research. Identification and categorization of tree species are critical for both monitoring ecosystem health and addressing climate change. Researchers proposed techniques to use UAV photos to assess the distribution of tree types in a large forest region using images from SENTINEL-2 (Chung et al., 2023; Ma et al., 2021). They helped with forest planning and preparedness for the effects of climate change by clearly identifying grassland regions, deciduous trees, and evergreen trees. To enhance textural feature separation between different tree species, Ma et al. (Ma et al., 2021) employed satellite photos and a random forest classifier. Tree distribution was impacted by aspect, slope, and altitude, yet overall classification accuracy was 86.49% with a 0.83 Kappa value. These results were significant for monitoring biodiversity, classifying species, and guiding inventory estimation.

17.7　CHALLENGES AND LIMITATIONS

As PA continues to revolutionize the farming industry with its cutting-edge technology and data-driven approaches, several challenges and limitations have emerged that threaten to hinder its widespread adoption and effectiveness. Here are the challenges and limitations of implementing GNSS in agriculture.

17.7.1 Technical Challenges

PA, while promising significant advancements in farming efficiency and productivity, faces several technical challenges that must be addressed for its full potential to be realized. The successful implementation of PA relies on overcoming a range of technical challenges, including issues with signal interference, accuracy, connectivity, and data management, which can hinder the effectiveness of this innovative approach to farming.

1. Signal Interference: GPS technology is crucial for PA but is susceptible to signal interference from tall buildings, trees, and atmospheric conditions. This interference can lead to inaccuracies in location tracking, which are critical for operations like planting, spraying, and harvesting (Yui, 2024).
2. Accuracy Issues: Although GPS provides high accuracy for many agricultural tasks, achieving sub-meter precision can be challenging. This level of accuracy is essential for certain applications, such as seed spacing and fertilizer application, where even slight deviations can affect yield and resource efficiency (Zhongshan Samli Drones Co., 2024).
3. Data Management: Large volumes of data are produced by PA from a variety of sources, including sensors, drones, and satellites. One major technological challenge is organizing, processing, and interpreting this data effectively so that judgments may be made (Taylor, 2023).
4. Connectivity Issues: Many PA technologies rely on stable internet connections, which can be problematic in rural or remote farming areas. Inconsistent connectivity can hinder real-time data transfer and communication between devices (Fakhruddin, 2017).
5. Integration with Existing Systems: It might be difficult to integrate new PA technology with current machinery and methods. This necessitates cross-platform and cross-system compatibility, which is frequently absent and results in inefficiencies and higher expenses (Fakhruddin, 2017).

17.7.2 Economic Barriers and Cost Considerations

Among the many advantages of PA are higher crop yields and a less environmental footprint. However, several economic barriers and cost considerations hinder its widespread adoption; the upfront acquisition costs of PA technologies are often prohibitive for many farmers. This includes the cost of advanced equipment, software, and training, which can be substantial for small- to medium-sized farms. Farmers, particularly those in rural areas, may have limited access to financial resources or credit facilities, making it difficult to invest in precision farming technologies. The return on investment (ROI) for PA technologies can be uncertain due to variable factors such as market prices, weather conditions, and the learning curve associated with new technologies. This uncertainty makes farmers hesitant to invest heavily. The cost of managing and analyzing large datasets generated by PA technologies can be high. This includes expenses related to data storage, software, and skilled personnel needed to effectively interpret the data. In many regions, there is a lack of government subsidies or incentives to support the adoption of PA technologies, which could help offset the initial costs for farmers.

17.7.3 Adoption Challenges Among Farmers

PA is a potent technique for meeting the problems of rising demands on limited resources. Aside from the benefits, farmers, researchers, scientists, and agriculturalists encounter a variety of challenges when putting it into practice. The adoption of PA technologies, such as IoT and wireless sensors, presents several challenges for farmers. While these technologies offer numerous benefits, including data accumulation and revenue support, farmers face issues with data ownership and exploitation by service providers.

The distribution of IoT systems requires careful consideration of various factors, including the ability of devices to connect anywhere, anytime with minimal configuration, support for seamless connectivity across different networks, and ensuring consistent performance and accuracy. Moreover, the increased reliance on technology raises concerns about system failures, which can have severe consequences, such as food safety compromises, and single-point failures, which can lead to downtime and losses.

17.7.4 REGULATORY AND POLICY ISSUES

PA presents numerous opportunities for enhancing agricultural efficiency and sustainability, but it also introduces several regulatory and policy challenges:

1. Data Privacy and Security: PA involves collecting vast amounts of data, including farm-specific and personal information. There are concerns about how this data is stored, used, and shared. Farmers are often worried that their data might be accessed or used without their consent, which could lead to regulatory actions or competitive disadvantages (Szira et al., 2023).
2. Lack of Standardized Regulations: There is no unified regulatory framework for PA, leading to inconsistent policies across regions. This lack of standardization can hinder the widespread adoption of precision technologies and create legal uncertainties for farmers and technology providers (Fusonie & LaRocco, 2021).
3. Intellectual Property Issues: PA technologies often involve proprietary software and algorithms. Farmers may face restrictions on their use, modifications, or transfer of these technologies, raising concerns about intellectual property rights and licensing agreements (Fusonie & LaRocco, 2021).
4. Environmental Regulations: PA aims to promote environmental sustainability, but it must align with existing environmental laws and policies. This includes regulations related to pesticide use, water management, and land conservation (Szira et al., 2023).
5. Policy Support and Incentives: Government policies play a crucial role in supporting PA adoption. However, inadequate policy frameworks or lack of financial incentives can impede the adoption of these technologies. Effective policies are needed to encourage research, innovation, and adoption while addressing farmers' concerns (Lencucha et al., 2020).

Addressing these regulatory and policy issues is essential for maximizing the benefits of PA and ensuring its sustainable integration into the agricultural sector.

17.8 FUTURE TRENDS AND INNOVATIONS

The advancement of GNSS technology is driving PA forward through several key developments (illustrated in Figure 17.7), which have significantly improved the accuracy, reliability, and functionality of positioning systems through various innovative developments, including improved signal accuracy and reliability, increased satellite constellations and signals, enhanced receiver sensitivity, and the integration of GNSS with other sensors, ultimately leading to more precise and efficient farming practices.

17.8.1 INTEGRATION WITH EMERGING TECHNOLOGIES

17.8.1.1 Modern Remote Sensing-based PA Solutions with GNSS

Light detection and ranging (LiDAR), UAV, harvesting robots, and NDVI were the most prevalent remote sensing-based options. The use of NDVI and LiDAR in agricultural and PA studies has been

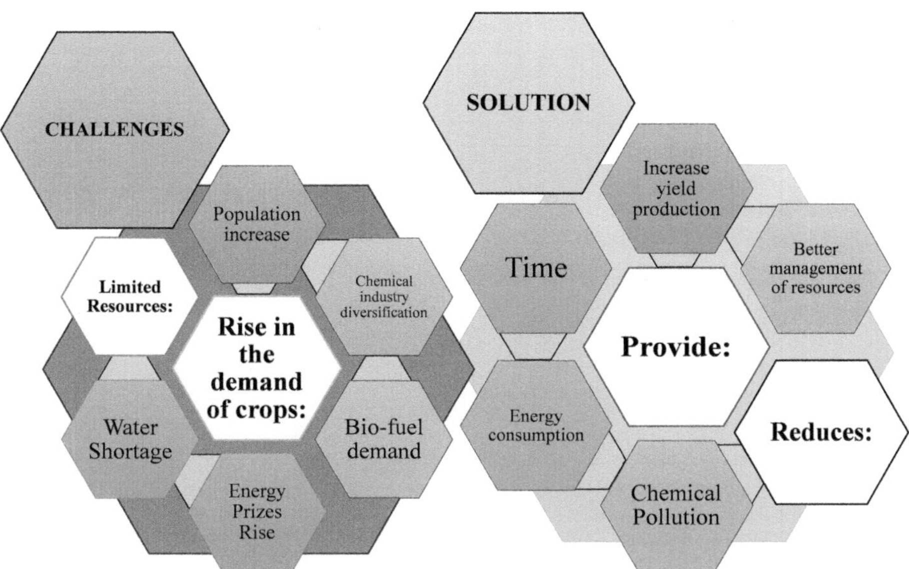

FIGURE 17.7 Answers for agriculture challenges through PA.

gradually increasing, but since 2010 and 2016, respectively, research on UAVs and harvesting robots has accelerated. All these solutions depend on GNSS; however, only a little more than ten percent of the examined papers for NDVI, LiDAR, and harvesting robots have GNSS as their main area of study (Table 17.2).

17.8.1.2 Modern Computer Processing-based PA Solutions with GNSS

Geo-statistics is a prominent and widely recognized field when it comes to computer processing-based solutions connected to GNSS and PA; however, the following solutions are fresh innovations. Computer processing-based solutions, such as accurate point positioning, synchronous localization and mapping, and geo-statistics, were considerably more closely tied to GNSS than remote sensing-based solutions (Table 17.3). Supported by the GNSS technology, the IoT and deep learning (DL) have expanded rapidly, while their main attention has been diverted to other advancements. Despite this, current trends indicate that the usage of GNSS in PA will probably improve because of the significant tailwinds from the IoT and DL research.

These future trends and innovations will further transform the agricultural industry, enabling more efficient, sustainable, and productive farming practices. GNSSs will play a critical role in this transformation, providing the precision and accuracy needed to support emerging technologies and sustainable agriculture practices.

17.8.2 POTENTIAL FUTURE APPLICATIONS

In industries including surveying, remote sensing, monitoring objects or animals, and aviation, GNSS technology is now preferred for its location and timing capabilities. Systems for global navigation satellites provide previously unheard-of levels of positional precision, 24/7 worldwide coverage, and comparatively cheap infrastructure requirements. As a result, it is anticipated that GNSSs will be crucial to the next wave of applications that emerge in the upcoming years (Table 17.4). It is important to note that because of their technological limitations, classic GNSS receivers are mostly employed in open-sky environments. Indoor positioning is one of the biggest problems facing modern GNSS technology, yet GNSSs can be utilized practically anywhere with the use of more sophisticated algorithms and better hardware.

TABLE 17.2
Studies of PA with GNSS

Sensor/Satellite/ System	GNSS Technology-enabled Ends/ Findings	Outcomes/ Result:		Reference
		Before GNSS is enabled	**After GNSS is enabled**	
Multispectral sensors (MS) (Landsat, Sentinel); MS-mounted unmanned aerial vehicles (UAVs); handheld or tractor-mounted radiometer	Georeferencing of measurements enables spatially explicit crop status monitoring and analysis.	Plant health and vegetation vigor	Time-series Analysis and Vegetative Growing Season(Radočaj et al., 2020)	(Pallottino et al., 2019; J. Zhang et al., 2020)
Airborne LiDAR sensors: e.g., lasers, and detectors, installed UAVs Ground-based LiDAR sensors	GNSS is used by ground-based LiDAR systems to precisely georeference the data they gather, establishing a precise connection between the three-dimensional measurements and their locations in the field.	Provide information about the three-dimensional composition of crops. Plant canopy, terrain height, and crop structural elements are all captured on camera by the laser beams.	Enables accurate LiDAR data georeferencing by coordinating the orientation and location of the sensor with the measured values.	(Banerjee et al., 2020; Hariz et al., 2023; Ravi et al., 2018)
RGB cameras sensor (Harvesting Robot)	Enables precise geolocation information to be used for these robots' navigation and operation, leading to more accurate and efficient harvesting operations.	Improve agricultural harvesting by automating labor-intensive operations.	Precise localization, RGB camera computer vision for crop detection and identification.	(Karkee et al., 2021; Kootstra et al., 2021; Mavridou et al., 2019)
Advanced locating methods using GNSS technology, such as RTK and Post-Processing Kinematic (PPK) using Unmanned Aerial Vehicles	By lowering the possibility of spraying outside the approved area, aerial spraying technologies minimize environmental damage and maximize resource use.	Affordable. Effective methods for gathering and processing data.	Give users the ability to geolocate with extreme precision, which boosts the effectiveness of PA's data collection and processing.	(Aslan et al., 2022; Gašparović et al., 2020; Lu et al., 2022; Merz et al., 2022)

TABLE 17.3

List of Computer-based Options Where GNSS Approach Is Used in PA

Computer-based Option	GNSS Technology/ Advanced Techniques/Approach	Function for PA	GNSS Active/Focused Results	Reference
1. **Geo-statistics**	Kriging 80, Variable Rate Technology (VRT)	Draw a map showing the variations in soil, vegetation, and terrain.	Digital elevation models are created using GNSS data (DEMs). Gather information on the increase and fluctuation of agricultural output.	(Oliver & Webster, 2014)
2. **Precise Point Positioning (PPP)**	VRT, Inertial Measurement Units (IMUs) and cameras	Enhances the field's ability to identify spatial heterogeneity.	Allow users more positional flexibility in PA.	(Hu et al., 2022; X. Zhang et al., 2022)
3. **Simultaneous Localization and Mapping (SLAM)**	LiDAR, cameras, and IMUs	Technology that can both map out a space and pinpoint a robot's or vehicle's location within it.	Used to guide agricultural machines precisely.	(Ding et al., 2022; Li et al., 2022)
4. **Internet of Things (IoT)**	IoT sensors	Collecting meteorological information, such as temperature, humidity, and soil moisture.	Provides the accurate position and timing data needed by several Internet of Things applications.	(Ayaz et al., 2019; Pandey et al., 2021)
5. **Deep Learning**	IoT sensors, Convolutional Neural Networks (CNNs), UAV imaging, satellite imagery analysis, and livestock monitoring	Precision agricultural data analysis.	Accurate geolocation information for satellite photos, allowing deep learning systems to track crop progress over time.	(Ampatzidis et al., 2017; Nguyen et al., 2020)

17.9 CONCLUSION

PA uses GNSSs to enhance crop yields and reduce waste. GNSSs provide location and time information, enabling PA to track equipment, livestock, and crops in real time; monitor soil conditions and manage crops with precision; enable autonomous farming, smart farming, precision irrigation, and predict yields; monitor livestock movement and health; and create accurate farm maps.

The principle of GNSS is based on trilateration, where a receiver uses signals from multiple satellites to determine its location. The GNSS constellation consists of various satellite systems, including GPS (USA), GLONASS (Russia), BeiDou (China), QZSS (Japan), Galileo (Europe), and NavIC (India), providing global coverage and accurate location information.

PA is a farm management strategy that maximizes agricultural production while cutting waste by utilizing cutting-edge technologies. The idea behind PA is to use precise data on crops, soil, and weather to inform data-driven farm management decisions. In PA, GNSSs are essential because they help farmers make accurate decisions regarding planting, irrigation, fertilization, and harvesting. PA is undergoing a revolution thanks to the GNSS's integration with other technologies like AI and IoT,

TABLE 17.4

GNSS Advancement, and Status, with Future Possibilities in PA

Advancement	Current Status	Recent Development/ Measurements	Future Possibilities in PA	Reference
1. **GNSS network and satellite constellations**	GNSS network stations: 1000 regional continuous GPS stations and the 300 permanent IGS stations located across the world. For instance, China has around 300 continuous GPS stations, while Japan has 1000 GPS Earth observation networks.	GLONASS, GALILEO, and Beidou, developed with multi frequencies. Increase of future GNSS satellite constellations and LEO satellite missions.	At both global and regional scales, it is feasible to follow more monitor processes and advances in the ionospheric profiles. Moreover, GPS networks with multiple pathways from the ground may be able to provide beneficial predictions of the moisture content of the local soil.	(Wu et al., 2010)
2. **Advanced GNSS receivers**	The Blackjack GPS receiver was created by the Jet Propulsion Laboratory (JPL) and used in several projects, including JASON-1 (2000), ICESat (2001), GRACE (2002, 2003), etc.	Surrey Satellite Technology Limited (SSTL) is working with colleagues from the National Oceanography Centre (NOC), the University of Bath, and the Surrey Space Centre (SSC) of the University of Surrey to develop a next-generation Space GPS Receiver-Remote Sensing Instrument (SGR-ReSI).	Designed for Reflectometry and Radio Occultation applications, SGR-ReSI expands the use of GNSS for remote sensing in land, ocean, cryosphere, and atmospheric research applications.	(Unwin et al., 2010)
3. **New and emerging applications**	GNSS-R, GNSS-RO	Space-borne GPS reflectometry and refractometry research will be conducted using future GNSS Receiver-Remote Sensing Instruments (such as Shuttle measurements).	German-Indonesian tsunami early warning system, will expand to the Mediterranean and the Atlantic Ocean with new space-based GNSS reflectometry and scatterometry utilizing updated GPS signal, restored GLONASS, establishing Galileo, and the upcoming Compass GNSS. Analogous to Synthetic Aperture Radar (SAR), used in combination with other sensors to monitor crustal deformation and examine the global-scale geodynamic processes.	(Unwin et al., 2010)

which empower farmers to optimize their operations and make data-driven decisions. With the increasing accuracy of GNSS, farmers can now precision-map their fields, monitor crop health, and apply precise amounts of water and fertilizer, reducing waste and environmental impact.

The future of GNSS-enabled PA looks promising, with possibilities including increased accuracy, indoor positioning, integration with IoT and AI, autonomous farming, and global food security. GNSSs will become more and more significant in determining the direction of agriculture as technology develops, guaranteeing efficient and sustainable food production for an expanding world population. Furthermore, novel applications like precision irrigation, which allows farmers to optimize water consumption based on soil moisture levels, and precision animal husbandry, which allows farmers to follow the movement and health of their cattle in real time, are made possible by GNSS-enabled PA. Furthermore, GNSSs are also being used in agricultural research, enabling scientists to study crop yields, soil health, and weather patterns with greater precision.

Overall, the integration of GNSS with PA is transforming the agriculture industry, enabling farmers to produce more with less, reducing waste and environmental impact, and ensuring global food security. We may anticipate seeing even more cutting-edge GNSS uses in agriculture as technology develops, which will result in a more effective and sustainable food system.

REFERENCES

Alwis, S. De, Hou, Z., Zhang, Y., Na, M. H., Ofoghi, B., & Sajjanhar, A. (2022). A survey on smart farming data, applications and techniques. In *Computers in Industry* (Vol. 138, p. 103624). Elsevier. https://doi.org/10.1016/j.compind.2022.103624

Amani, M. A., & Marinello, F. (2022). A Deep Learning-Based Model to Reduce Costs and Increase Productivity in the Case of Small Datasets: A Case Study in Cotton Cultivation. *Agriculture (Switzerland), 12*(2), 267. https://doi.org/10.3390/agriculture12020267

Ampatzidis, Y., Bellis, L. De, & Luvisi, A. (2017). iPathology: Robotic applications and management of plants and plant diseases. In *Sustainability (Switzerland)* (Vol. 9, Issue 6, p. 1010). Multidisciplinary Digital Publishing Institute. https://doi.org/10.3390/su9061010

Aslan, M. F., Durdu, A., Sabanci, K., Ropelewska, E., & Gültekin, S. S. (2022). A Comprehensive Survey of the Recent Studies with UAV for Precision Agriculture in Open Fields and Greenhouses. In *Applied Sciences (Switzerland)* (Vol. 12, Issue 3, p. 1047). Multidisciplinary Digital Publishing Institute. https://doi.org/10.3390/app12031047

Ayaz, M., Ammad-Uddin, M., Sharif, Z., Mansour, A., & Aggoune, E. H. M. (2019). Internet-of-Things (IoT)-based smart agriculture: Toward making the fields talk. *IEEE Access, 7,* 129551–129583. https://doi.org/10.1109/ACCESS.2019.2932609

Banerjee, B. P., Spangenberg, G., & Kant, S. (2020). Fusion of spectral and structural information from aerial images for improved biomass estimation. *Remote Sensing, 12*(19), 1–22. https://doi.org/10.3390/rs12193164

Barna, R., Tóth, K., Nagy, M. Z., & Solymosi, K. (2020). Technical characteristics of global navigation satellite systems and their role in precision agriculture. *Journal of Agricultural Informatics, 11*(1), 52–66. https://doi.org/10.17700/jai.2020.11.1.573

Berger, G. S., Teixeira, M., Cantieri, A., Lima, J., Pereira, A. I., Valente, A., Castro, G. G. D., & Pinto, M. F. (2023). Cooperative Heterogeneous Robots for Autonomous Insects Trap Monitoring System in a Precision Agriculture Scenario. *Agriculture (Switzerland), 13*(2), 239. https://doi.org/10.3390/agriculture13020239

Chen, Y., & Zhan, X. (2021). GNSS vulnerability reliable assessment and its substitution with visual–inertial navigation. In *Aerospace Systems* (Vol. 4, Issue 3, pp. 179–189). Springer Science and Business Media B.V. https://doi.org/10.1007/s42401-021-00099-6

Chung, Y. S., Yoon, S. U., Heo, S., Kim, Y. S., Kim, Y. H., Han, G. D., & Ahn, J. (2023). Classification of Tree Composition in the Forest Using Images from SENTINEL-2: A Case Study of Geomunoreum Forests Using NDVI Images. *Applied Sciences (Switzerland), 13*(1). https://doi.org/10.3390/app13010303

Ding, H., Zhang, B., Zhou, J., Yan, Y., Tian, G., & Gu, B. (2022). Recent developments and applications of simultaneous localization and mapping in agriculture. *Journal of Field Robotics, 39*(6), 956–983. https://doi.org/10.1002/rob.22077

Fakhruddin, H. (2017). *Precision Agriculture: Top 15 Challenges and Issues*. Teknowledge Software. https://teksmobile.com/precision-agriculture-top-15-challenges-and-issues/

Fusonie, T. H., & LaRocco, C. A. (2021). *Understanding the Legal Aspects of Precision Farming Data*. Vorys News and Insight. https://www.vorys.com/publication-Understanding-the-Legal-Aspects-of-Precision-Farming-Data

Gašparović, M., Zrinjski, M., Barković, Đ., & Radočaj, D. (2020). An automatic method for weed mapping in oat fields based on UAV imagery. *Computers and Electronics in Agriculture, 173*, 105385. https://doi.org/10.1016/j.compag.2020.105385

Gawande, V., Saikanth, D. R. K., Sumithra, B. S., Aravind, S. A., Swamy, G. N., Chowdhury, M., & Singh, B. V. (2023). Potential of Precision Farming Technologies for Eco-Friendly Agriculture. *International Journal of Plant & Soil Science, 35*(19), 101–112. https://doi.org/10.9734/ijpss/2023/v35i193528

Ghafar, A. S. A., Hajjaj, S. S. H., Gsangaya, K. R., Sultan, M. T. H., Mail, M. F., & Hua, L. S. (2021). Design and development of a robot for spraying fertilizers and pesticides for agriculture. *Materials Today: Proceedings, 81*(2), 242–248. https://doi.org/10.1016/j.matpr.2021.03.174

Hadi Ishak, A., Hajjaj, S. S. H., Rao Gsangaya, K., Thariq Hameed Sultan, M., Fazly Mail, M., & Seng Hua, L. (2021). Autonomous fertilizer mixer through the Internet of Things (IoT). *Materials Today: Proceedings, 81*(2), 295–301. https://doi.org/10.1016/j.matpr.2021.03.194

Hariz, F., Bouslimani, Y., & Ghribi, M. (2023). High-Resolution Mobile Mapping Platform Using 15-mm Accuracy LiDAR and SPAN/TerraStar C-PRO Technologies. *IEEE Journal on Miniaturization for Air and Space Systems, 4*(2), 122–135. https://doi.org/10.1109/JMASS.2023.3240892

Hibbert, L. (2011). Satellite alert: we've become increasingly reliant on global navigation systems. So what would happen if they became disrupted by jamming or bad weather in space? *Professional Engineering Magazine, 24*(4). https://go.gale.com/ps/i.do?id=GALE%7CA254482568&sid=go.gleScholar&v=2.1&it=r&linkaccess=abs&issn=09536639&p=AONE&sw=w

Hu, W., Neupane, A., & Farrell, J. A. (2022). Using PPP Information to Implement a Global Real-Time Virtual Network DGNSS Approach. *IEEE Transactions on Vehicular Technology, 71*(10), 10337–10349. https://doi.org/10.1109/TVT.2022.3187416

Iacomi, C., & Popescu, O. (2015). A New Concept for Seed Precision Planting. *Agriculture and Agricultural Science Procedia, 6*, 38–43. https://doi.org/10.1016/j.aaspro.2015.08.035

Iftikhar, S., Gill, S. S., Song, C., Xu, M., Aslanpour, M. S., Toosi, A. N., Du, J., Wu, H., Ghosh, S., Chowdhury, D., Golec, M., Kumar, M., Abdelmoniem, A. M., Cuadrado, F., Varghese, B., Rana, O., Dustdar, S., & Uhlig, S. (2023). AI-based fog and edge computing: A systematic review, taxonomy and future directions. In *Internet of Things (Netherlands)* (Vol. 21, p. 100674). Elsevier. https://doi.org/10.1016/j.iot.2022.100674

Javaid, M., Haleem, A., Singh, R. P., & Suman, R. (2022). Enhancing smart farming through the applications of Agriculture 4.0 technologies. *International Journal of Intelligent Networks, 3*, 150–164. https://doi.org/10.1016/j.ijin.2022.09.004

Jin, X., Li, Q. W., Zhao, K. X., Zhao, B., He, Z. T., & Qiu, Z. M. (2019). Development and test of an electric precision seeder for small-size vegetable seeds. *International Journal of Agricultural and Biological Engineering, 12*(2), 75–81. https://doi.org/10.25165/j.ijabe.20191202.4618

Kahveci, M. (2017). Contribution of GNSS in precision agriculture. *Proceedings of 8th International Conference on Recent Advances in Space Technologies*, RAST 2017, 513–516. https://doi.org/10.1109/RAST.2017.8002939

Karkee, M., Zhang, Q., & Silwal, A. (2021). *Agricultural Robots for Precision Agricultural Tasks in Tree Fruit Orchards* (pp. 63–89). https://doi.org/10.1007/978-3-030-77036-5_4

Kitchen, N. R., Snyder, C. J., Franzen, D. W., & Wiebold, W. J. (2002). Educational needs of precision agriculture. *Precision Agriculture, 3*(4), 341–351. https://doi.org/10.1023/A:1021588721188

Kootstra, G., Wang, X., Blok, P. M., Hemming, J., & van Henten, E. (2021). Selective Harvesting Robotics: Current Research, Trends, and Future Directions. *Current Robotics Reports, 2*(1), 95–104. https://doi.org/10.1007/s43154-020-00034-1

Lencucha, R., Pal, N. E., Appau, A., Thow, A. M., & Drope, J. (2020). Government policy and agricultural production: A scoping review to inform research and policy on healthy agricultural commodities. In *Globalization and Health* (Vol. 16, Issue 1). BioMed Central Ltd. https://doi.org/10.1186/s12992-020-0542-2

Li, Y., Li, J., Zhou, W., Yao, Q., Nie, J., & Qi, X. (2022). Robot Path Planning Navigation for Dense Planting Red Jujube Orchards Based on the Joint Improved A* and DWA Algorithms under Laser SLAM. *Agriculture (Switzerland), 12*(9), 1445. https://doi.org/10.3390/agriculture12091445

Liu, Y., Ma, X., Shu, L., Hancke, G. P., & Abu-Mahfouz, A. M. (2021). From Industry 4.0 to Agriculture 4.0: Current Status, Enabling Technologies, and Research Challenges. *IEEE Transactions on Industrial Informatics*, *17*(6), 4322–4334. https://doi.org/10.1109/TII.2020.3003910

Lu, W., Okayama, T., & Komatsuzaki, M. (2022). Rice Height Monitoring between Different Estimation Models Using UAV Photogrammetry and Multispectral Technology. *Remote Sensing*, *14*(1), 78. https://doi.org/10.3390/rs14010078

Ma, M., Liu, J., Liu, M., Zeng, J., & Li, Y. (2021). Tree species classification based on sentinel-2 imagery and random forest classifier in the eastern regions of the qilian mountains. *Forests*, *12*(12), 1736. https://doi.org/10.3390/f12121736

Mavridou, E., Vrochidou, E., Papakostas, G. A., Pachidis, T., & Kaburlasos, V. G. (2019). Machine vision systems in precision agriculture for crop farming. In *Journal of Imaging* (Vol. 5, Issue 12, p. 89). Multidisciplinary Digital Publishing Institute. https://doi.org/10.3390/jimaging5120089

Merz, M., Pedro, D., Skliros, V., Bergenhem, C., Himanka, M., Houge, T., Matos-Carvalho, J. P., Lundkvist, H., Cürüklü, B., Hamrén, R., Ameri, A. E., Ahlberg, C., & Johansen, G. (2022). Autonomous UAS-Based Agriculture Applications: General Overview and Relevant European Case Studies. *Drones*, *6*(5), 128. https://doi.org/10.3390/drones6050128

Misara, R., Verma, D., Mishra, N., Rai, S. K., & Mishra, S. (2022). Twenty-two years of precision agriculture: a bibliometric review. *Precision Agriculture*, *23*(6), 2135–2158. https://doi.org/10.1007/s11119-022-09969-1

Nguyen, T. T., Hoang, T. D., Pham, M. T., Vu, T. T., Nguyen, T. H., Huynh, Q. T., & Jo, J. (2020). Monitoring agriculture areas with satellite images and deep learning. *Applied Soft Computing Journal*, *95*, 106565. https://doi.org/10.1016/j.asoc.2020.106565

Oliver, M. A., & Webster, R. (2014). A tutorial guide to geostatistics: Computing and modelling variograms and kriging. In *Catena* (Vol. 113, pp. 56–69). Elsevier. https://doi.org/10.1016/j.catena.2013.09.006

Ortega, L., Vilà-Valls, J., & Chaumette, E. (2022). Theoretical Evaluation of the GNSS Synchronization Performance Degradation under Interferences. *35th International Technical Meeting of the Satellite Division of the Institute of Navigation, ION GNSS+ 2022*, *5*, 3726–3735. https://doi.org/10.33012/2022.18564

Pallottino, F., Antonucci, F., Costa, C., Bisaglia, C., Figorilli, S., & Menesatti, P. (2019). Optoelectronic proximal sensing vehicle-mounted technologies in precision agriculture: A review. In *Computers and Electronics in Agriculture* (Vol. 162, pp. 859–873). Elsevier. https://doi.org/10.1016/j.compag.2019.05.034

Pandey, P. C., Tripathi, A. K., & Sharma, J. K. (2021). An evaluation of GPS opportunity in market for precision agriculture. In *GPS and GNSS Technology in Geosciences* (pp. 337–349). Elsevier. https://doi.org/10.1016/B978-0-12-818617-6.00016-0

Papadopoulos, G., Arduini, S., Uyar, H., Psiroukis, V., Kasimati, A., & Fountas, S. (2024). Economic and environmental benefits of digital agricultural technologies in crop production: A review. In *Smart Agricultural Technology* (Vol. 8, p. 100441). Elsevier. https://doi.org/10.1016/j.atech.2024.100441

Pini, M., Marucco, G., Falco, G., Nicola, M., & De Wilde, W. (2020). Experimental Testbed and Methodology for the Assessment of RTK GNSS Receivers Used in Precision Agriculture. *IEEE Access*, *8*, 14690–14703. https://doi.org/10.1109/ACCESS.2020.2965741

Radočaj, D., Jurišić, M., Gašparović, M., & Plaščak, I. (2020). Optimal soybean (Glycine max L.) land suitability using gis-based multicriteria analysis and sentinel-2 multitemporal images. *Remote Sensing*, *12*(9), 1463. https://doi.org/10.3390/RS12091463

Ramachandran, V., Ramalakshmi, R., Kavin, B. P., Hussain, I., Almaliki, A. H., Almaliki, A. A., Elnaggar, A. Y., & Hussein, E. E. (2022). Exploiting IoT and Its Enabled Technologies for Irrigation Needs in Agriculture. In Water (Switzerland) (Vol. 14, Issue 5, p. 719). Multidisciplinary Digital Publishing Institute. https://doi.org/10.3390/w14050719

Ravi, R., Shamseldin, T., Elbahnasawy, M., Lin, Y. J., & Habib, A. (2018). Bias impact analysis and calibration of UAV-based mobile LiDAR system with spinning multi-beam laser scanner. *Applied Sciences (Switzerland)*, *8*(2), 297. https://doi.org/10.3390/app8020297

Saba, T., Rehman, A., Haseeb, K., Bahaj, S. A., & Lloret, J. (2023). Trust-based decentralized blockchain system with machine learning using Internet of agriculture things. *Computers and Electrical Engineering*, *108*, 108674. https://doi.org/10.1016/j.compeleceng.2023.108674

Saranya, T., Deisy, C., Sridevi, S., & Anbananthen, K. S. M. (2023). A comparative study of deep learning and Internet of Things for precision agriculture. In *Engineering Applications of Artificial Intelligence* (Vol. 122, p. 106034). Pergamon. https://doi.org/10.1016/j.engappai.2023.106034

Senapaty, M. K., Ray, A., & Padhy, N. (2023). IoT-Enabled Soil Nutrient Analysis and Crop Recommendation Model for Precision Agriculture. *Computers*, *12*(3), 61. https://doi.org/10.3390/computers12030061

Shafi, U., Mumtaz, R., García-Nieto, J., Hassan, S. A., Zaidi, S. A. R., & Iqbal, N. (2019). Precision agriculture techniques and practices: From considerations to applications. In *Sensors (Switzerland)* (Vol. 19, Issue 17, p. 3796). Multidisciplinary Digital Publishing Institute. https://doi.org/10.3390/s19173796

Szira, Z., Varga, E., Csegődi, T. L., & Milics, G. (2023). The Benefits, Challenges and Legal Regulation of Precision Farming in the European Union. *EU Agrarian Law*, *12*(1), 1–7. https://doi.org/10.2478/eual-2023-0001

Taylor, J. A. (2023). Precision agriculture. In *Encyclopedia of Soils in the Environment, Second* Edition (pp. V4-710–V4-725). John Wiley & Sons, Ltd. https://doi.org/10.1016/B978-0-12-822974-3.00261-5

Tegedor, J. (2015). Multi-Constellation Satellite Navigation: Precise Orbit Determination and Point Positioning. https://nmbu.brage.unit.no/nmbu-xmlui/handle/11250/2496562

Unwin, M., De Vos Van Steenwijk, R., Gommenginger, C., Mitchell, C., & Gao, S. (2010). The SGR-ReSI - A new generation of space GNSS receiver for remote sensing. *23rd International Technical Meeting of the Satellite Division of the Institute of Navigation 2010, ION GNSS* 2010, 2, 1061–1067. https://www.ion.org/publications/abstract.cfm?articleID=9221

Wu, Y., Jin, S. G., Wang, Z. M., & Liu, J. B. (2010). Cycle slip detection using multi-frequency GPS carrier phase observations: A simulation study. *Advances in Space Research*, *46*(2), 144–149. https://doi.org/10.1016/j.asr.2009.11.007

Yasyukevich, Y. V., Zhang, B., & Devanaboyina, V. R. (2024). Advances in GNSS Positioning and GNSS Remote Sensing. In *Sensors* (Vol. 24, Issue 4, p. 1200). Multidisciplinary Digital Publishing Institute. https://doi.org/10.3390/s24041200

Yui, S. (2024). *What are the challenges of GPS in precision agriculture?* LinkedIn. https://www.linkedin.com/pulse/what-challenges-gps-precision-agriculture-sisi-yui-ffcye/

Zhang, J., Wang, C., Yang, C., Jiang, Z., Zhou, G., Wang, B., Shi, Y., Zhang, D., You, L., & Xie, J. (2020). Evaluation of a UAV-mounted consumer grade camera with different spectral modifications and two handheld spectral sensors for rapeseed growth monitoring: performance and influencing factors. *Precision Agriculture*, *21*(5), 1092–1120. https://doi.org/10.1007/s11119-020-09710-w

Zhang, X., Ren, X., Chen, J., Zuo, X., Mei, D., & Liu, W. (2022). Investigating GNSS PPP–RTK with external ionospheric constraints. *Satellite Navigation*, *3*(1). https://doi.org/10.1186/s43020-022-00067-1

Zhongshan Samli Drones Co L. (2024). *What are the challenges of GPS in precision agriculture?* LinkedIn. https://www.linkedin.com/pulse/what-challenges-gps-precision-agriculture-5e1ve/

Zidan, J., Adegoke, E. I., Kampert, E., Birrell, S. A., Ford, C. R., & Higgins, M. D. (2021). GNSS Vulnerabilities and Existing Solutions: A Review of the Literature. In *IEEE Access* (Vol. 9, pp. 153960–153976). https://doi.org/10.1109/ACCESS.2020.2973759

Section 4

GNSS Applications for Space Weather

18 Role of GNSS-derived TEC for Space Weather Studies

Gopi Krishna Seemala

18.1 INTRODUCTION

Modern society's ever-growing dependence on technology requires careful review of reliability of the technologies used. Most technologies these days either use or depend on satellite-based communication and navigation systems. These satellite signals must pass through the ionosphere, which is a part of Earth's upper atmosphere. The ionosphere is a region from about 60 km to about 2000 km, where the neutral atmosphere is partially ionized by radiation (UV and x-rays) from the Sun—producing enough electrons and ions that can noticeably affect radio signals traversing through. It is known that the ionosphere reaches peak ionization or maximum electron density at around 300 km, but yet only a very small fraction (<0.1%) of the neutral gas is ionized. After this height of maximum ionization, the density of neutral gas decreases faster, so the ionization ratio will increase with altitude. The ionospheric regions can absorb or damp radio signals, or they can bend radio waves, as well as reflect radio signals.

Interest in the study of the ionosphere began soon after long-distance radio communication was demonstrated by Guglielmo Marconi in December 1901. Solar radiation passing through the atmosphere is absorbed and produces ionization. This absorption means less radiation reaches the lower levels of the atmosphere, and the level of ionization is reduced. Also, as the density of the atmosphere decreases with altitude, there is less gas to ionize and less ionization. This results in a peak in the level of ionization known as the Chapman layer. The real ionosphere is more complicated than this, thus resulting in different layers of ionization densities known as D, E, F, or further F1 and F2 layers in daytime; see Figure 1.2 in Kelley (2009). The typical variation of electron density and different ionospheric layers is shown in Figure 18.1, which also depicts the variation of density profiles for day and night. Also, the distinct density variation between high and low solar active times can be noticed in this figure. The IRI model (Bilitza and Reinisch, 2008) was used for deriving these density profiles, which is available at http://irimodel.org/.

The ionosphere is highly dynamic, as it is formed due to ionization of gases because of the Sun's radiation. The ionospheric density changes drastically from Earth's day side to night side, as can be seen from Figure 18.1. At night the ionosphere density becomes less as the ions and electrons will recombine back to form neutral gas. It is observed from past observations that ionospheric electron density varies both with altitude and spatially, substantially more with latitude. It is obvious that the ionization density of the ionosphere is proportional to the solar flux or the radiation received. Thus, the electron density will vary with time of day and season and also exhibits 11-year solar cycle variations as it follows solar radiation. Apart from these variations, there are more unpredictable changes in ionospheric density caused by factors arising from the space above and from the Earth below. This makes it harder to know or predict the density of ionosphere at a given time.

It is known that in the early days, research on the ionosphere was done to improve the radio or wireless communication for long distances. But today, the focus of research has shifted towards understanding the ionosphere and its effects on satellites as they orbit in the ionosphere. Thus, we have huge datasets generated from satellite remote sensing owing to the progress in laboratory

DOI: 10.1201/9781032712444-22

"""

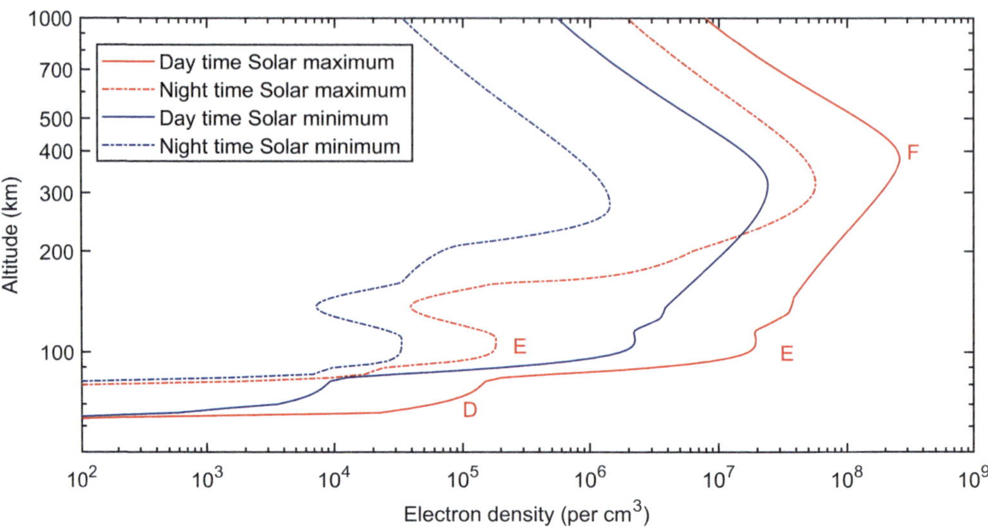

FIGURE 18.1 The variation of ionospheric density during day and night obtained from the IRI 2016 model, also representing low and high solar active conditions (Seemala, 2023).

experimental techniques and theory, which led to vast knowledge on ionosphere processes and climatology (Seemala, 2023 and references therein).

18.1.1 TOTAL ELECTRON CONTENT

It is known that the density of the ionosphere is mostly dependent on the flux of received solar radiation. As discussed in the previous section, the electron density of the ionosphere and its variability varies temporarily with time of the day, day of the year, and activity of the solar cycle.

Radio signals experience group delay when they traverse through the ionosphere, which affects the modulation or the information carried on these radio waves. The ionosphere is a dispersive medium, i.e., the higher frequencies travel faster than lower frequencies through this medium. This dispersion results in a spreading out or separation of different frequency components of a signal, which contributes to group delay.

The other effects on the ionosphere are carrier phase advance, Faraday rotation of plane of polarization, Doppler shift, angular refraction, distortion of transmitted pulses, and amplitude and phase scintillations. All these effects listed here except scintillation are proportional to the integrated or total electron content (TEC), at least to first order, that is encountered by the traversing radio wave. TEC is defined as the total number of free electrons in a column of unit cross-sectional area along the raypath or line of sight between the receiver and satellite or transmitter. TEC is expressed in TEC units (TECU), where 1 TECU = 10^{16} electrons per square meter. When TEC is measured along an oblique line of sight, it is called a Slant TEC or STEC; it becomes Vertical TEC or VTEC if the elevation angle is zenith or 90 degrees.

TEC is typically measured using received dual frequency signals from Global Navigation Satellite System (GNSS) satellites. The ionosphere is a dispersive medium for radio frequencies (Davies, 1990), so the signals or radio waves experience different delays in the ionosphere for different frequencies due to this dispersive nature. Common errors or other atmospheric delays are eliminated by calculating the difference of range delays between two transmitted frequencies leaving the residual differential delay of the ionosphere. This differential delay is proportional to TEC. Thus, using the group delay and phase advance data of the signals from two or more frequencies received on the ground, the TEC values can be derived (Lanyi and Roth, 1988; Coco et al., 1991; Seemala and Valladares, 2011; Seemala, 2023).

18.1.2 Space Weather

The Sun is the main source of energy for the Earth; thus, it is the primary driver of this planet's weather. The term Space weather usually defines changes of the space environment between the sun and Earth (Nagai, 1988), which gained popularity during the mid-1990s. However, attempts or forecasts of Space weather phenomena were made at least a couple of centuries before this time. The first 'Space weather' or auroral activity prediction was made by the Swedish astronomer and mathematician Pehr Wargentin, which was noted after the fact by him in a 1750 paper. He used the linkage between geomagnetic disturbance and auroral occurrence that was reported by earlier Swedish scientists in a 1747 paper. In application point of view, the term Space weather often refers to phenomena that could impact spaced-based systems and technologies either in orbit or on Earth. For more details, refer to the website of the Space Weather Prediction Center, NOAA (www.swpc. noaa.gov/phenomena).

Space weather includes solar radiation and also the charged particles that emanate from the Sun known as solar wind, which carries the interplanetary magnetic field (IMF), with the other end of those field lines attached to the Sun spiraling outward. Just like changes in temperature, rainfall, and winds make up the weather on Earth, changes in the sun's radiation, solar wind density/speed, IMF, and other factors collectively make up Space weather (Hanslmeier, 2007; Moldwin, 2008). Variations in the Sun's radiation can induce dramatic effects in the near-Earth or surrounding space environment. It is interesting to note that Space weather has existed since the beginning of the Earth. However, only since the last decade of the 20th century has it become a problem, as society has become increasingly dependent on space-based infrastructure.

Activities of the Sun such as Coronal Mass Ejections (CMEs) and solar flares emit radiation and significant magnetic fields into our solar system. Most of them miss the Earth entirely, but some do collide with the planet. The concern is about local influence of Space weather, meaning those CMEs or solar flares directed at Earth and influencing it. When a CME is directed towards Earth, the magnetosphere of Earth and solar wind interact very complexly, creating a storm in near-Earth space and on Earth. The interaction of Earth's magnetosphere with the IMF and solar wind dictates the intensity of a geomagnetic storm and thus the effects of Space weather on it.

The performance and reliability of space-borne and ground technological systems are affected by Space weather, which in turn could affect general public activities. It is obvious that satellites orbiting in near-Earth space are easily affected by Space Weather. In particular, the higher orbiting satellites such as geosynchronous satellites are more vulnerable to these Space weather effects and high-energy particles emitted from the Sun. These energetic particle effects may damage spacecraft electronics, and increased atmospheric drag may shorten satellites' lifetimes, affecting low-orbit satellites more. The energetic radiation can harm astronauts if they are not shielded properly.

Also, people in high-altitude aircraft are affected by this high energetic radiation and particles, as the radiation exposure is doubled every 2.2 km in altitude above ground. Solar flares can increase the radiation 20–30 times. It is known that intense Space weather storms can induce global electric circuit currents, causing the failure of electricity power grids over large areas, particularly in high-latitude regions.

The magnetospheric ring current intensity that is indicated by the Dst index is usually used as a measure of strength of a geomagnetic storm. This index gives a rough estimate of the amount of energy that is coupled into the upper atmosphere. CMEs are known to cause the strongest geomagnetic storms, but for these CMEs to be geo-effective, the solar wind and energetic particles must be directed towards Earth, and its IMF must have a suitable configuration with its Bz component to be southward and strong for long duration (Akasofu et al., 1985). Interaction of interconnection of the Earth's magnetic field and southward IMF leads to more energetic particle precipitation in the polar and high-latitude regions.

The solar activity variation and the 11-year-long sunspot cycle affects Space weather activity. The real-time forecast of Space weather response in Earth's space environment will be possible by having a better understanding of the physical mechanism of the Sun-Earth connection in addition to the solar wind observations before they reach the Earth.

18.1.3 Measure of Space Weather Activity

Space weather activity can be measured using various indices that monitor different aspects of solar and geomagnetic activity. Solar indices include the Sunspot Number (SSN) and Solar Flux Index (F10.7). The Solar Flux index measures the intensity of the radio wavelength of 10.7 cm emitted from the Sun, indicating solar activity levels. The SSN quantifies the intensity of solar activity by counting the visible sunspots on the solar disk; it has a vast database extending back to 17th century with annual values, while daily values exist from the year 1818.

The Geomagnetic Indices—Kp, ap, AE, and Dst—give a measure of strength of the disturbances caused in Earth's magnetosphere due to solar wind interactions (Rostoker, 1972). The DST index specifically measures the intensity or magnitude of the ring current, which is closely related to geomagnetic activity in the equatorial region. In general, the presence of a geomagnetic storm is identified when the Dst index goes below a minimum or critical value of -50 nT or -100 nT (Sugiura & Chapman, 1960). Even the strength of geomagnetic storms is classified based on the minimum values observed (Gonzalez et al., 1994). It is known that different geomagnetic storms will result in various magnitudes of ring currents, enhancements in auroral and radiation-belt currents, plasma sheet temperature, etc. As observed from the earlier studies, the Dst index is a poor identifier of high-speed-stream-driven geomagnetic storms and of other types of geospace storms (Borovsky and Shprits, 2017). The Sym-H is a newer index designed to provide more accurate and symmetric measure of geomagnetic activity compared to Dst.

And lastly, the Ionospheric Indices include TEC and Sudden Ionospheric Disturbance (SID) Index. As discussed in the previous section, TEC is a measure of the total number of free electrons in a column of unit area from the surface of Earth to top of the ionosphere, which is also affected by geomagnetic activity and solar parameters. The SID indicates rapid changes in the ionospheric electron density due to solar flares or storms.

These indices collectively provide a comprehensive view of the state of Space weather as a result of solar activity and geomagnetic disturbances to ionospheric changes and radiation hazards. These indices are used to forecast and monitor Space weather events, and each index serves a specific purpose in understanding different aspects of Space weather phenomena.

18.2 TEC AND GLOBAL IONOSPHERIC MAPS

We know the ionosphere is made up of ionized gas, and these charged particles are very sensitive to the variation in electric and magnetic conditions in space. Also, the ionosphere plays an important role in our day-to-day navigation and communication systems. GNSS's and communication radio signals have to traverse this atmospheric layer or depend on it for bouncing off it (in case of terrestrial to terrestrial communication) to arrive at its destination. In either of these cases, the density and composition of the ionosphere can disturb these signals. For any satellite-based system, radio signals must pass through the ionosphere; thus, the TEC becomes an important parameter for any satellite-based systems.

The Ionosphere Working Group of the International GNSS Service (Iono-WG) (https://igs.org/products/#ionosphere) was created in 1998 to generate reliable global vertical TEC maps from IGS data. For more than 20 years, the availability of GNSS receivers was a great opportunity to observe ionospheric variability. Now, with better connectivity, we can observe the ionosphere in real time from GNSS-derived TEC.

The global ionospheric map (GIM) of vertical TEC routinely generated and archived by the International GNSS Service (IGS) has become a comprehensive tool to visualize the global variation of the ionosphere (Dow et al., 2009). It provides detailed information about global ionospheric conditions and is instrumental for Space weather monitoring, GNSS operations, and scientific research. The GIM gives the distribution of ionospheric parameters like TEC, electron density, and other related metrics across the globe in IONEX format (https://files.igs.org/pub/data/format/ionex1.pdf). By visualizing and interpreting TEC and other ionospheric parameters, GIMs support a wide range of applications, from Space weather prediction to GNSS system performance. The GIM maps can be used to correct single frequency GNSS receiver errors for more accurate positioning. GIMs help in understanding the impact of geomagnetic storms on the ionosphere.

Figure 18.2 shows the GIM TEC variation over the entire region covering all longitudes and latitudes from -87.5 to 87.5 degrees. Figure 18.2(a) shows a typical TEC variation during a quiet day (22 April 2023) depicting an Equatorial Ionization Anomaly (EIA) feature over the Indian sector. The day 22nd April 2023 is a geomagnetically quiet day, as listed in WDC Kyoto. Figure 18.2(b) presents a disturbed day scenario (24th April 2023) showing reduced TEC over the mid latitudes and absence of EIA showing increased TEC over the magnetic equator. Thus, these maps offer a snapshot of ionospheric conditions at a specific time or provide temporal data for analysis over time (Li et al., 2020).

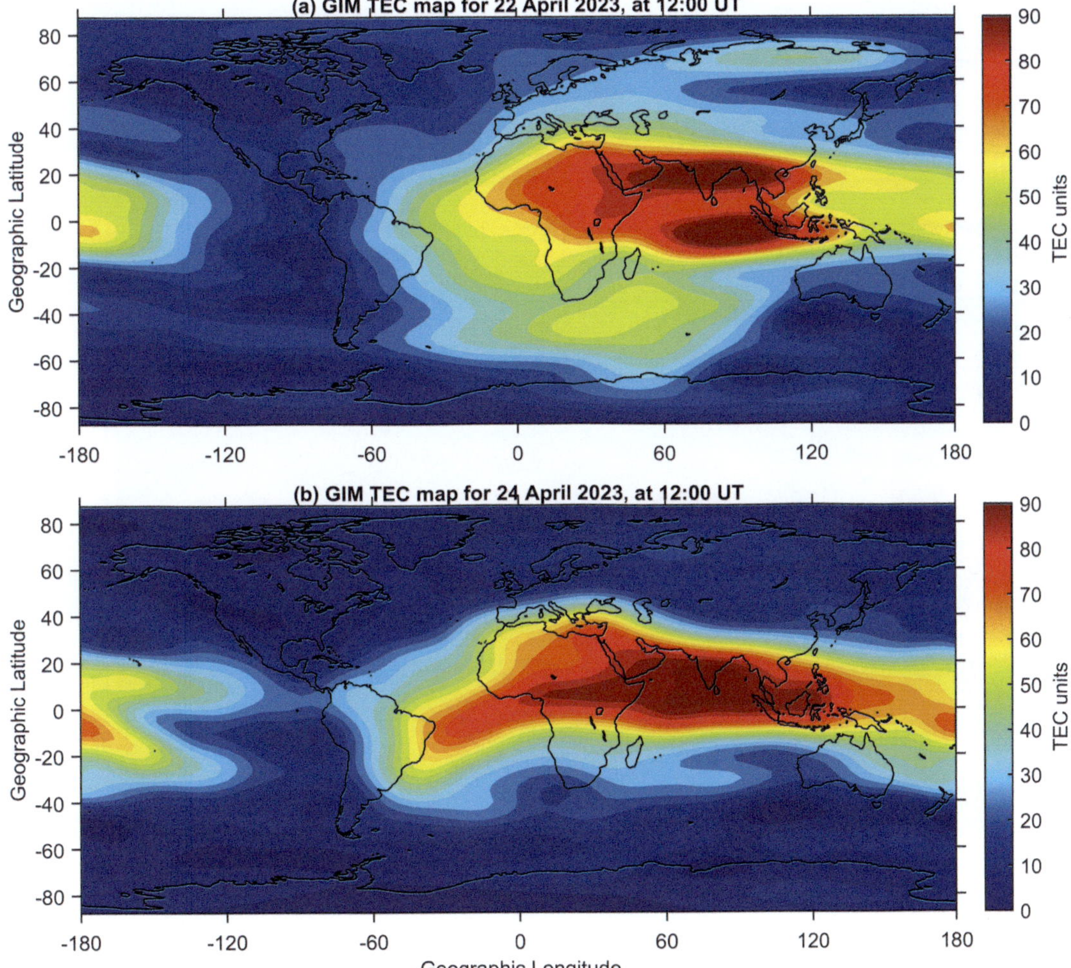

FIGURE 18.2   GIM TEC maps at 12 UT hours for the day of a) 22 April 2023 depicting a quiet day TEC variation, and b) 24 April 2024 depicting a disturbed day variation of TEC across the globe.

18.3 APPLICATION OF TEC IN SPACE WEATHER

As discussed, TEC is a direct measure of the ionosphere's electron density. Variations in TEC can indicate changes in ionization levels due to solar activity, solar flares, and CMEs. Thus, long-term TEC observations can help in studying trends related to the solar cycle and understanding how forces from below, like climate change, might affect ionosphere. During Space weather events like geomagnetic storms, TEC values can be increased or decreased depending on the storm-time-induced electric fields and the increased energy deposition from the solar wind.

In this chapter, a storm that occurred on 23–24 April, 2023 is showcased to describe the effect on the ionosphere as observed by TEC. On April 21 at about 18:12 UT, an active solar region released a CME that was directed towards Earth, leading to this severe geomagnetic storm.

Figure 18.3 shows the various solar wind parameters such as velocity and density variations in panel (a). On 23rd April, the IMF Bz component started turning southward at 09:00 UT, which is associated with the beginning of the main phase of this geomagnetic storm (Habarulema et al., 2024). It is known that the north or south direction of IMF-Bz controls the extent of solar wind energy coupled to the magnetosphere through reconnection of the Earth's magnetic field and IMF-Bz (Dungey, 1961).

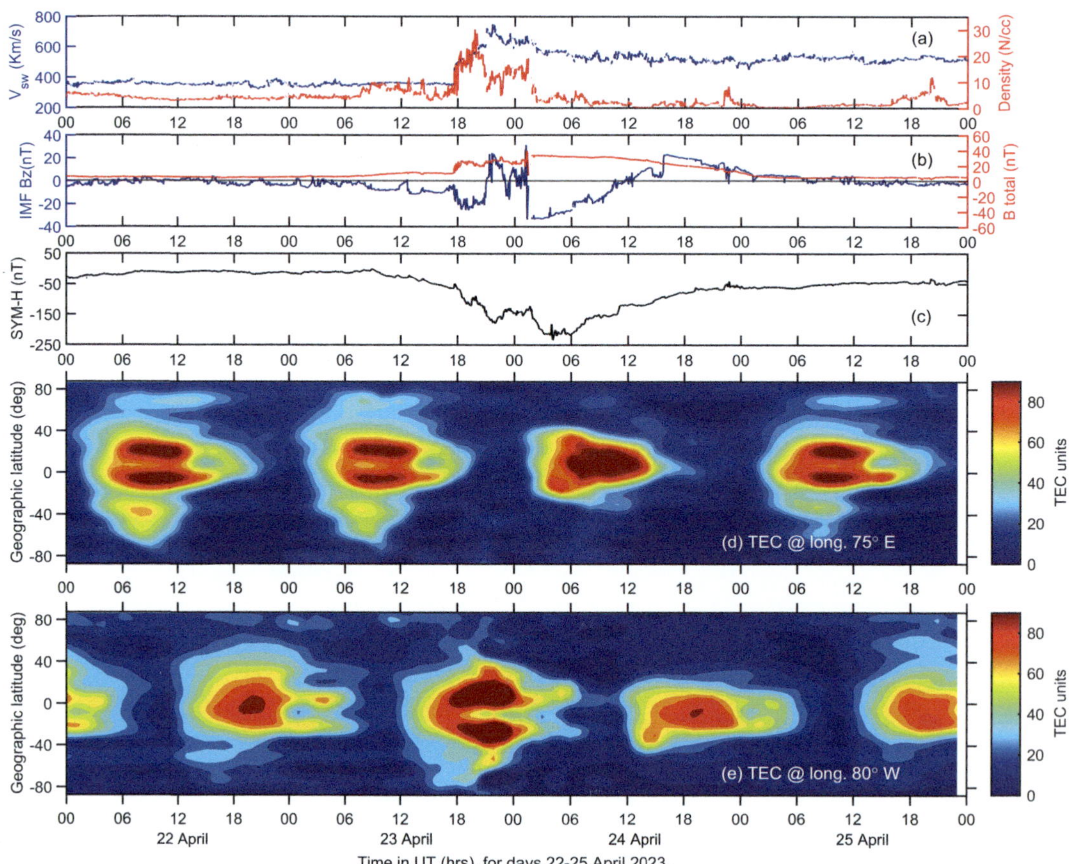

FIGURE 18.3 This figure shows the various parameters of the solar wind and their effect on TEC for the geomagnetic storm of 22–25 April 2023. The panels (a) shows variation of solar wind speed and proton density, (b) IMF Bz and IMF total field, (c) geomagnetic index SYM-H, (d) latitudinal variation of TEC in the Indian sector, and (e) latitudinal variation of TEC in the American sector.

Thus, this event has IMF-Bz turned southward for most of the time, which resulted in more energy coupling from the solar wind. On 23 April at 17:45 UT, the rapid increase of the speed of the solar wind (Figure 18.3a) from 350 to 500 km/s is observed, and the solar wind proton density has also shown a sharp increase. These fluctuations in IMF-Bz resulted in a two-step main phase for this storm, with Sym-H (shown in Figure 18.3c) reaching ~179 nT at ~22:00 UT on 23 April and ~233 nT at 04:03 UT on 24 April, signifying the intense geomagnetic storm.

Figures 18.3(d & e) shows TEC latitudinal variation with time at the Indian sector (75° E longitude chosen here) and the American sector (80° W chosen here), respectively. The main phase of the storm occurred around local midnight of 23 April for the Indian sector, during which VTEC values dropped below their quiet-time levels. This reduction was more significant near the equator and at low latitudes. Increased TEC was observed in the American sector during the local night hours; this could be due to the reverse fountain effect. There are no significant changes observed at mid-latitude regions. In the Indian sector on 24 April, TEC values were higher compared to quiet time except in the mid-latitude region. The Equatorial Ionization Anomaly (EIA) feature is suppressed in both the Indian and American sectors on 24th April due to disturbed thermospheric winds.

Thus, TEC can reveal disturbances like ionospheric storms, which affect different regions in unique ways. For instance, at the equator, TEC can show equatorial ionization anomalies or the absence of them, while in Polar Regions, it can show polar cap absorption effects. So, by observing real-time changes in TEC, one can predict the influence of solar events on the ionosphere. For example, sudden increases in TEC can signal the arrival of a solar storm. TEC data is used in models to simulate the response of the ionosphere to the solar and geomagnetic activity. Thus, TEC is a crucial tool in Space weather observation; it will help understand and predict how solar and geomagnetic activities impact the Earth's ionosphere and related technologies.

18.4 SUMMARY

The term Space weather refers to conditions on the Sun and its variability of parameters that can affect the reliability and performance of space-borne and ground technological systems, thus also affecting the general public that depends on these technologies. Modern technology is becoming vulnerable day by day to these Space weather disturbances outside Earth due to increased reliance on satellite-based technologies. The economic consequences of the Space weather effects are enormous. Recently, Space weather and its forecasts have gained significant attention, not only among researchers but also among the general public. The vast amount of observational data was recorded/ archived over the last two decades from a large fleet of space-based instruments. This data has allowed us to improve our understanding of the physical processes involved on the Sun, the near-Earth environment, and interplanetary space.

Thus, TEC has become an essential parameter in Space weather observation and research. TEC data is crucial for understanding and optimizing HF and VHF radio communications. Satellites equipped with ionospheric sounders or GNSS receivers can also measure TEC. TEC data is usually collected from the networks of GNSS ground receivers. Examples include the IGS and other regional networks.

ACKNOWLEDGEMENTS

The geomagnetic indices and solar wind parameters were downloaded from the NASA GSFC SPDF OMNI website (http://omniweb.gsfc.nasa.gov). Global ionosphere maps data was downloaded from the NASA CDDIS website (https://cddis.nasa.gov/archive/gnss/products/ionex/). The lists of international quiet days are taken from the World Data Centre (WDC) for Geomagnetism, Kyoto (http://wdc.kugi.kyoto-u.ac.jp/qddays/index.html).

REFERENCES

Akasofu, S.I., Olmsted, C., Smith, E.J., Tsurutani, B., Okida, R., and Baker, D.N. (1985). Solar wind variations and geomagnetic storms: A study of individual storms based on high time resolution ISEE 3 data, J. Geophys. Res., 90(A1), 325–340, doi:10.1029/JA090iA01p00325

Bilitza, D., and Reinisch, B.W. (2008). International reference ionosphere 2007: improvements and new parameters. Adv Space Res., 42(4), 599–609, doi:10.1016/j.asr.2007.07.048

Borovsky, J. E., and Shprits, Y. Y. (2017). Is the Dst index sufficient to define all geospace storms? *J. Geophys. Res. Space. Phys.*, 122(11), 11,543–11,547. https://doi.org/10.1002/2017JA024679

Coco, D. S., Coker, C., Dahlke, S. R., and Clynch, J. R. (1991). Variability of GPS satellite differential group delay biases, *IEEE Trans. Aerosp. Electron. Sys.*, 27, 931–938.

Davies, K. (1990). Ionospheric Radio. IEEE Electromagnetic Waves Series 31, Peter Peregrinus, London.

Dow, J.M., Neilan, R.E., Rizos, C., (2009). The international GNSS service in a changing landscape of global navigation satellite systems. *J. Geod.* 83(3–4), 191–198.

Dungey, J. W. (1961), Interplanetary magnetic field and the auroral zones, *Phys. Rev. Lett.*, 6, 47–48, doi:10.1103/PhysRevLett.6.47

Gonzalez, W., Joselyn, J.A., Kamide, Y. et al. (1994). What is a geomagnetic storm? *J. Geophys. Res. Space. Phys.* 99(A4), 5771–5792. https://doi.org/10.1029/93JA02867

Hanslmeier, A. (2007). *The Sun and Space Weather, ASSL 347*, Springer, doi.org/10.1007/978-1-4020-5604-8

Habarulema, J.B., Zhang, Y., Matamba, T., Buresova, D., Lu, G., Katamzi-Joseph, Z., et al. (2024). Absence of high frequency echoes from ionosondes during the 23–25 April 2023 geomagnetic storm; what happened? *J. Geophys. Re.s Space Phys.*, 129, e2023JA032277. https://doi.org/10.1029/2023JA032277

Kelley, M. C. (2009). *The Earth's Ionosphere: Plasma Physics and Electrodynamics, volume 96 of International geophysics series*. Elsevier, second edition, 10.1016/S0074-6142(09)60212-6

Lanyi, G. E. and Roth, T. (1988). *A comparison of mapped and measured total ionospheric electron content using Global Positioning System and beacon satellites observations, Radio Sci.*, 23, 483–492.

Li, Z., Wang, N., Hernández-Pajares, M. et al. (2020). IGS real-time service for global ionospheric total electron content modeling. *J Geod* 94, 32. https://doi.org/10.1007/s00190-020-01360-0

Moldwin, M. (2008). *An Introduction to Space Weather*, Cambridge Univ. Press, Cambridge, U. K., 10.1017/9781108866538

Nagai, T. (1988). "Space weather forecast": Prediction of relativistic electron intensity at synchronous orbit, *Geophys. Res. Lett.*, 15(5), 425–428, 10.1029/GL015i005p00425

Rostoker, G. (1972), Geomagnetic indices, *Rev. Geophys.*, 10(4), 935–950, doi:10.1029/RG010i004p00935

Seemala, G.K., (2023). Chapter 4 - Estimation of ionospheric total electron content (TEC) from GNSS observations, Editor(s): Singh, A.K., Tiwari, S., In *Earth Observation, Atmospheric Remote Sensing*, Elsevier, 63–84, 10.1016/B978-0-323-99262-6.00022-5

Seemala, G.K., and Valladares, C.E. (2011). Statistics of total electron content depletions observed over the South American continent for the year 2008, *Radio Sci.*, 46, RS5019, doi:10.1029/2011RS004722

Sugiura, M., and Chapman, S. (1960). The average morphology of geomagnetic storms with sudden commencement. In *Abhandlungen der Akademie der Wissenschaften zu Göttingen* (pp. 1–53). Göttingen: Göttingen Math. Phys. Kl., Sondernheft Nr.4.

19 Characteristic of Ionospheric Irregularities Using GNSS Measurements over Indian EIA Region Varanasi

Mini Rajput, Mukulika Mondal, and Abhay Kumar Singh

19.1 INTRODUCTION

19.1.1 THE EARTH'S IONOSPHERE

The discovery of the ionosphere marked a quantum leap in the development of satellite navigation and radio wave communication systems. Since its discovery, it has fascinated researchers worldwide, opening the door to studying plasma physics and contributing to the beginning of a new era in radio wave communication.

Early research goes back to the 19th century, when Samuel P. Langley first proposed the presence of an atmospheric layer that could affect radio wave propagation, but his idea was not accepted at the time. Later, in 1901, Guglielmo Marconi conducted his famous experiment where he sent a radio signal from Cornwall, England, to Newfoundland, Canada. This long distance radio transmission, which was not possible via direct line-of-sight, reignited interest in the phenomenon. Further, subsequent suggestions were made by Heaviside [1] and Arthur E. Kennelly [2] that a conductive layer was present in the atmosphere, and radio waves must have bounced back from this conductive layer because of the curvature of the Earth.

In 1920 Edward V. Appleton [3] was the first to illustrate direct experimental confirmation of the existence of charged particles in the high altitude region. He introduced the concept of ionospheric layers, the E layer and F layer. The term "ionosphere" for this ionized layer was later introduced by Robert Watson-Watt [4]. In the 1920s Sydney Chapman gave a detailed explanation about the formation of the ionosphere and the part solar radiation plays in the process. Chapman's work introduced a new chapter in ionospheric studies by detailing the association between the Sun and the ionosphere, providing physicists with ample information to explore, both experimentally and through modeling [5].

19.1.2 FORMATION OF THE IONOSPHERE

Ninety-nine point nine percent of the entire mass of the atmosphere is situated below the stratopause, which is about 50 km altitude; above the stratopause the air is extremely thin and only one-thousandth of the whole atmosphere's total molecules. As the constitution of the atmosphere varies with height, at ground level the atmosphere is relatively dense, and density declines with altitude. Thus, in this rarefied region of the atmosphere, the lighter gases, mainly hydrogen, become more dominant. When solar radiation penetrates into this upper rarefied region and strikes the atoms and molecules of gases of this region, then the energetic proton ionizes the atoms or molecules and alters the atmospheric composition [6–8]. This process is called "ionization."

DOI: 10.1201/9781032712444-23

For example, the electron is ejected from the atom, when helium is heated by the solar radiation. Thus, it gets ionized, causing the free electrons and the positive ions. Since the atmospheric pressure is quite low above 100 km, the ionized particles do not recombine quickly, and the permanent numbers of ions and free electrons are present in this region. The existence of these ions and free electron governs the properties and behavior of this upper region. This ionized part expanding from 60 km to the altitude beyond 1000 km, which has a sufficient number of free ions and free electrons, is known as the 'ionosphere.' The ionosphere affects the radio wave propagation through it [9].

The intensity of the radiation decreases since energy is expended during each ionization event. Consequently, most of the ionization in the upper regions is caused by UV radiation, while at lower altitudes, more ionization is caused by the radiation, which can penetrate deeper, such as extreme UV radiation and X-rays. As altitude decreases, the intensity of radiation is reduced, leading to variations in the level of ionization across different layers of the ionosphere. This variation is also reflected in the density of gases, which changes with altitude. Moreover, the monoatomic forms of gases are prominent at higher altitudes, resulting in the difference in proportions of monoatomic and molecular forms of gases. Therefore, with the dominance of monatomic ions at higher altitudes, the composition of ionized constituents varies depending on altitude.

Solar activity, the time of day, and the seasons are other factors that influence the ionization level of the ionosphere. While atoms and molecules break out into free electrons and positive ions from the Sun's radiation, they may, on the other hand, recombine if a positive ion encounters a negative electron. Thus, two opposite effects—splitting as well as recombination —have occurred. This is known as the state of dynamic equilibrium. Consequently, the level of ionization depends upon both: ionization and recombination rates [10]. After ionization, the ions distribute themselves according to their weight; the heavier ions settle at the bottom with the lighter ones at the top. The ionosphere, therefore, is not a continuous region—it is layered.

In the daytime, the ionosphere segregates into several distinct regions, mainly the D-region, E-region, F1-region, and F2-region. In the nighttime, the D-region disappears, and F1 and F2 merge to form the F region. These ionospheric regions are defined by a peak in density at a particular altitude [10].

19.1.3 D-Region

The D-region is the bottommost part of the ionosphere, which varies from about 50 km to 90 km above the Earth's surface. This region is ionized primarily by the Lyman-α radiation with wavelength at 121.6 nm, which ionizes Nitric Oxide (NO) present there. At night, ionization in the D-region decreases greatly, because nitrogen and oxygen molecules recombine in the absence of sunlight to form neutral particles. While the layer significantly reduces after sunset, it is still present due to galactic cosmic rays [11]. Absorption of radio waves is mainly made by the D layer, with absorption peak at midday and decreasing at night. Since, in the D layer, the recombination rate is high, the net ionization effect is reduced, due to which, High Frequency (HF) radio waves are not returned by the D-region.

19.1.4 E-Region

The E-region is the middle region of the ionosphere, which varies from approximately 90 km to 150 km above the surface. The E-region is mainly ionized by X-rays in the range of 8–140 Å and UV radiation of 796–1027 Å. This region preliminary comprises molecular ions such as N_2^+, O_2^+, and NO^+. Generally, this layer reflects the radio waves of frequencies less than 10 MHz, while for frequencies above this, has an opposite effect. But, it can reflect high frequencies like 250 MHz during severe sporadic E events. The highest electron density in the E region is about 10^{11} electrons per cubic meter in the daytime, occurring at around 100 km altitude. At night, due to the absence of a primary source of ionization, the E layer dissipates [11].

19.1.5 F-Region

The F layer is the ionosphere's last and topmost layer, extending from approximately 150 km to 600 km. The ionization of atomic oxygen in this layer is dominated by Extreme Ultraviolet (EUV) solar radiation. At night, the F-region contains a single layer, but in the day, due to more ionization by solar radiation, the F-region splits into F_1 and F_2 layers. Due to the presence of major ion species O^+, the F_2 layer is able to sustain even at night, as these species have longer lifetimes. This region is of great interest because most of the sky wave propagation is possible due to this layer only, and thus, it helps in HF radio communication over long distances.

19.1.6 Influence of Ionosphere in Radio Wave Propagation

When radio waves propagate through the ionosphere, they experience phenomena such as refraction, reflection, diffraction, absorption, polarization, and scattering [12]. Radio propagation refers to the behavior of radio waves as they transmitted or propagated from one point to another on Earth or through various layers of the atmosphere. The ionosphere spreads the radio propagation to distant locations on the Earth, making it of great practical importance.

The ionosphere acts as an opaque barrier to the electromagnetic waves. It absorbs electromagnetic waves with a frequency less than 2 MHz and allows those with frequencies greater than 30 MHz to pass through. When low frequency waves are transmitted through the ionosphere, they are reflected back due to total internal reflection (TIR). The refraction of a radio signal in the ionosphere relies upon the frequency of radio waves and on the dispersive nature of the ionospheric plasma. Since the ionospheric refractive index varies with the density of plasma and the frequency of radio waves, the propagation speed of the radio waves changes accordingly. The ionospheric refractive index is given by:

$$n = 1 - \frac{40.3N}{f^2}$$

where N is the electron number density in m^{-3} and f is the frequency of the wave.

When waves travel along the ionosphere, the electron density is lower in the lower regions, causing the wave to propagate with the speed of light. However, as the altitude increases, the electron density also increases but the frequency remains constant. This causes the refractive index to decrease (as per the equation above), and at a particular condition it becomes zero, and the wave is bounced back to the Earth. This reflective property of the ionosphere is utilized for radio wave communication for long distances. The effectiveness of the ionosphere to reflect the transmitted radio wave depends upon the frequency of the wave and, therefore, to know about the behavior of the ionosphere is important in order to choose the good frequency [10, 13].

Radio wave propagation across the ionosphere is more complex to predict and analyze during the disturbed ionospheric conditions termed as 'ionospheric irregularities' than in free space [6]. There is a strong association between the ionospheric radio wave propagation and Space weather. When solar flares emit X-rays that ionize the D-region of the ionosphere, a sudden ionospheric disturbance or short-wave is fadeout. Consequently, the absorption of radio signals increases in this region due to enhancement in ionization. During the period of high solar activity, strong solar X-ray flares lead to the absorption of virtually all ions. Thus, the accuracy of GPS (Global Positioning System) is disturbed, as these solar flares disrupt HF radio propagation.

Radio wave propagation is influenced by multiple factors, such as ionization of the ionosphere caused by the Sun and daily variations in tropospheric water vapor. Additionally, the path of radio wave propagation from point to point can vary, including over-the-horizon paths influenced by ionospheric refraction or direct line-of-sight paths. Ionospheric variability, such as spordiac-E, spread-F, ionospheric layer tilts, and solar activities like geomagnetic storms, solar flares, and solar proton events, also affect ionospheric radio signal propagation. Due to different frequencies, radio waves propagate in different modes. For example, the wavelength of extra low frequencies (ELF) and very low frequencies (VLF) are immensely larger than the separation between the Earth's surface and the ionospheric D region. Consequently, electromagnetic waves can propagate as a waveguide in this region.

The wave propagates as a single waveguide for frequencies less than 20 KHz, with horizontal magnetic field and vertical electric field [10]. Thus, the interaction of ionized regions of the atmosphere with radio waves is complex to predict and examine than free space. Since radio propagation cannot be predictable, many services have been moved to communication satellites. For instance, services like emergency locator transmitters in flight communication such as ocean-crossing aircraft and television broadcasting had been using satellite communication. Despite the expense, satellite links offer highly stable and predictable line-of-sight coverage for specific areas.

19.2 IONOSPHERIC IRREGULARITIES

The ionospheric irregularities are disruptions or anomalies that can affect the density and composition of the ionosphere. These variations are collectively referred as ionospheric irregularities. It is crucial to understand these irregularities because they can influence radio signal propagation, satellite communication, and GPS accuracy. "Non-frozen irregularity" is a rapid fluctuation in an electron density with respect to time and space, and "frozen irregularity" is the one that does not change with time [14]. The plasma irregularities are categorized into various scale length as follows:

(a) **Large scale – (>10 km)**
(b) **Intermediate scale – (10–0.1 km)**
(c) **Transitional scale – (100–10 m)**
(d) **Small scale – (<10 m)**

The long distance radio communication navigation system encountered various difficulties like errors in positioning accuracy, fading of signal, etc., due to the presence of these irregularities, particularly, the small-scale irregularities [15].

19.2.1 Types of Ionospheric Irregularities

19.2.1.1 Equatorial Spread F (ESF)

This irregularity is common around the magnetic equator. The solar radiations are maximum at the equatorial region, which enhances the ionization during the day, and just after sunset the ionospheric E-region experiences a rapid depletion in the electron density due to recombination of ions and electrons. In contrast, the F-region has higher electron density compared to the E-region because of a slow recombination rate. This creates a vertical slope in plasma density between the F-region and E-region, which in the presence of a gravitational field, induces Rayleigh-Taylor instability, which enhances the eastward electric field [16]. When this eastward electric field grows sufficiently strong, the vertical $\mathbf{E} \times \mathbf{B}$ shifts the plasma irregularity upward. These ionospheric irregularities moving towards the F-region grow in size, forming immense structures of plasma-depleted regions known as equatorial plasma bubbles (EPB). These plasma bubbles spread downward along geomagnetic field lines, and this process creates varying electron density and contributes to the dynamic behavior of the ionosphere [17].

19.2.1.2 Auroral Irregularities

Auroral irregularities in the ionosphere are instigated by disturbances associated with auroras. When solar winds, which consist of charged particles, arrive at the Earth, they interact with the planet's magnetic field. Due to this interaction, a current flow in the magnetosphere accelerates the charged particles towards the polar regions. Upon reaching the upper atmosphere, these accelerated particles strike the neutral atoms and molecules, resulting in excitation and ionization. Due to this process, the ionization in the auroral region is amplified, leading to the formation of auroral irregularity.

19.2.1.3 Polar Patches

Polar patch irregularity in the ionosphere occurs in the polar regions (especially at high latitude regions), 70°–80° north and south of the poles. These irregularities are characterized by regions of enhanced plasma density within the polar ionosphere. They form during a high geomagnetic activity period, compelled by the interaction of the solar wind with Earth's magnetic field, leading to the inoculation of high-energy solar particles into the polar regions. This interaction enhances the convection electric field, resulting in the transfer of plasma from the dayside to the nightside of the polar cap, leading to the formation of polar patches. These patches cause irregularities such as scintillation, signal delay, etc., which can affect communications.

19.3 IONOSPHERIC SCINTILLATION

Scintillation is caused by rapid fluctuation in radio signals as they pass through irregularities in the ionosphere. Ionospheric irregularities are responsible for producing scintillation by degrading the radio signals [18]. Scintillation affects GPS and satellite communication systems, as they rely on satellite signals. Effect of scintillation on radio waves:

Amplitude Scintillation: When radio signals pass through the irregular ionosphere, the fluctuation in its amplitude is known as amplitude scintillation. The amplitude scintillation leads to the potential loss of data in navigation and communication signals.

The scintillation occurrence and its characteristics are analyzed using the amplitude scintillation index S_4. The S_4 index is expressed in relation to the intensity of the received signal as follows:

$$S_4^2 = \frac{\langle I^2 \rangle - <I>^2}{<I>^2}$$

where S_4 is the amplitude scintillation index, $<I^2>$ is the mean of the square of the received signal intensity, and $<I>^2$ is the mean of the received signal intensity.

Phase Scintillation: Similarly, the fluctuation in the phase of a radio signal is referred to as phase scintillation. It affects the phase information, such as GPS satellite signals [19].

Ionospheric scintillation is an important tool in studying the characteristics of ionospheric irregularities. Scintillation is influenced by various factors such as: latitude, longitude, local time, season, solar activity, solar cycle, solar flares, geomagnetic activity, geomagnetic storms, etc. [20]. The extent of the amplitude and phase scintillation is quantified by the S4 index and $\sigma\phi$ index, respectively. The S4 index represents the standard deviation of the normalized carrier-to-noise (C/N) ratio of the signal. The $\sigma\phi$ index is derived from the variance of measurement of the carrier phase [21].

Scintillation is usually witnessed over polar and low latitudes during the period of solar maximum. In low latitude regions, plasma instabilities are accountable for scintillation, whereas in the polar region, precipitation of particles are accountable [22–24]. In low latitude regions, amplitude scintillation is more recurrent than the phase scintillation; in contrast, in polar regions phase scintillation is more frequent [25].

19.4 GPS SCINTILLATION

The Global Navigation Satellite System (GNSS) is a navigation system based on satellites that provide global coverage of three-dimensional positioning. The existing GNSSs are the USA 'GPS (NAVSTAR),' Russian 'GLONASS,' 'GALILEO' of the European Union, the proposed 'COMPASS' navigation system of China, and India's 'Indian Regional Navigation Satellite System (IRNSS).' Among these, the fully functional GNSS is the United States' GPS (NAVSTAR) with a constellation of 31 working satellites. The GPS delivers user position, velocity, and time information with good accuracy regardless of weather conditions. GPS was initially created by the U.S. Department of Defense (DoD) only for military users, but later in the 1980s it was made accessible for civilian use. In order to give global coverage, GPS is made up of a constellation of 24 satellites with specific spacing in six orbits, such that if you are anywhere on the Earth's surface, you have a minimum number of four satellites observable to you at any time at any location on Earth.

GPS satellites convey L-band signals on two carrier frequencies: L1 (1575.42 MHz) and L2 (1227.60 MHz). These signals are modulated with digital codes, the precision (P) code and the Coarse Acquisition (C/A) code, as well as a navigation message that modulates only onto the L1 signal. The P code is longer and more precise than the C/A code. The P code and the C/A code are referred to as Pseudo Random Noise (PRN) codes since they are random binary sequences. To keep minimum cross-correlation, different PRN codes are assigned to different satellites. Each satellite has a unique PRN number, which helps in easily identifying the satellites that transmit a particular code [26].

As the GPS satellite signal traverses the dispersive ionosphere, it experiences both a delay in time and an advance in phase of the signal. This time delay of the GPS signal is influenced by the ionospheric electron density. Therefore, by the estimation of the delay of GPS signal, we can determine ionospheric total electron content (TEC). This GPS TEC and its rate of change of TEC index (ROTI) is also used for the study of ionospheric irregularities. The GPS also provides amplitude scintillation S4 index. Thus, both S4 index and ROTI are used to study the different characteristics of ionospheric irregularities.

19.4.1 OCCURRENCE OF IONOSPHERIC SCINTILLATION OVER INDIAN LOW LATITUDE STATION VARANASI

In this chapter, the occurrence of amplitude scintillations is studied from January 2021 to December 2022 (during the ascending phase of the 25th solar cycle) using the amplitude scintillation index S_4. The S_4 index is derived from the multi-frequency GNSS receiver Septentrio PolaRx5s installed at Banaras Hindu University, Varanasi (latitude: 25.3176, longitude: 82.9739). The study has been performed to examine the diurnal, monthly, and seasonal occurrence characteristics of ionospheric irregularities.

19.4.1.1 Monthly Percentage Occurrence of S4 Index

The monthly variation in the percentage occurrence of S4 index with monthly mean sunspot number from January 2021 to December 2022 over Varanasi is depicted in Figure 19.1. It is observed that during the periods of low solar activity, scintillation occurrence in March and September (equinox) is lower than May (summer). However, the difference in scintillation occurrence between March (an equinox month) and May (a summer month) is minimal. In contrast, during the period of moderate solar activity, the highest scintillation occurrence is observed during the equinox. Further, scintillation events are slightly less frequent in March–April compared to September, suggesting equatorial asymmetry. The trend of variation of both parameters is almost similar.

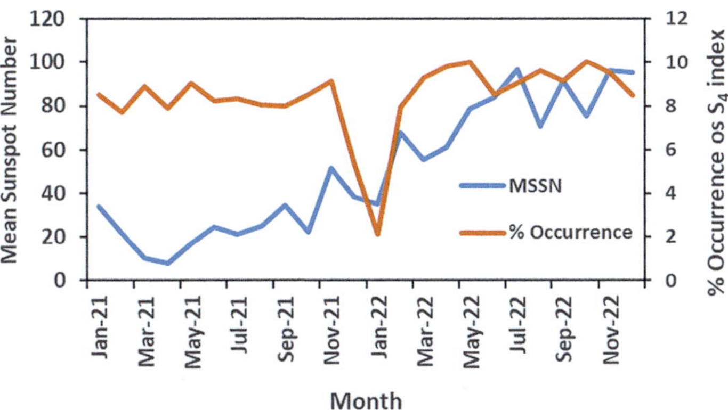

FIGURE 19.1   Monthly variation of percentage occurrence of scintillation at Varanasi for the years 2021–2022.

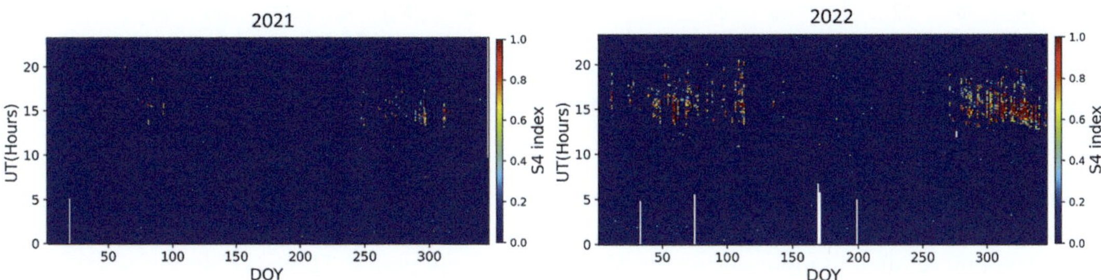

FIGURE 19.2   Diurnal variation of S_4 index over Varanasi for years (a) 2021 and (b) 2022. The color bar specifies the intensity of the S_4 index.

19.4.1.2   Diurnal S4 Index

To observe the occurrence of scintillations during the entire study period, the variation of the daily S4 index of two years is plotted in Figure 19.2. This figure illustrates the diurnal variation of the S4 index over Varanasi for (a) 2021 and (b) 2022. The x-axis represents the day of the year, while the y-axis shows Universal Time (UT) in hours. The color bar specifies the intensity of the S4 index. The S4 index data is collected from all PRNs during their passage over Varanasi, recorded at an elevation of 10° to minimize data loss due to higher elevation angles.

Figure 19.2 shows that the overall scintillation occurrence characteristics over Varanasi were less frequent in 2021 during minimum solar activity. For 2022, scintillation occurrences increased, correlating with moderate solar activity. In all cases, the majority of the diurnal S4 values are below 0.2, with values under 0.1 delineated as noise. The figure also shows that scintillation occurrences are mostly seen during nighttime and equinoctial months.

19.4.1.3   Seasonal Percentage Occurrence

The occurrence and intensity of ionospheric scintillation vary with several factors, including latitude, local time, solar activity, and seasons. In equatorial and low latitude regions, scintillation tends to be more pronounced around the equinoxes (March and September). This is due to the alignment of the geomagnetic equator and the solar terminator, which enhances the development of ionospheric irregularities. Scintillations over low latitudes are most intense after sunset due to the formation of equatorial plasma bubbles and spread F, which are intermediate-scale ionospheric irregularities.

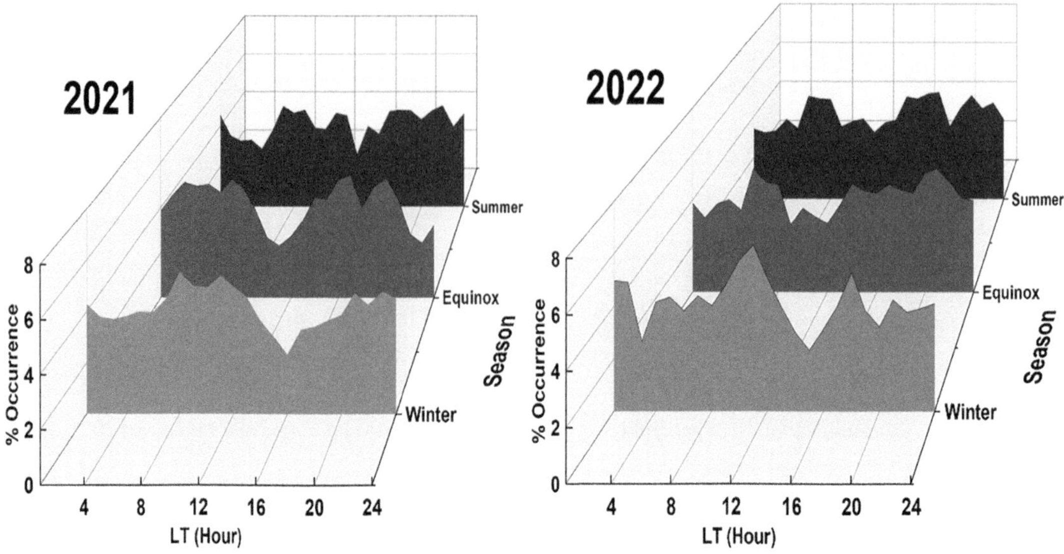

FIGURE 19.3 Seasonal variation of percentage occurrence of scintillation over Varanasi during 2021–2022.

Figure 19.3 illustrates the seasonal percentage scintillation occurrence concerning local time in Varanasi. It shows scintillation occurrence during nighttime as well as daytime. Naturally, scintillation occurs more often at night during the high solar activity period. However, during years of low solar activity, daytime scintillation occurrences are also observed.

Several studies have been reported about the monthly and seasonal variation of scintillation occurrence over different ionospheric regions using GPS measurements [27–28]. Our above results matched well with that of different low latitude scintillation occurrence characteristics.

19.5 CONCLUSION

In this chapter, we presented the morphology of ionosphere, summary of ionospheric irregularities, and scintillation. A study of characteristic of scintillation occurrence over Indian EIA station Varanasi for the period of two years from January 2021 to December 2022 has been conducted. Based on these studies, the following conclusions were reached:

(1) Throughout the modest solar activity period, the highest percentage occurrence of ionospheric irregularities is witnessed during the equinox months, moderate occurrence during winter months, and lowest occurrence during the summer months. On the other hand, during the minimal solar activity period, scintillation percentage occurrence in May is greater than the equinox months.

(2) The presence of equinoctial asymmetry has been noted, i.e., scintillation occurrence in March–April is slightly lesser than that of September. This equinoctial asymmetry might be attributed to the inter-hemispheric neutral winds.

(3) The diurnal occurrence of scintillation is most common throughout the equinoctial months and at nighttime.

(4) The dependency of occurrences of scintillations on solar activity has been found. The scintillation occurrence frequency at Varanasi during low solar activity year 2021 was weaker. However, the scintillation occurrence was more frequent corresponding to the modest solar activity year 2022.

REFERENCES

[1] O. Heaviside, *Telegraphy, Britannica 33*, 10th ed., Encyclopedia Britannica, Chicago, 1902, pp. 213–218.

[2] A.E. Kennelly. On the elevation of the electrically conducting strata of the earth's atmosphere, Electr. *World Eng.* 39 (1902) 473.

[3] E.V. Appleton and M.A.F. Barnett, *On some direct evidence for downward atmospheric reflection of electric rays, Nature* 115 (1925) 333.

[4] Hargreaves, J.K., 1992. *The solar-terrestrial environment: an introduction to geospace-the science of the terrestrial upper atmosphere, ionosphere, and magnetosphere.* Cambridge University Press.

[5] Anduaga, A., 2009. Sydney chapman on the layering of the atmosphere: Conceptual unity and the modelling of the ionosphere. *Annals of Science*, 66(3], pp. 333–344.

[6] Chapagain, N.P., 2015. Dynamics ionospheric plasma bubbles measured by optical imaging system. *Journal of Institute of Science and Technology*, 20(1), pp. 20–27.

[7] Chapman, S. (1931a), The absorption and dissociative or ionizing effect of monochromatic radiation in an atmosphere on a rotating Earth, Proc. *Phys. Soc. (London)*, 43, 26–45.

[8] Chapman, S. (1931b), The absorption and dissociative or ionizing effect of monochromatic radiation in an atmosphere on a rotating Earth II, Proc. *Phys. Soc (London)*, 43, 483–501.

[9] Hargreaves, J.K., 1979. *The upper atmosphere and solar-terrestrial relations-An introduction to the aerospace environment.* New York.

[10] Chapagain, N.P. and Patangate, L., 2016. *Ionosphere and its influence in communication systems. an annual publication of Central Department of Physics*, 10.

[11] Rishbeth, H. and O.K. Garriott, 1969. *Introduction to ionospheric physics*, Academic press, New York & London, 1–5.

[12] Booker, H.G. and Wells, H.W., 1938. Scattering of radio waves by the Foregion of the ionosphere. *Terrestrial Magnetism and Atmospheric Electricity*, 43(3), pp. 249–256.

[13] Bora, S., 2017. Ionosphere and radio communication. *Resonance*, 22(2), pp. 123–133.

[14] G. Haerendal, Theory of equatorial Spred-F, Report, Max-Planck-Institute, Space Plasma Physics of Near-Earth Environment, MPI for Extraterrestrial Physics, Max Planck Society, ou_159897, Fur. Phys. Und. Astrophys. Garching, West Germany (1974).

[15] Kelley, M.C., Siefring, C.L., Pfaff, R.F., Kintner, P.M., Larsen, M., Green, R., Holzworth, R.H., Hale, L.C., Mitchell, J.D. and Le Vine, D., 1985. Electrical measurements in the atmosphere and the ionosphere over an active thunderstorm: 1. Campaign overview and initial ionospheric results. *Journal of Geophysical Research: Space Physics*, 90(A10), pp. 9815–9823.

[16] Abdu, M.A., 2005. Equatorial ionosphere–thermosphere system: Electrodynamics and irregularities. *Advances in Space Research*, 35(5), pp. 771–787.

[17] B. Basu, On the linear theory of the equatorial plasma instability: comparison of different descriptions, *J. Geophys. Res.* 107 (A8) (2002).

[18] Mondal, M., Kumar, S., Banola, S. and Singh, A.K., 2024. Occurrence of ionospheric scintillations under different solar and geomagnetic conditions over low latitude station Varanasi. *Advances in Space Research*, 73(7), pp. 3658–3674.

[19] Crane, R.K., 1977. Ionospheric scintillation. *Proceedings of the IEEE*, 65(2), pp. 180–199.

[20] Fukao, S., Ozawa, Y., Yokoyama, T., Yamamoto, M. and Tsunoda, R.T., 2004. First observations of the spatial structure of F region 3-m-scale field-aligned irregularities with the equatorial atmosphere radar in Indonesia. *Journal of Geophysical Research: Space Physics*, 109(A2).

[21] Van Dierendonck, A.J. and Enge, P., 1994, April. RTCA SC-159 Wide Area Integrity Broadcast/ Wide Area Differential GPS Status. In *Proceedings of 1994 IEEE Position, Location and Navigation Symposium-PLANS'94* (pp. 613–620). IEEE.

[22] Rastogi, R.G. and Klobuchar, J.A., 1990. Ionospheric electron content within the equatorial F 2 layer anomaly belt. *Journal of Geophysical Research: Space Physics*, 95(A11), pp. 19045–19052.

[23] Basu, S., Basu, S., Groves, K.M., MacKenzie, E., Keskinen, M.J. and Rich, F.J., 2005. Near-simultaneous plasma structuring in the midlatitude and equatorial ionosphere during magnetic superstorms. *Geophysical Research Letters*, 32(12).

[24] Basu, S., Basu, S., Rich, F.J., Groves, K.M., MacKenzie, E., Coker, C., Sahai, Y., Fagundes, P.R. and Becker-Guedes, F., 2007. Response of the equatorial ionosphere at dusk to penetration electric fields during intense magnetic storms. *Journal of Geophysical Research: Space Physics*, 112(A8).

[25] Azeem, I., Crowley, G., Reynolds, A., Santana, J. and Hampton, D., 2013, April. First results of phase scintillation from a longitudinal chain of ASTRA's SM-211 GPS TEC and scintillation receivers in Alaska. In *Proceedings of the ION 2013 Pacific PNT Meeting* (pp. 735–742).

[26] Shreedevi, P.R. and Choudhary, R.K., 2017. Impact of Oscillating IMF Bz During 17 March 2013 Storm on the Distribution of Plasma Over Indian Low-Latitude and Mid-Latitude Ionospheric Regions. *Journal of Geophysical Research: Space Physics*, 122(11), pp. 11–607.

[27] Rathore, V.S., Kumar, S., Singh, A.K., 2018. Occurrence characteristics of ionospheric irregularities over Indian low-latitude region Varanasi during ascending phase of solar cycle 24, *Annales of Geophysics*, 61 (1), PA113.

[28] Sahithi, K., Sridhar, M., Kotamraju, S. K., Kavya, K. C. S., Sivavaraprasad, G., Ratnam, D. V., & Deepthi, C., 2019. Characteristics of ionospheric scintillation climatology over Indian low-latitude region during the 24th solar maximum period. *Geodesy and Geodynamics*, 10(2), pp. 110–117.

20 GNSS as a Cost-effective Tool for Continuous Monitoring of Solar Flare Effect on the Ionosphere

Suniti Saharan, Ajeet Kumar Maurya,
Mahesh N. Shrivastava, and Rajesh Singh

20.1 INTRODUCTION

GNSS (Global navigation satellite system) is a large term that refers to any satellite constellation supplying global or regional positioning, navigation, and timing services. Several countries posses GNSS technology: GPS (Global Positioning System) from United States, IRNSS (Indian Regional Navigation Satellite System) from India, BeiDou (BeiDou Navigation Satellite System) from China, Galileo (European Global Navigation Satellite System) from Europe, GLONASS (Global Navigation Satellite System) from Russia, and QZSS (Quasi-Zenith Satellite System) from Japan. GNSSs supply a global dataset critical for geodetic reference frame analysis, automotive engineering, cadastral surveying, crustal deformation monitoring, and even Space weather. However, the high cost of GNSS receivers has been a big drawback for researchers.

Fortunately, significant advances in GNSS technology in recent decades have enabled the evolution of low-cost single- and double-frequency sensors for ionosphere studies (e.g., U-BLOX, ScintPi 2.0 and 3.0, u-blox ZED-F9P, Tersus BX305) that scientific interests and various investigations have used to examine ionospheric variations, compared to high-cost receivers (Okoh et al., 2021; Gomez Socola & Rodrigues, 2022; Kogogin et al., 2021; Bramanto et al., 2018; Dan et al., 2021). GNSS satellites orbit approximately 20,000 km above the Earth's surface and send signals through the ionosphere. As signals disperse when they pass through the ionosphere, they are delayed and cause errors in positioning, navigation, and timing services (Rao & Singh, 2021).

The ionized upper atmosphere above 50 km altitude that has enough electrons and ions to obstruct radio wave transmissions is known as the ionosphere. Its elasticities form about 80 to 1000 km above the Earth's surface. The Earth's higher ambiance is where sun-powered Extreme Ultraviolet (EUV) and X-ray radiations are captured, creating the ionosphere, resulting in significant photoionization and a partially ionized atmosphere. The ionosphere is vertically classified and separated into interconnecting layers, known as the F, E, and D layers. F1 and F2 layers are further divisions of the F layer, which are only visible during the day. Vertical classification is thought to be caused by gravity and the structure of the atmosphere. Figure 20.1 represents the ionospheric vertical structure, with a clear day-night disparity in the ionosphere density profile. The profile also differs with the solar cycle, with higher densities during periods of high solar activity. Other processes influence the ionosphere profile depending on latitudinal location (Rao & Singh, 2021).

DOI: 10.1201/9781032712444-24

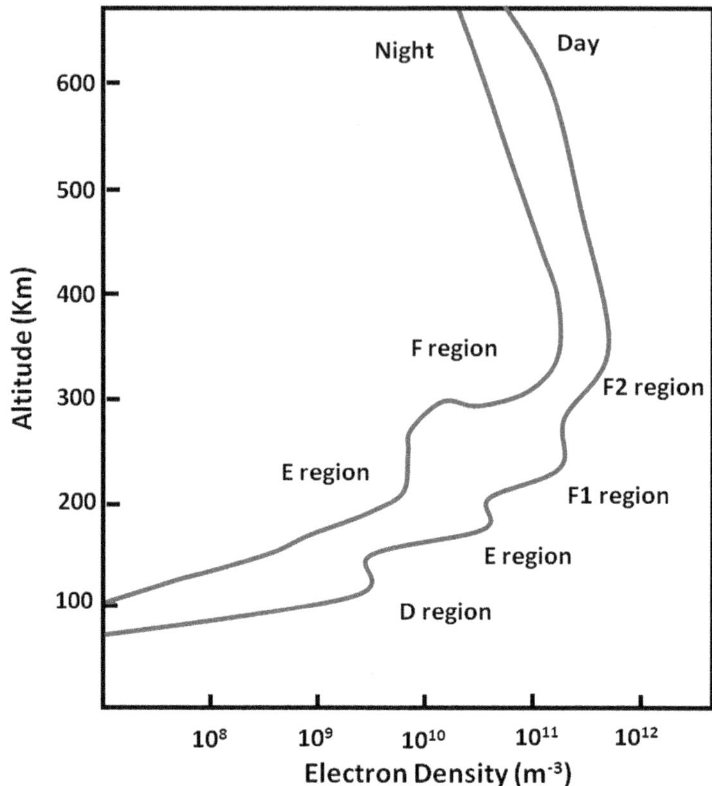

FIGURE 20.1 Variation of electron density during the day and nighttime.

20.2 IMPORTANT PARAMETER OF IONOSPHERE

The different parameters of the ionosphere that are affected due to solar events are ionization density, F layer critical frequency, total electron content, etc. However, a crucial ionosphere characteristic known as electron density has been regularly assessed using tools from the ground, spacecraft, and satellites (Rao & Singh, 2021). A signal's transmission pathway integral electron number density (in m^{-3}) is proportional to the difference in delay between two signals; this is known as total electron content (1 TEC unit = 1×10^{16} electrons/m²) (Gomez Socola & Rodrigues, 2022). The first step in calculating TEC is to estimate Slant Total Electron Content (STEC), which measures the quantity of free electrons in a square meter within the satellite and the receiver's line of sight. Subsequently, STEC is converted into Vertical Total Electron Content (VTEC); this VTEC value is equivalent to the TEC measurement (Afraimovich et al., 2000). Figure 20.2 illustrates the transformation process from STEC to VTEC.

The TEC on the slant path between a ground-based receiver (Rx) and a satellite (Tx), referred to as STEC, as per Hofmann-Wellenhof et al. (2012), can be expressed as:

$$\text{STEC} = \int_{Rx}^{Tx} N \, dl \tag{20.1}$$

In this equation, N represents the electron density, which is expressed in electrons per cubic meter (electron/m³) and varies depending on the location. The 'l-axis' denotes the path taken by the satellite and the receiver.

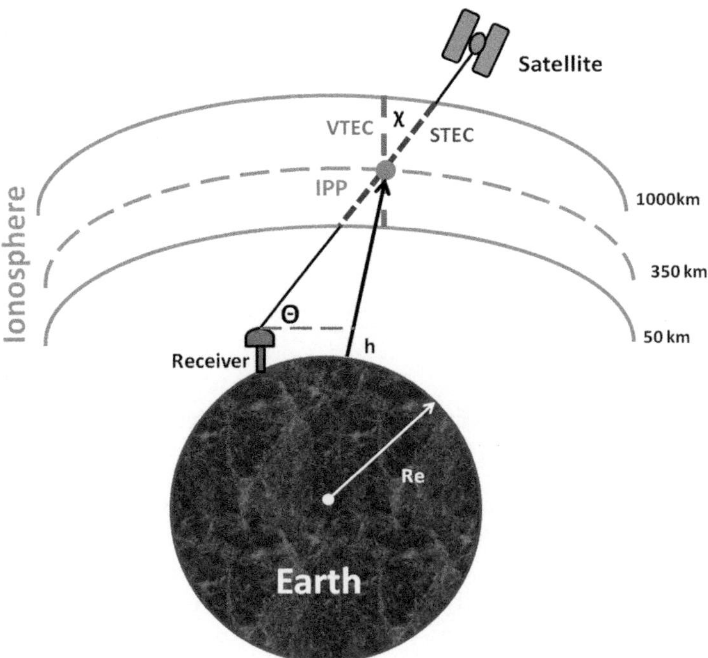

FIGURE 20.2 VTEC derived from STEC for Precise Atmospheric Monitoring.

Each GPS satellite transmits at two carrier L-band frequencies: L1 (1,575.42 MHz) and L2 (1,227.60 MHz). A dual-frequency GPS receiver detects the variation in ionospheric delay between the L1 and L2 signals as the GPS signal passes through the ionosphere. The TEC value is then calculated using this discrepancy (Bagiya et al., 2009). Phase and pseudo-range data are used to calculate the TEC:

$$\text{STEC} = \frac{1}{40.3} \times \left[\frac{1}{L1^2} - \frac{1}{L2^2} \right]^{-1} \times \left(P1 - P2 \right) + \text{TEC}_{\text{cal}} \tag{20.2}$$

Here, TEC_{CAL} stands for bias error correction, and P1 and P2 stand for the pseudo range at L1 and L2, respectively, as described in Hofmann-Wellenhof et al. (2012) and Fayose et al. (2012).

Utilizing the ray path, the STEC depends on the geometric arrangement throughout the ionosphere. A corresponding VTEC is computed, which is unaffected by the ray path's elevation angle. The thin shell model is used to project from the slant to the vertical, with an assumed height of 350 kilometers, to calculate the VTEC, following the methodology outlined by Klobuchar, (1987).

$$\text{VTEC} = \text{STEC} \times \cos \chi \tag{20.3}$$

$$\sin \chi = \frac{R_e}{R_{e+h}} \cos \cos \Theta$$

$$\cos \cos \chi = \sqrt{1 - \left(\frac{R_e}{R_{e+h}} \cos \Theta \right)^2}$$

$$\text{VTEC} = \text{STEC} \times \cos \left[\sin^{-1} \left(\frac{R_e}{R_{e+h}} \cos \cos \Theta \right) \right] \tag{20.4}$$

The Ionospheric Pierce Point (IPP) is a crucial concept in this context, and it depends on various parameters. Among these parameters, we have the Zenith angle (χ) and the satellite's elevation angle (θ), both measured in degrees. The Earth's radius (R_e), approximately 6,378 km, also plays a key role in these calculations. Additionally, it's assumed that the ionospheric layer's height (h) is fixed at 350 km. The IPP is determined and characterized using the relationship between these variables in the context of satellite-based navigation systems and ionospheric modeling (Silwal et al., 2021).

The TEC can be affected by many Space weather events (Liu et al., 2011), i.e., solar flares, geomagnetic storms, coronal mass ejection, solar wind, etc., and various tropospheric-lithospheric events (Chernigovskaya et al., 2015), like cyclones, earthquakes, hurricanes, etc. The impact of solar flares on TEC variation in the ionosphere is the main topic of this chapter. We also discuss the best way to calculate TEC.

20.3 IMPACT OF SOLAR FLARES AND SOLAR FLUXES ON IONOSPHERIC CONDITIONS

In the sun's active regions, solar flares are abrupt outbursts of solar energy in the twisted magnetic field lines. These abrupt events generate electromagnetic waves across a broad spectrum, with a primary contribution in the X-ray and EUV ranges, which raises the ionization rate and, consequently, the ionosphere's electron density. The emission of the X-ray spectral band between 0.1 and 0.8 nm is used to classify solar flares, the data obtained by the Geostationary Operational Environmental Satellite (GOES). A, B, C, M, and X letters are used to categorize solar flares in order of increasing X-ray flux. The most powerful solar flare is one that emits 10^{-4} Wm-^{2} flux and belongs to the X class (Mitra, 1974). The solar flares in Table 20.1 are classified based on their X-ray flux emissions measured by the GOES satellite.

Furthermore, a more detailed scale ranging from 1 to 9 exists within each letter class. A and B classes of flares are usually too weak to have an impact on the ionosphere. M-class flares, on the other hand, have the potential to cause short-term radio blackouts in polar regions as well as minor radiation storms that can endanger astronauts. It's important to emphasize that, even though "X" represents the highest letter in the classification, there are flares with intensities exceeding 10 times that of an X1, indicating that X-class flares can surpass the upper limit of 9.

20.3.1 CORRELATION BETWEEN SOLAR FLUXES

The dissimilarities in X-ray and EUV flux with altitude are known as Sudden Ionospheric Disturbances (SID) (Donnelly, 1969; Davies, 1990); Sudden Frequency Deviation (SFD) (Donnelly, 1971; Liu et al., 1996); Sudden Phase Anomaly (SPA) (Jones, 1971; Ohshio, 1971); Short Wave Fadeout (SWF) (Stonehocker, 1970); Sudden Cosmic Noise Absorption (SCNA) (Deshpande & Mitra, 1972); Sudden

---

TABLE 20.1

Solar Flare Classes and Associated X-ray Flux Levels

Classes of solar flare	X-ray flux 0.1–0.8 nm (W/m^2)
A	10^{-8}–10^{-7}
B	10^{-7}–10^{-6}
C	10^{-6}–10^{-5}
M	10^{-5}–10^{-4}
X	$>10^{-4}$

Enhancement/decrease of atmospherics (SES) (Sao et al., 1970). Many SID studies have been conducted, including a number of comprehensive reviews (Mitra, 1974; Davies, 1990).

The optically thin sun corona is the primary source of EUV emissions and soft X-rays with wavelengths shorter than 25 nm. These emissions rapidly intensify during solar flares and affect the lower ionosphere of the Earth (150 km) (Qian et al., 2011). Atomic oxygen in the F region is mainly ionized by EUV emissions from the optically thick transition layer and chromosphere at wavelengths of 26–34 nm at altitudes greater than 200 km (Qian et al., 2011; Richards et al., 1994; Woods et al., 2011). As a result, the magnitude of solar X-ray and EUV flux determines the magnitude of change in the ionosphere. Therefore, to evaluate the risk of solar fluxes, the connection between X-ray and EUV flux should be examined. Figure 20.3 displays the emissions of EUV flux and X-rays during a solar flare of X1.8 class in 2014. The last graph depicts the temporal evolution of VTEC in the context of an X-class solar flare.

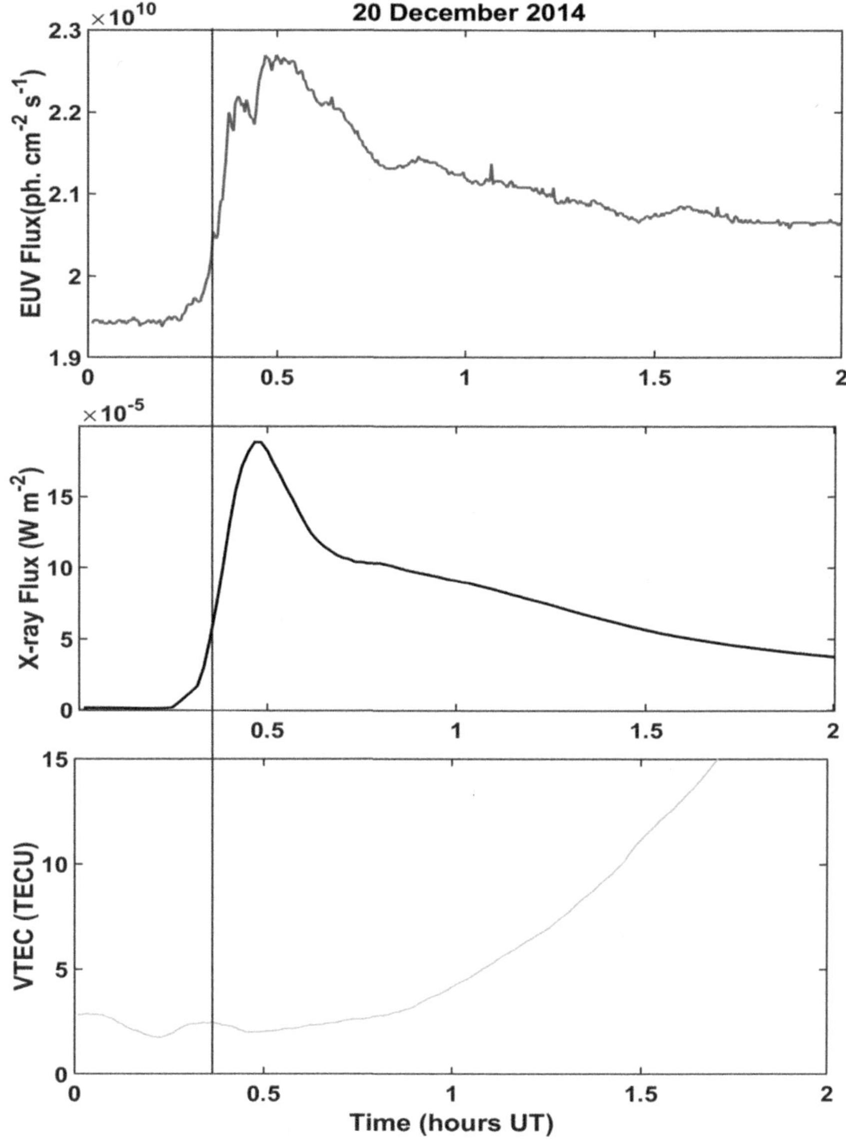

FIGURE 20.3 Time series of X-ray flux, EUV flux, and VTEC for an X-class of solar flare.

Over the past few decades, the impact of solar flare-induced EUV and X-ray fluxes on the ionosphere has been studied (Garriott et al., 1967; Horan & Kreplin, 1981; Bauer, 1973; Horan et al., 1982; Horan et al., 1983; Donnelly, 1976). It was determined that solar fluxes caused an increase in photoionization in the ionosphere. Furthermore, some research has been conducted to investigate the dependence of ionospheric TEC upon solar activity, and it was also discovered that TEC is dependent on solar fluxes (Dabas et al., 1993; Chakerbarty et al., 2015; Patel et al., 2017). However, this chapter focuses on the influence of solar flares on the TEC of the ionosphere as well as their relationship with solar fluxes.

20.3.2 Correlation between TEC and Solar Fluxes

As indicated by the findings of preceding studies, the ionospheric TEC is influenced by solar flux. Liu et al. (2006) observed the ionospheric TEC change due to solar flares, and they reported the strong linear association between ΔTEC (change in TEC) with X-ray flux. Leonovich et al. (2010) investigated how ionospheric TEC responds to C, M, and X-class solar flares using both GPS data and theoretical models. They discovered an empirical connection between the TEC response's amplitude and the solar flares X-ray peak output. This relationship is approximated by the function $\Delta I = 649.4\varphi^{0.7}$, where ΔI represents TEC, and φ corresponds to X-ray flux.

Mahajan et al. (2010) inspected the X-class solar flares and discovered a correlation in the EUV and X-ray flux. They investigated the association among peak EUV and peak X-ray fluxes and the number of solar flares. With 10 X class solar flares and 70 X class solar flares, they discovered a 0.38 and 0.44 correlation, respectively. When the peak fluxes are changed to ΔX-ray and ΔEUV, the correlation improves to 0.51 and 0.59, respectively. They also looked into the connection between solar fluxes and TEC. There is a 0.85 correlation between EUV and TEC and a 0.53 correlation between X-ray flux and TEC. Further, they reported the effect of solar flare location. Using solar fluxes, Zhang et al. (2011) calculated the ionospheric TEC. They discovered a strong relationship between TEC and EUV flux at 26-34 nm. They also investigated the TEC variation in relation to the position effect of solar flares. However, this chapter does not address the location effect of solar flares.

Le et al. (2011) examined the upper atmosphere's response to a solar flare. They discovered a significant correlation among X-ray and EUV flux for different types of solar flares. Correlations for C-class flares were 0.54, M-class flares were 0.58, and X-class flares were 0.66 observed. When the location of solar flares was also considered, the correlation increased, and multiplication cos (CMD) in X-ray was performed. Le et al. (2013) conducted a study of X class solar flares using GPS TEC data. They found a 0.67 correlation among X-ray and TEC. Furthermore, they discovered a 0.91 and 0.84 correlation among TEC and EUV flux (at different wavelengths) using EUV flux.

Hazarika et al. (2016) explored the solar cycle 24 ascending phase. They studied an X-class solar flare that happened around this time. Their findings revealed several significant correlations: a correlation of 0.47 between ΔX-ray and ΔEUV flux, ΔX-ray & ΔTEC have a correlation of 0.68, and a strong correlation of 0.86 between ΔEUV flux and ΔTEC. Saharan et al. (2023) looked into the impact of various solar flare classes that coincided with sun cycle 24's height. Notably, they observed a correlation of 0.53 between ΔX-ray and ΔEUV flux in X-class solar flares, and as the class of solar flare decreased, this correlation value tended to decrease. For X-class, they reported correlations among ΔTEC and ΔX-ray of 0.6, and between ΔTEC and ΔEUV flux of 0.67 using a baseline method. Using a mean method, they found correlations of 0.52 between ΔTEC and ΔX-ray, and 0.81 between ΔTEC and ΔEUV flux. However, these correlations tended to diminish with decreasing solar flare class. Additionally, their research considered the impact of the location of solar flares, which appeared to have a noticeable effect in X-class solar flares but was negligible in M and C-class flares.

20.4 TECHNIQUES OF TEC CALCULATION: ADVANTAGES AND DISADVANTAGES

Researchers have used GNSS TEC data to study ionospheric behavior over time, tracking variations that occur daily, seasonally, and during solar events. This information aids in understanding the ionosphere's reaction to solar flares, geomagnetic storms, and other Space weather phenomena. Monitoring TEC changes with time is not only important for precise satellite navigation but also for Space weather forecasting and ionospheric research, contributing to our knowledge of the vibrant nature of Earth's higher ambiance. The TEC calculation methods have evolved to leverage the increased availability of GNSS data, provide higher spatial and temporal resolution, incorporate ionospheric models, offer real-time updates, utilize advanced filtering techniques, and integrate Space weather information. These modifications reflect ongoing efforts to improve the accuracy and utility of TEC calculations for a variety of applications ranging from navigation to Space weather monitoring and ionospheric research.

GNSS data is used for the investigation of ionospheric variation; for that, researchers use dual-frequency measurements. This included measurements of phase and pseudo-range. Both measurements have their own set of advantages and disadvantages. While the pseudo-range method provides absolute values, it is more susceptible to noise. In contrast, phase measurements provide more precise TEC measurements but are ambiguous due to the unknown number of phase cycles. As a result, researchers employ a hybrid of the two methods to increase the accuracy in the TEC calculations. However, this approach allows for the estimation of ionospheric TEC changes while mitigating the drawbacks associated with each individual method (Opperman et al., 2007). By merging phase and pseudo range measurements, researchers can obtain more comprehensive and reliable insights into ionospheric disturbances in GNSS data analysis.

In starting, Lanyi & Roth, (1988) measured TEC using Jet Propulsion Laboratory (JPL) GPS receivers of dual-frequency. These assessments integrated data from six GPS satellites for the duration of multiple 5-hour sessions, utilizing P-code and carrier phase data. To map GPS TEC measurements to the Faraday beacon satellite's line of sight, a statistical fitting process was applied with a simple ionospheric model. The accuracy of this approach, including the simultaneous estimation of satellite and receiver offsets alongside TEC values, can yield GPS-derived line-of-sight TEC with an error as low as 1×10^{16} el/m^2. This accuracy level matches that of the Faraday rotation technique while offering superior sky coverage. Mannucci et al. (1998) employed a method where they partitioned the ionosphere into triangular regions, akin to sections of a sphere, for TEC calculation. This technique integrates high-resolution pixel-based methods with effective gradient estimation. To address data gaps, they incorporated predictions from models. Over time, they continually updated their TEC maps with fresh data, employing Kalman filtering to enhance their smoothness and estimate errors. Preliminary comparisons with other data sources indicated that their maps effectively captured ionospheric variations from the equator to approximately 65 degrees latitude.

20.4.1 PRESENT TECHNIQUES OF TEC CALCULATION

Over time, researchers have increasingly tried several approaches to extract the impact of solar flares in the VTEC signal. Among the approaches currently employed by many researchers are the Global Detection Technology (GLOBDET), rate change of TEC method (rTEC), baseline method, and mean method, which are mentioned below.

The first method of TEC calculation is GLOBDET, which is recommended by Afraimovich et al. (2000). It is a novel approach dependent on TEC phase measurements in the ionosphere. This technique uses a global GPS network to identify atmospheric disturbances caused by humans as well as natural causes. This method is based on the rapid change in the electron content of the ionosphere at the time of a solar flare. They collect data from all stations with a time resolution range of 30s to

0.1s; for fine detailing please follow cited paper. Ionospheric TEC based on phase measurement is given by the following formula:

$$I = \frac{1}{40.308} \frac{f_1^2 f_2^2}{f_1^2 - f_2^2} \left[\left(L_1 \lambda_1 - L_2 \lambda_2 \right) + \text{const} + nL \right]$$

Where $L_1 \lambda_1$ and $L_2 \lambda_2$ are the increments of the radio signal phase path brought on by the phase delay in the ionosphere (m), const is some unknown initial phase path (m), and nL is the error in determining the phase path (m). The values of λ_1 and λ_2 represent the wavelengths (m) for the frequencies f_1 and f_2, respectively, and L_1 and L_2 also denote the total number of phase rotations. Now, the GPS STEC data converted to GPS standard RINEX file format. The ionospheric response to solar flares was investigated via the coherent summation of time derivatives of the series of fluctuations of the VTEC value. Coherent summation formula is:

$$sd = \sum_{i=1}^{n} \frac{dI(t)}{dti} \times K_i$$

$$K_i = \cos \left[arcsin\ arcsin \left(\frac{R_z}{R_z + h_{\max}} \cos \theta i \right) \right]$$

where LOS has a value of n. To convert the STEC to a comparable VTEC value, the correction coefficient K_i is needed, ionospheric F2-layer maximum height is denoted by $h_{\max}$, and Earth's radius is represented by R_z.

$$\Delta I(t) = \int_{t1}^{t2} Sd_{ex}(t)\, dt$$

The trend, which is expressed as a polynomial on the associated time interval, is then subtracted by the coherent summation of the time derivatives result (normalized to the number of LOS). The computed time dependence $(Sd_{ex}(t))$ is then integrated to yield the mean integral TEC increment $\Delta I(t)$ on the specified time interval $[t_1 - t_2]$.

Further, using the same technology they examined the effect of bright and faint solar flares. They conclude that this approach serves as a valuable tool for determining the ionospheric reactions triggered by weak solar flares, specifically those categorized as X-ray class C. It becomes especially pertinent when the amplitude of TEC response to individual line-of-sight connections with GPS satellites is similar to the background variations. Furthermore, the research delves into examining how the location of bright solar flares on the Sun affects the TEC variation response's amplitude (Afraimovich et al., 2002).

Many researchers employ the second method of TEC calculation, known as the rate change of TEC method (dTEC/dt or rTEC); in this method for each satellite and receiver pair, first convert the STEC data into VTEC data for analysis. Usually, a solar flare causes a brief change in TEC; however, other factors were considered when analyzing the VTEC data for each satellite receiver pair. When calculating the TEC change, the instrumental bias of the satellite is eliminated, but not that of the receiver.

$$\text{TEC}_i = 2.853 \left(B_{rs} + dt_{pi} - dt_s - dt_r \right) \cos \chi_i,$$

$$B_{rs} = \frac{\sum_{i=1}^{i=N} \left(dt_{ri} - dt_{pi} \right)}{N},$$

$$\chi_i = \arcsin \arcsin \left[\left(\frac{R}{R + h_{ion}} \right) \cos \left(el_i \right) \right],$$

where the satellite and receiver's instrument biases are denoted by dt_s and dt_r, and the time delays, dt_{ri} and dt_{pi}, are calculated using the carrier phase and two frequency pseudo range measurements. The offset between absolute and relative TEC is called B_{rs}. The variable el_i is the satellite's elevation angle. The satellite zenith angle at IPP at the estimated 400 km altitude of the ionospheric shell is denoted by χ_i. R is the radius of the Earth, and N is the number of measurements in a continuous phase track. The rate of change of the TEC method employed by Zhang and Xiao (2003) was utilized to investigate the solar flare that occurred on April 15, 2001, using GPS TEC data.

They aimed to isolate and study the TEC changes solely attributed to the solar flare, while eliminating the influence of other factors, such as background TEC variations. To achieve this, they employed the following expression:

$$\text{DTEC} = \sum_{t_i}^{t_{i+n}} \frac{\partial TEC}{\partial t} - \frac{n}{m} \times \sum_{t_{i-m}}^{t_i} \frac{\partial TEC}{\partial t}$$

In this expression:

- "t_i" represents the starting time of TEC raise in the TEC curves produced via the solar flare.
- "t_{i+n}" signifies the time when the TEC reaches its highest increase due to the solar flare.
- "t_{i-m}" is a reference time carefully chosen just before the onset of the solar flare.

On the right, the second term of the equation is used to compute the TEC changes associated with the background ionosphere in the timeframe of the solar flare. This approach allowed Zhang & Xiao (2003) to isolate and analyze the specific TEC variations induced by the solar flare while accounting for the baseline ionospheric conditions. Zhang et al. (2011) delved into the impact of solar flare irradiation on the ionosphere. They adopted the method proposed by Zhang & Xiao, (2003) to minimize the influence of background TEC. By averaging the TEC increment values taken from several VTEC curves gathered in the subsolar region, the final TEC improvement value is calculated.

Further, Liu et al. (2004) used GPS data to explore the impact of a solar flare that happened on July 14, 2000, via rate of change of TEC (rTEC) method. Using the dual frequency method, they examined the TEC of the ionosphere. To improve TEC calculation accuracy, they subtracted each TEC value from the 30-second TEC value before it, essentially using a two-point differentiation approach. More precision is obtained from GPS data during the analysis of ionospheric TEC using the two-point differentiation approach. However, they concluded that by using GPS data, the rTEC could detect the sudden change in the ionosphere begun by a solar flare. Liu et al. (2006) looked two major solar flares that happened on October 28 and November 4, 2003. They examined the position effect of these solar flares by computing rTEC with a dual frequency method. They concluded that the position of solar flares has the noteworthy impact over the TEC of ionosphere TEC. Mahajan et al. (2010) utilized a two-point differentiation method to study an X-class solar flare over the upper ionosphere.

In recent studies, Saharan et al. (2023) employed the baseline method and the mean method to investigate various techniques for calculating total electron content (TEC) in order to explore the fluctuations in ionospheric TEC. They employed the mean approach and the baseline method as their two TEC methodologies. For both approaches, Gopi Seemala's RINEX GPS TEC analysis software was used to convert STEC values into VTEC values. In addition, a trend removal technique was triggered to eliminate the daily oscillations and parabolic trend caused by the satellite route. Initially, the temporal range of the given PRN must be reduced using the X-ray flux and VTEC figure. Furthermore, the solar flare VTEC data must be removed from the short time range VTEC data. Then, the optimal lowest RMS (root mean square) polynomial should be fitted to the remaining curve. Furthermore, the TEC change (DVTEC) is calculated by fitting this optimal polynomial to the VTEC data once more, which includes solar flare data. To compute the TEC change in the baseline procedure, a line was drawn in the DVTEC curve of a suitable PRN prior and following the peak time of solar flare. Figure 20.4 explains the baseline method in step via diagram.

Employing a uniform methodology for each PRN, the mean technique was utilized to calculate the DVTEC for solar flares, as illustrated in Figure 20.5. Subsequently, the mean was determined.

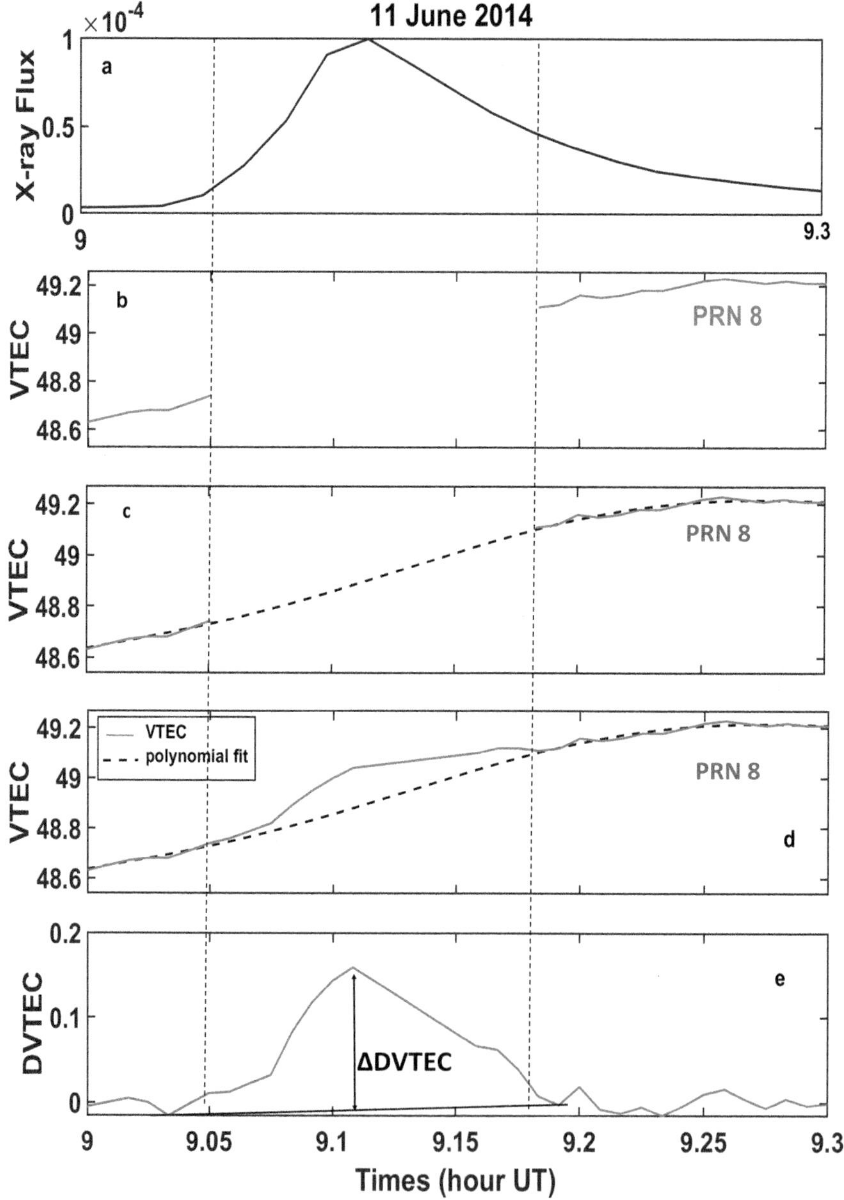

FIGURE 20.4 Baseline method applied on 11 June 2014, X class solar flare. (a) Shows the X-ray flux emission during X class and (b), (c) the trimming of solar flare and polynomial fitting, (d) and (e) display the polynomial fit with solar flare data and ΔDVTEC. Vertical dash line shows the trimmed timing of solar flare.

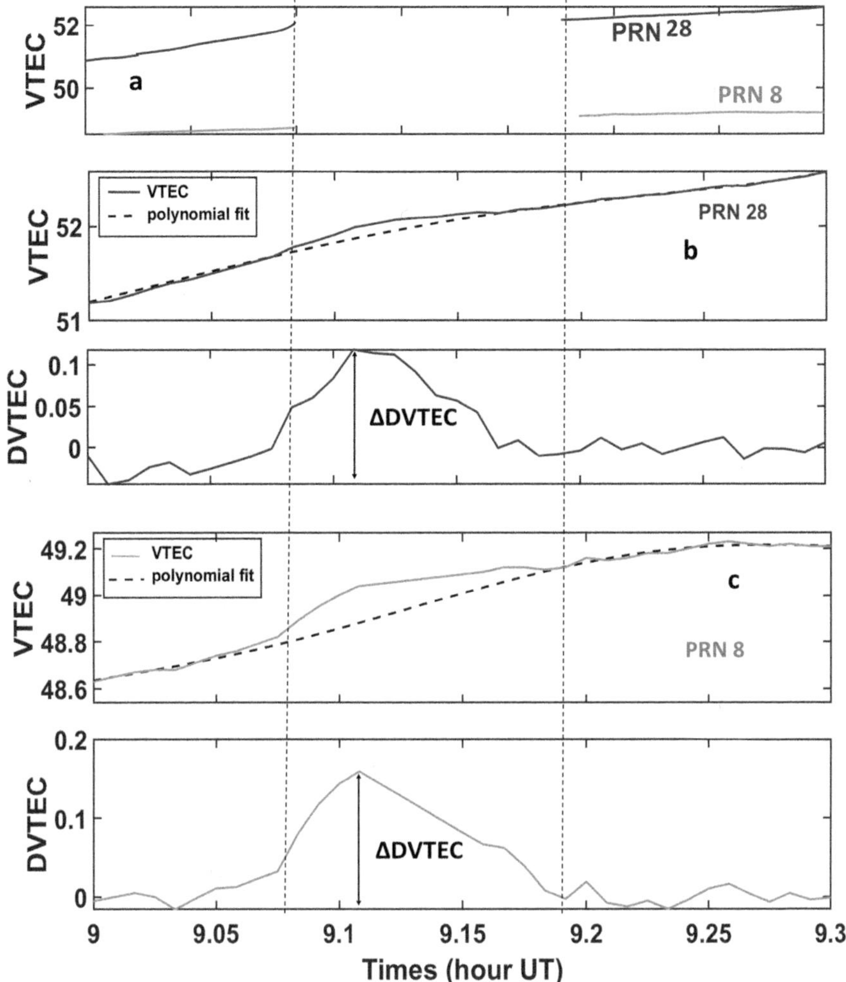

FIGURE 20.5 Mean method applied on 11 June 2014, X class solar flare. (a) Shows the visible PRN (number mention on the panels) and (b), (c) the display the polynomial fit with solar flare data and ΔDVTEC.

A solar flare's minimum and maximum DVTEC values were deducted from each other. Additionally, we compute the mean of all chosen PRNs and determine the ΔDVTEC of a solar flare. In comparison to the mean method, the baseline method demonstrated superior correlation for solar flares in the X, M, and C classes, even though it seemed to work less well for the M and C classes.

GPS data were utilized by Leonovich et al. (2010) to quantify the ionosphere's elevated TEC during solar flares. They choose dual frequency to diminish the effect of the Earth's shadow on the signal path from satellite to ground receiver. They add a suitable polynomial over the VTEC data curve after having VTEC. Furthermore, the polynomial was subtracted from VTEC to calculate the TEC change caused by a solar flare.

20.5 SUMMARY

In summary, GNSS data give positioning, navigation, and timing information over the global coverage. However, one major obstacle for researchers is the expensive cost of GNSS receivers. Additionally, as science progresses, inexpensive GNSS receivers support the study of ionospheric fluctuations. In order to provide high spatial resolution, real-time monitoring, and an investigation of ionospheric variation resulting from Space weather events, various TEC calculation techniques were employed.

Ionospheric TEC computations are performed using phase and pseudo-range observations. Pseudo-range methods provide noise with an absolute value, while phase approaches yield an indeterminate number of phase cycles with more accurate values. Nonetheless, there are benefits and drawbacks to both phase and pseudo-range measurements. Researchers integrated both approaches to improve the accuracy of TEC calculations to address the problems. They blend phase and pseudo-measurements in their dual frequency measurements. Additional adjustments were made to increase the accuracy of the TEC computation to eliminate extraneous influences and remove trends from the satellite. In addition, researchers employ the GLOBDET, rTEC, baseline, and mean methods to increase the precision of ionospheric TEC calculations. Nonetheless, scientists employ these techniques based on the type of solar flare.

REFERENCES

Afraimovich, E. L., Altynsev, A. T., Grechnev, V. V., & Leonovich, L. A. (2002). The response of the ionosphere to faint and bright solar flares as deduced from global GPS network data. *Annals of Geophysics*, *45*(1), 31–40.

Afraimovich, E. L., Kosogorov, E. A., & Leonovich, L. A. (2000). The use of the international GPS network as the global detector (GLOBDET) simultaneously observing sudden ionospheric disturbances. *Earth, planets and space*, *52*, 1077–1082.

Bagiya, M. S., Joshi, H. P., Iyer, K. N., Aggarwal, M., Ravindran, S., & Pathan, B. M. (2009, March). TEC variations during low solar activity period (2005–2007) near the equatorial ionospheric anomaly crest region in India. In *Annales Geophysicae* (Vol. 27, No. 3, pp. 1047–1057). Copernicus GmbH.

Bauer, S. J., & Bauer, S. J. (1973). Sources of Ionization. *Physics of Planetary Ionospheres*, 45–66.

Bramanto, B., Gumilar, I., Sidiq, T. P., Kuntjoro, W., & Tampubolon, D. A. (2018, May). Sensing of the atmospheric variation using Low Cost GNSS Receiver. In *IOP Conference Series: Earth and Environmental Science* (Vol. 149, No. 1, p. 012073). IOP Publishing.

Chernigovskaya, M. A., Shpynev, B. G., & Ratovsky, K. G. (2015). Meteorological effects of ionospheric disturbances from vertical radio sounding data. *Journal of Atmospheric and Solar-Terrestrial Physics*, *136*, 235–243.

Dabas, R. S., Lakshmi, D. R., & Reddy, B. M. (1993). Solar activity dependence of ionospheric electron content and slab thickness using different solar indices. *pure and applied geophysics*, *140*, 721–728.

Dan, S., Santra, A., Mahato, S., Koley, C., Banerjee, P., & Bose, A. (2021). On use of low cost, compact GNSS receiver modules for ionosphere monitoring. *Radio Science*, *56*(12), 1–11.

Davies, K. (1990). Ionospheric Radio. Peter Peregrinus Ltd, UK.

Deshpande, S. D., & Mitra, A. P. (1972). Ionospheric effects of solar flares—IV. Electron density profiles deduced from measurements of SCNA's and VLF phase and amplitude. *Journal of Atmospheric and Terrestrial Physics*, *34*(2), 255–266.

Donnelly, R. F. (1969). Contribution of X-ray and EUV bursts of solar flares to sudden frequency deviations. *Journal of Geophysical Research*, *74*(7), 1873–1877.

Donnelly, R. F. (1971). Extreme ultraviolet flashes of solar flares observed via sudden frequency deviations: experimental results. *Solar physics*, *20*, 188–203.

Donnelly, R. F. (1976). Empirical models of solar flare X ray and EUV emission for use in studying their E and F region effects. *Journal of Geophysical Research*, *81*(25), 4745–4753.

Fayose, R. S., Babatunde, R., Oladosu, O., & Groves, K. (2012). Variation of Total Electron Content [TEC] and their effect on GNSS over Akure, Nigeria. *Applied Physics Research*, *4*(2), 105.

Garriott, O. K., da Rosa, A. V., Davis, M. J., & Villard Jr, O. G. (1967). Solar flare effects in the ionosphere. *Journal of Geophysical Research*, *72*(23), 6099–6103.

Gomez Socola, J., & Rodrigues, F. S. (2022). ScintPi 2.0 and 3.0: low-cost GNSS-based monitors of ionospheric scintillation and total electron content. *Earth, Planets and Space*, *74*(1), 185.

Hazarika, R., Kalita, B. R., & Bhuyan, P. K. (2016). Ionospheric response to X-class solar flares in the ascending half of the subdued solar cycle 24. *Journal of Earth System Science*, *125*, 1235–1244.

Hofmann-Wellenhof, B., Lichtenegger, H., & Collins, J. (2012). *Global positioning system: theory and practice*. Springer Science & Business Media.

Horan, D. M., & Kreplin, R. W. (1981). Simultaneous measurements of EUV and soft X-ray solar flare emission. *Solar Physics, 74*, 265–272.

Horan, D. M., Kreplin, R. W., & Dere, K. P. (1983). Direct measurements of the gradual extreme ultraviolet emission from large solar flares. *Solar Physics, 85*, 303–312.

Horan, D. M., Kreplin, R. W., & Fritz, G. G. (1982). Direct measurements of impulsive extreme ultraviolet and hard X-ray solar flare emission. *The Astrophysical Journal, 255*, 797–805.

Jones, T. B. (1971). VLF phase anomalies due to a solar X-ray flare. *Journal of Atmospheric and Terrestrial Physics, 33*(6), 963–965.

Klobuchar, J. A. (1987). Ionospheric time-delay algorithm for single-frequency GPS users. *IEEE Transactions on aerospace and electronic systems*, (3), 325–331.

Kogogin, D. A., Nasyrov, I. A., Sokolov, A. V., Shindin, A. V., Ryabov, A. V., Maksimov, D. S., & Zagretdinov, R. V. (2021, July). Capacities of TEC measurements by the low-cost GNSS receiver based on the u-blox ZED-F9P for ionospheric research. In *Journal of Physics: Conference Series* (Vol. 1991, No. 1, p. 012020). IOP Publishing.

Lanyi, G. E., & Roth, T. (1988). A comparison of mapped and measured total ionospheric electron content using global positioning system and beacon satellite observations. *Radio science, 23*(4), 483–492.

Le, H., Liu, L., Chen, Y., & Wan, W. (2013). Statistical analysis of ionospheric responses to solar flares in the solar cycle 23. *Journal of Geophysical Research: Space Physics, 118*(1), 576–582.

Le, H., Liu, L., He, H., & Wan, W. (2011). Statistical analysis of solar EUV and X-ray flux enhancements induced by solar flares and its implication to upper atmosphere. *Journal of Geophysical Research: Space Physics, 116*(A11).

Leonovich, L. A., Tashchilin, A. V., & Portnyagina, O. Y. (2010). Dependence of the ionospheric response on the solar flare parameters based on the theoretical modeling and GPS data. *Geomagnetism and Aeronomy, 50*, 201–210.

Liu, J. Y., Chiu, C. S., & Lin, C. H. (1996). The solar flare radiation responsible for sudden frequency deviation and geomagnetic fluctuation. *Journal of Geophysical Research: Space Physics, 101*(A5), 10855–10862.

Liu, J. Y., Lin, C. H., Chen, Y. I., Lin, Y. C., Fang, T. W., Chen, C. H., … & Hwang, J. J. (2006). Solar flare signatures of the ionospheric GPS total electron content. *Journal of Geophysical Research: Space Physics, 111*(A5).

Liu, J. Y., Lin, C. H., Tsai, H. F., & Liou, Y. A. (2004). Ionospheric solar flare effects monitored by the ground-based GPS receivers: Theory and observation. *Journal of Geophysical Research: Space Physics, 109*(A1).

Liu, L., Wan, W., Chen, Y., & Le, H. (2011). Solar activity effects of the ionosphere: A brief review. *Chinese Science Bulletin, 56*, 1202–1211.

Mahajan, K. K., Lodhi, N. K., & Upadhayaya, A. K. (2010). Observations of X-ray and EUV fluxes during X-class solar flares and response of upper ionosphere. *Journal of Geophysical Research: Space Physics, 115*(A12).

Mannucci, A. J., Wilson, B. D., Yuan, D. N., Ho, C. H., Lindqwister, U. J., & Runge, T. F. (1998). A global mapping technique for GPS-derived ionospheric total electron content measurements. *Radio science, 33*(3), 565–582.

Mitra, A. P. (1974). *Ionospheric effects of solar flares* (Vol. 46). Boston: Reidel.

Ohshio, M. (1971). Negative sudden phase anomaly. *Nature Physical Science, 229*(8), 239–240.

Okoh, D., Obafaye, A., Rabiu, B., Seemala, G., Kashcheyev, A., & Nava, B. (2021). New results of ionospheric total electron content measurements from a low-cost global navigation satellite system receiver and comparisons with other data sources. *Advances in Space Research, 68*(9), 3835–3845.

Opperman, B. D., Cilliers, P. J., McKinnell, L. A., & Haggard, R. (2007). Development of a regional GPS-based ionospheric TEC model for South Africa. *Advances in Space Research, 39*(5), 808–815.

Patel, N. C., Karia, S. P., & Pathak, K. N. (2017). GPS-TEC variation during low to high solar activity period (2010–2014) under the northern crest of Indian equatorial ionization anomaly region. *Positioning, 8*(2), 13–35.

Qian, L., Burns, A. G., Chamberlin, P. C., & Solomon, S. C. (2011). Variability of thermosphere and ionosphere responses to solar flares. *Journal of Geophysical Research: Space Physics, 116*(A10).

Rao, S. S., & Singh, A. K. (2021). Probing the upper atmosphere using GPS. In *GPS and GNSS Technology in Geosciences* (pp. 135–153). Elsevier.

Richards, P. G., Fennelly, J. A., & Torr, D. G. (1994). EUVAC: A solar EUV flux model for aeronomic calculations. *Journal of Geophysical Research: Space Physics, 99*(A5), 8981–8992.

Saharan, S., Maurya, A. K., Dube, A., Patil, O. M., Singh, R., & Sharma, H. (2023). Low latitude ionospheric TEC response to the solar flares during the peak of solar cycle 24. *Advances in Space Research, 72*(9), 3890–3902.

Sao, K., Yamashita, M., Tanahashi, S., Jindoh, H., & Ohta, K. (1970). Sudden enhancements (SEA) and decreases (SDA) of atmospherics. *Journal of Atmospheric and Terrestrial Physics, 32*(9), 1567–1576.

Silwal, A., Gautam, S. P., Poudel, P., Karki, M., Adhikari, B., Chapagain, N. P., ... & Migoya-Orue, Y. (2021). Global positioning system observations of ionospheric total electron content variations during the 15th January 2010 and 21st June 2020 solar eclipse. *Radio Science, 56*(5), 1–20.

Stonehocker, G. H. (1970). Advanced telecommunication forecasting technique. In *Ionospheric Forecasting, AGARD Conf. Proc* (Vol. 29, pp. 27–31).

Woods, T. N., Hock, R., Eparvier, F., Jones, A. R., Chamberlin, P. C., Klimchuk, J. A., ... & Tobiska, W. K. (2011). New solar extreme-ultraviolet irradiance observations during flares. *The Astrophysical Journal, 739*(2), 59.

Zhang, D. H., & Xiao, Z. (2003). Study of the ionospheric total electron content response to the great flare on 15 April 2001 using the International GPS Service network for the whole sunlit hemisphere. *Journal of Geophysical Research: Space Physics, 108*(A8).

Zhang, D. H., Mo, X. H., Cai, L., Zhang, W., Feng, M., Hao, Y. Q., & Xiao, Z. (2011). Impact factor for the ionospheric total electron content response to solar flare irradiation. *Journal of Geophysical Research: Space Physics, 116*(A4).

21 Geomagnetic Storms and Their Effect on Global Positioning System

A Supplication of Global Navigation Satellite System Receivers

Kalpana Patel and Sunil Kumar Chaurasiya

21.1 INTRODUCTION

Due to its dispersive nature, the ionosphere influences radio signal propagation, which contributes to range inaccuracy within GPS transmission, and it is connected with the ionosphere's Total Electron Content (TEC). A thorough comprehension of the physical behavior attained by TEC is crucial, resulting in the growing number of applications for the Global Positioning System (GPS) and the need for greater positional precision. TEC is a quantity that is majorly reliant on numerous geomorphology and solar occurrences, such as geomagnetic storms and solar flares, as well as distinct geographic locations. A magnetic storm occurring geomagnetically, also known as magnetic disturbances, causes a brief disruption of the magnetosphere on Earth due to the relationships between Earth's field of magnetism, solar and shock waves of the sun. Bulk energy is sent to the magnetic domain by the solar wind's magnetic field, which increases plasma movement and creates an overabundance of electric current.

On the basis of magnetometer locations, a geomagnetic storm is quantified by the shift in the Earth's magnetic field's parallel factor as calculated by D_{st}-pointer, a globally averaged (Hz) of the Earth's magnetic equator. Geomagnetic disturbances have a direct or indirect relationship with many Space weather events. These phenomena include auroral displays at lower latitudes than usual; radio, radar, and compass scintillation; geomagnetic-induced currents; solar energetic particle events; and ionosphere storms. Earth's low-latitude and equatorial ionosphere is subjected to geomagnetic storms, which regulate the currents and fields in that area [1]. The ionosphere's plasma dispersion is impacted by field perturbations.

21.2 GEOMAGNETIC STORMS MECHANISM

According to the work of [2] evaluated, a geomagnetic active time begins when the Sun's disturbance-carrying wind crosses through it. According to [3] area representing the boundary in this instance is the magnetosphere of Earth, specifically between the solar wind's dayside interface and the transition between the magnetic tail's open and closed lines. The mechanics by which the magnetic field of the solar wind compresses the diurnal border of the magnetospheric crater is dependent on unit configuration, concentration, swiftness, and route. The geosphere's magnetic field is strengthened

DOI: 10.1201/9781032712444-25

on a route equivalent with geomagnetic league as a result of flow of a magnetopause current. The constructive track of the confident D_{st} index is increased by an initial increase in the magnetic field. The constant interaction of high-energy elements within the near-Earth magnetosphere with solar wind constituent parts in the magneto-tail on Earth's nightside generates the ring current that circles the planet. This ring current produces significant changes in the area of the magnetogram from a nearby observatory [4]. As the storm recovers, we might observe a decrease in this ring current throughout the preliminary days of the disturbed duration.

Magnetic disturbances are the primary indicator of geographic effects [5, 6]. As reported by Chakborty et al., rapid diffusion of electric field, being directed towards east during diurnal hours, strengthens the dynamo electric field. When the plasma lifts towards higher altitudes, the generated electric field amplifies the perpendicular E × B plasma gist [7]. Because of higher production to loss ratio at these altitudes, the dayside sector's electron density is enhanced. As a result, the rapid penetration electric field is linked to both significant dayside sector TEC enhancement [8] and nightside sector TEC depletion [9].

21.3 GEOMAGNETIC INDICES

To inquire about complexity and variability related to Earth's magnetic area as a single number, geomagnetic indices have been developed. Numerous indices exist, each of which describes a distinct facet or temporal span of the dynamic geomagnetic field. The K index is a measurement of the disturbance in the Earth's magnetic field's horizontal component. (https://geomagnetism. ga.gov.au/geomagnetic-indices/other-indices; https://www.ncei.noaa.gov/products/geomagnetic-indices)

There are many different magnetic activity indices.

21.3.1 Kp, K Index

Intended as a sole geomagnetic test center location, K index is the virtual-arithmetic general indicator of a three-hour array in magnetic activity in the direction of an adopted quiet-diurnal arc. It was first presented by J. Bartels in 1938 and uses unit number between 0 and 9, every three hours of universal time (UT) diurnally.

The designed identical K-value of "13" geomagnetic measuring rooms located geomagnetically at 44 and 60 degrees northern or southern latitude is known as planetary 3-hour-range index, or Kp. In thirds of a unit, the scale runs from 0 to 9. For example, 5- is 4.67, 5- is 5, and 5+ is 5.33. This planetary index's objective is to measure sun particle radiation by means of its magnetic effects.

21.3.2 Amplitude Antipodal Value (AA)

The 3-hour amplitude antipodal value is represented by this index, aa. Ap and daily average AA can be calculated in a similar way. By using magnetograms, from earlier observations, to scale out the magnetic activity, these indices have been extended further back in time using AA rather than Ap. This has historically been advantageous. The 3-hour corresponding amplitude antipodal value is represented as AA. Ap and daily average AA can be calculated in a similar way.

Numerically, AA* and Ap* are not alike due to the dissimilar presentation units; therefore, for the two series, separate large magnetic storm onset and end threshold values are applied. However, comparing the years with overlapping coverage, we find that the annual frequency of big storm occurrence is rather similar. An additional explanation for the variations is that an index derived from magnetic perturbation values at only two observatories is likely to experience larger extreme values if either input site is well positioned to the overhead ionospheric or field-aligned current systems producing the magnetic storm effects.

21.3.3 AP INDEX

When a geomagnetic storm occurs, Ap is the earliest occurring maximum 24-hour value that can be determined by calculating an 8-point running average of consecutive 3-hour ap indices. This computation is done without taking into account the beginning and ending times of the UT-day. It has a special connection to the storm occurrence. Values of Ap* over many years offer a maximum disturbance measure helpful in determining the chronological order (date and start time) and amplitude (largest to smallest) of major geomagnetic storms.

21.3.4 DISTURBED STORM TIME INDEX (DST)

The globally symmetrical equatorial electrojet, or "ring current," determined by the DST value, quantifies magnetic activity attained through a grid of near-equatorial geomagnetic stations.

Equatorial magnetic disturbance indices comparable to DST (Disturbance Storm Time) are obtained for every hour scale of low-latitude parallel magnetic changes. Others demonstrate the impact from global and balanced high height latitudinal ring current flowing westward, resulting in "main phase" depressing H-factor field throughout significant magnetic storms. A global array of low-latitude observatories' records is examined to eliminate annual secular change trends using hourly H-component magnetic variations. The factor known as DST index is isolated through harmonic analysis, and residual variations are transformed to their equatorial equivalents through cosine value of located latitude.

21.3.5 AURAL ELECTROJET INDEX (AE)

Auroral Electrojet Index (AE) quantifies quantifiable, worldwide measurement for auroral region magnetic action generated through increased ionospheric currents that flow both inside and beneath auroral ovate. It is ideally the entire range of deviation from quiet day values of the horizontal magnetic field (h) around the auroral oval at a given moment in time.

Compared to other geomagnetic indices, AE has advantages, or at least shares some of their benefits:

- It can be obtained instantly or by averaging the variances calculated over a chosen period of time.
- It is a quantitative measure generally connected to the mechanisms causing the magnetic variations that are seen.
- Its derivation process is fairly straightforward, objective, and digital, making it a good fit for current computer processing methods.
- It can be applied to statistical aggregates or individual events.

21.4 PHASES OF GEOMAGNETIC STORMS

According to [10], geomagnetic storms are classified in three stages, which are:

1. **Onset or Early Stage**: A sudden onset of magnetic field activity may be associated with a sudden commencement (SC). The time between the SC and the main phase is known as early stage.
2. **Primary Stage**: The main phase is the period of gradual decline of Earth's magnetic field strength.
3. **Recapture Stage**: Recapture stage, a period where the Earth's magnetic field strength returns to normal.

21.5 EFFECTS AND STUDY OF GEOMAGNETIC STORMS

As per [1], magnetic storms have an impact on the ionospheric regions of Earth that are equatorial and low latitude. These regions experience similar notable singularities, such as Equatorial Ionization Anomaly (EIA), Equatorial Electrojet (EEJ), Equatorial spread-F (ESF), and others. Various technology-based instruments are utilized for calculating the impact of geomagnetic action for the disturbances in the ionosphere based on their geomagnetic indices.

GPS, an application of the GNSS, is a well-known instrument. These days, GPS is used to determine the Total Electron Content (TEC) as a vital instrument to understand the ionosphere. The geomagnetic storm disturbed dynamo electric field (DDEF) is diurnal-edge, which points west, and this results in a reduction of the dynamo electric field. Consequently, it decreases vertical drift of E $\times$ B. TEC is being depleted in the dayside sector as a result, and EIA is being suppressed. Constructive and destructive storm outcomes terms are used for describing the detected rise and downfall of TEC in addition to electron concentrations of F-region, individually [11].

Latitude, local time, and storm phase all affect the existence and strength of both constructive plus destructive storm consequences [12]. For the past several years, researchers have examined the results of storm stretch electrodynamics, neutral winds plus compositional changes [13]. A great deal of research has also been done both theoretically and empirically on storm time ionospheric responses [14, 15]. Moreover, GPS measurements are widely used to investigate how geomagnetic storms affect TEC at different latitudes. Low latitudes [16, 17] and mid-latitudes [18–20] have both reported TEC enhancements linked to geomagnetic storms. [14] demonstrated a long-term, significant (12-hour) increase in TEC value over the Pacific Ocean from low-latitudes towards mid-latitudes throughout the first part of a disturbed main phase on December 15, 2006. [21] investigated how GPS-based TEC varied in Indian segments and through intense storms that occurred between November 8 and November 12, 2004, at various latitudes.

Seasons, magnetic local times, and magnetic latitudes all affect how long it takes for TEC responses to occur after geomagnetic disturbances, according to [22]. Recently, [23] conducted an investigation into the TEC variation over Indian stations, such as Lucknow and Bangalore, for the duration of the solar cycle 24. Several TEC variations have been carried out with the aid of spectral regression and analysis. There has been discussion of TEC from IRI-2016 and GPS as a comparative study. [24] also demonstrated that while ionospheric TEC can be purely estimated all through the recapture stage of geomagnetic storms, neither NN nor Ne Quick models can reliably forecast it throughout intense geomagnetic storms for the equatorial stations Bangalore and Equatorial Ionization Anomaly (EIA) Lucknow.

21.6 INSTRUMENTATION AND ANALYSIS

GPS navigation arrangement contains three separate "segments," as per information provided by [23–27]. The 24 satellites that make up the "space segment" of the global positioning system are arranged in six trajectorial levels round the Earth at an altitude of 20,200 km, plus a trajectorial duration of 12 hours. Every satellite broadcast for 2 rate of recurrence (f1 = 1575.42 MHz; $f2$ = 1227.60 MHz) through binary distinct encryptions (P1 or C/A) plus P2, as well as dual carrier stages (L1 and L2), originated from a joint oscillator working at 10.23 MHz. Secondly, the "control segment," containing ground positions that are customized to track satellites, transmits signs towards each satellite's transmitted codes and waveforms as well as engineering control. Thirdly, the operator section uses GPS receivers and satellite-transmitted signals. These are the three segments of the Global Positioning System.

GPS signals can be impacted by the ionosphere in a variety of methods in a way of their movement towards the ground recipient due to its refractive index at radio rate of recurrences, not equal to 1 [28, 29]. GPS waves passing through the ionosphere experience an extra interruption comparative to TEC, which is the overall quantity of unrestricted electrons along a tube of one meter squared cross section between two points. This is one of the major effects.

An effective method for estimating TEC values with more temporal and spatial coverage is provided by the GPS data [30–32]. GPS signals are not significantly impacted by the Earth's magnetic field and ionospheric absorption because the frequencies used in the GPS system are high enough. This is true mutually for rapid- and extended-time disparities in the ionospheric configuration.

Many GPS stations all over the world are linked to IGS stations. The acquired GPS data is in RINEX FORMAT and is processed using Seemala software called "GOPI GPS ver 2.9.5" to obtain the TEC data. On the way to yield rectified slant total electron content (STEC), Seemala program system measures the satellite (b_S) and recipient (b_R) inaccuracies for the unaltered pseudo-range, including moving stage information on L1, L_2 rate of recurrence. The software uses the translation method provided by [33] to calculate vertical total electron content (VTEC).

$$\text{VTEC} = \left(\text{STEC} - \left[b_R + \text{bs}\right]\right) / S(\text{El}) \tag{21.1}$$

To obtain a job that plots out alongwith sign S (El), one computes the reciprocal of cosine of solar zenith angle (χ). Φ, reliant on degree-based angle of elevation (El) of satellite. It also depends on the ionosphere's effective height 'h' ('h' = 350 km) of shrill case model in addition to mean Earth's radius R_E [34]. Another name for the VTEC is vertical total electron content. The phase, group velocities of satellite signal, are disturbed by disparities in TEC.

The equation yields the STEC.

$$S(\text{El}) = 1 / \cos(\chi) = 1 / \sqrt{(1 - \{(R_E \cos(\text{El}) / (R_E + h)\}^2} \tag{21.2}$$

Here, El, h, and R_E depict mean Earth's radius in kms, solar zenith angle, elevation angle of satellite, plus the ionospheric effective altitude, respectively.

The following formula provides ionospheric time delay experienced by GPS signal along with rate of occurrences 'f'

$$t = 40.3 \times \text{TEC} / \text{cf}^2 \tag{21.3}$$

where in c represents light rapidity in vacuum.

The World Data Center (WDC) (https://wdc.kugi.kyoto-u.ac.jp/) provides access to D_{st} index.

21.7 RESULTS ON EFFECT OF GEOMAGNETIC STORMS ON GPS-DERIVED TEC

Further related to the consequences of geomagnetic storms on the GPS, following are certain significant cases as obtained by [35] given in the chapter.

21.7.1 THE GEOMAGNETIC STORMS AND THEIR SOLAR GEOPHYSICAL IMPACTS

The Dst-index, Kp index, IMF Bz, IEF Ey, and velocity of solar wind Vx during March 17–21, 2015, are all displayed in Figure 21.1. St. Patrick's storm began on 17th March 2015, at approximately 06:00 UT, as observed by the change within the Dst index (Figure 21.1).

Figure 21.1 shows the chronology of the geomagnetic storm from 17–21 March 2015 plotted against the following variables: Dst-index, AE-index, Kp index, IMF BZ, IEF Ey, as well as solar wind speed. The main phase onset (MPO) of the storm began at about 07:00 on 17th March 2015, as shown in Figure 21.1. The sudden commencement (SSC) of the storm happened at approximately 04:00 UT, and throughout the main phase, the value of Dst index fell as low as −234 nT at 23:00 UT on 17th March 2015. Following this, the recovery phase started 20th March 2015, and the storm went into quiet time conditions. This storm is a two-step development storm, with the first step ending at 10:00 UT and the second step ending at 23:00 UT.

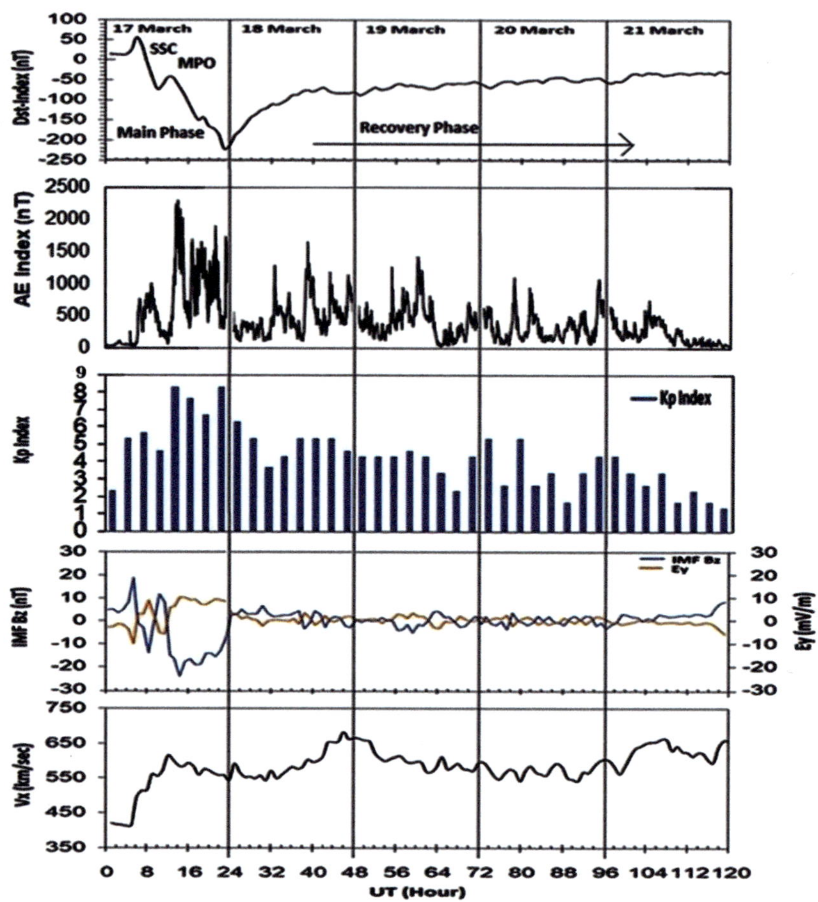

FIGURE 21.1 The Dst-index, Kp Index, IMF (Bz), IEF (Ey), and solar wind velocity Vx fluctuated from top to bottom for 17–21 March 2015.

Several factors, such as the intricacy of solar activity, interactions amid the solar wind and Earth's magnetosphere, and the continuing influence of a CME as it moves across space, can lead to the two-step development of geomagnetic storms. The strength and duration of these storms can vary, and they can cause breathtaking auroras at higher latitudes as well as disturbances to satellite operations, electrical grids, and communication systems [36, 37].

The highest value of the Kp index (8.5), which was detected on March 17 between 21:00 UT and 24:00 UT, further confirms the end of the main phase on that day. The AE-index displayed notable spikes over 1000 nT throughout the main and recovery phases during 17–20 March, demonstrating that heating in the auroral zone can initiate a gravity wave's spectrum, equatorward winds. On March 17, IMF-Bz made two moves toward the south, turning south at 06:00 UT, and staying south until 15:00 UT. IEF Ey altered through 20.27 mV/m throughout this time, going via -9.89 to 10.38 mV/m. There were no further notable fluctuations within IEF Ey. On 18th March 2015, the speed of the solar wind maximized at 700 km/s during the storm.

21.7.2 Effect of the Geomagnetic Storm at the Equatorial Station

The fluctuation of quiet mean TEC and VTEC at the equatorial station Bangalore over the storm period of 17–21 March 2015, is shown in Figure 21.2. Here, the TEC for the five quietest days of

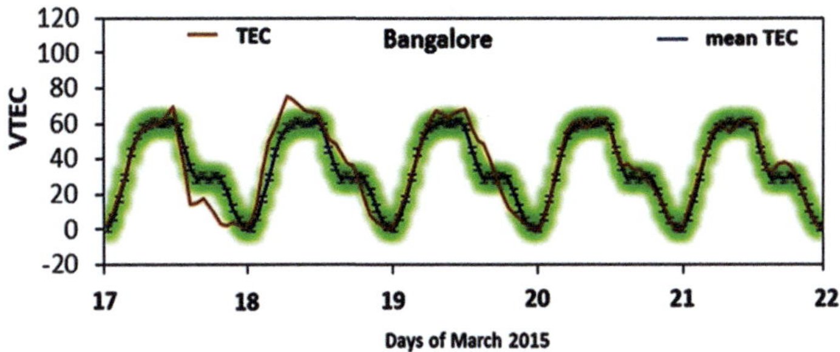

FIGURE 21.2 VTEC (average of all PRN) fluctuation for Bangalore during the storm on 17–21 March 2015, showing the variation. In the shaded area, the vertical bar represents the standard deviation of the predicted mean VTEC.

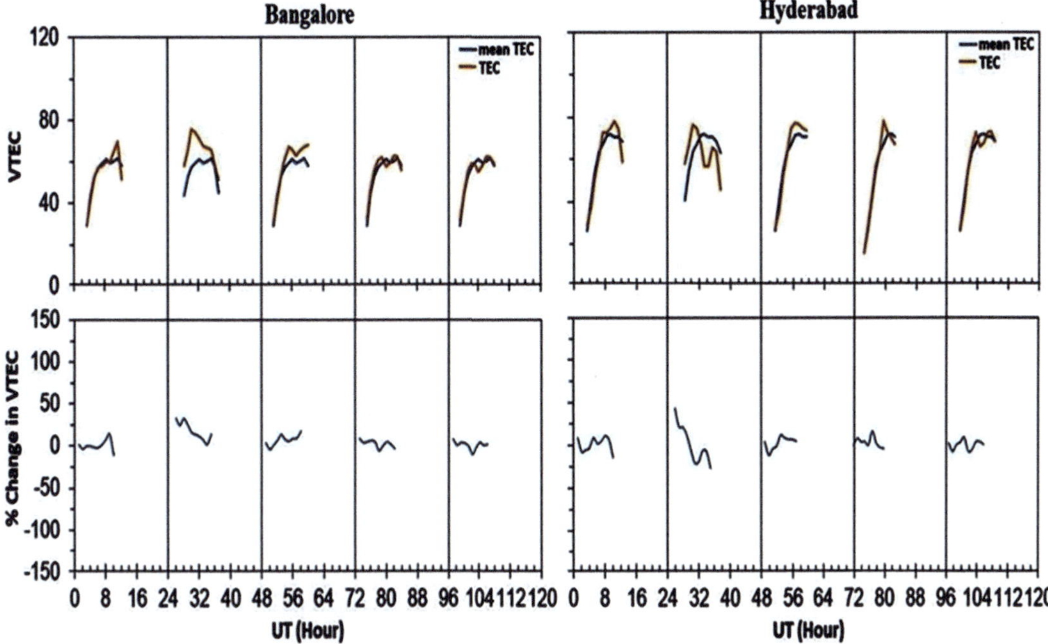

FIGURE 21.3 VTEC of a single PRN (PRN 17), as well as the percentage variation in VTEC relative to the mean VTEC over Bangalore and Hyderabad stations, displayed from top to bottom between March 17 and March 21, 2015.

March 2015 is averaged to determine the mean TEC. On March 17, between 14:00 UT and 21:00 UT, the equatorial station Bangalore recorded a similar decrease in TEC data. Here, the TEC for the five quietest days of March 2015 is averaged to determine the mean TEC. On March 17, between 14:00 UT and 21:00 UT, the equatorial station Bangalore recorded a similar decrease in TEC data.

Figure 21.3 shows the variation in TEC via PRN 17 and quiet mean TEC data at the equatorial stations Bangalore and Hyderabad. In this instance, it is also evident that the difference in TEC on March 17 is not noticeable and is within 20% over equatorial stations. This occurs because the primary stage of a geomagnetic storm is frequently connected to the initial disruption of the magnetic field of the Earth brought on through the arrival of a coronal mass ejection (CME) or a solar wind stream having high speed.

While these phenomena have the potential to create geomagnetic disturbances, the ionospheric response might not be noticeable or immediate during the storm's early stages. According to [38] and [39], it frequently takes some time for the ionosphere to react to changes in the geomagnetic field. The equatorial station Bangalore indicated improvement on 18th March 2015, i.e., positive storm effects, in contrast to the EIA station. This occurred because of the ionosphere's increased ionization that might occur throughout a geomagnetic storm's early stages. Temporarily raising TEC values in the ionosphere due to this enhanced ionization can have a positive impact on GPS TEC.

According to [38] and [39], it frequently takes some time for the ionosphere to react to changes in the geomagnetic field. Equatorial station Bangalore indicated improvement on 18th March 2015, i.e., positive storm effects, in contrast to the EIA station. This occurred because of the ionosphere's increased ionization that might occur throughout a geomagnetic storm's early stages. Temporarily raising TEC values in the ionosphere due to this enhanced ionization can have a positive impact on GPS TEC. According to [40] and [23], the increasing electron density may cause GPS signal delays to be more pronounced. Even though the Hyderabad station is a little higher in latitude than Bangalore, it exhibits a negative storm effect, although one of very slight intensity. It is possible that events during a geomagnetic storm, particularly at equatorial latitudes, lead to electron density depletions.

One of these processes is the penetration electric field, which can increase the ionospheric plasma's altitude and reduce its electron density at lower altitudes [24, 41]. Additionally, Hyderabad (27%) and Bangalore (33%), two equatorial stations, also show a negative storm effect according to TEC data from a single PRN 17. At equatorial stations, there were no noticeable impacts on March 19. Both the equatorial stations Bangalore and Hyderabad remained unaltered (within 20%) throughout the recovery period of 20–21 March 2015.

Compared to higher latitudes, Earth's magnetic field lines are more stable throughout the magnetic equator. The TEC levels in equatorial regions can be less affected by magnetic storms due to this stability. The local time also influences the TEC levels for equatorial locations. The largest TEC values occur between local noon and local midnight in the equatorial ionosphere, which experiences daily changes in ionization. These daily changes could assist in returning TEC values to standard levels throughout the recovery phase of the storm [39, 42]. Figure 21.4 displays the daily average GPS-TEC over the equatorial station.

Figure 21.4 shows that the GPS TEC grows over the equatorial station Bangalore through the ascending portion of the solar cycle 24, reaches its highest value throughout the solar maximum year, and gradually falls through the solar cycle's declining phase. Figure 21.5 displays the contour plot of the GPS TEC over the equatorial station Bangalore from 2007 to 2017. The color bar displays the GPS TEC's magnitude. In Figure 21.5, the peak GPS TEC across Bangalore is depicted as being between 07:00 UT and 13:00 UT.

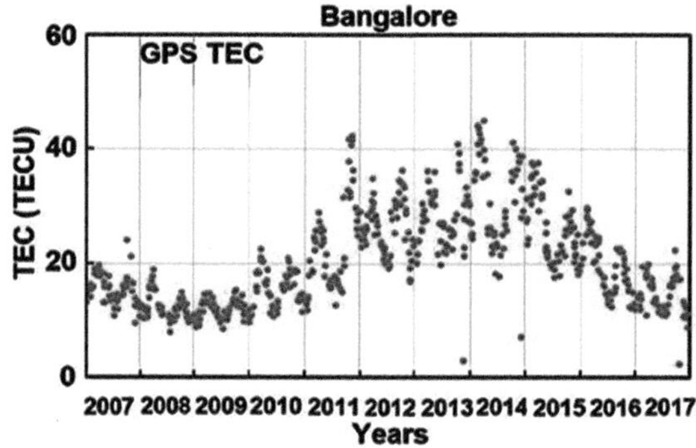

FIGURE 21.4　Daily average GPS-TEC over Bangalore from 2007–2017.

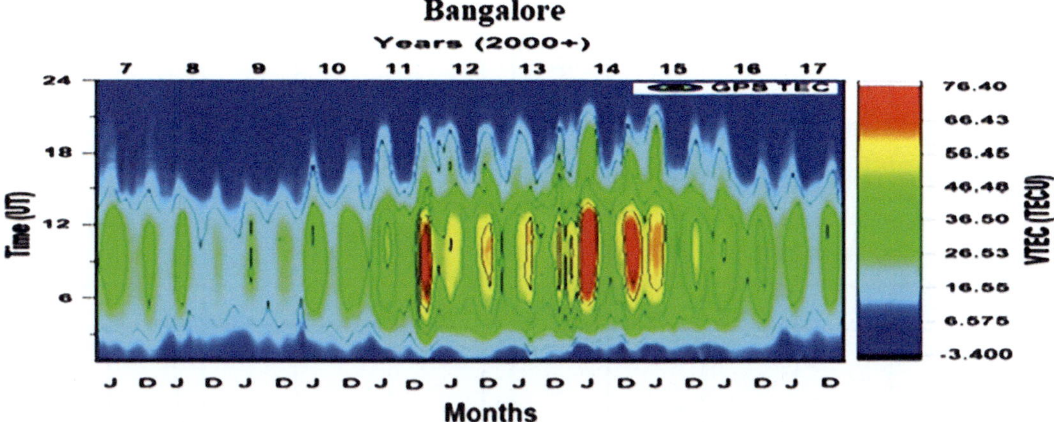

FIGURE 21.5 Indian station Bangalore is represented in a contour plot of monthly average GPS data from 2007 to 2017. The abscissa indicates the months of the year, and the ordinate depicts the time in UT. The value scale of TEC is a vertical color bar on the right side of the picture.

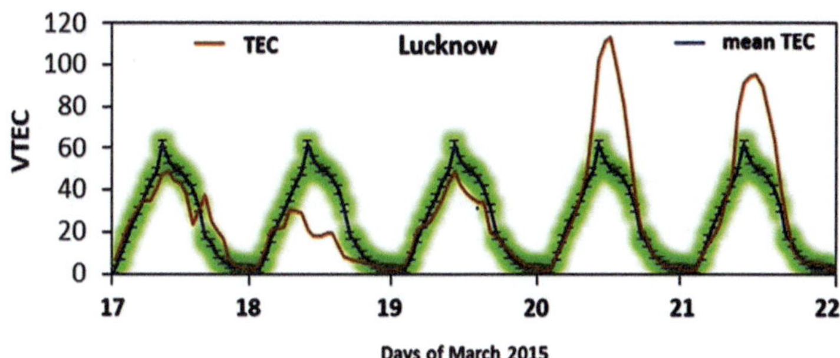

FIGURE 21.6 VTEC (average of all PRN) fluctuation for Lucknow during the storm on 17–21 March 2015, showing the variation. In the shaded area, the vertical bar represents the standard deviation of the predicted mean VTEC.

21.7.3 EFFECT OF A GEOMAGNETIC STORM AT AN EIA STATION

Figure 21.6 shows the fluctuation in quiet mean TEC and VTEC at the EIA station Lucknow throughout the storm period of 17–21 March 2015. The TEC for the five quietest days of March 2015 has been averaged to determine the mean TEC in this case. At 13:00 UT and 15:00 UT on March 17, the EIA station Lucknow recorded the same decrease in TEC values. On March 17, the storm did not impact the EIA station Lucknow. TEC data from a single PRN from 17–21 March 2015 have been studied to better understand the fluctuation in TEC.

TEC fluctuation via PRN 17 and quiet average TEC data from Varanasi and Lucknow are displayed in Figure 21.7. In this instance, it is also obvious that the variance in TEC on March 17 is not noticeable and was found to be under 20% across EIA stations. This occurs because the primary phase of the storm is often linked to the initial disturbance of magnetic field of the Earth brought on via the development of a CME or a rapid solar wind stream. Although such occurrences have the potential to generate geomagnetic disruptions, it is possible that the ionospheric response will not be noticeable or immediate throughout the storm's initial phases.

According to [25], the ionosphere often needs some time to react to fluctuations in the geomagnetic field. Over EIA stations Varanasi and Lucknow, negative storm effects were noted on March 18

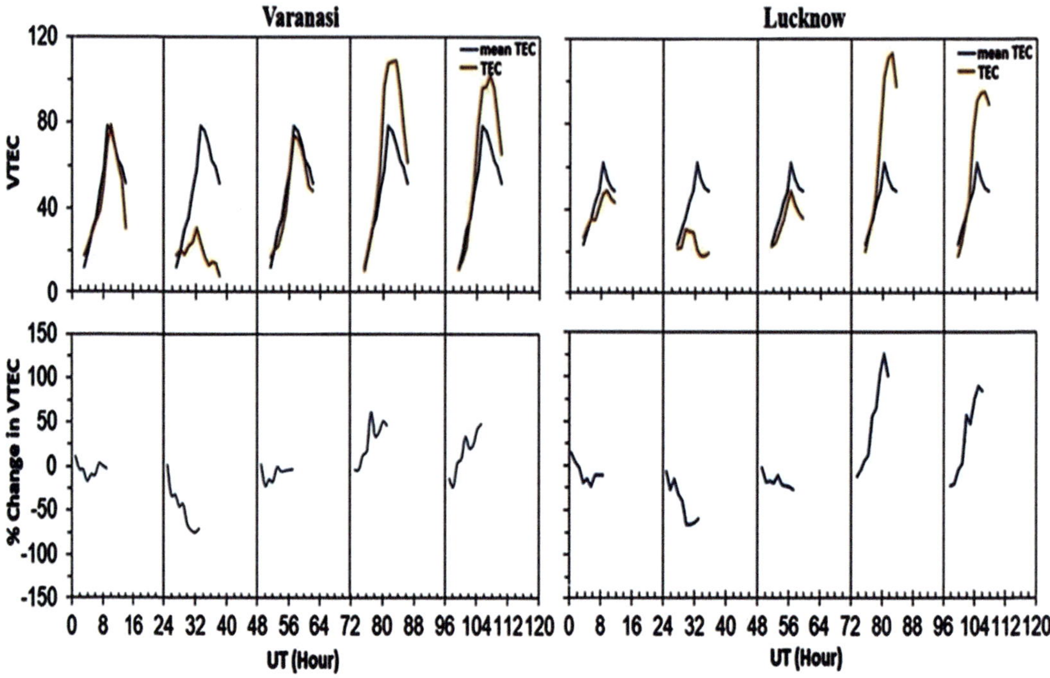

FIGURE 21.7 VTEC of a single PRN (PRN 17), as well as the percentage variation in VTEC relative to the mean VTEC over Varanasi and Lucknow stations, displayed from top to bottom between 17–21 March 2015.

throughout the afternoon hour, resulting in a considerable fall in TEC values. Varanasi (74%) and Lucknow (66%) both experienced a negative storm effect, according to TEC data from a single PRN 17. The F2 layer in the ionosphere may shift toward the equator because of the mid-latitude magnetic storms becoming more intense. According to [39], this transfer of electrons may result in a drop in electron density at the EIA locations. At EIA stations, no notable impacts were noticed on March 19.

Daytime positive storm effects were noted across Varanasi and Lucknow on the 20th and 21st of March 2015 throughout the recovery period. According to TEC data via PRN 17, Varanasi experienced the greatest rise within TEC of 61% on 20th March 20 and 48% on 21st March, while Lucknow experienced the greatest rise in TEC data of 144% on March 20 and 80% on March 21. The ionosphere across the EIA stations can temporarily see an increase within TEC throughout the recovery phase, which causes a positive storm impact. According to [43], this enhancement is probably the result of ion and electron recombination, as well as other ionospheric mechanisms that occur as the ionosphere reaches a state of equilibrium.

Figure 21.8 displays the GPS-TEC daily average for the EIA station (Lucknow). Figure 21.8 shows that the GPS TEC rises over the EIA station during the ascending phase, reaches its highest value throughout the solar maximum year, and gradually falls throughout the solar cycle 24's falling phase.

In Figure 21.9, the GPS TEC contour plot for the EIA station Lucknow from 2007 to 2017 is depicted. Every hour of every day for every month has been averaged for the GPS TEC. This is an hourly illustration of the TEC variation as averaged over a given month. The color bar displays the measurement of the GPS TEC. Figure 21.9 shows that the peak GPS TEC over EIA station Lucknow is between 07:00 UT and 11:00 UT.

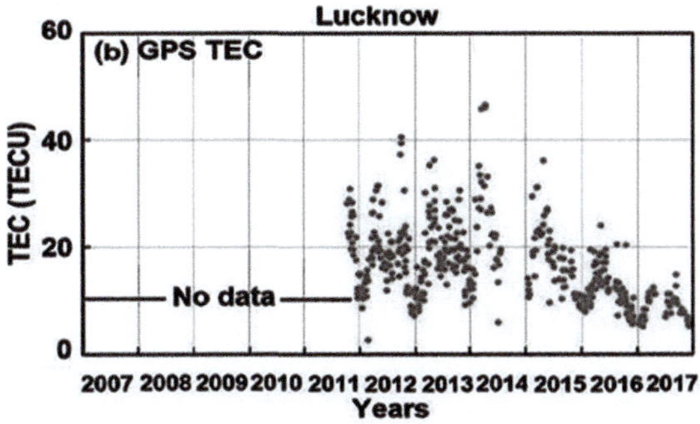

FIGURE 21.8 Daily average GPS-TEC over Lucknow from 2007–2017.

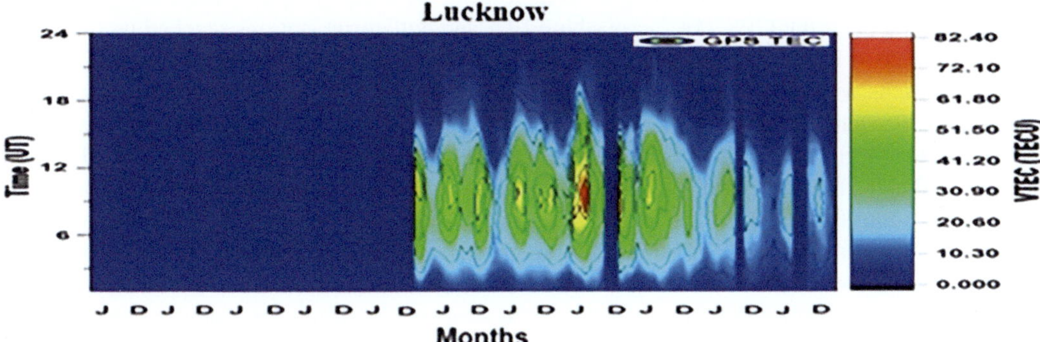

FIGURE 21.9 Indian station Lucknow is represented in a contour plot of monthly average GPS data from 2007 to 2017. The abscissa indicates the months of the year, and the ordinate depicts the time in UT. The value scale of TEC is a vertical color bar on the right side of the picture.

21.7.4 COMPARATIVE ANALYSIS OF THE EFFECT OF A GEOMAGNETIC STORM AT EQUATORIAL EIA STATIONS

Based on Figures 21.3 and 21.7, on March 17, both equatorial and EIA locations recorded TEC variations that were not statistically significant and within 20%. This demonstrates that the geomagnetic storm had no appreciable impact on the TEC's main phases beginning (17 March 2015) at either equatorial or EIA stations. This occurs due to the primary phase of the storm being linked to the initial disturbance of the magnetic field of the Earth brought on via the development of a CME or an intense solar wind stream. Despite such occurrences having the potential to create geomagnetic disruptions, the ionospheric effect might not be noticeable or immediate throughout the storm's initial phases. According to [23], fluctuations in the geomagnetic field frequently cause the ionosphere to take some time to react. Both EIA stations Varanasi and Lucknow reported large decreases in TEC data throughout the daytime hour on March 18, indicating negative storm effects; however, equatorial locations in Bangalore showed enhancements, indicating positive storm effects on 18th March 2015. A negative storm effect can be seen at the Hyderabad station, which is somewhat greater in latitude than Bangalore, but its magnitude is extremely tiny.

The single PRN 17's TEC data also shows a negative storm effect on both EIA stations, while showing a positive storm effect over Bangalore and a negative storm effect over Hyderabad. This occurs because of possible increased ionization in the ionosphere throughout the early stages of a geomagnetic storm. Positive storm effects on GPS TEC can emerge from this enhanced ionization, which can momentarily raise TEC values within the ionosphere. According to [43], the increased electron density can result in more pronounced GPS signal delays.

GPS signals moving through the ionosphere encounter intensified scintillation and refraction as an outcome of ionospheric disruptions. This may result in a decrease in the observed TEC values, which would harm TEC measurements [23]. From this, it can be shown that both EIA stations throughout the storm's main phase (18th March 2015) only displayed negative storm effects, while equatorial stations displayed both negative and positive storm effects. Over both EIA and equatorial locations, no noteworthy impacts were seen on 19th March 2015, the first day of the recovery phase of the storm. Daytime positive storm effects were noted across both EIA locations throughout the recovery period (20–21 March 2015); however, neither equatorial station was influenced.

The ionosphere progressively returns to its previous state throughout the recovery phase that follows the primary phase of a geomagnetic storm. The ionosphere continues to settle down during this period, and the particle's density begins to decline. The detected positive storm effect in GPS TEC is therefore caused by TEC possibly increasing momentarily [39, 44]. According to Figures 21.4 and 21.8, the GPS TEC rises across both EIA and equatorial locations, reaches its highest value during the year of solar maximum, and then begins to fall during the solar cycle 24's declining phase. Also studied is the fact that the GPS TEC across both EIA and equatorial stations reached its first maximum value in 2014 but reached its second-highest value in 2012. Because of the rise in ionization and electron density within the ionosphere during solar maximum, the total quantity of electrons along a particular path, or TEC, typically reaches its maximum at that time. As a result, TEC values are often greater during the solar cycle's solar maximum phase [24, 45]. Figures 21.4 and 21.8 peculiar result is that the maximum GPS TEC value for 2014 across the EIA station Lucknow was higher than the maximum GPS TEC value for 2014 across the equatorial station Bangalore. This occurs because TEC is more common in higher latitudes and less common at lower ones. In contrast to Bangalore, Lucknow is situated in India at a greater latitude.

The angle at which solar radiation penetrates the atmosphere affects the electron density within the ionosphere, and this angle varies with latitude [46]. This means that greater latitude stations typically have higher TEC levels. From 2007 to 2017 for both EIA and equatorial stations, the maximum TEC is calculated from Figures 21.5 and 21.9 to be approximately between 7:00 UT and 13:00 UT.

Ionization in the Earth's ionosphere is typically triggered by the Sun. Solar radiation reaches its peak intensity during the day, particularly around noon. Due to the increased ionization of the ionospheric plasma caused by this intensive solar radiation, the electron density increases. This is so that free electrons and positive ions can be produced in the ionosphere by the solar ultraviolet (UV) light that can ionize neutral particles there. These free electrons help increase TEC levels [23, 41]. TEC has followed a consistent trend during Solar Cycle 24, growing in the ascending phase (2007–2012), reaching its extreme value during 2013–2014, and falling throughout its descending phase (2015–2017).

Additionally, across both EIA and equatorial stations, GPS-TEC rises steadily from January until it attains a peak in March or April, after which it starts to fall until July. Subsequently, between September and October, it increases to a high number earlier, then progressively drops between November and December. The EIA is a region of enhanced ionization. The EIA intensifies and moves toward the magnetic equator at the equinoxes. As a result, TEC within the equatorial region is significantly improved. The "equatorial fountain effect" is a special ionospheric phenomenon that only occurs in the equatorial regions. This effect is stronger in the equinoxes and helps explain why the TEC levels are higher. According to [40, 44], this impact entails the upward transit of ionized plasma through the F-region to the topside ionosphere.

21.8 CONCLUSIONS

At both the equatorial and EIA locations, the geomagnetic storm had no discernible effect on the TEC's main phase start date (17 March 2015). On 18th March 2015, equatorial sites in Bangalore saw improvements, suggesting favorable storm impacts. Although the amount of the negative storm impact was quite modest, it may be observed over Hyderabad, which is a little higher in latitude than

Bangalore. The TEC for the single PRN 17 indicates a positive storm impact across Bangalore and a negative storm impact across Hyderabad, along with a negative storm impact across both EIA locations.

From this, it can be shown that both EIA stations throughout the storm's main phase (18 March 2015) only displayed negative storm effects, while equatorial stations displayed both negative and positive storm effects. Over both EIA and equatorial locations, no noteworthy impacts were seen on 19th March 2015, the first day of the recovery phase of the storm. Daytime positive storm effects were noted across both EIA locations throughout the recovery period (20–21 March 2015); however, neither equatorial station was influenced. The GPS TEC increases across both equatorial and EIA sites, peaks within the year of solar maximum, and subsequently starts to decline within the decreasing portion of the solar cycle 24. In comparison to the highest GPS TEC value over the equatorial location Bangalore, the highest GPS TEC value over the EIA location Lucknow in 2014 was greater. The strongest solar radiation occurs throughout the day, especially at noon. There is more ionization in the EIA. At the equinoxes, the EIA becomes stronger and travels closer to the magnetic equator.

REFERENCES

[1] Chakraborty, M., Kumar, S., De, B.K., Guha, A. 2015. Effects of geomagnetic storm on low latitude ionospheric total electron content: A case study from Indian sector. *J Earth Syst Sci* 124, 1115–1126. https://doi.org/10.1007/s12040-015-0588-3

[2] Sastri, J.H., M.A. Abdu, and J.H.A. Sobral. 1997. Response of Equatorial Ionosphere to Episodes of Asymmetric Ring Current Activity. *Annales Geophysicae* 15, no. 10 (October 31): 1316–1323. https://doi.org/10.1007/s00585-997-1316-3

[3] Ondede, G.O., A.B. Rabiu, D. Okoh, P. Baki, J. Olwendo, K. Shiokawa, and Y. Otsuka. 2022. Relationship between Geomagnetic Storms and Occurrence of Ionospheric Irregularities in the West Sector of Africa during the Peak of the 24th Solar Cycle. *Frontiers in Astronomy and Space Sciences* 9 (November 17). https://doi.org/10.3389/fspas.2022.969235

[4] Campbell, W.H. 1979. Occurrence of AE and Dst Geomagnetic Index Levels and the Selection of the Quietest Days in a Year. *Journal of Geophysical Research: Space Physics* 84, no. A3 (March): 875–881. https://doi.org/10.1029/ja084ia03p00875

[5] Burch, J.L., T.E. Moore, R.B. Torbert, and B.L. Giles. 2016. Magnetospheric Multiscale Overview and Science Objectives. *Space Science Reviews* 199, no. 1–4 (May 30): 5–21. https://doi.org/10.1007/s11214-015-0164-9

[6] Kendall, P.C., S. Chapman, S.-I. Akasofu, and P.N. Swartztrauber. 1966. The Computation of the Magnetic Field of Any Axisymmetric Current Distribution?With Magnetospheric Applications. *Geophysical Journal International* 11, no. 3 (November): 349–364. https://doi.org/10.1111/j.1365-246x.1966.tb03088.x

[7] Richardson, I.G., and H.V. Cane. 2011. Galactic Cosmic Ray Intensity Response to Interplanetary Coronal Mass Ejections/Magnetic Clouds in 1995–2009. *Solar Physics* 270, no. 2 (May 12): 609–627. https://doi.org/10.1007/s11207-011-9774-x

[8] Gopalswamy, N. 2009. Halo Coronal Mass Ejections and Geomagnetic Storms. *Earth, Planets and Space* 61, no. 5 (May): 595–597. https://doi.org/10.1186/bf03352930

[9] Rastogi, R.G., and J.A. Klobuchar. 1990. Ionospheric Electron Content within the Equatorial F2 Layer Anomaly Belt. *Journal of Geophysical Research: Space Physics* 95, no. A11 (November): 19045–19052. https://doi.org/10.1029/ja095ia11p19045

[10] Tsurutani, B. 2004. Global Dayside Ionospheric Uplift and Enhancement Associated with Interplanetary Electric Fields. *Journal of Geophysical Research* 109, no. A8. https://doi.org/10.1029/2003ja010342

[11] Abdu, M.A., T. Maruyama, I.S. Batista, S. Saito, and M. Nakamura. 2007. Correction to "Ionospheric Responses to the October 2003 Superstorm: Longitude/Local Time Effects over Equatorial Low and Middle Latitudes". *Journal of Geophysical Research: Space Physics* 112, no. A12 (December): n/a. https://doi.org/10.1029/2007ja012908

[12] Tsutomu, N., Geomagnetic Storms, *Journal of the Communications Research Laboratory* Vol. 49 No. 3 2002.

[13] Buonsanto, M. 1999. Ionospheric Storms — A Review. *Space Science Reviews* 88, 563–601. https://doi. org/10.1023/A:1005107532631

[14] Pedatella, N.M., J. Lei, K.M. Larson, and J.M. Forbes. 2009. Observations of the Ionospheric Response to the 15 December 2006 Geomagnetic Storm: Long-Duration Positive Storm Effect. *Journal of Geophysical Research: Space Physics* 114, no. A12 (December). https://doi.org/10.1029/2009ja014568

[15] Fuller-Rowell, T.J., G.H. Millward, A.D. Richmond, and M.V. Codrescu. 2002. Storm-Time Changes in the Upper Atmosphere at Low Latitudes. *Journal of Atmospheric and Solar-Terrestrial Physics* 64, no. 12–14 (August): 1383–1391. https://doi.org/10.1016/s1364-6826(02)00101-3

[16] Fejer, B.G., J.W. Jensen, T. Kikuchi, M.A. Abdu, and J.L. Chau. 2007. Equatorial Ionospheric Electric Fields During the November 2004 Magnetic Storm. *Journal of Geophysical Research: Space Physics* 112, no. A10 (October). https://doi.org/10.1029/2007ja012376

[17] Kumar, S., and A.K. Singh. 2011. GPS Derived Ionospheric TEC Response to Geomagnetic Storm on 24 August 2005 at Indian Low Latitude Stations. *Advances in Space Research* 47, no. 4 (February): 710–717. https://doi.org/10.1016/j.asr.2010.10.015

[18] Kumar, S., Singh, A.K. 2011. Storm time response of GPS-derived total electron content (TEC) during low solar active period at Indian low latitude station Varanasi. *Astrophysics and Space Science* 331, 447–458. https://doi.org/10.1007/s10509-010-0459-y

[19] Basu, S., S. Basu, J.J. Makela, R.E. Sheehan, E. MacKenzie, P. Doherty, J.W. Wright, et al. 2005. Two Components of Ionospheric Plasma Structuring at Midlatitudes Observed during the Large Magnetic Storm of October 30, 2003. *Geophysical Research Letters* 32, no. 12 (April 29). https://doi. org/10.1029/2004gl021669

[20] Basu, S., S. Basu, J.J. Makela, E. MacKenzie, P. Doherty, J.W. Wright, F. Rich, M.J. Keskinen, R.E. Sheehan, and A.J. Coster. 2008. Large Magnetic Storm-Induced Nighttime Ionospheric Flows at Midlatitudes and Their Impacts on GPS-Based Navigation Systems. *Journal of Geophysical Research: Space Physics* 113, no. A3 (March). https://doi.org/10.1029/2008ja013076

[21] Rama Rao, P.V.S., S. Gopi Krishna, J. Vara Prasad, S.N.V.S. Prasad, D.S.V.V.D. Prasad, and K. Niranjan. 2009. Geomagnetic Storm Effects on GPS Based Navigation. *Annales Geophysicae* 27, no. 5 (May 7): 2101–2110. https://doi.org/10.5194/angeo-27-2101-2009

[22] Liu, J., B. Zhao, and L. Liu. 2010. Time Delay and Duration of Ionospheric Total Electron Content Responses to Geomagnetic Disturbances. *Annales Geophysicae* 28, no. 3 (March 18): 795–805. https:// doi.org/10.5194/angeo-28-795-2010

[23] Kumar Chaurasiya, S., K. Patel, S. Kumar, and A. Kumar Singh. 2023a. Analysis of GPS-TEC and IRI Model over Equatorial and EIA Stations during Solar Cycle 24. *Advances in Space Research* (September). https://doi.org/10.1016/j.asr.2023.09.014

[24] Chaurasiya, S.K., K. Patel, and A.K. Singh. 2023b. Total Electron Content Forecasting with Neural Networks during Intense Geomagnetic Storms of the Solar Maximum and Moderate Years of Solar Cycle 24 in the Low Latitude Indian Region. *Astrophysics and Space Science* 368, no. 9 (September). https:// doi.org/10.1007/s10509-023-04237-8

[25] Kumar, S., and A.K. Singh. 2010. The Effect of Geomagnetic Storm on GPS Derived Total Electron Content (TEC) at Varanasi, India. *Journal of Physics: Conference Series* 208 (February 1): 012062. https://doi.org/10.1088/1742-6596/208/1/012062

[26] G.J. Sonnenberg, *Radar and Electronic Navigation* (Elsevier, 2013).

[27] N. Ackroyd and R. Lorimer, *Global Navigation: A GPS User's Guide* (Lloyd's of London Press, 1990).

[28] Coco, D. (1991). GPS satellites of opportunity for ionospheric monitoring. *GPS World* 2, no. 9, 47–50.

[29] Klobuchar, J. A. 1996. Ionospheric Effects on GPS in: Global Positioning System: Theory and Applications. Vol 2, edited by: Parkinson B W and Spilker J J *Progress in Astronautics and Aeronautics* 164 p. 485.

[30] Davies, K., and G.K. Hartmann. 1997. Studying the Ionosphere with the Global Positioning System. *Radio Science* 32, no. 4 (July): 1695–1703. 10.1029/97rs00451

[31] Hocke, K., and A.G. Pavelyev. 2001. General Aspect of GPS Data Use for Atmospheric Science. *Advances in Space Research* 27, no. 6–7 (January): 1313–1320. https://doi.org/10.1016/s0273-1177(01)00141-7

[32] Rama Rao, P.V.S., S. Gopi Krishna, K. Niranjan, and D.S.V.V.D. Prasad. 2006. Temporal and Spatial Variations in TEC Using Simultaneous Measurements from the Indian GPS Network of Receivers during the Low Solar Activity Period of 2004–2005. *Annales Geophysicae* 24, no. 12 (December 21): 3279–3292. https://doi.org/10.5194/angeo-24-3279-2006

[33] Birch, M.J., J.K. Hargreaves, and G.J. Bailey. 2002. On the Use of an Effective Ionospheric Height in Electron Content Measurement by GPS Reception. *Radio Science* 37, no. 1 (January). https://doi.org/10.1029/2000rs002601

[34] Mannucci, Anthony J., Wilson, Brian D., Edwards, Charles D. 1993. A New Method for Monitoring the Earth's Ionospheric Total Electron Content Using the GPS Global Network. *Proceedings of the 6th International Technical Meeting of the Satellite Division of The Institute of Navigation (ION GPS 1993)*, Salt Lake City, UT, pp. 1323–1332.

[35] Biqiang, Z., W. Weixing, L. Libo, and M. Tian. 2007. Morphology in the Total Electron Content under Geomagnetic Disturbed Conditions: Results from Global Ionosphere Maps. *Annales Geophysicae* 25, no. 7 (July 30): 1555–1568. https://doi.org/10.5194/angeo-25-1555-2007

[36] Astafyeva, E., I. Zakharenkova, and M. Förster. 2015. Ionospheric Response to the 2015 St. Patrick's Day Storm: A Global Multi-instrumental Overview. *Journal of Geophysical Research: Space Physics* 120, no. 10 (October): 9023–9037. https://doi.org/10.1002/2015ja021629

[37] Nava, B., J. Rodríguez-Zuluaga, K. Alazo-Cuartas, A. Kashcheyev, Y. Migoya-Orué, S.M. Radicella, C. Amory-Mazaudier, and R. Fleury. 2016. Middle- and Low-latitude Ionosphere Response to 2015 St. Patrick's Day Geomagnetic Storm. *Journal of Geophysical Research: Space Physics* 121, no. 4 (April): 3421–3438. https://doi.org/10.1002/2015ja022299

[38] Astafyeva, E.I. 2009. Dayside Ionospheric Uplift during Strong Geomagnetic Storms as Detected by the CHAMP, SAC-C, TOPEX and Jason-1 Satellites. *Advances in Space Research* 43, no. 11 (June): 1749–1756. 10.1016/j.asr.2008.09.036

[39] Chaurasiya, S.K., K. Patel, S. Kumar, and A.K. Singh. 2022. Ionospheric Response of St. Patrick's Day Geomagnetic Storm over Indian Low Latitude Regions. *Astrophysics and Space Science* 367, no. 10 (October). https://doi.org/10.1007/s10509-022-04137-3

[40] Kumar, S. 2016. Performance of IRI-2012 Model during a Deep Solar Minimum and a Maximum Year over Global Equatorial Regions. *Journal of Geophysical Research: Space Physics* 121, no. 6 (June): 5664–5674. https://doi.org/10.1002/2015ja022269

[41] Chen, Y., L. Liu, H. Le, W. Wan, and H. Zhang. 2016. Equatorial Ionization Anomaly in the Low-latitude Topside Ionosphere: Local Time Evolution and Longitudinal Difference. *Journal of Geophysical Research: Space Physics* 121, no. 7 (July): 7166–7182. https://doi.org/10.1002/2016ja022394

[42] Singh, A., V.S. Rathore, S. Kumar, S.S. Rao, S.K. Singh, and A.K. Singh. 2021. Effect of Intense Geomagnetic Storms on Low-Latitude TEC during the Ascending Phase of the Solar Cycle 24. *Journal of Astrophysics and Astronomy* 42, no. 2 (August 23). https://doi.org/10.1007/s12036-021-09774-8

[43] Wang, W., J. Lei, A.G. Burns, S.C. Solomon, M. Wiltberger, J. Xu, Y. Zhang, L. Paxton, and A. Coster. 2010. Ionospheric Response to the Initial Phase of Geomagnetic Storms: Common Features. *Journal of Geophysical Research: Space Physics* 115, no. A7 (July). https://doi.org/10.1029/2009ja014461

[44] Zhang, W., X. Zhao, S. Jin, and J. Li. 2018. Ionospheric Disturbances Following the March 2015 Geomagnetic Storm from GPS Observations in China. *Geodesy and Geodynamics* 9, no. 4 (July): 288–295. https://doi.org/10.1016/j.geog.2018.02.001

[45] Rao, S.S., M. Chakraborty, and A.K. Singh. 2023. Observed (GPS) and Modeled (IRI and TIE-GCM) TEC Trends over Southern Low Latitude during Solar Cycle-24. *Advances in Space Research* 71, no. 8 (April): 3394–3407. https://doi.org/10.1016/j.asr.2022.12.030

[46] Chaurasiya SK, Patel K, Singh AK (2022a) Equatorial plasma bubbles for the solar maximum and moderate year of solar cycle 24. *Advances in Space Research* 70, 2856–2866. https://doi.org/10.1016/j.asr.2022.07.036

22 GPS TEC Response During the Solar Flare

Sardar Singh Rao and Abhay Kumar Singh

22.1 GENERATION OF SOLAR FLARES

A solar flare is a dynamic event involving the sun's magnetic field, charged particles, and the explosive release of pent-up solar energy. The magnetic fields, created by the churning motion of charged particles, create a dynamic landscape. As they reassemble and reconnect, they release an immense amount of stored magnetic energy, triggering a cascade of events leading to the initiation of the flare. High-energy particles, primarily electrons and protons, are accelerated to near the speed of light, releasing energy across the electromagnetic spectrum, including a burst of X-rays. The intensity of the X-ray emission is directly linked to the flare's scale and energy release, allowing scientists to categorize flares based on their X-ray flux. In some cases, solar flares are accompanied by coronal mass ejection (CME), where a massive amount of solar material, including charged particles and magnetic fields, is ejected into space at high speeds. The major solar flares could be classified in the following manner (Table 22.1).

22.2 WHY STUDYING SOLAR FLARES IS NECESSARY

Solar flares, despite their beauty, pose a significant threat to our technologically advanced world. They can cause massive auroras and disastrous effects on modern infrastructure. As technology's reliance increases, so does our vulnerability to solar flares, which can disrupt power lines, pipelines, satellites, communication networks, and the interconnected world.

Proactive measures to protect our technological oasis from solar flare impacts include improved Space weather forecasting, resilient infrastructure development, and understanding the solar cycle to predict disturbance activity periods.

22.3 IMPACT OF SOLAR FLARES ON THE EARTH'S IONOSPHERE

The ionosphere is a dynamic layer of charged particles created by high-energy solar radiation. Solar flares release intense radiation, causing electron density elevation and ionospheric disturbances.

22.3.1 EFFECTS ON RADIO COMMUNICATION

One of the most visible effects of solar flare-induced ionospheric disruptions is their influence on radio communication. Radio signals at high frequencies (HF) bounce off the ionosphere, allowing for long-distance communication. Signal propagation can, however, be interrupted during periods of disturbed ionospheric activity. The higher electron density can absorb or reflect radio waves, resulting in signal deterioration or, in severe situations, blackout.

22.3.2 GLOBAL NAVIGATION SATELLITE SYSTEMS (GNSSs) ANOMALIES

The ionosphere is also critical to the operation of Global Navigation Satellite Systems (GNSSs), such as GPS. The increased electron density caused by solar flare-induced disturbances can cause

DOI: 10.1201/9781032712444-26

TABLE 22.1

Classification of the Solar Flare

Flare Class	*C-Class Flares*	*M-Class Flares*	*X-Class Flares*
Based on intensity	Peak fluxes of less than 10^{-6} Watts per square meter. It may not cause significant technology interruptions; they add to the enthralling display of auroras at higher latitudes.	Peak flux between 10^{-6} and 10^{-5} Watts per square meter. It can produce temporary radio signal blackouts in the Polar Regions and represent a moderate hazard to satellite operations.	Peak flux levels exceeding 10^{-5} Watts per square meter. It can cause severe disruptions in radio communication and navigation systems, perhaps creating global technological chaos.
Based in location	*Limb Flares*; Limb flares occur close to the sun's limb, or edge, as viewed from Earth. Often ejecting material into space, these flares produce coronal mass ejections (CMEs), which are aesthetically magnificent. The related CMEs may not face Earth directly, yet they might nevertheless have indirect consequences on our globe.	*Center-disk Flare* Flares erupting from the center of the sun's disk pose a more direct threat to Earth. These events are more likely to launch charged particles and radiation in our direction, increasing the potential for technological disruptions.	
Based on duration of X-ray emission	*Impulsive Flare* Impulsive flares are defined by a sudden increase in X-ray emission, quickly followed by a sharp decrease.	*Long-duration Flares* Long-duration flares occur over longer timespans and exhibit a progressive rise and fall in X-ray emission.	

delays and errors in satellite signals as they transit through the ionosphere. This can influence navigation system accuracy, affecting various applications ranging from aircraft to everyday GPS-based navigation.

22.3.3 AURORAS AND IONOSPHERIC CHEMISTRY

While ionosphere disturbances might bring technological hurdles, they also contribute to the beautiful visual display of auroras. These bright displays in the Polar Regions are caused by solar flare-induced energetic particles interacting with charged particles in the ionosphere. Furthermore, the enhanced ionization caused by solar flares can change the chemistry of the ionosphere, altering its composition and thermal structure.

22.4 DIFFERENT FACTORS AFFECTING SOLAR FLARE INFLUENCE ON EARTH'S IONOSPHERE

22.4.1 SOLAR ZENITH ANGLE

The solar zenith angle, which is at its minimum at solar noon and maximum during sunrise or sunset, is influenced by the observer's geographical location and time of day, creating a dynamic canvas for solar flares. Solar flares emit high-energy X-rays that affect Earth's ionosphere, affecting absorption and potential disruptions. The solar zenith angle determines the angle at which these X-rays penetrate the atmosphere, with higher angles causing increased absorption and disruption.

The solar zenith angle affects radio wave propagation through the ionosphere, affecting the distance and angle at which waves encounter ionospheric layers. Higher angles can increase signal path lengths, potentially causing signal degradation, particularly in the high-frequency band. Satellite communication, heavily reliant on signals traversing the ionosphere, is also subject to the influence of solar zenith angles during flare events. Higher angles can lead to increased signal path lengths, potentially affecting the reliability and performance of satellite communication systems. Solar flares may affect the ionosphere at different latitudes, although they tend to have a more noticeable effect on auroral displays at higher latitudes. The sun's zenith angle influences aurora visibility; greater angles typically result in more extensive and colorful displays in the Polar Regions during geomagnetic storms caused by solar flares.

The ionospheric response to a solar flare is primarily determined by the solar zenith angle, with changes more noticeable when small. The time rate of flare-induced TEC increase is proportional to efficient flare radiation flux, so the TEC derivative should be proportional to the cosine of the solar zenith angle. Studies by Zhang et al. (2002) and Sripathi et al. (2013) found a significant enhancement and correlation of dayside TEC values during flare events with small solar zenith angles.

Barta observed a solar-zenith-angle-dependency of the enhancement of the ionosonde echo under a different solar zenith angle, indicating that the ionospheric electron density enhancement during a solar flare has a solar zenith angle dependency.

The solar zenith angle varied between 12 and 13 March 2015, with the M1.8 flare occurring with a local time of around 1200 LT at Varanasi, India. The flare caused a 30 TECU increase in TEC over Varanasi, India (Figure 22.1), with the most pronounced effects occurring during noon when the angle was close to zero (see lower panel of Figure 22.2). On the other hand, the M1.6 flare occurred on 12 March 2015 with a 90° solar zenith angle (see upper panel of Figure 22.2) at Varanasi India, resulting in a less TEC increase of ~15 TECU (Figure 22.1) due to the flare occurring in the dusk sector.

22.4.2 Location of the Solar Active Region

Solar flare effects on Earth and its surrounding space environment are strongly influenced by the position of these active regions on the solar surface. The complex magnetic fields in active regions can give rise to intense events that send charged particles and bursts of radiation flying into space, acting as the solar flare's origin. The location of solar active regions is a key factor in determining the potential impact of solar flares on Earth. If an active region is Earth-facing, solar flares originating from that region are more likely to have a direct influence on our planet. Conversely, if the active region is situated on the far side of the sun, the impact on Earth is diminished. The ionospheric TEC response sufficiency decreases as the center meridian distance (CMD) increases. Locations of flares on the sun affect TEC variabilities.

Whereas EUV predominates in ionization above 150 km, soft X-rays dominate the E region. A solar flare's location impacts the thermosphere ionosphere reactions in the F region.

The thermospheric-ionosphere response in the F region is 2–3 times larger for disk-center flares than limb flares due to flare location influences on EUV intensification. Flares near the solar disk center significantly affect the TEC compared to those near the solar limb region. The longitude effect on TEC makes it challenging to filter the effect of solar flare location in TEC response. The more significant enhancement in EUV for disk flares is due to the difference in optical thickness of soft X-rays and EUV in the solar atmosphere. Optically dense emissions are absorbed more if a flare is on the limb.

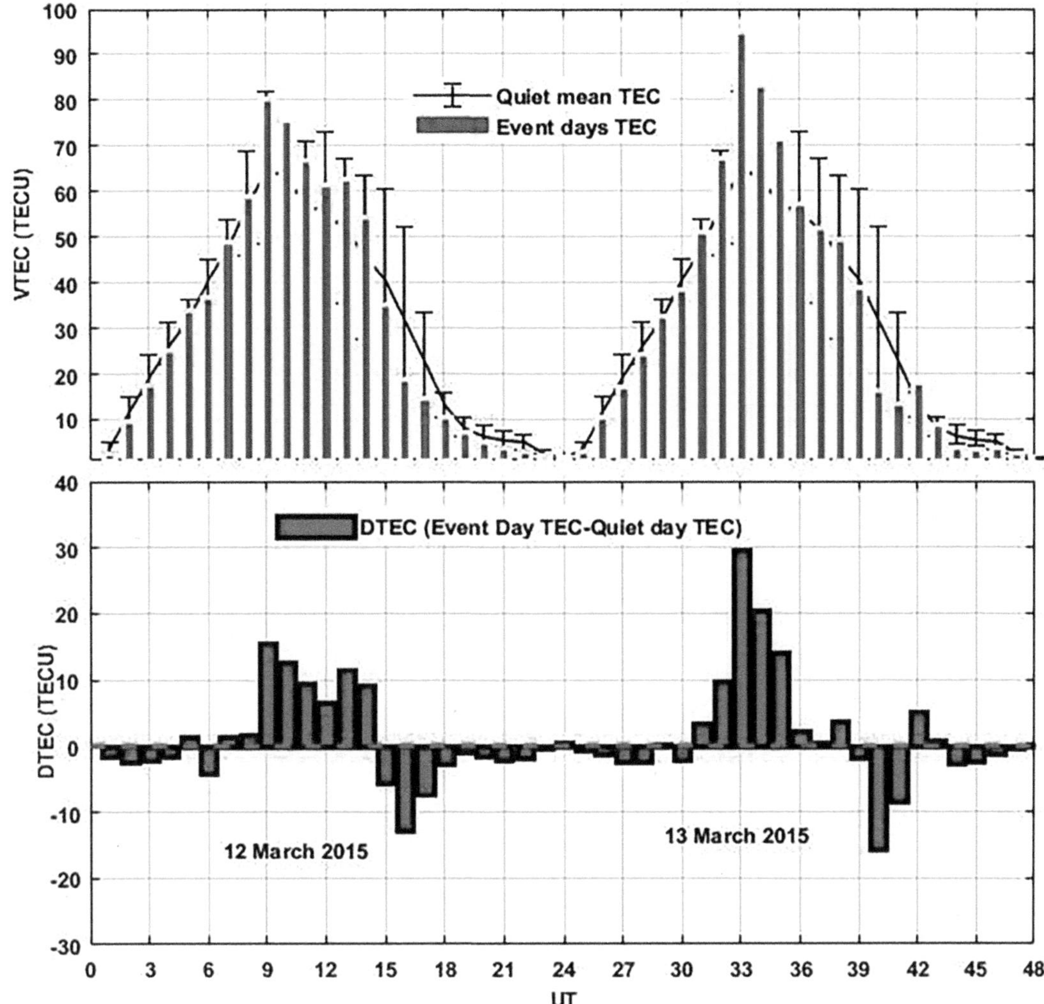

FIGURE 22.1 The study shows TEC variation over Varanasi during M1.6 and M1.8 solar flares, with the difference to the quiet mean TEC (ΔTEC) in the lower panel, calculated using VTEC of the five quietest days in a given month.

The soft X-ray enhancement, which dominates D/E region ionization, is not significantly affected by the location of a flare on the sun. The extreme position of the flare affects more EUV radiation than X-ray radiation. Qian et al. (2019) suggested that solar EUV enhancement, which dominates ionization above ~150 km, is weaker for limb flares compared to disk flares of the same magnitude. Qian et al. (2019) found that disk flares have a more significant impact on the ionospheric response to solar flares above 150 km, with the effect being influenced by local time and longitude. The intensity of disk flare X9.3 and limb flare X8.2 were similar (shown in Figures 22.3 and 22.4). The X9.3 flare had a larger enhancement in the EUV due to the difference in optical thickness of the soft X-ray and EUV in the solar atmosphere. It can be seen from Figure 22.5 that the maximum TEC increase during the disk flare X9.3 was ~15-20 TECU, whereas the maximum TEC increase during the limb flare X8.2 was ~10 TECU.

The X8.2 flare significantly reduced optically thick chromospheric and transition region emissions, resulting in a significant decrease in center-to-limb variability.

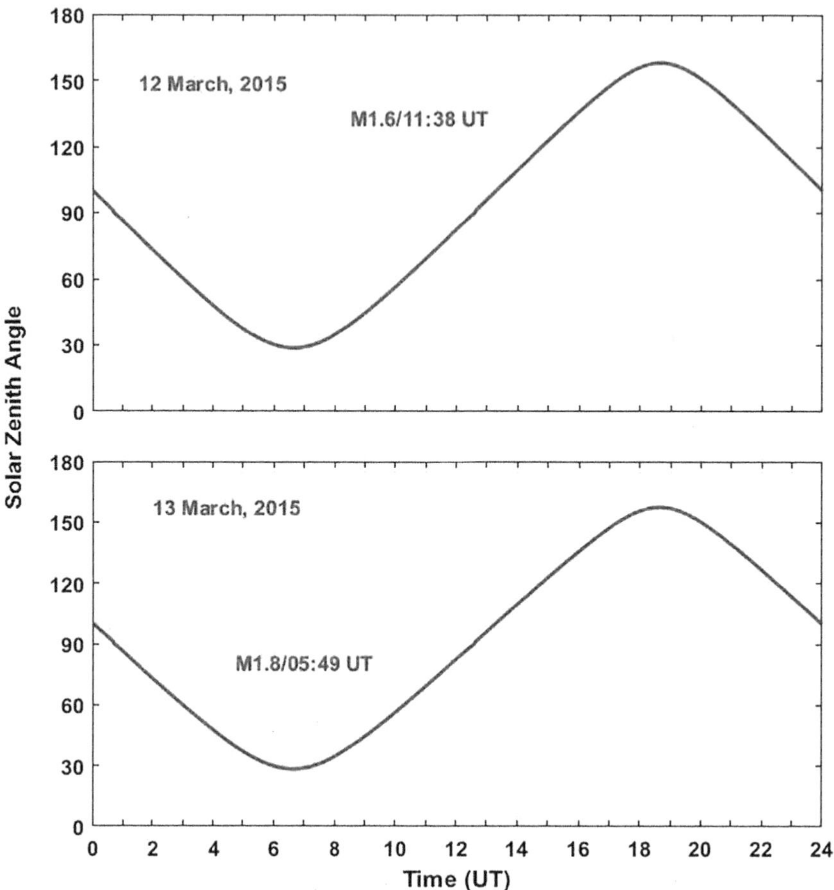

FIGURE 22.2 The variations of the solar zenith angle at the location of Varanasi on 12 and 13 March 2015 are given in the upper and lower panels, respectively.

22.4.3 SEASONAL EFFECT

It is generally recognized that stronger ionospheric reactions occur during equinoxes than at solstices. This might be explained by the seasonal variation of the neutral density. Previous research revealed that the ratio of atomic oxygen to molecular nitrogen (O/N2) varies significantly semi-annually, with higher levels at equinoxes than at solstices (Qian et al., 2009). The semi-annual change in the O/N2 ratio has been proposed as a cause of TEC variation (Le et al., 2013). Le et al.'s 2013 study found that solar flares cause summer-winter asymmetry in total electron content, which is linked to O/N2 asymmetry. The average ΔTEC values for the March, June, September, and December solstices were 1.26 TECU, 0.78 TECU, 1.33 TECU, and 1.01 TECU, with September having the most impact.

22.4.4 RATE OF RISE AND DECAY PHASE OF THE FLARE

The Neupert Effect was introduced by American astrophysicist Robert Neupert in 1968. It describes the observational correlation between the time profiles of microwave and hard X-ray emissions during the impulsive phase of solar flares. The idea is that microwave emission, associated with non-thermal processes, occurs earlier than hard X-ray emission, associated with thermal processes. In simple terms, the Neupert Effect suggests that as non-thermal particles (such as energetic electrons)

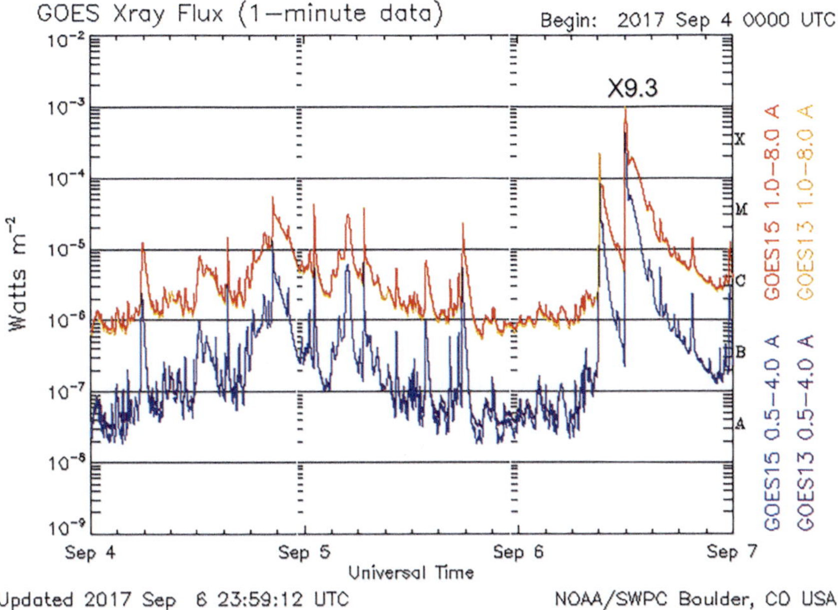

FIGURE 22.3 Variation of X-ray flux during 4–7 September 2017. The X9.3 disk solar flare that occurred on 6th September 2017 at 11:53 UT peaked at 12:02 UT and ended at 12:10 UT.

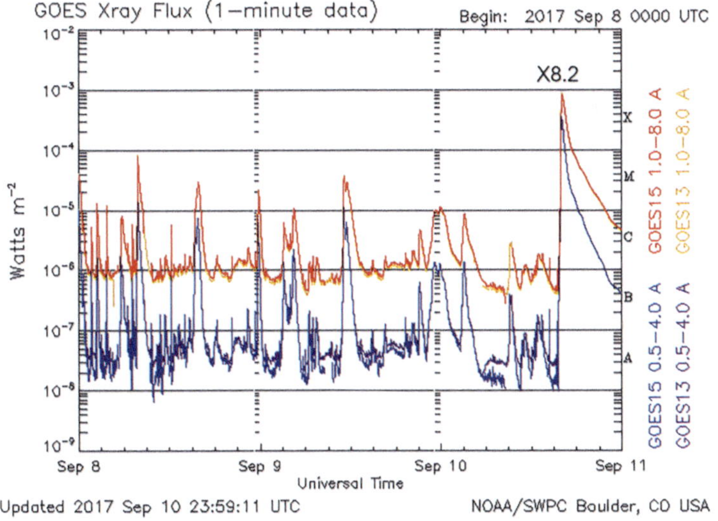

FIGURE 22.4 Variation of X-ray flux during 8–11 September 2017. The X8.2 disk solar flare that occurred on 10th September 2017 at 15:35 UT peaked at 16:06 UT and ended at 16:31 UT.

are accelerated during a solar flare, they produce microwave emission first. Subsequently, these accelerated particles can heat the ambient plasma, leading to the production of thermal X-rays. Neupert (1968, 1989) demonstrated that the positive time derivative of the GOES XRS X-ray irradiance (DXRS/Dt) is critical in calculating impulsive phase EUV emissions.

This is now referred to as the "Neupert Effect." It was demonstrated by Chamberlin et al., 2008 and Qian et al. (2011) that the rise and decay rates of solar flare s affect their ionospheric and thermospheric impacts.

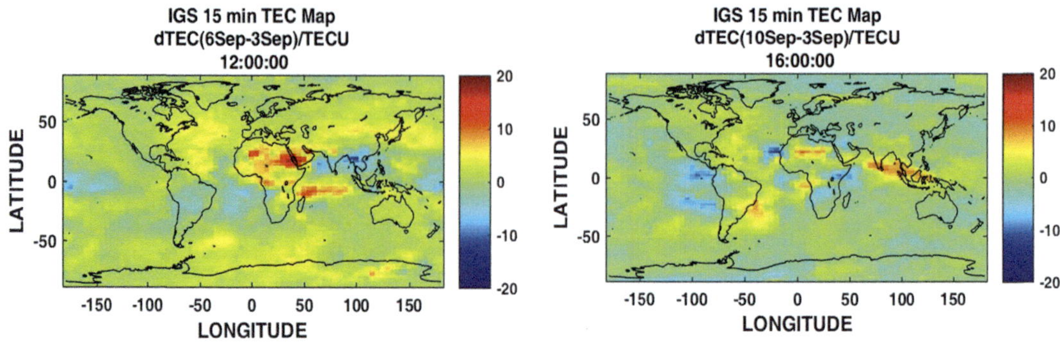

FIGURE 22.5 Global TEC maps during the solar flare X9.3 (disk flare) on 6th September 2017 and solar flare X8.2 (limb flare) on 10th September 2017.

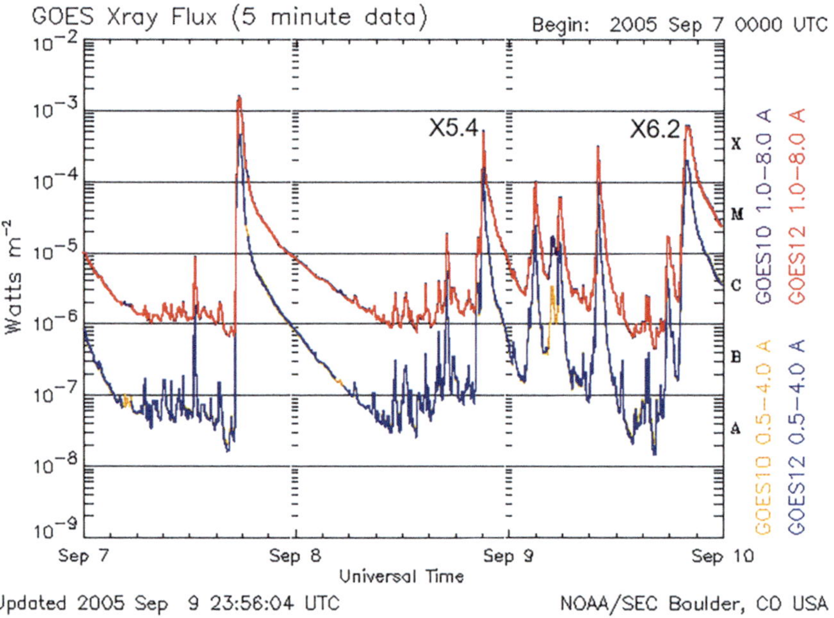

FIGURE 22.6 Variation of X-ray flux during 8th September to 10th September 2005. The X5.4 limb solar flare occurred on 8th September 2005 at 20:52 UT peaked at 21:06 UT and ended at 21:17 UT. The X6.2 limb solar flare occurred on 9th September 2005 at 19:13 UT peaked at 20:04 UT and ended at 20:36 UT.

At wavelengths less than 27 nm, these impulsive phase EUV radiations are comparatively small in coronal emissions, but quite high in EUV emissions generated in the transition area.

The Neupert Effect, a measure of the strength of EUV during impulsive phase, is a key factor in understanding how neutral density and TEC responses are affected.

The Neupert Effect can easily be understood from the solar flare events of September 2005. An X5.4 flare occurred on 8 September 2005 and an X6.2 flare occurred on 9 September 2005 (see Figure 22.6). The two flares had approximately similar soft X-ray peak intensities and the same location on the solar limb. The rise times for the X5.4 was 15 min and decay time for the X5.4 was 2 h 35 min. The rise times for the X6.2 was 40 min and decay time for the X5.4 was 5 h. Thus, the X6.2 flare had longer rise and decay times. It can be seen from Figure 22.7 that the maximum total electron content (TEC) enhancements from the X6.2 and X5.4 flares were ~20 TECU (right panel) and ~10 TECU (left panel) at sub-solar point, respectively.

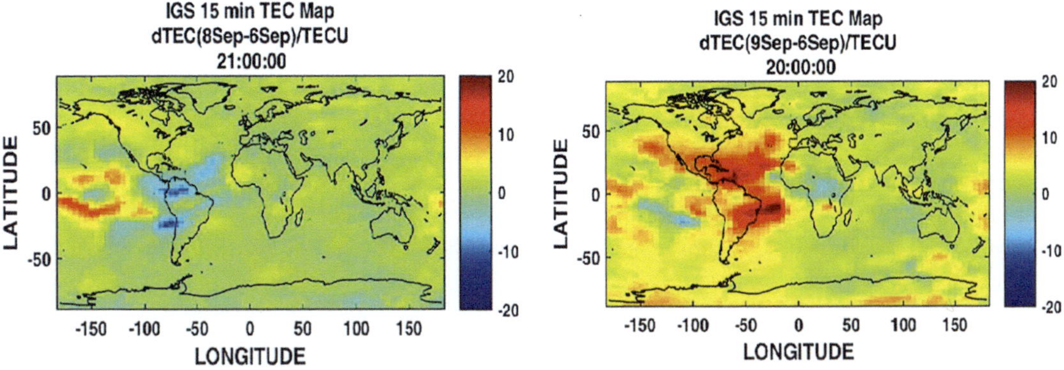

FIGURE 22.7 Global maps of TEC during the X5.4 flare on 8 September 2005 (left panel) and X6.2 flare on 9 September 2005 (right panel).

The ionospheric response to a solar flare can be delayed, particularly for flare-induced TEC changes. Zhang et al. (2011) study found that the ionospheric TEC response to a flare lagged behind the HF Doppler shift records, suggesting that a solar flare does not instantaneously influence the TEC. The peak solar flare effect is constantly delayed with increasing solar zenith angle values in the sunlit hemisphere. Li et al. (2018) observed several hours of delay between the decay of EUV emission and TEC recovery due to the long-term decay of flare emissions and contributions from higher altitudes (over 300 km, where recombination is weaker), which have been observed to induce delays in reaching the pre-flare level of TEC (Qian et al., 2011; Le et al., 2016). Additionally, the ionospheric response and solar flare may be delayed due to thermospheric-ionospheric interaction, which is significantly disrupted by solar flares (Tsugawa et al., 2006).

22.5 SUMMARY

1. A larger flare-induced TEC increase is found at sub-solar locations (when the solar zenith angle tends towards 0° or noon times) and a lower enhancement when the sun zenith angle tends towards 90°. This clearly reveals the solar zenith angle dependency of TEC when responding to solar flares.
2. The study reveals that solar extreme ultraviolet enhancement, which dominates ionization above 150 km, is less intense for limb flares than disc flares—and soft X-ray enhancement, which dominates E region ionization, is not affected by an Active Region's position.
3. Numerous observational and modeling studies show that the rise and decay times of flares with identical magnitudes and active region locations significantly influence the responses of the thermosphere and ionosphere.
4. Solar flare-induced TEC enhancements revealed seasonal variation, which might be related to seasonal changes in background neutral density.

22.6 FUTURE DIRECTION

The ionosphere, a dynamic canvas of charged particles, responds to the sun's energetic outbursts with a symphony of disturbances that, while challenging, also contribute to the mesmerizing dance of auroras. As technology advances, so too does our ability to understand and predict these interactions, allowing us to adapt and safeguard our interconnected world against the cosmic forces at play. Efforts to mitigate the impact of solar flare-induced ionospheric disturbances involve advanced Space weather forecasting. By monitoring solar activity and predicting periods of increased flare activity, scientists and technologists can implement measures to safeguard communication and navigation systems.

ACKNOWLEDGEMENTS

The Council of Scientific and Industrial Research (CSIR), New Delhi, funded this research through the Pool Scientist Scheme awarded to Senior Research Associate Dr. Sardar Singh Rao (File No. 13(9203-A)/2021-POOL dated 28/12/2021).

The GOES X-ray plots were taken from the link (https://www.spaceweatherlive.com/en/solar-activity/top-50-solar-flares.html). The TEC data for station Varanasi, India, was obtained from a Trimble 5700 GPS receiver at Banaras Hindu University, Varanasi, India, and global maps were retrieved from the Crustal Dynamic Data Information System's website (https://cdaweb.gsfc.nasa.gov/cgi-bin/eval2.cgi).

REFERENCES

Chamberlin, P. C., T. N. Woods, and F. G. Eparvier (2008), Flare Irradiance Spectral Model (FISM): Flare component algorithms and results, *Space Weather*, **6**, S05001, doi:10.1029/2007SW000372

Le, H, Liu, L, Chen, Y, Wan W (2013) Statistical analysis of ionospheric responses to solar flares in the solar cycle 23; *J Geophys Res. Space Phys*, 118, 576–582.

Le H, Liu L, Ren Z, Chen, Y, Zhang, H, Wan W (2016) A modelling study of global ionospheric and thermospheric responses to extreme solar flare; *J Geophys Res. Space Phys*, 121, 832–840.

Li N, Lei J, Luan X, Chen J, Zhong J, Wu Q, Xu Z, Lin L (2018), Responses of the D region ionosphere to solar flares revealed by MF radar measurements; *J. Atmos. Sol.-Terr. Phys.*, 182, 211–216.

Neupert, W. N. (1968), Comparison of solar X-ray line emission with microwave emission during flares, *Astrophys. J.*, 153, L59–L64, doi:10.1086/180220

Neupert, W. N. (1989), Transient coronal extreme ultraviolet emission before and during the impulsive phase of a solar flare, *Astrophys. J.*, 344, 504–512, doi:10.1086/167819

Qian, L, Burns A G, Chamberlin P C, Solomon S C (2011), Variability of thermosphere and ionosphere responses to solar flares; *J. Geophys. Res. Space Phys*, 116, A10309.

Qian, L., S. C. Solomon, and T. J. Kane (2009), Seasonal variation of thermospheric density and composition, *J. Geophys. Res.*, 114, A01312, doi:10.1029/2008JA013643

Qian, L, Wang, W, Burns, A G, Chamberlin, P C, Coster, A, Zhang, S R, Solomon, S C (2019) Solar *Flare and Geomagnetic Storm Effects on the Thermosphere and Ionosphere During 6–11 September 2017*; 124(3) 2298–2311.

Sripathi, S., Balachandran, N., Veenadhari, B., Singh, R., and Emperumal, K. (2013), Response of the equatorial and low-latitude ionosphere to an intense X-class solar flare (X7/2B) as observed on 09 August 2011. *J. Geophys. Res. Space Phys*. 118, 2648–2659. doi:10.1002/jgra.50267

Tsugawa T, Sadakane T, Sato J, Otsuka Y, Ogawa T, Shiokawa K, Saito A (2006), Summer-winter hemispheric asymmetry of sudden increase in ionospheric total electron content induced by solar flares: A role of O/N2 ratio; *J. Geophys. Res. Space Phys*, 111(A11), A11316.

Zhang, D. H., X. H. Mo, L. Cai, W. Zhang, M. Feng, Y. Q. Hao, and Z. Xiao (2011), Impact factor for the ionospheric total electron content response to solar flare irradiation; *J. Geophys. Res.*, 116, A04311, doi:10.1029/ 2010JA016089

Zhang D H, Xiao Z, Igarashi K and Ma G Y (2002), GPS derived ionospheric total electron content response to a solar flare that occurred on 14 July 2000; *Radio Sci.* 37, 1086.

23 A Brief Review of Low-latitude Ionosphere

With Multi-frequency Observation Using GPS, NavIC, and GMRT

*Abhirup Datta, Bhuvnesh Brawar, Sarvesh Mangla,
Deepthi Ayyagari, and Sumanjit Chakraborty*

23.1 INTRODUCTION

The ionosphere is the upper part of the atmosphere, which spans above Earth's surface from approximately 60 km height to beyond 1000 km. In that region, there is an abundance of charged particles and electrons (Kelley, 1989). It affects transionospheric radio communication and radio observations and limits the performance of radio interferometers and navigation systems by degrading the amplitude and phase of the signals. In the equatorial and low-latitude regions, the horizontal alignment of the magnetic field lines creates an electro-dynamic effect. This leads to complex phenomena like equatorial ionization anomaly (EIA), equatorial electrons (EEJs), plasma bubbles, vertical electron drift, and electron fountains.

During extreme geomagnetic conditions, ionosphere effects are more pronounced and unpredictable. Signals traversing the ionosphere experience face scattering due to irregularities in the ionospheric F region, known as spread-F (Fejer & Kelley, 1980; Mangla, Chakraborty, Datta, & Paul, 2023; Woodman, Chau, Aquino, Rodriguez, & Flores, 1999). The ionosphere has been thoroughly studied for decades using ionosondes and the Global Navigation Satellite System (GNSS) at a rough temporal resolution of 2 hours and a spatial resolution of $5° \times 2.5°$ in longitude and latitude, respectively.

The total electron content (TEC), that is, the column density of electrons along the Line of Sight (LoS), is commonly used to quantify the ionosphere where a TEC unit or $1TECu = 10^{16}$ electrons/m^2. The incoming transionospheric signals of far-field astronomical sources are refracted by these electrons, causing a delay frequently seen as a phase shift in the radio observation. An interferometer cannot determine the total phase contribution caused by the ionosphere because it measures the phase difference across baselines.

Alternatively, the δTEC resulting from the ionosphere's spatial fluctuations is measured by an interferometer (Mangla et al., 2023; Mangla & Datta, 2022). The effects of scintillation on communication systems were investigated in the equatorial anomaly region by looking at the variations in decorrelation time, amplitude cumulative distribution functions, fade duration, phase and intensity rate distribution, and diffractive scattering-induced depolarization effects (Basu, Fougere, MacKenzie, Basu, & Costa, 1987; Basu & Whitney, 1983; Franke & Liu, 1985). Phase scintillation has been studied less thoroughly than amplitude scintillation, even though both concepts are well-developed (Priyadarshi, 2015).

This chapter focuses on the multiple observations and results of the ionosphere using GPS, NavIC, and GMRT. It provides comprehensive insights into the ionosphere at different frequencies of radio signals using GMRT (235MHz, 610MHz), GPS (L1, L2), and NavIC (L5, S1) (Ayyagari

DOI: 10.1201/9781032712444-27

et al., 2020; Ayyagari, Datta, & Chakraborty, 2022; Mangla et al., 2023) and references. GPS and NavIC data reported in this text were collected from the receivers housed in IIT Indore from September 01, 2017, to September 30, 2019 (25 months). GMRT observation was taken on the night of August 5 and 6, 2012, at 235 and 610 MHz frequencies. NavIC and GPS show similar patterns of scintillation events, and signal fading was modeled as Nakagami-m distribution and $\alpha - \mu$ distribution. GMRT provides fluctuation values in TEC gradients with a precision up to 1×10^{-3}TECu. TEC surface was reconstructed using the polynomial fitting method. Information about Medium Scale Traveling Ionospheric Disturbances (MSTID) was also derived using GMRT data.

23.2   PROBES OF LOW-LATITUDE IONOSPHERE

The ionosphere has an abundance of electrons and charged particles; there, the medium behaves like plasma. The ionosphere affects radio signals by reflecting, refracting, and scattering them. The ionospheric refractive index 'n' can be estimated exactly for a collisionless, cold, magnetized plasma (Davies, 1990). Signals with frequencies $\nu >> \nu_p$ (Plasma frequency; ~10 MHz for ionosphere), 'n' can be expanded (Datta-Barua, Walter, Blanch, & Enge, 2008) using third-order Taylor approximation, preserving terms up to ν^{-4}, as:

$$n \approx 1 - \frac{q^2}{8\pi^2 m_e \epsilon_0} \cdot \frac{n_e}{\nu^2} \pm \frac{q^3}{16\pi^3 m_e^2 \epsilon_0} \cdot \frac{n_e B \cos\theta}{\nu^3} - \frac{q^4}{128\pi^4 m_e^2 \epsilon_0^2} 16 \cdot \frac{n_e^2}{\nu^4} - \frac{q^4}{64\pi^4 m_e^3 \epsilon_0} \cdot \frac{n_e B^2 \left(1 - \cos^2\theta\right)}{\nu^4}, (23.1)$$

where m_e is the electron mass, q is the electron charge, n_e is the number density of free electrons, ϵ_0 is the electric permittivity in vacuum, the magnetic field strength is denoted by B, and θ is the angle between B and the propagation direction of an electromagnetic wave.

When radio signals traverse the ionosphere, they face propagation delay due to the dispersive nature of the medium. The total propagation delay integrated along the LoS at frequency ν results in phase rotation given by:

$$\phi_{\text{ion}} = -\frac{2\pi\nu}{c} \int_{\text{LoS}} \left(n - 1\right) dx \tag{23.2}$$

In equation (23.1), the first term, which is related to dispersive delay proportional to the TEC along the LoS, is the dominant term. When considering frequencies higher than a few hundred MHz, higher terms can be disregarded. In the second term, the positive and negative signs are connected to left- and right-hand polarized signals, respectively, which results from Faraday rotation. The final two terms are typically disregarded, but they are considered for observational frequencies lower than 40 MHz.

$$\phi_{\text{ion}} = \frac{q^2}{4\pi c m_e \epsilon_0 \nu} \int_{\text{LoS}} n_e \, dx$$

$$\phi_{\text{ion}} = 84.36 \left(\frac{\nu}{100\text{MHz}}\right)^{-1} \left(\frac{\text{TEC}}{1\text{TECU}}\right) \text{radians} \tag{23.3}$$

23.2.1   GNSS-GPS/NavIC

The Global Navigation Satellite System (GNSS) is a group of different satellite constellations such as the Global Positioning System (GPS), Navigation with Indian Constellation (NavIC), Globalnaya Navigatsionnaya Sputnikovaya Sistema (GLONASS), etc. These systems work for positioning services, timing services, and ionospheric observations. GPS provides services around the globe with more than 30 satellites in Medium Earth Orbit (MEO). The GPS system utilizes multiple frequencies for signal transmission. The primary frequencies are the L1 band (1575.42 MHz), L2 band (1227.60 MHz), and the L5 band (1176.45 MHz), all with bandwidths of 24 MHz. These frequencies support various signals, including the civilian-accessible C/A code on the L1 band and the P(Y) code and M-code used by authorized users on both L1 and L2 bands.

Additionally, the L5 band is designed for higher performance and improved reliability, particularly for safety-of-life applications. Contrary to GPS, NavIC is a regional navigation system. It has a combination of three geosynchronous orbits (GSO) and three geostationary Earth orbits (GEO) in its space segment. It is developed and being operated by the Indian Space Research Organization (ISRO). Its primary service area is designated as the Indian region and a region of 1500 km from its boundary, and it was created to give Indian users positional accuracy information. NavIC has a provision for an extended service area that lies between the primary service area, and the area spans between the rectangular grid from 30°S to 50°N in latitude to 30°E to 130°E in longitude. Furthermore, these satellites broadcast signals in a 24 MHz spectrum bandwidth in the L5 and S1 bandwidth carrier frequencies of 1176.45 MHz and 2492.028 MHz, respectively.

GNSS observation provides us with ionospheric conditions and responses to Space weather conditions by measuring the ionosphere's Total Electron Content (TEC). The TEC is measured using the frequency dependence effect (Arikan, Erol, & Arikan, 2003; Arikan, Nayir, Sezen, & Arikan, 2008; Julien, Priya, Issler, & Lestarquit, 2013; Klobuchar, 1983; Komjathy & Langley, 1996; Lanyi & Roth, 1988; Mitch, Psiaki, & Tong, 2013) as:

$$\rho_{\text{iono},\nu} = \int_{\text{LoS}} (n-1)\,dx = 40.3\,\frac{\text{STEC}}{\nu^2} \tag{23.4}$$

Here, ν represents the satellite signal's operational frequency in Hz, ρ_{iono},ν is the iono-delay in m. The iono-delay is approximated up to the first order of the ionospheric term. Using the pseudorange measurements of dual frequencies observations, TEC can be calculated as follows. The standard estimation procedure of pseudorange measurement equations at f_1 5 and f_2 6 dual frequencies composed of the true range r includes the ionospheric delay term ρ_{iono},ν indicated in equation (23.4) and other delays like the receiver and satellite clock bias Δt_r, Δt_s, the ephemeris error component e along the satellite-receiver line-of-sight, the tropospheric delay T_t, the satellite and receiver instrument delays k^s and k^r, the code phase multipath error M_p, and the random noise v.

$$\tilde{P}_{f1} = r + c\left(\ddot{A}t_r - \ddot{A}t_s\right) + e + \rho_{\text{iono},f1} + T_t + ck_{f1}^r - ck_{f1}^s + M_{pf1} + v_{f1} \tag{23.5}$$

$$\tilde{P}_{f2} = r + c\left(\Delta t_r - \Delta t_s\right) + e + \rho_{\text{iono},f2} + T_t + ck_{f2}^r - ck_{f2}^s + M_{pf2} + v_{f2} \tag{23.6}$$

In order to eliminate satellite instrument delays for dual-frequency users, the difference between equations (23.5) and (23.6) is utilized to form iono-free pseudoranges, which are given as follows:

$$\rho_{IF} = \frac{\rho_{\text{iono},f1} - \left(\gamma\right)\rho_{\text{iono},f2}}{1 - \left(\gamma\right)} \tag{23.7}$$

where $\gamma = \left(\dfrac{F_{f2}}{F_{f1}} \right)^2$. GNSS satellite constellations provide a wide coverage from a single receiver.

This coverage can be measured by calculating Ionospheric Pierce Points (IPPs) as follows:

$$\psi_{pp} = \frac{\pi}{2} - E - \sin^{-1}\left[\frac{R_e \cos(El)}{R_e + h_I} \right]$$

$$\phi_{pp} = \sin^{-1}\left[\sin\phi_u \cos\psi_{pp} + \cos\phi_u \sin\psi_{pp} \cos(Az) \right]$$

$$\lambda_{pp} = \lambda_u + \sin^{-1}\left[\frac{\sin\psi_{pp} \sin(Az)}{\cos\phi_{pp}} \right] \tag{23.8}$$

(ϕ_u, λ_u) are the user's Earth coordinates in latitude and longitude, El is the elevation to the satellite, ψ_{pp} is the angle from the user to the IPP as measured from the center of the Earth, R_e is the radius of the Earth, and $(\phi_{pp}, \lambda_{pp})$ are the Earth coordinate of the IPP (Figure 23.1).

In this report, the data used from the GPS and NavIC receivers are housed at IIT Indore (Lat: 22.52°N, Lon: 75.92°E; Magnetic dip: 32.23°N). Figure 23.2 provides the coverage of the IPPs over central India.

23.2.2 Radio Interferometer – GMRT

A radio interferometer measures the "spatial coherence function" directly with antenna pairs aimed skyward (Cornwell & Fomalont, 1999). By executing the Fourier inversion of the function, which characterizes the far-field radiation pattern, one can obtain an image of the sky signal's intensity distribution. The van Cittert-Zernike theorem yields the observed visibility (V_{ij}) for a baseline (i, j) as following:

$$V_{ij}(u, v) = G_{ij} \int e^{-i\left(\phi_i(l,m) - \phi_j(l,m)\right)} I(l, m) e^{-2\pi i\left(u_{ij}l + v_{ij}m\right)} dldm \tag{23.9}$$

where $G_{ij}(t)$ is "instrumental gain," and $I(l, m)$ is the brightness distribution of the sky. The phase of a point source at sky position (l, m) for an antenna i can be expressed as $\phi_i(l, m)$, which is the effect of the ionosphere by simply adding an excess path length. Nonetheless, the ionospheric disturbances' scale size is approximately a few hundred kilometers. Consequently, in cases where the field of view (FoV) is small, $\phi_i(l, m)$ is often constant throughout the primary beam.

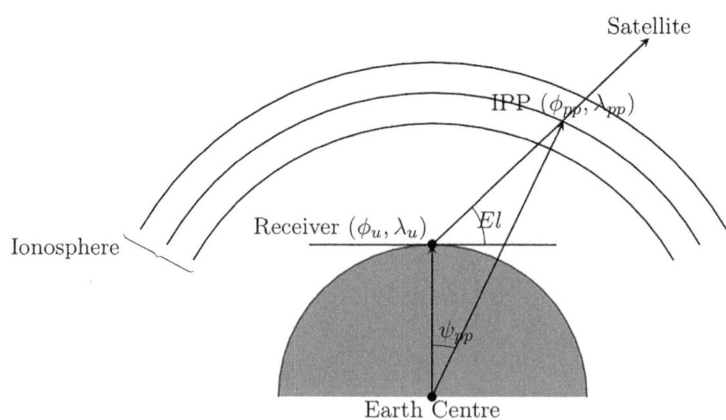

FIGURE 23.1 Geometry for Ionospheric Pierce Points.

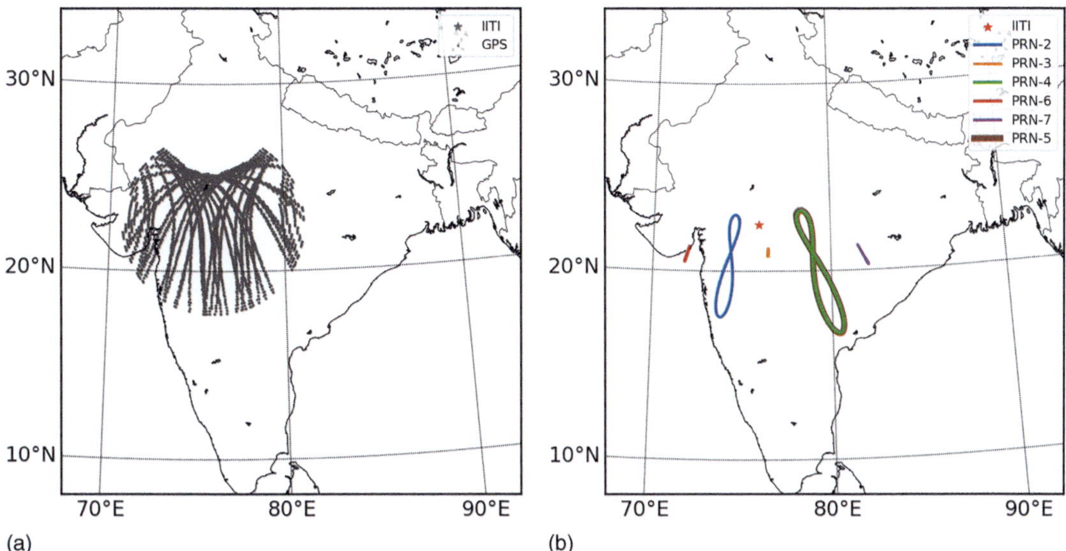

FIGURE 23.2 The figure represents the observational sites used in the work. The location of the receiver is indicated by the red stars on each map. The first map shows the location of the ionospheric pierce points, and the second map's eight-shaped plots represent the locations of the ionospheric pierce points of NavIC satellites based on the receiver positions for Indore (Ayyagari et al., 2022).

A constant 'n' in time and space will cause a constant phase error in the measurement, resulting in a constant spatial shift of the observed sources compared to the true sky. However, when 'n' strongly depends on space and time, things get more complicated.

The GMRT (Swarup et al., 1991) is among the world's largest and fully operational sensitive telescopes that work at low radio frequencies. The GMRT has 30 dishes, each 45 m diameter, and these antennas span over 25 km, providing a total collecting area of about 30,000 m^2 at meter wavelengths, with a moderately good angular resolution (~arcsec). Out of 30 antennas, 14 antennas are randomly distributed in a central square, which is 1.4×1.4 km^2 in extent. The other 16 antennas are placed along three arms, each about 14 km long, similar to a 'Y'-shaped configuration. The array layout is shown in Figure 23.3.

A cosmic radio source (3C68.2) was scanned for nearly nine hours to observe the ionosphere with GMRT. The observations were taken simultaneously on the night of August 5 and 6, 2012, at 235 and 610 MHz. The observations were divided into "scans" (blocks of time), each lasting roughly an hour, with the exception of scan number four, which was a nearly five-hour continuous observation period from midnight to post-sunrise with 0.5-second temporal sampling. Furthermore, a moderate level of solar activity (F10.7 = 137.9 SFU; 1 SFU = 10^{-22} $Wm^{-2}Hz^{-1}$) and a moderate level of geomagnetic activity (Kp index ~1-3) were both present throughout the observation.

23.3 GNSS OBSERVATIONS

23.3.1 TEC Derived from NavIC vs GPS

Based on the IPP locations of NavIC satellites at every time instant, the IPP values of GPS are considered in a grid of $1° \times 1°$ surrounding the NavIC ray path. Hence, for each time instant corresponding to a TEC value estimated by NavIC, few TEC values estimated from GPS satellites have been observed. Here, the assumption is that the ionosphere is invariant (Paul, Chakraborty, Das, DasGupta, & Mitra, 2005) in a $1° \times 1°$ grid, and the same ionosphere is sampled by NavIC and GPS. Of course, this assumption may not always be true in a strict sense; however, reducing the grid

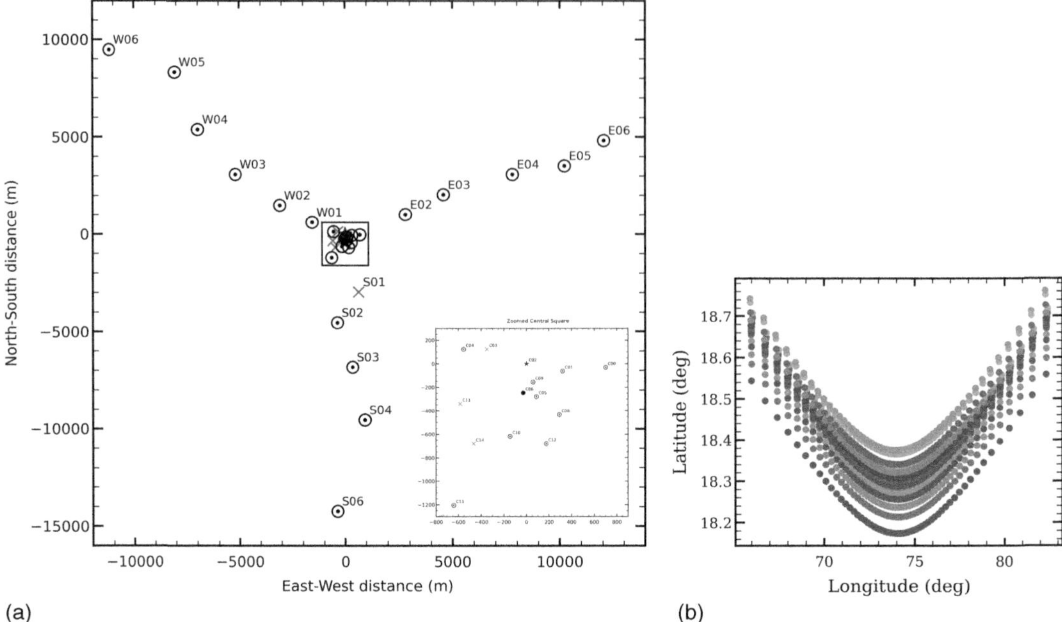

FIGURE 23.3 Left Panel: shows the layout of the full GMRT array during the astronomical source (3C68.2) observation run; box inside left panel shows the zoomed-in configuration of the central box from the left panel. Blue star ("C02") is the array center; the reference antenna ("C06") is highlighted in the black circle. The red crosses represent malfunctioning antennas during the observation. Right Panel: shows the IPP of the radio source for each of the antennas (Mangla & Datta, 2022).

size further for the present purpose will be impractical due to the limited availability of GPS ray paths over any location. The VTEC values estimated by these two navigation systems are then compared for a period of one year, starting from September 1, 2017, to September 30, 2018.

To get a broader view of the estimation process of VTEC from these constellations of satellites under varied ionospheric conditions, the total period is divided into two parts, namely quiet and disturbed days based on the K_p index. K_p index indicates the disturbances in the horizontal component of the Earth's magnetic field in the range from 0 to 9. K_p index value up to 4 signifies a calm period, and 5 or more indicates a geomagnetic storm After synchronizing the corrected VTEC estimates from both receivers based on the same instant in a $1° \times 1°$ grid, the NavIC VTEC is plotted against GPS VTEC estimates in Figure 23.4 for the whole period, quiet and disturbed period, respectively.

23.3.2 Amplitude Scintillations

Scintillation properties were checked using GPS and NavIC receiver data located in central India at IIT Indore. Scintillation is particularly abrupt fluctuations in the amplitude or phase of the signal. In the case of Amplitude scintillation, it is measured as S_4. Total scintillation index S_4 (Yeh & Liu, 1982) can be defined as the fractional variance of received signal intensity-

$$S_4^2 = \frac{\left\langle \left(A^2 - \langle A^2 \rangle \right)^2 \right\rangle}{\langle A^2 \rangle^2} = \frac{\langle I \rangle^2 - \langle I \rangle^2}{\langle I \rangle^2}, \tag{23.10}$$

where the Intensity $I \propto$ Power $P \propto A^2$, $I = 10^{0.1 \times C/No}$. Here, A is the amplitude of the received signal, C/No is the carrier-to-noise ratio of the signal, and $\langle \cdot \rangle$ represents the average values for the period of time. Here, GNSS (GPS) data was averaged for each 1 minute, and S_4 was calculated.

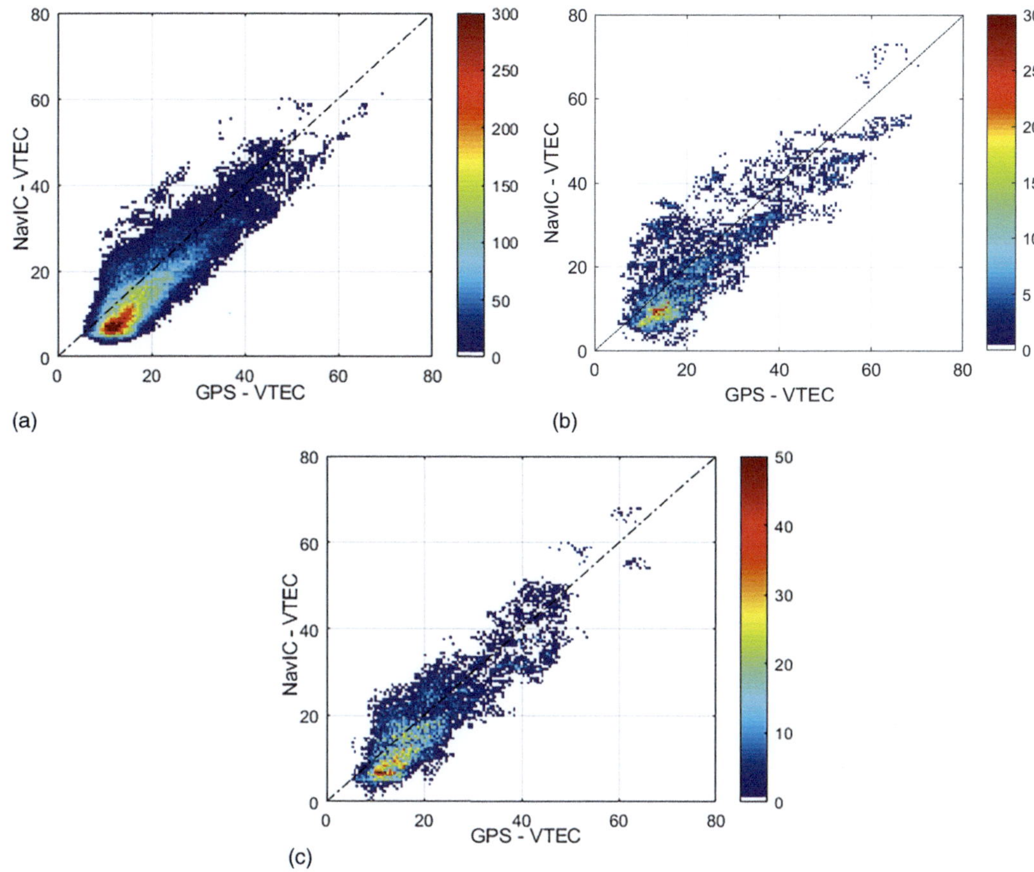

FIGURE 23.4 The scatter plot between NavIC and GPS-derived VTEC. VTEC estimates after making diurnal minimum value corrections with the removal of anomalous data. The color bar indicates the number of data points, (a) for a period of one year, (b) for the quiet days, (c) for the disturbed days (Ayyagari et al., 2020).

The S_4 index calculated above has ambient noise. To exclude the effect of ambient noise, we calculate *corrected* S_4, i.e., S_{4c}, as follows (Ayyagari et al., 2022; Van Dierendonck, Klobuchar, & Hua, 1993):

$$S_{4C}^2 = \begin{cases} S_4^2 - S_{AN}^2, & \text{if } S_4^2 - S_{AN}^2 \geq 0 \\ 0, & \text{if } S_4^2 - S_{AN}^2 < 0. \end{cases} \tag{23.11}$$

The ambient thermal noise can be calculated as follows:

$$S_{AN} = \sqrt{\frac{100}{C/No}\left[1 + \frac{500}{19C/No}\right]}. \tag{23.12}$$

The ionospheric scintillations can also be classified based on their strength or severity. The classification of ionospheric scintillations based on strength or severity can vary depending on the application or system of interest. However, here is a general classification scheme (Ayyagari et al., 2022; Beniguel & Hamel, 2011; "Septentrio PolaRx5S User Manual," n.d.) based on the amplitude scintillation index (S_4) (Table 23.1).

TABLE 23.1

Severity of scintillations as a measure of scintillation index (S_4)

Scintillation Class	S_4
Strong	$S_4 \geq 0.6$
Medium	$0.6 > S_4 \geq 0.3$
Weak	$S_4 < 0.3$

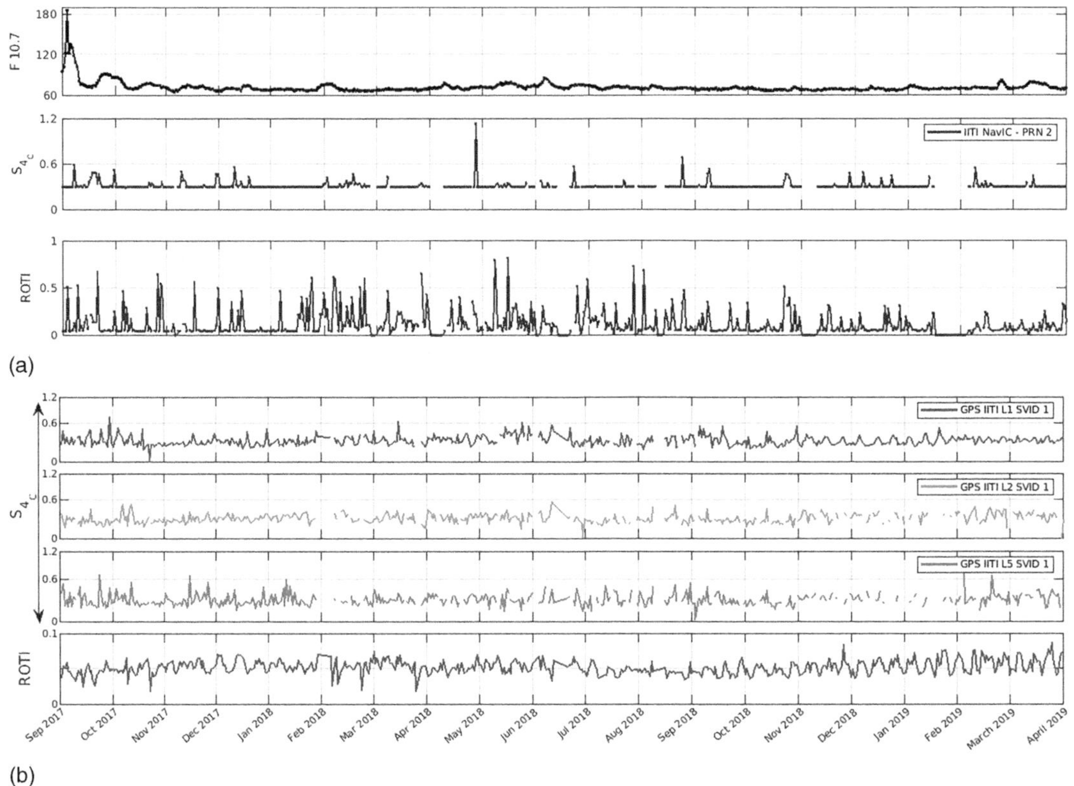

(a)

(b)

FIGURE 23.5 The fluctuation of F10.7 along with NavIC's PRN 2 and GPS SVID 1 recorded values of peak S_{4C} and ROTI from the Indore station for the accessible L band frequencies during the September 2017 through April 2019 analysis period (Ayyagari et al., 2022).

23.3.2.1 A Typical Scintillation Event

This subsection describes a typical scintillation event observed on September 08, 2017. The S_{4C}, the ROT (dTEC/dt), the ROTI, and the scattering coefficients (Goswami, Paul, & Paul, 2017) (see A3) are plotted as a function of LT (h) in Figure 23.6 for NavIC and GPS observations over Indore, respectively (Figure 23.5).

The dominant mechanism, in the onset of amplitude scintillations, is the Pre-reversal Enhancement generated as a result of the enhancement in the vertical **E×B** drift due to the eastward electric field at the sunset terminator (where the terminator implies a moving boundary between the day side and night side regions on the Earth) generating EPBs and ESFs around this time (Fejer, Scherliess, & de

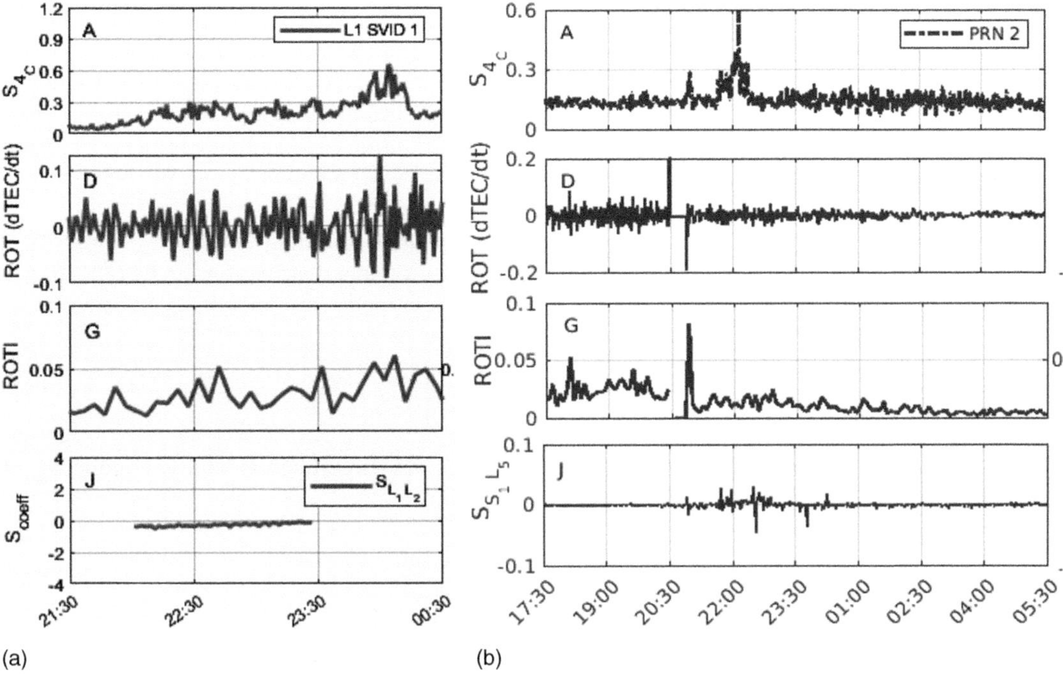

FIGURE 23.6 The S_{4C}, ROT, ROTI, and S_{coeff} variations as observed from Indore station during the night of September 8–9, 2017, for GPS SVID 1 for L1 frequency and for NavIC's PRN 2 (Ayyagari et al., 2022).

Paula, 1999; Ghosh, Otsuka, Mani, & Shinagawa, 2020; Woodman, 1970). Figure 23.6 depicts the variation of scattering coefficients (S_{coeff}) from two frequencies L_1 and L_2 based on the deviations estimated from each of the frequencies of GPS. A clear deviation from the mean value is observed, supporting the fact that signal strength is affected during S_{4C} events. Similarly, Figure 23.6 shows the onset times observed for NavIC's PRNs 2 over Indore. The onset times were observed to be around 20:30 LT (h), and peak values of S_{4C} at this location vary between 22:00 and 23:30 LT (h) for the PRNs 2 NavIC. The oscillating values observed in the $S_{S1}\,L_5$ match with the fluctuations of S_{4C}.

23.3.3 Characteristics of Amplitude Scintillation

Previous studies (E. J. Fremouw, Livingston, & Miller, 1980; E. J. Fremouw et al., 1978) found evidence to evaluate statistical models to characterize amplitude scintillations at low latitudes. Ayyagari et al. (2022) and references therein offer two statistical models for signal fading: Nakagami-m distribution and $\alpha - \mu$ distribution. The model is described by the coefficients α and μ of the normalized amplitude envelope of the received signal (r), where α is the modulus of the sum of the multipath components and μ is the number of multipath components and the $\alpha - \mu$ probability density function, assuming that the average signal power (or intensity) r^2 is equal to 1 for $E[R^2] = 1$ is defined as:

$$f(r) = \frac{\alpha r^{\alpha\mu-1}}{\zeta^{\alpha\mu/2}\Gamma(\mu)}\exp\left(-\frac{r^\alpha}{\zeta^{\alpha/2}}\right), \tag{23.13}$$

Here

$$\zeta = \frac{\Gamma(\mu)}{\Gamma(\mu+2/\alpha)} \tag{23.14}$$

$\Gamma(.)$ is the gamma function. (Yacoub, 2007) used this model to describe the mobile communication channel, assuming that the received signal is the consequence of a cluster of multipath waves propagating in a heterogeneous environment. The most interesting feature of this model is that it can simulate a number of different scenarios depending on the values of its variables; for instance, the $\alpha - \mu$ distribution can be transformed into a Nakagami-m distribution ($\alpha = 2$ and $\mu = $ m, where m is the single parameter of the distribution), a Rayleigh distribution ($\alpha = 2$ and $\mu = 1$), or a Weibull distribution (with two parameters α and $\gamma = \zeta^{1/2}$, for $\mu = 1$).

As described, S_4 index characterizes the strength of amplitude scintillation where the intensity of received signal $I = |r^2|$. The Nakagami-m parameter in the Nakagami distribution can be related to S_4 index by the following notation where $m = 1 / S_4^2$ and the relation to estimate this is given below:

$$m = \frac{E^2\left(r^2\right)}{E\left(r^4\right) - E^2\left(r^2\right)} \tag{23.15}$$

The α and μ coefficients can be calculated based on an experimental study conducted by (Yacoub, 2007).

$$\frac{E^2\left(r^\beta\right)}{E\left(r^{2\beta}\right) - E^2\left(r^\beta\right)} = \frac{\Gamma^2\left(\mu + \beta / \alpha\right)}{\Gamma\left(\mu\right)\Gamma\left(\mu + 2\beta / \alpha\right) - \Gamma^2\left(\mu + \beta / \alpha\right)} \tag{23.16}$$

The left-hand side of the equation (23.16) can be derived from field data for randomly chosen values of the parameter β, which provides the order of the system of r to be determined. In fact, when the left-hand side of the equation (23.16) for $\beta = 2$ is compared to the right-hand side of the equation, the following results are obtained:

$$S_4^2 = \frac{\Gamma\left(\mu\right)\Gamma\left(\mu + 4 / \alpha\right) - \Gamma^2\left(\mu + 2 / \alpha\right)}{\Gamma^2\left(\mu + 2 / \alpha\right)} \tag{23.17}$$

If $\alpha = 2$ in the above equation (23.17) and the properties of the gamma function are used, the value of the left-hand side of the same equation implies $1/\mu = 1/m$, which is the condition for the Nakagami-m distribution (de Oliveira Moraes et al., 2018).

The input discusses the analysis of fading occurrences in signal intensity for different frequencies using empirical data. In Figure 23.7, values of amplitude scintillation peaks are marked to calculate the fading sample. The corresponding values of $\alpha - \mu$ and Nakagami-m distribution for the scintillation occurrences are also shown. The severity of fading occurrences increases with higher values of S_{4C}, and the α parameter increases with the increase in S_{4C}. Understanding the statistics of these fades is important for developing signal-processing techniques and positioning algorithms to mitigate their effects. The observations presented will be used for further studies on positional accuracy using NavIC under different ionospheric conditions.

23.3.3.1　Phase Scintillation: Phase Screen Model

Equatorial ionospheric irregularities have been studied in the past and have produced interesting insights about ionospheric physics and processes. Equatorial Plasma Bubbles (EPBs) are the post-sunset plasma instability in the equatorial ionosphere, generating irregularities and large-scale depletion in the density of the electrons. When radio waves propagate through these irregularities, they experience scattering and diffraction, causing random fluctuations or scintillations of the Very High Frequency (VHF) signal amplitude, phase, polarization, and propagation direction. These scintillation-producing irregularities are generally present in and around the F-layer of the ionosphere, where the plasma density is maximum. The primary mechanism responsible for generating the EPBs, with scale sizes from a few hundred meters to several kilometers, is the Rayleigh-Taylor (R-T) instability (Carrano, Groves, & Caton, 2012b; Carrano, Groves, Caton, Rino, & Straus, 2011;

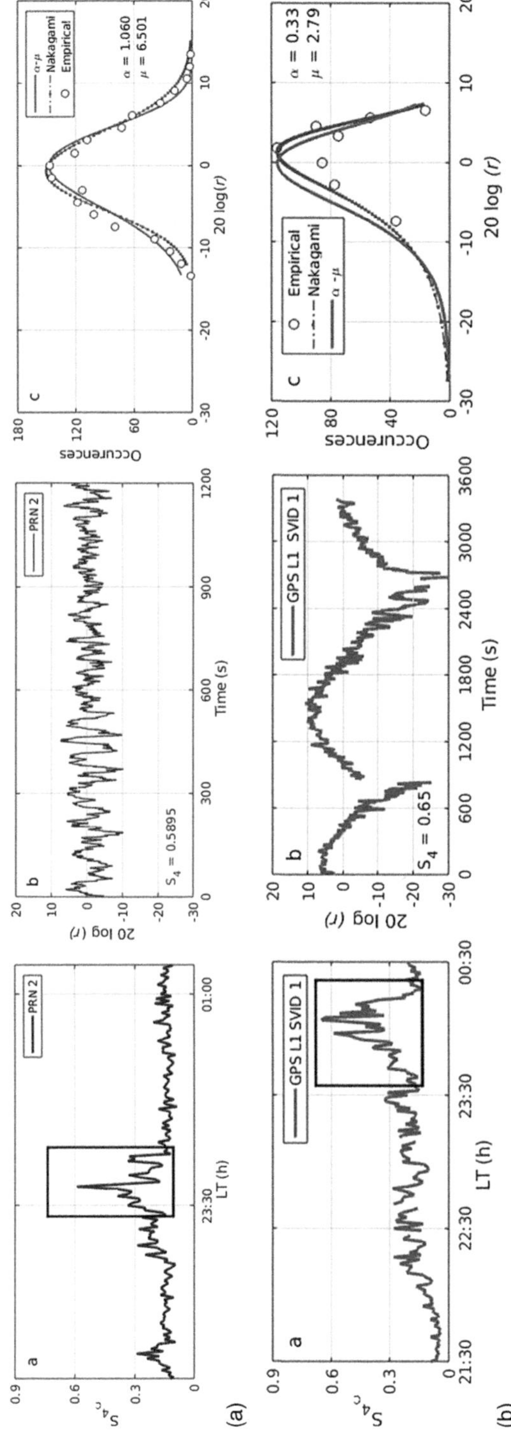

FIGURE 23.7 Scintillation event on September 08, 2017, as observed by NavIC's PRN 2 with L_5 signal and GPS SVID 1 L_1 signal over Indore. Column a shows variations of S_{4c} for a portion of the day. The black rectangular box denotes the epochs when scintillation was detected. Column b shows fading signal intensity for the period marked with a rectangular box in column a. Column c shows the calculated occurrences along with the $\alpha - \mu$ distributions and the Nakagami-m distribution (Ayyagari et al., 2022).

Carrano, Valladares, & Groves, 2012a; Priyadarshi, 2015). Scintillations are frequent when solar activity is high, and they occur near the magnetic equator in the post-sunset to midnight sector (Basu, Basu, Weber, & Coley, 1988b; Basu, MacKenzie, & Basu, 1988a). As these scintillations are demonstrations of Space weather effects that affect the performance of space-based navigation systems (de Oliveira Moraes et al., 2017), it becomes crucial to understand the occurrence of scintillation and its aftereffects.

Since the early 1950s and following the works by (Booker, Ratcliffe, Shinn, & Bragg,1950; Hewish & Bragg, 1951), characterization of ionospheric scintillation has been performed by several researchers who developed numerous phase screen theories. Their fundamental approach was the use of wave propagation theory in random media to obtain the exiting wave parameters and further solve the problem using the theory of Fresnel diffraction that involves radio wave propagation between the ionosphere and the Earth, which finally would replace the ionosphere by an equivalent random phase screen (Taylor & Infosino, 1976). They considered the ionospheric irregularities as phase objects with the ground pattern produced by the propagating waves through this phase screen derived from the Fresnel diffraction theory (Bhattacharyya, 1999).

Next, we show a case study related to ionospheric studies near the Equatorial Ionization Anomaly (EIA) using the Navigation with the Indian Constellation (NavIC). We have characterized the ionospheric irregularities in terms of the Power Spectral Density (PSD) at different dynamical frequencies (f). The formalism is similar to what was suggested by (Rino, 1979a, 1979b; Strangeways, 2009) using the phase screen modeling of the ionosphere. Here, we have used data from the same NavIC receiver.

According to the NOAA, a G1 (Kp= 5, minor) level geomagnetic storm was observed on December 4, 2017, due to the passage of a High-Speed Solar Wind (HSSW) stream around the Earth, arising from a positive polarity CH on the Sun. The minimum Dst index showed -45 nT at 22:00 UT. Figure 23.8 shows the C/No variation (dB-Hz) for the entire day of December 4, 2017, as observed by the L5 signal of NavIC. Drops in the C/No can be observed over multiple time stamps throughout the day by all the PRNs. It is to be noted that the C/No values had dropped to zero in these time stamps and that they are designated in the figure as drop-downs to 30 dB-Hz. However,

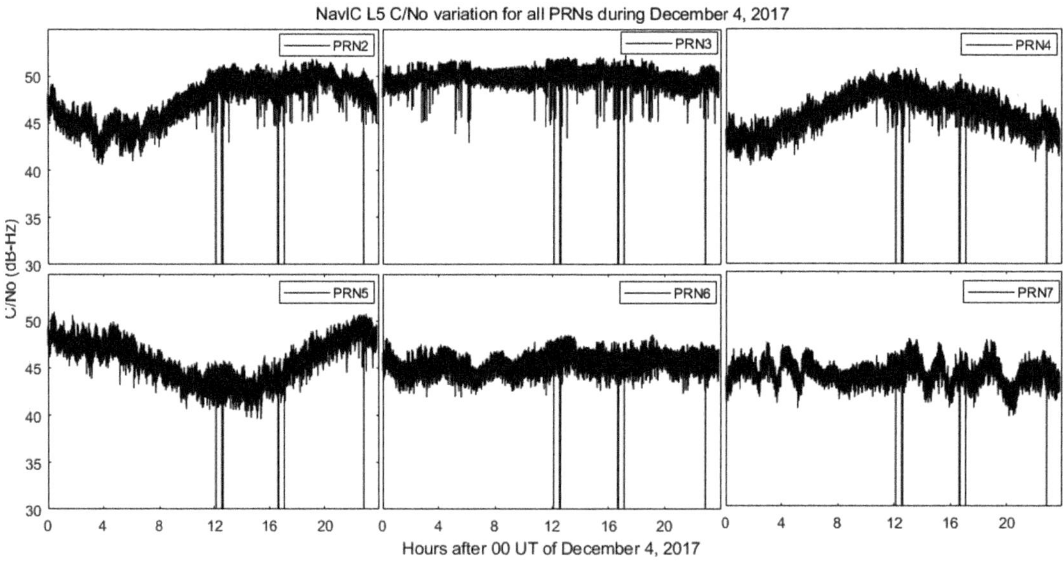

FIGURE 23.8 The C/No (dB-Hz) variation during the entire day of December 4, 2017, as observed by the L_5 signal of NavIC satellite PRNs 2-7. The C/No values that had dropped to zero are designated here as drops to 30 dB-Hz (Chakraborty & Datta, 2021).

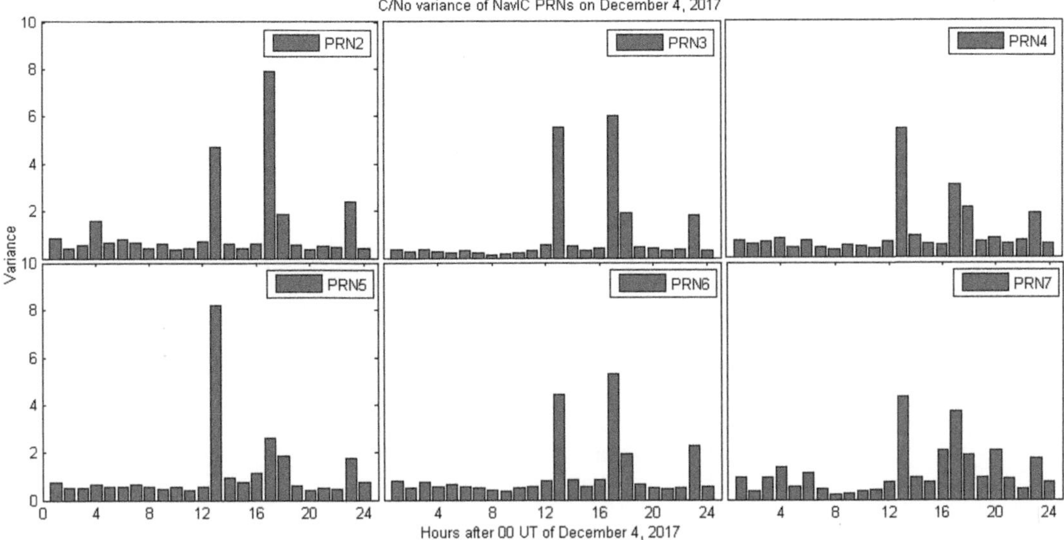

FIGURE 23.9 The hourly binned variance plots of C/No for all PRNs of NavIC on December 4, 2017. Most significant rises in variance values as observed from PRN 2 (16-17 UT bin) and PRN 5 (12-13 UT bin) (Chakraborty & Datta, 2021).

to verify whether the C/No drops are significant, Figure 23.9 shows the hourly binned variance plots of all the NavIC PRNs. The bin of 16-17 UT for PRN 2 and that of 12-13 UT for PRN 5 show the most significant rise and, hence, maximum variation among all the bins of the day.

Following the formalism in Rino (1979a, 1979b); Strangeways (2009) and in its simplified form, the PSD is given by:

$$P(f) = \alpha f^{-\beta} \tag{23.18}$$

where α is the spectral strength and β is the spectral slope.

Figure 23.10 depicts the PSD variation from one of the PRNs (PRN 5), which shows a power-law variation in the PSD and thus is taken into consideration for the calculation of the β, which is found to be 0.46 ± 0.04, showing the occurrence of weak scintillation detected by NavIC. To summarize, this initial study has been performed to demonstrate the capability of NavIC for the detection of ionospheric irregularities using the phase screen theory and the corresponding PSD analysis approach.

23.4 INTERFEROMETRIC OBSERVATIONS

Using the antenna-based method (Mangla et al., 2023) [and references therein], one observes a single bright source at the phase center and measures the varying phase on each baseline (a baseline refers to the distance between a pair of antennas) as a function of time.

From equation (23.3), one can infer that the difference in TEC along the LoS to the two antennas is directly proportional to the measured ionospheric-induced phase. This method will be referred to as the *antenna-based method*, which can study the temporal variation in differential TEC (δTEC) corresponding to the projected array onto the ionosphere. Several studies have used this method to probe the ionosphere.

To estimate the additional phase introduced due to the radio signal passing through the Earth's ionosphere, we follow the procedure outlined in (Helmboldt, Lazio, Intema, & Dymond, 2012a).

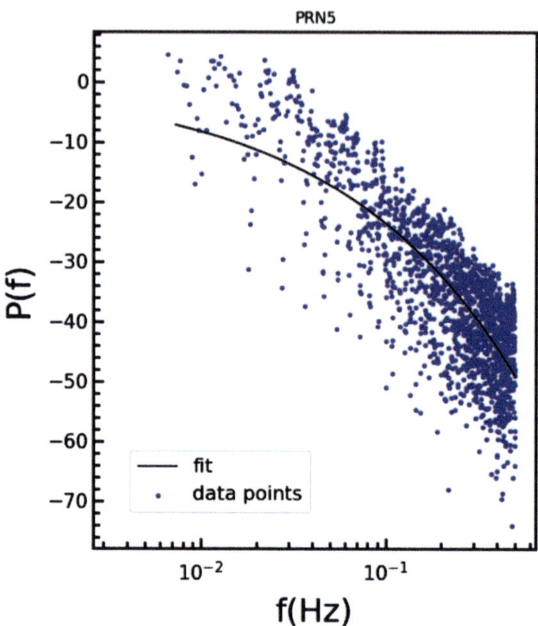

FIGURE 23.10   The PSD variation with the least square fit (black solid line) and the smoothed data (red solid line) corresponding to the C/No variation designated in horizontal bars in Figure 23.9 as observed from PRN 5 on December 4, 2017 (Chakraborty & Datta, 2021).

After calculating the phase terms from CASA, several steps are performed on phase data to get the ionospheric information. The phase terms contain other effects and the ionospheric contribution, which is highest in low-frequency regimes.

The phase difference between the two antenna elements (Helmboldt et al., 2012a; Intema et al., 2009) is given by

$$\Delta\phi = \Delta\phi_{\text{ion}} + \Delta\phi_{\text{other}} \tag{23.19}$$

where $\Delta\phi_{\text{ion}}$ represents the difference in the ionospheric phases of the two antennas along the line of sight given by equation (23.3), $\Delta\phi_{\text{other}}$ denotes the combined differences in the instrumental effects between the two antennas, 2π ambiguities and the phase difference from the observed source structure.

The $\Delta\phi_{\text{other}}$ phase delay effect is removed or reduced as the method mentioned in (Mangla & Datta, 2022). From equation (23.3), one can infer that the difference in TEC along the LoS to the two antennas is directly proportional to the measured ionospheric induced phase. This method will be referred to as the *antenna-based method*, which can study the temporal variation in differential TEC (δTEC) corresponding to the projected array onto the ionosphere. Several studies have used this method to probe the ionosphere.

The resulting δTEC values are plotted for each antenna in the central square in Figure 23.11, the arms in Figure 23.12, along with the Median Absolute Deviation (MAD) values (plotted in red) to demonstrate the relative accuracy to which δTEC is estimated. The uncertainty in δTEC expressed by the MAD calculations is of the order of 1×10^{-3} TECU, showing the remarkable ability of the GMRT to detect small fluctuations while measuring TEC. These results suggest that the variation in δTEC for the southern arm is more than the other two arms and the central square configuration.

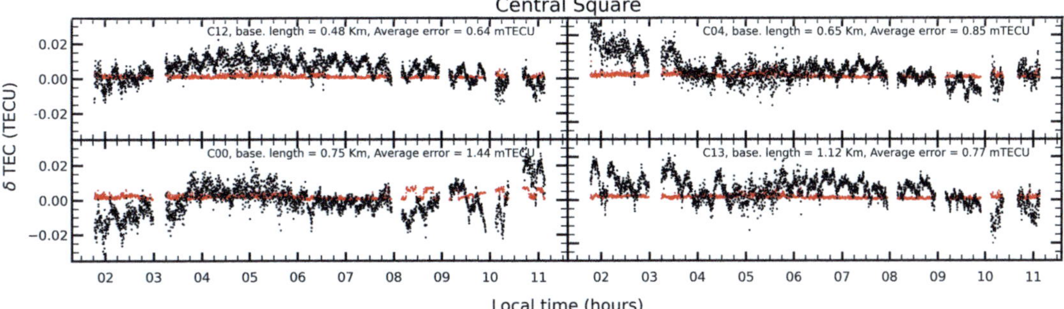

FIGURE 23.11 Differential TEC (δTEC) is plotted with respect to time for each antenna in the compact central square of the GMRT. This δTEC was estimated along the antenna's line of sight with respect to the reference antenna line of sight. In each panel, the estimated uncertainty is also plotted in red (Mangla & Datta, 2022).

Differential TEC along three arms (Figure 23.12) shows a different pattern, which may be because a dominant wave is propagating with smaller-scale wave(s) or several waves are propagating in different directions. Spectral analysis of such patterns will be demonstrated in our subsequent work. Further, the larger amplitude and longer period fluctuation (known behavior of TIDs [Hernandez-Pajares, Juan, & Sanz, n.d.]) are present throughout the observation in the southern arm, but the same characteristic fluctuations are present only during the nighttime (before dawn) in the remaining two arms. Also, antennas located at different distances along the same direction or the same azimuth describe that δTEC appears to be proportional to the baseline length. Measuring the gradient is, therefore, necessary to understand the behavior of any ionospheric phenomena.

The two-dimensional TEC surface over the array can be computed using the second-order two-dimensional Taylor series (maximum baseline for the array is smaller than transient ionospheric waves), which has the following form:

$$\text{TEC} = p_0\, x + p_1\, y + p_2\, x^2 + p_3\, y^2 + p_4\, xy + p_5 \tag{23.20}$$

where x and y are antenna positions along with north-south and east-west directions. p_0 to p_5 are the polynomial coefficients. By calculating the difference of δTEC between antenna pairs (120 baselines) at each time step, one can increase the accuracy of these polynomial coefficients/parameters. Thus, equation (23.20) transforms into the following form:

$$\delta\text{TEC}_i - \delta\text{TEC}_j = p_0\left(x_i - x_j\right) + p_1\left(y_i - y_j\right) + p_2\left(x_i^2 - x_j^2\right)$$
$$+ p_3\left(y_i^2 - y_j^2\right) + p_4\left(x_i y_i - x_j y_j\right) \tag{23.21}$$

where subscripts i and j correspond to different antenna pairs (where $i = j$). It is important to note that the fitting is done independently at each time step, thus conserving the temporal and spatial TEC variation over the array. The obtained polynomial coefficients are shown in Figure 23.13, along with the standard error for each coefficient. One can easily notice that the amplitude for the p_1 coefficient (along the east-west direction) is significantly high during the local nighttime. The same phenomenon is observed for p_0 coefficient (along north-south direction) during the sunrise hour (around 6:00 a.m.), which is a known behavior of MSTIDs, commonly detected around sunrise and sunset (Hernandez-Pajares et al., n.d.). Variations in higher-order coefficients (p_2 to p_4) are more significant during the nighttime, suggesting unanticipated ionospheric changes in and around the EIA region during these local times.

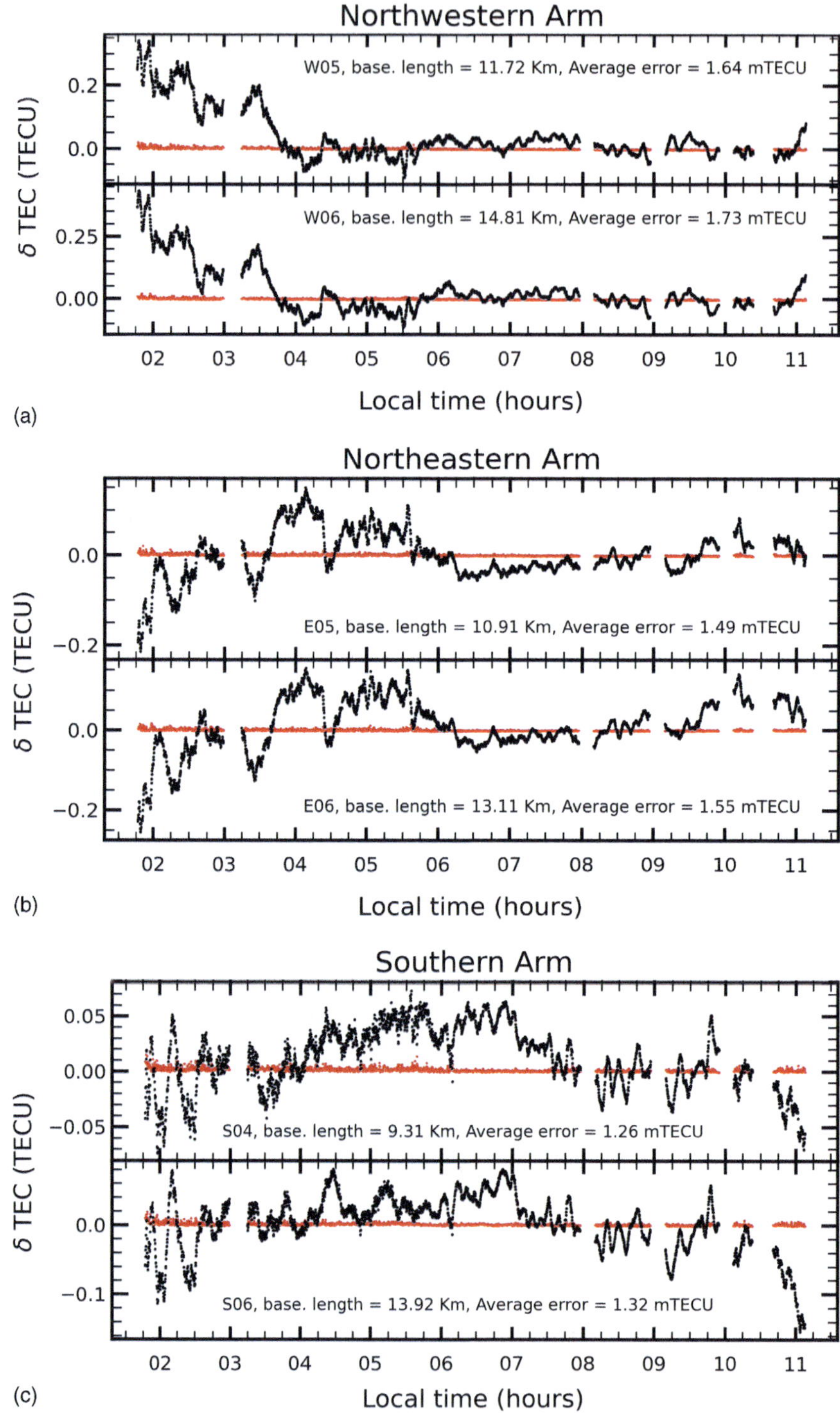

FIGURE 23.12 Same as Figure 23.11 but along the arms of the GMRT (Mangla & Datta, 2022).

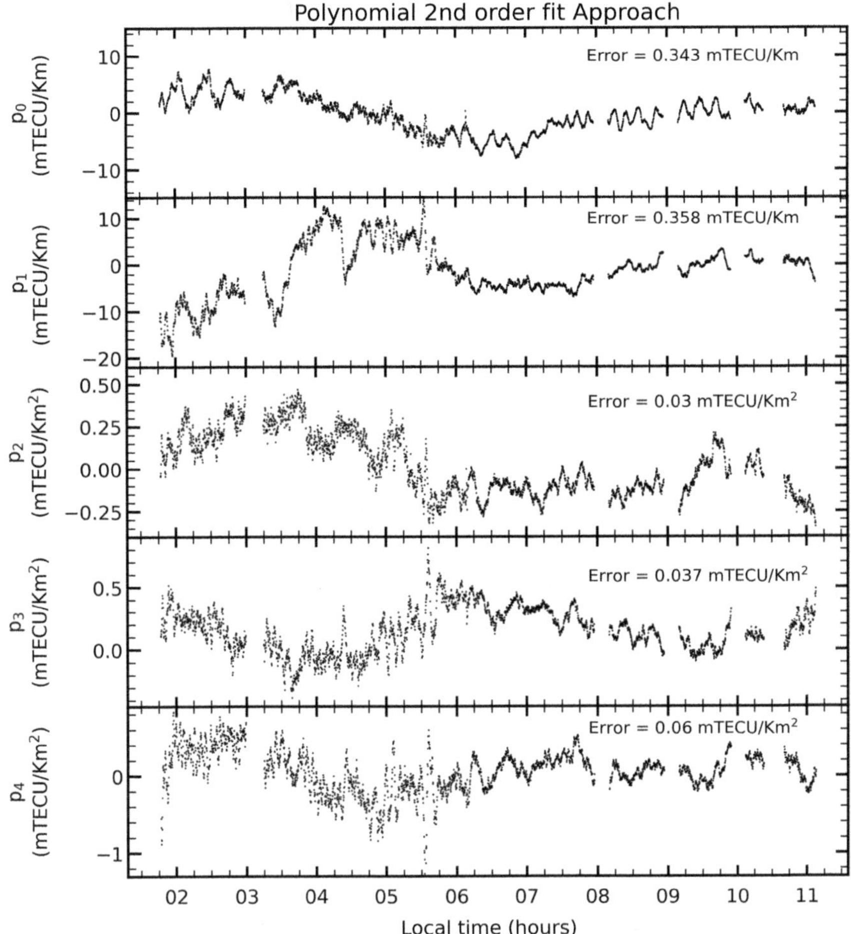

FIGURE 23.13 The fitted coefficients using the second-order polynomial equation (23.21) as a function of local time. The coefficients were estimated using the difference between δTEC for each of 325 antenna pairs at each time step. The estimated standard-error for each coefficient is mentioned in the respective panel (Mangla & Datta, 2022).

23.4.1 Derived Wave Properties from GMRT Data

The input discusses the use of polynomial-based TEC gradient fits to compute power as a function of time and frequency at two separate frequency bands. This is done by performing discrete Fourier transforms on the time series of polynomial coefficients. To enable the use of DFTs, missing time steps are filled with zeros through a process called "zero-padding." The DFTs are performed with a sliding window of 1 hour for frequencies up to $10hr^{-1}$. The power is computed for a period of 30 minutes before and after the observing run. The DFTs are used to calculate the power for the two frequency bands, and significant wave detections above the background are observed. To isolate these detections, a mask is produced by determining the median and median absolute deviation within elliptical "annuli" around each pixel. A pixel exceeding two times the MAD over the median for its annulus is considered a detection above the background.

Based on the highly correlated coefficient observed in Figure 23.14(d), three events have been selected, indicated by A, B, and C in Figure 23.14(a). Calculated properties of the events are presented in Table 23.2. Events B and C show agreement in wave speed and azimuth angle, while event A only partially agrees. The wave speed in the two bands for event A is approximately the same, but

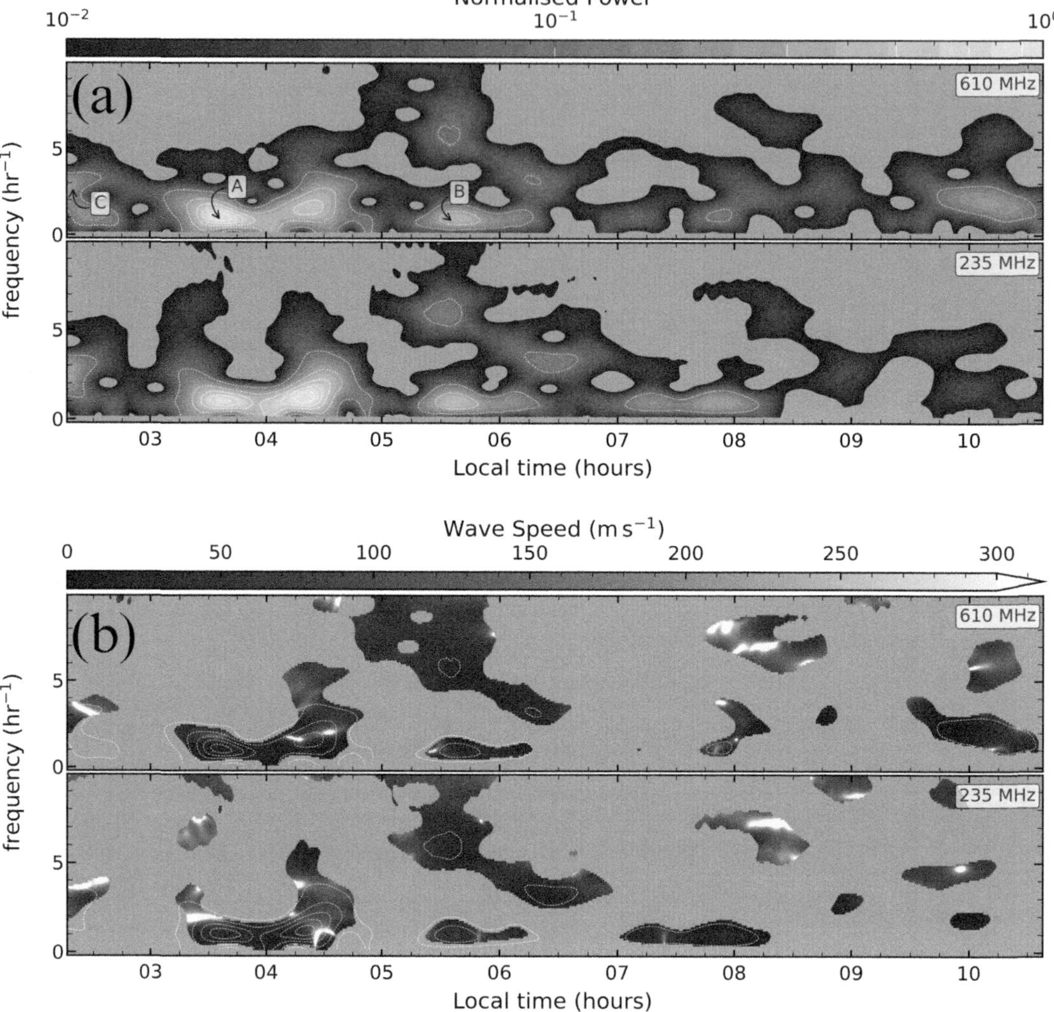

FIGURE 23.14 Spectral properties at 235 and 610 MHz. In both panels (a and b), the white contours are of spectral power. (a) The normalized power in the direction of the wave number vector as a function of temporal frequency (y-axis) and local time (x-axis). Similar structures are present in both bands; in addition, smaller structures are visible in 235 MHz band compared to 610 MHz band. (b) The wave speed of each Fourier mode with power significantly larger than background noise (Mangla & Datta, 2023).

TABLE 23.2

Wave parameters derived from GMRT measurements at 235 and 610 MHz (Mangla & Datta, 2023)

Events (m/s)	Period (km)	Wave speed		Wavelength	
		610 MHz	235 MHz	610 MHz	235 MHz
A	1 hr 10 min	46.55 ± 5.16	48.02 ± 6.54	195.51 ± 21.67	201.68 ± 27.48
B	1 hr 33 min	54.91 ± 3.45	56.78 ± 5.23	263.57 ± 16.56	272.54 ± 25.10
C	24 min	95.47 ± 9.20	106.20 ± 8.02	137.19 ± 13.25	152.93 ± 11.55

the propagation direction is out of phase by approximately $180°$, which may be due to scintillation observed in the higher order terms of polynomial fitting at 235 MHz data (Mangla et al., 2023).

23.5 DISCUSSION

The reliability of NavIC has been proven in studying the ionosphere (Ayyagari et al., 2020; Chakraborty, Datta, Ray, Ayyagari, & Paul, 2020a). The variation of S_{4C} registered from the GPS closely matches that of NavIC S_{4C}. During the post-sunset period S_{4C} index of NavIC PRNs 2, 5, and 6 provide clear signatures of the occurrence of ionospheric scintillation. The study shows a strong correlation between NavIC signals from L_5 and S_1. The amplitude distribution is statistically well-modeled as Nakagami-m and $\alpha - \mu$ distributions.

Analysis of GMRT data unveils its ability to measure fluctuations in TEC gradients at 1×10^{-3}TECu level with 1σ uncertainty. With a polynomial fitting approach, 2D TEC gradient surface is reconstructed over the array. The results closely match with GNSS observations by (Ayyagari et al., 2020; Chakraborty, Ray, Sur, Datta, & Paul, 2020b). The polynomial-based method is used previously by (Helmboldt et al., 2012a) to study the ionosphere structure with the VLA. The second-order polynomial fit approximated the curvature needed to recover much of the observed structure, but small-scale disturbances were missed during the nighttime. Our study shows that a second-order polynomial fit fails to recover structures for antennas that are beyond 6km from the array center. However, much of the observed larger structures are recovered by a third-order polynomial fit, but small-scale fluctuations are lost during the night.

Therefore, to approximate the curvature in the 2D TEC surface and study the ionosphere over the GMRT, an additional higher-order fit is needed, as it is located near the magnetic equator. The unique positional advantage of the GMRT unveils more about a number of structures (small-scale to large-scale) present, compared to VLA ($\sim34°N$), LOFAR ($\sim53°N$), and MWA ($\sim26°S$), even during low solar activity and a moderate amount of geomagnetic activity. In events A, B, and C, MSTIDs were observed. Event A shows MSTID with a wave speed of about $50ms^{-1}$; Event B was observed during dusk time. It has eastward waves along with many small waves.

The observed ionospheric structure has the period and wavelength of 1 hr 33 min and ~300 km, respectively, propagating with a speed between 50 and 60 ms^{-1}, which falls within the typical range of equatorial spread F (ESF). Event C was observed during nighttime with a northward moving MSTID having a period of 24 min and wave speed of about 100 ms^{-1}. This is a rare event in the northern hemisphere and is suggested to be caused by gravity waves. Similar events have been observed in the southern hemisphere, also attributed to gravity waves.

APPENDIX A CALCULATION OF ROT, ROTI AND SCATTERING COEFFICIENTS

The ROT and ROTI values (Pi, Mannucci, Lindqwister, & Ho, 1997) are calculated as given in equations below where in equation (A1) $STEC_{r+1}$ and $STEC_r$ are slant TEC at $r+1$ and r time epochs; Δt_r time interval; usually the unit of ROT is TECU/min and in equation (A2) where $\langle ROT \rangle$ denotes averaging ROT during N epochs.

$$ROT = \frac{STEC_{r+1} - STEC_r}{\Delta t_r}. \tag{A1}$$

$$ROTI = \sqrt{\langle ROT^2 \rangle - \langle ROT \rangle^2} \tag{A2}$$

A scattering coefficient (Goswami et al., 2017) across a pair of frequencies was later defined as the difference of C/No fluctuations normalized with respect to the amount of those fluctuations, based on these computed C/No deviations. The formula for scattering coefficient is given by A3, where a

is the *C/No* deviations calculated from one frequency and *b* is the *C/No* deviations calculated from the second frequency, respectively. Here, $S_{a,b}$ is a dimensionless quantity with values near zero, suggesting a good similarity between the signals. Here, the *C/No* deviations were determined by subtracting the moving averaged values over a 90-minute running time period from the instantaneous *C/No* calculation.

$$S_{a,b} = \left[\frac{a-b}{a+b} \right] \tag{A3}$$

APPENDIX B HIGHER ORDER POLYNOMIAL TO FIT TEC AND WAVE SPEED

The Fourier-based approach uses TEC gradient measurements obtained from the polynomial method to detect and analyze individual or groups of waves passing over the array. This method was previously utilized by (Helmboldt, Lazio, Intema, & Dymond, 2012b) to account for wavefront distortions and detect multiple wavefronts simultaneously.

The primary objective is to improve the array's spatial coverage by considering the source's apparent position and movements of ionospheric disturbances, which have different speeds, wavelengths, and directions. To estimate the properties of each structure, the TEC fluctuations are approximated as a combination of multiple oscillating modes, each having a specific form

$$\mathrm{TEC}(t) = \tau_\omega \exp\left[i\left(\omega t - k_x x - k_y y \right) \right] \tag{B1}$$

where $\tau\omega$ is the amplitude that varies perpendicular to wavefront distortions, ω is the oscillation frequency, and k_x and k_y are the wave number for position x (North-South) and y (East-West), respectively. For a single Fourier mode, the Fourier transforms of the partial derivatives with respect to x and y are:

$$D_x\left(\omega;x,y\right) = \left[\frac{\partial \tau_\omega}{\partial x} - ik_x\tau_\omega \right] \times \exp\left[-i\left(k_x x + k_y y \right) \right]$$

$$D_y\left(\omega;x,y\right) = \left[\frac{\partial \tau_\omega}{\partial y} - ik_y\tau_\omega \right] \times \exp\left[-i\left(k_x x + k_y y \right) \right] \tag{B2}$$

Mangla and Datta (2022) employed the polynomial series method to estimate TEC in the plane transverse to the line of sight. This method assumes that TEC can be represented by a two-dimensional, third-order polynomial, such that the difference in TEC between two antennas i and j can be represented as

$$\begin{aligned}
\Delta\mathrm{TEC}_{ij} = &\, p_0\left(x_i - x_j\right) + p_1\left(y_i - y_j\right) + p_2\left(x_i^2 - x_j^2\right) + p_3\left(y_i^2 - y_j^2\right) + p_4\left(x_i y_i - x_j y_j\right) \\
&+ p_5\left(x_i^3 - x_j^3\right) + p_6\left(y_i^3 - y_j^3\right) + p_7\left(x_i^2 y_i - x_j^2 y_j\right) + p_8\left(x_i y_i^2 - x_j y_j^2\right)
\end{aligned} \tag{B3}$$

where x and y are the antenna positions projected onto the transverse plane in the north-south and east-west directions, respectively. At any given time, this method has the capability to replicate most of the structure observed. Subsequently, the polynomial coefficients time series can then be Fourier transformed to decompose the transverse gradient in TEC into the Fourier modes (Helmboldt, Haiducek, & Clarke, 2020), such that

$$D_x\left(\omega;x,y\right) = P_0\left(\omega\right) + 2P_2\left(\omega\right)x + P_4\left(\omega\right)y + 3P_5\left(\omega\right)x^2 + 2P_7\left(\omega\right)xy + P_8\left(\omega\right)y^2$$

$$D_y\left(\omega;x,y\right) = P_1\left(\omega\right) + 2P_3\left(\omega\right)y + P_4\left(\omega\right)x + 3P_6\left(\omega\right)y^2 + P_7\left(\omega\right)x^2 + 2P_8\left(\omega\right)xy \tag{B4}$$

where D_x and D_y represent the Fourier transforms of the north-south and east-west components of the TEC gradient, respectively, while $P_n(\omega)$ corresponds to the Fourier transform of $p_n(t)$. By expanding equation (B2) using Taylor series and comparing the variables (x, y, x^2, y^2, xy) with equation (B4) for respective D_x and D_y, the wave number vector components (k_x and k_y) for each mode can be approximated as

$$k_x(\omega) \simeq -Im\left\{\frac{2P_2}{P_0}(\omega)\right\} \simeq -Im\left\{\frac{P_4}{P_1}(\omega)\right\}$$

$$k_y(\omega) \simeq -Im\left\{\frac{2P_3}{P_1}(\omega)\right\} \simeq -Im\left\{\frac{P_4}{P_0}(\omega)\right\} \tag{B5}$$

Since there are two estimators for both k_x and k_y, a weighted average is utilized, with the weights being $|P_0|^2$ and $|P_1|^2$. It is to be noted that in the above equation, P_5 to P_8 are not considered, as they will induce higher-order effects. The spectral power, wave speed, and azimuth (propagating wave direction) can be estimated respectively as

$$\mathcal{P} \equiv \left|\mathcal{F}\{\Delta TEC\}\right|^2 \tag{B6}$$

$$V \simeq \frac{\omega}{\sqrt{k_x^2 + k_y^2}} \tag{B7}$$

$$Az \simeq \tan^{-1}\left(\frac{k_y}{k_x}\right) \tag{B8}$$

REFERENCES

Arikan, F., Erol, C., & Arikan, O. (2003). Regularized estimation of vertical total electron content from global positioning system data. *Journal of Geophysical Research: Space Physics*, *108* (A12).

Arikan, F., Nayir, H., Sezen, U., & Arikan, O. (2008). Estimation of single station interfrequency receiver bias using gps-tec. *Radio Science*, *43* (04), 1–13.

Ayyagari, D., Chakraborty, S., Das, S., Shukla, A., Paul, A., & Datta, A. (2020, March). Performance of NavIC for studying the ionosphere at an EIA region in india. *Advances in Space Research*, *65* (6), 1544–1558. Retrieved from http://doi.org/10.1016/j.asr.2019.12.019

Ayyagari, D., Datta, A., & Chakraborty, S. (2022, October). Systematic study of ionospheric scintillation over the Indian low-latitudes during low solar activity conditions. *Advances in Space Research*, *70* (8), 2506–2521. Retrieved from http://doi.org/10.1016/j.asr.2022.07.026

Basu, S., Basu, S., Weber, E. J., & Coley, W. R. (1988b, July). Case study of polar cap scintillation modeling using DE 2 irregularity measurements at 800 km. *Radio Science*, *23* (4), 545–553. Retrieved from http://doi.org/10.1029/RS023i004p00545

Basu, S., Fougere, P. F., MacKenzie, E. M., Basu, S., & Costa, E. (1987, February). 250 MHz/GHz scintillation parameters in the equatorial, polar, and auroral environments. *IEEE Journal on Selected Areas in Communications*, *5*, 102–115.

Basu, S., MacKenzie, E., & Basu, S. (1988a, June). Ionospheric constraints on VHF/UHF communications links during solar maximum and minimum periods. *Radio Science*, *23* (3), 363–378.

Basu, S., & Whitney, H. E. (1983, April). The temporal structure of intensity scintillations near the magnetic equator. *Radio Science*, *18* (2), 263–271.

Beniguel, Y., & Hamel, P. (2011, October). A global ionosphere scintillation propagation model for equatorial regions. *Journal of Space Weather and Space Climate*, *1* (1), A04.

Bhattacharyya, A. (1999, January). Deterministic retrieval of ionospheric phase screen from amplitude scintillations. *Radio Science*, *34* (1), 229–240. Retrieved from http://doi.org/10.1029/98RS02351

Booker, H. G., Ratcliffe, J. A., Shinn, D. H., & Bragg, W. L. (1950). Diffraction from an irregular screen with applications to ionospheric problems. *Philosophical Transactions of the Royal Society of London. Series A, Mathematical and Physical Sciences*, *242* (856), 579–607. Retrieved from https://royalsocietypublishing.org/doi/abs/10.1098/rsta.1950.0011

Carrano, C. S., Groves, K. M., & Caton, R. G. (2012b, July). The effect of phase scintillations on the accuracy of phase screen simulation using deterministic screens derived from gps and altair measurements. *Radio Science, 47* (4). Retrieved from http://doi.org/10.1029/2011RS004958

Carrano, C. S., Groves, K. M., Caton, R. G., Rino, C. L., & Straus, P. R. (2011, July). Multiple phase screen modeling of ionospheric scintillation along radio occultation raypaths. *Radio Science, 46* (6). Retrieved from http://doi.org/10.1029/2010RS004591

Carrano, C. S., Valladares, C. E., & Groves, K. M. (2012a). Latitudinal and local time variation of ionospheric turbulence parameters during the conjugate point equatorial experiment in Brazil. *International Journal of Geophysics*, 2012, 1–16. Retrieved from http://doi.org/10.1155/2012/103963

Chakraborty, S., & Datta, A. (2021, August). Study of low-latitude ionospheric scintillation using navic. In *2021 XXXIVth General Assembly and Scientific Symposium of the International Union of Radio Science (URSI GASS)*. IEEE. Retrieved from http://doi.org/10.23919/URSIGASS51995.2021.9560192

Chakraborty, S., Datta, A., Ray, S., Ayyagari, D., & Paul, A. (2020a, August). Comparative studies of ionospheric models with GNSS and NavIC over the Indian longitudinal sector during geomagnetic activities. *Advances in Space Research, 66* (4), 895–910. Retrieved from http://doi.org/10.1016/j.asr.2020.04.047

Chakraborty, S., Ray, S., Sur, D., Datta, A., & Paul, A. (2020b, January). Effects of CME and CIR induced geomagnetic storms on low-latitude ionization over Indian longitudes in terms of neutral dynamics. *Advances in Space Research, 65* (1), 198–213. Retrieved from http://doi.org/10.1016/j.asr.2019.09.047

Cornwell, T., & Fomalont, E. B. (1999, January). Self-Calibration. In G. B. Taylor, C. L. Car Illi, & R. A. Perley (Eds.), *Synthesis Imaging in Radio Astronomy II* (Vol. 180, p. 187).

Datta-Barua, S., Walter, T., Blanch, J., & Enge, P. (2008, October). Bounding higher-order ionosphere errors for the dual-frequency GPS user. *Radio Science, 43* (5), RS5010.

Davies, K. (1990). *Ionospheric radio*. IET.

de Oliveira Moraes, A., Costa, E., Abdu, M. A., Rodrigues, F. S., de Paula, E. R., Oliveira, K., & Perrella, W. J. (2017, April). The variability of low-latitude ionospheric amplitude and phase scintillation detected by a triple-frequency GPS receiver. *Radio Science, 52* (4), 439–460. Retrieved from http://doi.org/10.1002/2016RS006165

de Oliveira Moraes, A., Muella, M. T., de Paula, E. R., de Oliveira, C. B., Terra, W. P., Perrella, W. J., & Meibach-Rosa, P. R. (2018, April). Statistical evaluation of glonass amplitude scintillation over low latitudes in the Brazilian territory. *Advances in Space Research, 61* (7), 1776–1789. Retrieved from http://doi.org/10.1016/j.asr.2017.09.032

Fejer, B. G., & Kelley, M. (1980). Ionospheric irregularities. *Reviews of Geophysics, 18* (2), 401–454.

Fejer, B. G., Scherliess, L., & de Paula, E. R. (1999, September). Effects of the vertical plasma drift velocity on the generation and evolution of equatorial spread *F*. *Journal of Geophysical Research: Space Physics, 104* (A9), 19859–19869. Retrieved from http://doi.org/10.1029/1999JA900271

Franke, S., & Liu, C. H. (1985). Modeling of equatorial multifrequency scintillation. *Radio Science, 20* (3), 403–415.

Fremouw, E., Livingston, R., & Miller, D. A. (1980, August). On the statistics of scintillating signals. *Journal of Atmospheric and Terrestrial Physics, 42* (8), 717–731. Retrieved from http://doi.org/10.1016/0021-9169(80)90055-0

Fremouw, E. J., Leadabrand, R. L., Livingston, R. C., Cousins, M. D., Rino, C. L., Fair, B. C., & Long, R. A. (1978, January). Early results from the DNA Wideband satellite experiment—Complex-signal scintillation. *Radio Science, 13* (1), 167–187. Retrieved from http://doi.org/10.1029/RS013i001p00167

Ghosh, P., Otsuka, Y., Mani, S., & Shinagawa, H. (2020, July). Day-to-day variation of pre-reversal enhancement in the equatorial ionosphere based on GAIA model simulations. *Earth, Planets and Space, 72* (1). Retrieved from http://doi.org/10.1186/s40623-020-01228-9

Goswami, S., Paul, K. S., & Paul, A. (2017, September). Assessment of gps multifrequency signal characteristics during periods of ionospheric scintillations from an anomaly crest location. *Radio Science, 52* (9), 1214–1222. Retrieved from http://doi.org/10.1002/2017RS006295

Helmboldt, J. F., Haiducek, J. D., & Clarke, T. E. (2020). The properties and origins of corotating plasmaspheric irregularities as revealed through a new tomographic technique. *Journal of Geophysical Research: Space Physics, 125* (3), e2019JA027483. Retrieved from https://doi.org/10.1029/2019JA027483

Helmboldt, J. F., Lazio, T. J. W., Intema, H. T., & Dymond, K. F. (2012a, February). High- precision measurements of ionospheric TEC gradients with the Very Large Array VHF system. *Radio Science, 47*, RS0K02.

Helmboldt, J. F., Lazio, T. J. W., Intema, H. T., & Dymond, K. F. (2012b, February). A new technique for spectral analysis of ionospheric TEC fluctuations observed with the Very Large Array VHF system: From QP echoes to MSTIDs. *Radio Science, 47*, RS0L02.

Hernandez-Pajares, M., Juan, J. M., & Sanz, J. (n.d.). Medium-scale traveling ionospheric disturbances affecting GPS measurements: Spatial and temporal analysis. *Journal of Geophysical Research: Space Physics*, *111* (A7). Retrieved from https://agupubs.onlinelibrary.wiley.com/doi/abs/10.1029/2005JA011474

Hewish, A., & Bragg, W. L. (1951). The diffraction of radio waves in passing through a phase-changing ionosphere. *Proceedings of the Royal Society of London. Series A. Mathematical and Physical Sciences*, *209* (1096), 81–96. Retrieved from https://royalsocietypublishing.org/doi/abs/10.1098/rspa.1951.0189

Intema, H. T., van der Tol, S., Cotton, W. D., Cohen, A. S., van Bemmel, I. M., & Ottgering, H. J. A. (2009, July). Ionospheric calibration of low frequency radio interferometric observations using the peeling scheme. I. Method description and first results. *Astronomy & Astrophysics*, *501* (3), 1185–1205.

Julien, O., Priya, L., Issler, J.-L., & Lestarquit, L. (2013). Estimating the ionospheric delay using GPS/Galileo signals in the e5 band. In *EWGNSS 2013, 6th European Workshop on GNSS Signals and Signal Processing*.

Kelley, M. C. (1989). Chapter 8 - instabilities and structure in the high-latitude ionosphere. In M. C. Kelley (Ed.), *The earth's ionosphere* (p. 345–423). Academic Press. Retrieved from https://www.sciencedirect.com/science/article/pii/B9780124040137500137

Klobuchar, J. A. (1983). *Ionospheric effects on earth-space propagation* (Vol. 84) (No. 4).Ionospheric Physics Division, Air Force Geophysics Laboratory.

Komjathy, A., & Langley, R. (1996). An assessment of predicted and measured ionospheric total electron content using a regional GPS network. In *Proceedings of the National Technical Meeting of the Institute of Navigation* (pp. 615–624).

Lanyi, G. E., & Roth, T. (1988). A comparison of mapped and measured total ionospheric electron content using global positioning system and beacon satellite observations. *Radio Science*, *23* (4), 483–492.

Mangla, S., Chakraborty, S., Datta, A., & Paul, A. (2023). Exploring earth's ionosphere and its effect on low radio frequency observation with the uGMRT and the SKA. *Journal of Astrophysics and Astronomy*, *44* (1), 2.

Mangla, S., & Datta, A. (2022, 04). Study of the equatorial ionosphere using the Giant Metrewave Radio Telescope (GMRT) at sub-GHz frequencies. *Monthly Notices of the Royal Astronomical Society*, *513* (1), 964–975. Retrieved from https://doi.org/10.1093/mnras/stac942

Mangla, S., & Datta, A. (2023). Spectral analysis of ionospheric density variations measured with the large radio telescope in the low-latitude region. *Geophysical Research Letters*, *50* (14), e2023GL103305. Retrieved from https://agupubs.onlinelibrary.wiley.com/doi/abs/10.1029/2023GL103305 (e2023GL103305 2023GL103305)

Mitch, R., Psiaki, M., & Tong, D. (2013). Local ionosphere model estimation from dual- frequency global navigation satellite system observables. *Radio science*, *48* (6), 671–684.

Paul, A., Chakraborty, S., Das, A., DasGupta, A., & Mitra, S. (2005). Estimation of satellite-based augmentation system grid size at low latitudes in the Indian zone. *Navigation*, *52* (1), 15–22.

Pi, X., Mannucci, A. J., Lindqwister, U. J., & Ho, C. M. (1997, September). Monitoring of global ionospheric irregularities using the worldwide GPS network. *Geophysical Research Letters*, *24* (18), 2283–2286. Retrieved from http://doi.org/10.1029/97GL02273

Priyadarshi, S. (2015). A review of ionospheric scintillation models. *Surveys in Geophysics*, *36*, 295–324.

Rino, C. L. (1979a, November). A power law phase screen model for ionospheric scintillation: 1. weak scatter. *Radio Science*, *14* (6), 1135–1145. Retrieved from http://doi.org/10.1029/RS014i006p01135

Rino, C. L. (1979b, November). A power law phase screen model for ionospheric scintillation: 2. strong scatter. *Radio Science*, *14* (6), 1147–1155. Retrieved from http://doi.org/10.1029/RS014i006p01147

Septentrio polarx5s user manual [Computer software manual]. (n.d.). Retrieved from https://www.septentrio.com/en/products/gnss-receivers/gnss-reference-receivers/polarx5s

Strangeways, H. J. (2009, February). Determining scintillation effects on GPS receivers. *Radio Science*, *44* (1). Retrieved from http://doi.org/10.1029/2008RS004076

Swarup, G., Ananthakrishnan, S., Kapahi, V. K., Rao, A. P., Subrahmanya, C. R., & Kulkarni, V. K. (1991, January). The Giant Metre-Wave Radio Telescope. *Current Science*, *60*, 95.

Taylor, L. S., & Infosino, C. J. (1976, May). On the strong phase screen theory of ionospheric scintillations. *Radio Science*, *11* (5), 459–463. Retrieved from http://doi.org/10.1029/RS011i005p00459

Van Dierendonck, A., Klobuchar, J., & Hua, Q. (1993). Ionospheric scintillation monitoring using commercial single frequency C/A code receivers. In *Proceedings of ION GPS* (Vol. 93, pp. 1333–1342).

Woodman, R. F. (1970, November). Vertical drift velocities and east-west electric fields at the magnetic equator. *Journal of Geophysical Research*, *75* (31), 6249–6259. Retrieved from http://doi.org/10.1029/JA075i031p06249

Woodman, R. F., Chau, J. L., Aquino, F., Rodriguez, R. R., & Flores, L. A. (1999). Low- latitude field-aligned irregularities observed in the E region with the Piura VHF radar: First results. *Radio Science, 34* (4), 983–990.

Yacoub, M. D. (2007, January). The α-μ distribution: A physical fading model for the Stacy distribution. *IEEE Transactions on Vehicular Technology, 56* (1), 27–34. Retrieved from http://doi.org/10.1109/TVT.2006.883753

Yeh, K. C., & Liu, C.-H. (1982). Radio wave scintillations in the ionosphere. *Proceedings of the IEEE, 70* (4), 324–360.

Section 5

Challenges and Future Prospects

24 Challenges and Future Developments in GNSS Technology in Disaster Management

Nitin Arora, Sakshi, and Sartajvir Singh

24.1 INTRODUCTION

Prior to the introduction of RS, GPS and other GNSS-based devices were used for mapping and monitoring.

The GNSS or Global Navigation Satellite System itself is a broader term that includes several navigation technologies such as GPS, GLONASS, and Galileo. While GNSSs are valuable for navigation and location-based applications such as tectonic plate monitoring, they fail to provide valuable information about environmental parameters like temperature, precipitation, or vegetation health, which are crucial for monitoring natural calamities.

Then came the concept of remote sensing. It usually involves the use of satellites, airborne sensors, and GB-SAR to measure the valuable data essential for monitoring the cause and effect of natural calamities [1].

24.1.1 Basics of GNSS

GNSS works on the basic mathematical principle of triangulation, which involves using the known positions of multiple satellites to determine the receiver's location on the Earth's surface. Considering GPS, it requires at least three satellites aligned to navigate the exact location, and the fourth satellite is responsible for synchronization between the timing and navigation data of the receiver. This synchronization is the most crucial part of the process of providing real-time data readings of utmost precision and accuracy [2, 3].

Figure 24.1 represents the basic methodology GNSS uses to gather information about the Earth's surface. Points A, B, and C are shown as three different altitudes, and the GNSS sensor—after capturing the information of all three points—sends the data to a Main Processing Unit (MPU). The task of the MPU is to process the data and compare the difference between the altitude readings and create a altitude map from it [4].

This map is useful in demonstrating the geographical structure of a region or to study about the movement of tectonic plates over a course of time.

Many natural calamities have occurred over the years, and GNSSs have been part of monitoring their causes. With the introduction of RS, GNSS technology has rapidly been used alongside RS for aerial mapping and understanding the change pattern of a particular region [5]. This changing environment is crucial for the development of prediction models, which are useful in determining the probability of occurrence of an event. Any event with an occurrence probability of more than 90 percent is certain to happen.

DOI: 10.1201/9781032712444-29

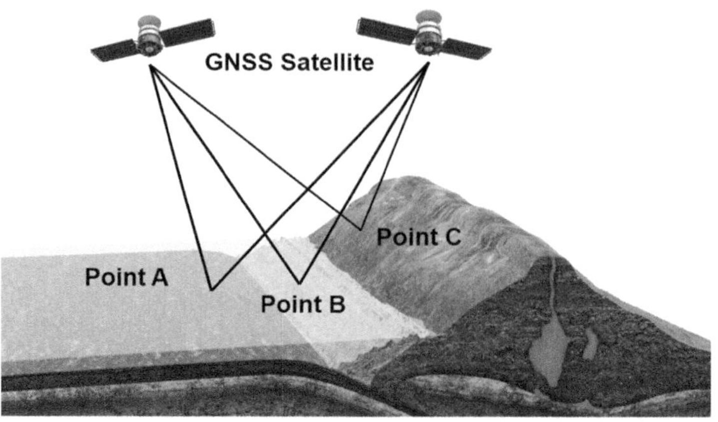

FIGURE 24.1   Representation of role of GNSS in monitoring the Earth's surface.

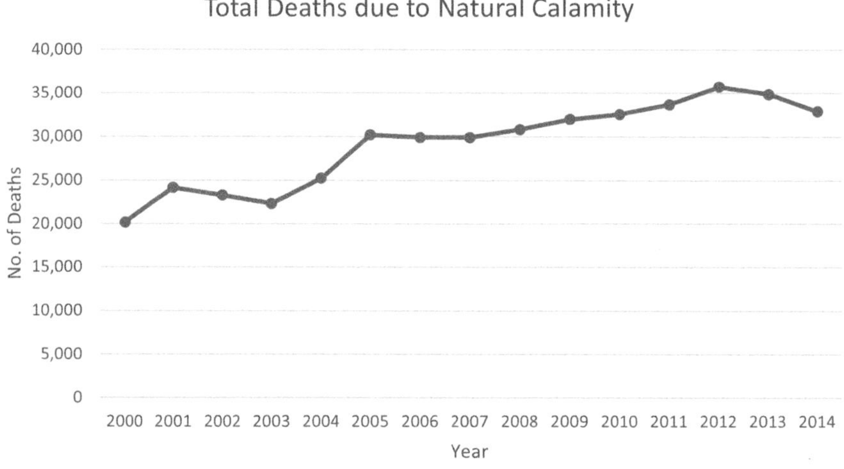

FIGURE 24.2   Representation of Deaths due to Natural Calamity from 2000–2024.

These predictions allow researchers to forecast the occurrence of an event well before it happens in real time, thus reducing damage to life forms to a significant extent [6, 7]. However, predictions are not always accurate; there are instances when false alarms are triggered due to internal or external factors, or due to spoofing from other sources or transmitters. The reasons behind these false alarms and their effects will be discussed further in the next sections of this chapter.

24.1.2   NEED OF GNSS AND RS IN CALAMITY PREDICTION

It is obvious that natural calamities are directly linked to casualties. See Figure 24.2 to obtain the statistical data of deaths occurring annually due to natural calamities. It is evident from the data that casualties have gradually decreased since 2012. This could be credited to the technical advancements in the field of RS and terrain mapping. Hence, investing in technological advancements seems to be a better choice, leading to a path of improvement [8, 9].

Similarly, GNSS technology was adopted in the 20th century by researchers to map regions, determining basic parameters such as surface roughness, altitude, and other navigation-based applications. This marked the onset of technical advancements in the field of remote sensing, paving the way for more precise and accurate readings.

24.1.3 HISTORICAL RELATIONSHIP BETWEEN GNSS AND REMOTE SENSING

The historical relationship between Global Navigation Satellite Systems (GNSSs) and remote sensing (RS) has evolved significantly over time, reflecting their complementary roles in spatial data collection and analysis. Initially, GNSS technology, epitomized by systems like GPS, focused primarily on providing precise positioning and navigation information for diverse applications, spanning military, civilian, and commercial sectors.

Satellite or airborne sensor technology that gathers information from the Earth's surface from a distance is called remote sensing [10]. Remote sensing experienced immense development with the advancement in science and technology [11, 12]. In its early days, GNSS was a sort of savior in remote sensing, providing navigation, geofencing, and coordinate fixing necessities, which helped data collection be more accurate. GNSS has thus enabled accurate data collection and georeferencing.

With the advancement of remote sensing technology, the integration of GNSS grew even deeper. Today, several highly sophisticated remote sensing technologies like High-resolution satellite imagery and LiDAR and others offer the basic positioning accuracy that allows remote sensing platforms to acquire high resolution images and data and simultaneously accurate to the location where the image was taken [13].

These applications, predominantly related to agriculture, forestry, urban planning, disaster management, and environmental monitoring, were developed with the coordination of GNSS and remote sensing [14]. In precision agriculture, for example, GNSS-guided equipment combines remote sensing data to help automate field practices such as irrigation and fertilizer application—from adjusting the number of rows of planters at the moment of passing the barrier to optimizing the use of resources, resulting in greater maximization of the crop area and less waste. GNSSs are used in disaster management for precise geological mapping based on remote sensing data, which helps gather timely and effective responses in relief efforts [15].

The interaction of GNSSs and remote sensing has developed with advancements in technology over the years [16]. Second, emerging multi-GNSS constellations including GPS, Galileo, GLONASS, and BeiDou greatly improve the accuracy and reliability of precise positioning information, augmenting remote sensing capabilities. Furthermore, advances in remote sensing technologies, such as small satellites and GNSS-equipped drones, have improved and expanded the utilization and availability of spatial data [17, 18].

In summary, the historical relationship between GNSSs and remote sensing is characterized by practical combinations that have greatly enhanced the capabilities of both technologies and applications—and contribute to how we live and care for the environment and Earth's resources.

24.2 ANALYSIS USING GNSS

The occurrence of natural calamities is divided into three phases: pre-disaster phase, occurrence during disaster, and post disaster phase (see Figure 24.3) [19]. This section will consider each phase separately and explain the role of GNSS during each phase.

24.2.1 PRE-DISASTER PHASE

Before the occurrence of a natural calamity, GNSSs play a critical role in monitoring and early warning systems. For example, GNSS technology can be used for:

1. Earthquake Prediction: By monitoring the movement of tectonic plates and the pattern they follow in the due course of time, GNSSs provide researchers an insight of the conditions that could potentially become contributory factors for land/underwater earthquakes.

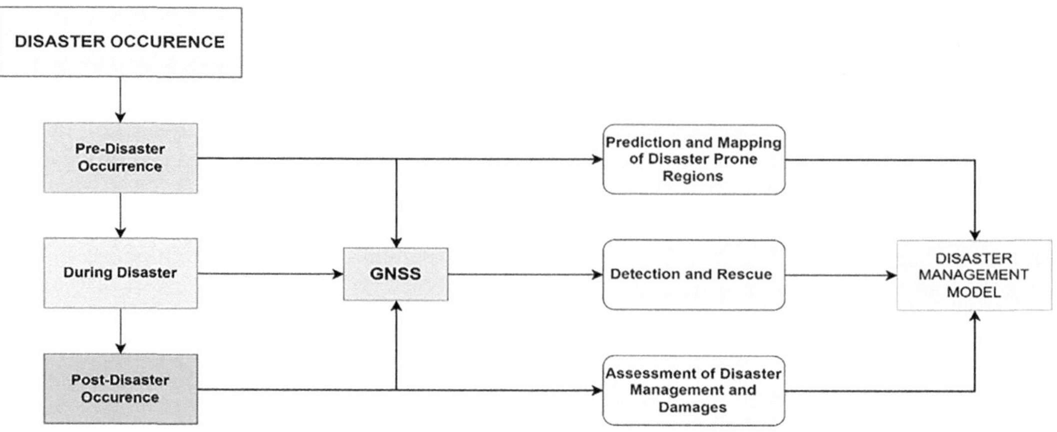

FIGURE 24.3 Representation of the role of GNSS in three phases of the Disaster Management system.

2. Weather Forecasting: GNSS data helps enhance the accuracy of weather prediction models by providing precise measurements of atmospheric parameters such as temperature, atmospheric pressure, and humidity.
3. Flood Prediction: GNSSs can monitor the changing water level across large water bodies such as oceans, deltas, and rivers. Any sudden change in water level could ultimately become a flood, and this information can be considered fruitful in saving goods and livestock.
4. Volcanic Activity Monitoring: GNSSs can track ground deformation around volcanoes, providing early warnings of potential eruptions. Apart from providing information about natural calamities, GNSSs are readily used to pinpoint hot spring water coordinates.

24.2.2 Occurrence During Disaster

1. Response to Natural Disasters: GNSS information is provided and applied in real time. The leading natural disaster response roles are as follows:
2. Positioning and Navigation: GNSSs help position and guide emergency response teams to affected spots quickly and precisely.
3. Disaster Mapping: Due to GNSSs, disaster mapping is now possible in real time, as it tells the locations of disasters, such as where the areas are flooded or the epicenter of an earthquake.
4. Search and Rescue Operations: GNSSs help search for people in hazardous situations or distress by overcoming challenges associated with tracing aspects.
5. Communication GNSS enables communication networks, which become very important during disaster management when traditional communication infrastructure may not be operational.

24.2.3 Post-disaster Phase

After the occurrence of a natural calamity, GNSSs continue to be essential in assessing damage and aiding recovery efforts [20, 21]. This phase includes:

1. Damage Assessment: GNSSs help survey affected areas to assess infrastructural, building, and environmental damage.
2. Reconstruction Planning: GNSS data is utilized to plan how to reconstruct damaged infrastructure so that the new infrastructure shall be resilient against future disasters.

3. Resource Allocation: GNSSs are also helpful in tracking the distribution of resources and materials, thus ensuring they are correctly allocated to their intended destinations.
4. Aftershocks Monitoring: In case of an earthquake, GNSSs are used to monitor aftershocks and ground stability to provide the most desirable information to institute further safety measures.

24.3 METHODOLOGY

The basic methodology of using GNSSs for prediction models is explained in Figure 24.4. Each step in the methodology is important and plays a crucial role in providing accuracy and robustness to the analyzed data and predictions made. As mentioned earlier, no system is 100% flawless, but a predictive model with more than 90% accuracy is considered an ideal model [22, 23]. Additionally, these models sometimes result in false predictions, but these instances are rare. The factors responsible for these false occurrences will be covered in further sections of this chapter [24].

Step 1: **Data acquisition and import**: GNSS data is collected either directly by satellite or by ground-based stations. Collected data consists of unwanted signal interruptions, spoofing, and timing irregularities.

Step 2: **Pre-processing**: The collected data, including interruptions and delays, are filtered using complex algorithms. The filtered data is then processed to extract important information about topography, wind speed, and surface roughness.

Step 3: **Prediction**: After filtering and data collection, it is trained to process every possible circumstance, making it robust and immune from external hindrances. This increases the dataset's reliability and is now preferable for monitoring and research works. After the collection of data is completed, it is then sent for further analysis and implementation.

Step 4: **Validation**: After more training samples, the information is compared to the database. Real-time data monitoring compares data from sensors both in the air and on the ground to forecast disasters. If the results are as expected, the model goes to the pollution phase.

Step 5: **Last stage: Execution**: The final stage is the execution of the prediction model into real-life scenarios and to monitor the real-time data for research and rescue purposes.

The system's precision can be increased by embedding machine learning algorithms. Machine learning lays the foundation for researchers to understand the changing environment better.

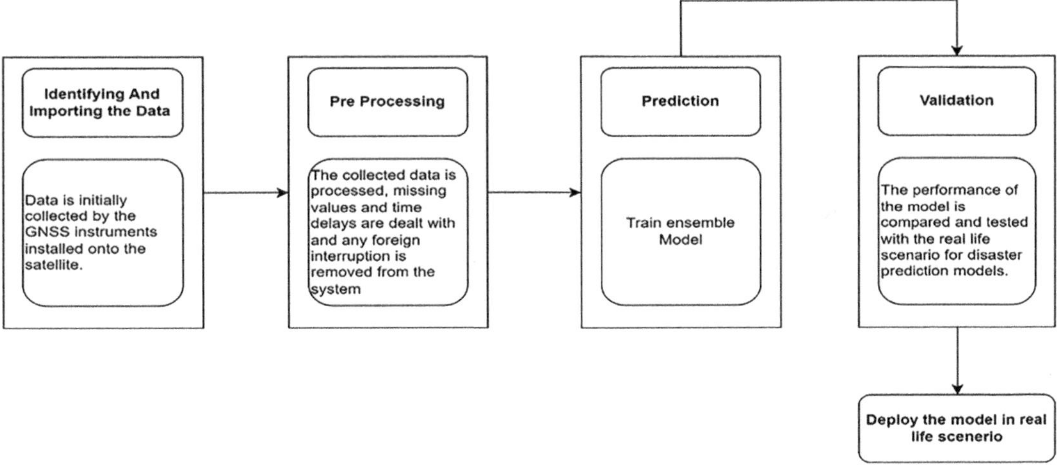

FIGURE 24.4 Basic methodology behind GNSS-based prediction models.

Embedding machine learning algorithms directly increases the accuracy and strengthens the predictive aspects of the system [25, 26].

24.4 LIMITATIONS OF GNSS

The Global Navigation Satellite System (GNSS) is an impressive technology, but like any system, it comes with its own set of limitations. These limitations are a result of both internal and external factors that directly or indirectly have a direct impact on GNSS signal accuracy and ultimately reduce the system's precision [4, 27]. Next, we cover these factors in detail.

24.4.1 INTERNAL FACTORS

24.4.1.1 Satellite Geometry

A satellite's accuracy plays a vital role in precise monitoring. Even the slightest change in a satellite's orientation directly impacts accuracy, causing non-reliable data. This data irregularity results in reducing the precision of the prediction model, making it practically useless [28].

24.4.1.2 Clock Timing

Synchronization of a satellite with the clock cannot be neglected in systems requiring precision. Thus, navigation systems must maintain a balance between real-time navigation data and accurate timing. GNSS satellite systems generally rely on atomic clocks for the most accurate reading possible.

24.4.1.3 Ionosphere and Tropospheric Effect

The accuracy of the GNSS system is also significantly influenced by the Earth's atmosphere. Ionospheric and tropospheric changes result in signal bending, ultimately causing errors in precise position calculation [29].

24.4.2 EXTERNAL FACTORS

Several external factors are capable of having a direct or indirect impact on GNSS system accuracy. The navigation system cannot operate normally in high-density areas like densely forested areas or cities with high-rise structures. Buildings, trees, and other physical obstructions may hinder GNSS signals, decreasing the number of satellites the receiver is able to detect and lowering accuracy [30–32]. Another name for this phenomenon is the "Urban Canyon Effect."

The other factor that reduces accuracy is spoofing. The act of intentionally inserting falsified signals into the system with the intent to take advantage of the original signals is commonly referred to as spoofing [33, 34]. The GNSS's limitations are a result of the previously mentioned factors [35]. It is necessary to acknowledge these constraints and acknowledge the need for enhancement to create more effective GNSS models.

These limitations include:

1. **Signal Blockage**: During natural disasters like earthquakes, hurricanes, or floods, the signals of the GNSS system are significantly obstructed, thus decreasing the system's overall accuracy.
2. **Infrastructure Dependency**: The GNSS system is dependent on a constellation of satellites for functioning; if even one satellite is out of order, the system as a whole becomes inefficient.
3. **Urban Accuracy**: Given that high-rise buildings cause multipath interference, GNSSs are unable to operate with high accuracy and precision. A limitation that becomes evident in

GNSS systems used for disaster management is that high accuracy and precision cannot be compromised in disaster conditions.

4. **Spoofing and Jamming**: GNSS signals can be intentionally interfered with by jamming or spoofing (fake signals). This process generates unwanted noise, which results in false coordinate reading and less accurate prediction models [36].
5. **Communication Latency**: GNSS data is transmitted to users through communication networks, which may experience latency or congestion during emergencies, thereby delaying the delivery of critical positioning information. This latency could be due to internal circuitry or external environmental factors.

24.5 AREA FOR IMPROVEMENT

While the authors and the literature focus on the utilization of GNSS technology in the management of disasters, several areas must be further improved to harness the potentials in the related area:

24.5.1 ACCURACY AND PRECISION

Improving the performance of GNSS signals is essential, especially on some rather critical and sensitive situations like the area provided by the tall building structures or forest regions that heavily absorb the GNSS signals emitted by the satellites. It can also be attained through advancements in the Multi Frequency and Multi Constellation GNSS receivers and advanced algorithms for correction of the errors.

24.5.2 REAL-TIME DATA PROCESSING

Optimize and improve the speed of the real-time analysis of the data with the view of providing a quick response during disastrous situations [37]. This would require modifications to the structure of the facilities, upgrade of the computation departments, and advancement of procedures in terms of speed in analyzing data and getting it out to the general public.

24.5.3 INTEGRATION OF OTHER TECHNOLOGIES

GNSS technologies must be integrated to other technologies like remote sensing, IoTs, and artificial intelligence so as to build up a better poised, a more elaborate, and more robust disaster management system. This helps in data gathering, information searching, making of conclusions, and significantly in decision-making processes.

24.5.4 ROBUSTNESS AND RELIABILITY

Ensuring the GNSS systems' effectiveness and their availability to continue operating during disasters contributes to service integrity regarding various satellite structures. Still, service integrity can be preserved by increasing the present satellite networks and ground-based augmentation systems at extreme situations.

24.5.5 USER TRAINING AND CAPACITY BUILDING

The effort will be directed towards the development of capacity and capability in the application of GNSS technology among the professionals in the field of disaster management. This can be accomplished through training, seminars and awareness sessions, workshops, education, and other activities.

24.6 CONCLUSION AND FUTURE SCOPE

This chapter incorporated the basic principle of GNSS, its history with the remote sensing, and real-life applications. The chapter serves as a foundation for researchers working in the field of geospatial data analysis and calamity prediction models with the integration of machine learning.

Thus, GNSS technology and remote sensing have provided the strongest intensification of the disaster management system by making it possible to predict disasters, monitor their effects, and respond to them appropriately. By using the accurate coordinates of GNSS integrated with the specifics of remote sensing data, events such as earthquakes, floods, and other disasters can be forecasted and predicted, so efficient responses can be conducted.

However, challenges remain. As a result, GNSS signals are relatively imprecise and dependent on satellite positioning, atmospheric conditions, and urban environments. The solution to these problems is upgrading the current technology; for example, multi-frequency receivers, and real-time data processing. Also, the merger of GNSS with other technologies such as Internet of Things (IoT) and Artificial Intelligence (AI) will help offer a wider-ranging disaster management response.

It is just as vital to train and educate disaster management professionals because of the critical role they play in responding to natural disasters. In so doing, capacity can be built up within these individuals to promote better preparedness and response scenarios to occur and, in effect, prevent loss of lives, as well as the high levels of impact from such disasters.

From a human orientation, these are the kinds of advancements necessary. Disasters are gruesome incidents that cause the loss of lives and properties; what matters most is to have the best interventions for the disasters. If we enhance our ability to forecast and predict the events and circumstances likely to lead to such failures, we can save lives and reduce losses to the economy. These technologies should be implemented appropriately and fairly to all, such that the disadvantaged persons have an equal opportunity to benefit from their availability.

It is clear that the fullest expression of GNSS and remote sensing in disaster management will remain an evolutionary process charted by research and development and a focus on people's lives. In conquering the present shortcomings and giving focus to these significant fields, it is possible to construct a global society that will be safer for all.

REFERENCES

[1] C. J. Hegarty and E. Chatre, "Evolution of the Global Navigation Satellite System (GNSS)," *Proceedings of the IEEE*, vol. 96, no. 12, pp. 1902–1917, Dec. 2008, doi: 10.1109/JPROC.2008.2006090

[2] X. Li et al., "Accuracy and reliability of multi-GNSS real-time precise positioning: GPS, GLONASS, BeiDou, and Galileo," *J Geod*, vol. 89, no. 6, pp. 607–635, Jun. 2015, doi: 10.1007/s00190-015-0802-8

[3] E. Schönemann, M. Becker, and T. Springer, "A new Approach for GNSS Analysis in a Multi-GNSS and Multi-Signal Environment," *Journal of Geodetic Science*, vol. 1, no. 3, Jan. 2011, doi: 10.2478/v10156-010-0023-2

[4] Y. Wang et al., "A Digital Closed-Loop Sense MEMS Disk Resonator Gyroscope Circuit Design Based on Integrated Analog Front-end," *Sensors*, vol. 20, no. 3, p. 687, Jan. 2020, doi: 10.3390/s20030687

[5] J. M. Dow, R. E. Neilan, and C. Rizos, "The International GNSS Service in a changing landscape of Global Navigation Satellite Systems," *J Geod*, vol. 83, no. 3–4, pp. 191–198, Mar. 2009, doi: 10.1007/s00190-008-0300-3

[6] S. M. Khan et al., "A Systematic Review of Disaster Management Systems: Approaches, Challenges, and Future Directions," *Land (Basel)*, vol. 12, no. 8, p. 1514, Jul. 2023, doi: 10.3390/land12081514

[7] L. A. Manfré et al., "An Analysis of Geospatial Technologies for Risk and Natural Disaster Management," *ISPRS Int J Geoinf*, vol. 1, no. 2, pp. 166–185, Aug. 2012, doi: 10.3390/ijgi1020166

[8] G. Johnston, A. Riddell, and G. Hausler, "The International GNSS Service," in *Springer Handbook of Global Navigation Satellite Systems*, Cham: Springer International Publishing, 2017, pp. 967–982. doi: 10.1007/978-3-319-42928-1_33

[9] N. Zhu, J. Marais, D. Betaille, and M. Berbineau, "GNSS Position Integrity in Urban Environments: A Review of Literature," *IEEE Transactions on Intelligent Transportation Systems*, vol. 19, no. 9, pp. 2762–2778, Sep. 2018, doi: 10.1109/TITS.2017.2766768

[10] S. K. Chaturvedi, "Disaster Management: Tsunami and Remote Sensing Technology," *Nature Environment and Pollution Technology*, vol. 20, no. 5, Dec. 2021, doi: 10.46488/NEPT.2021.v20i05.030

[11] D. K. Gupta, S. Prashar, S. Singh, P. K. Srivastava, and R. Prasad, "Introduction to RADAR remote sensing," in *Radar Remote Sensing*, Elsevier, 2022, pp. 3–27. doi: 10.1016/B978-0-12-823457-0.00018-5

[12] T. Angelopoulou, N. Tziolas, A. Balafoutis, G. Zalidis, and D. Bochtis, "Remote Sensing Techniques for Soil Organic Carbon Estimation: A Review," *Remote Sens (Basel)*, vol. 11, no. 6, p. 676, Mar. 2019, doi: 10.3390/rs11060676

[13] Z. Xu, L. Shen, J. Qian, and Z. Zhang, "Advanced Hand Gesture Prediction Robust to Electrode Shift with an Arbitrary Angle," *Sensors*, vol. 20, no. 4, p. 1113, Feb. 2020, doi: 10.3390/s20041113

[14] R. Dach et al., "GNSS processing at CODE: status report," *J Geod*, vol. 83, no. 3–4, pp. 353–365, Mar. 2009, doi: 10.1007/s00190-008-0281-2

[15] C. D. Bussy-Virat, C. S. Ruf, and A. J. Ridley, "Relationship between Temporal and Spatial Resolution for a Constellation of GNSS-R Satellites," *IEEE J Sel Top Appl Earth Obs Remote Sens*, vol. 12, no. 1, pp. 16–25, Jan. 2019, doi: 10.1109/JSTARS.2018.2833426

[16] Z. Yao and M. Lu, "Signal Multiplexing Techniques for GNSS: The Principle, Progress, and Challenges Within a Uniform Framework," *IEEE Signal Process Mag*, vol. 34, no. 5, pp. 16–26, Sep. 2017, doi: 10.1109/MSP.2017.2713882

[17] A. Pagrotra, N. Vyas, H. Dhiman, and V. Dutt, "Enhancing Land-Based Robotics Through the Development of Remotely Operated Vehicles for Military, Rescue and Industrial Applications," in *2023 3rd Asian Conference on Innovation in Technology (ASIANCON)*, IEEE, Aug. 2023, pp. 1–6. doi: 10.1109/ASIANCON58793.2023.10270756

[18] D. K. Singh et al., "Spatiotemporal Vegetation Variability and Linkage with Snow-Hydroclimatic Factors in Western Himalaya Using Remote Sensing and Google Earth Engine (GEE)," *Remote Sens (Basel)*, vol. 15, no. 21, p. 5239, Nov. 2023, doi: 10.3390/rs15215239

[19] A. Shukla, N. Adwani, T. Choudhury, and B. Dewangan, "Geospatial analysis for natural disaster estimation through arduino and node MCU approach," *GeoJournal*, vol. 88, no. S1, pp. 29–45, Sep. 2021, doi: 10.1007/s10708-021-10496-1

[20] M. Zhran and A. Mousa, "Planetary boundary layer height retrieval using GNSS Radio Occultation over Egypt," *The Egyptian Journal of Remote Sensing and Space Science*, vol. 25, no. 2, pp. 551–559, Aug. 2022, doi: 10.1016/j.ejrs.2022.03.013

[21] P. Dabove, A. M. Manzino, and C. Taglioretti, "GNSS network products for post-processing positioning: limitations and peculiarities," *Applied Geomatics*, vol. 6, no. 1, pp. 27–36, Mar. 2014, doi: 10.1007/s12518-014-0122-3

[22] F. Mahmood, A. Asar, and A. Mahmood, "GPS and Remote Sensing for Emergency Vehicle Navigation and Communication," in *2006 International Conference on Advances in Space Technologies*, IEEE, Sep. 2006, pp. 33–36. doi: 10.1109/ICAST.2006.313793

[23] S. Dey, I. Chakraborty, P. Banerjee, and A. Bose, "Anomalous GNSS Signal Strength Fluctuations during the Amphan Super Cyclone in Eastern India on May 20, 2020," *National Academy Science Letters*, vol. 45, no. 1, pp. 45–49, Feb. 2022, doi: 10.1007/s40009-021-01076-5

[24] F. Zangenehnejad and Y. Gao, "GNSS smartphones positioning: advances, challenges, opportunities, and future perspectives," *Satellite Navigation*, vol. 2, no. 1, p. 24, Nov. 2021, doi: 10.1186/s43020-021-00054-y

[25] C. Ghosh, "GIS and Geospatial Studies in Disaster Management," in *International Handbook of Disaster Research*, Singapore: Springer Nature Singapore, 2023, pp. 701–708. doi: 10.1007/978-981-19-8388-7_214

[26] A. Shukla, N. Adwani, T. Choudhury, and B. Dewangan, "Geospatial analysis for natural disaster estimation through arduino and node MCU approach," *GeoJournal*, vol. 88, no. S1, pp. 29–45, Sep. 2021, doi: 10.1007/s10708-021-10496-1

[27] M. Kahveci, "Recent advances in GNSS: Technical and legal aspects," in *2017 8th International Conference on Recent Advances in Space Technologies (RAST)*, IEEE, Jun. 2017, pp. 501–505. doi: 10.1109/RAST.2017.8002950

[28] P. Puricer and P. Kovar, "Technical Limitations of GNSS Receivers in Indoor Positioning," in *2007 17th International Conference Radioelektronika*, IEEE, Apr. 2007, pp. 1–5. doi: 10.1109/RADIOELEK.2007.371487

[29] M. M. Hoque and N. Jakowski, "Higher order ionospheric effects in precise GNSS positioning," *J Geod*, vol. 81, no. 4, pp. 259–268, Mar. 2007, doi: 10.1007/s00190-006-0106-0

[30] E. Andreou and K. Axarli, "Investigation of urban canyon microclimate in traditional and contemporary environment. Experimental investigation and parametric analysis," *Renew Energy*, vol. 43, pp. 354–363, Jul. 2012, doi: 10.1016/j.renene.2011.11.038

[31] E. Alexandri and P. Jones, "Temperature decreases in an urban canyon due to green walls and green roofs in diverse climates," *Build Environ*, vol. 43, no. 4, pp. 480–493, Apr. 2008, doi: 10.1016/j.buildenv.2006.10.055

[32] M. Tsakiri, A. Kealy, and M. Stewart, "Urban Canyon Vehicle Navigation with Integrated GPS/GLONASS/DR Systems," *Navigation*, vol. 46, no. 3, pp. 161–174, Sep. 1999, doi: 10.1002/j.2161-4296.1999.tb02404.x

[33] M. L. Psiaki and T. E. Humphreys, "GNSS Spoofing and Detection," *Proceedings of the IEEE*, vol. 104, no. 6, pp. 1258–1270, Jun. 2016, doi: 10.1109/JPROC.2016.2526658

[34] L. Junzhi, L. Wanqing, F. Qixiang, and L. Beidian, "Research Progress of GNSS Spoofing and Spoofing Detection Technology," in *2019 IEEE 19th International Conference on Communication Technology (ICCT)*, IEEE, Oct. 2019, pp. 1360–1369. doi: 10.1109/ICCT46805.2019.8947107

[35] A. Kumar, S. Kumar, P. Lal, P. Saikia, P. K. Srivastava, and G. P. Petropoulos, "Introduction to GPS/GNSS technology," in *GPS and GNSS Technology in Geosciences*, Elsevier, 2021, pp. 3–20. doi: 10.1016/B978-0-12-818617-6.00001-9

[36] L. Meng, L. Yang, W. Yang, and L. Zhang, "A Survey of GNSS Spoofing and Anti-Spoofing Technology," *Remote Sens (Basel)*, vol. 14, no. 19, p. 4826, Sep. 2022, doi: 10.3390/rs14194826

[37] K. Yu, C. Rizos, D. Burrage, A. G. Dempster, K. Zhang, and M. Markgraf, "An overview of GNSS remote sensing," *EURASIP J Adv Signal Process*, vol. 2014, no. 1, p. 134, Dec. 2014, doi: 10.1186/1687-6180-2014-134

25 Limitation, Improvement, and
Future Aspects of GNSS in
Climate Studies

*Anubhava Srivastava, Raziqa Masood, Raj Kumar Goel, and
Shikha Verma*

25.1 INTRODUCTION

In the field of climate research, Global Navigation Satellite Systems (GNSSs) have become crucial instruments because they offer accurate positioning and temporal data, both of which are necessary for monitoring changes in the environment. When used in conjunction with Geographic Information Systems (GIS), the integration of GNSS data improves the capabilities of spatial analysis, which in turn makes it easier to conduct comprehensive research and make decisions regarding climate-related issues. Despite its usefulness, Global Navigation Satellite System (GNSS) technologies have several drawbacks, including their vulnerability to atmospheric disturbances, their inability to block signals, and their inherent errors, all of which can affect the trustworthiness of climate-related research.

Using the paradigm of Geographic Information Systems (GIS) [1], this introduction lays the groundwork for a subsequent discussion on the limitations, improvements, and future aspects of GNSSs in climate studies. Researchers can harness the full potential of GNSS data by addressing these issues and exploiting the functions of GIS. This will allow them to advance our understanding of climatic dynamics and assist methods for sustainable environmental management. In this study, we investigate various methods that can be utilized to enhance the performance of Global Navigation Satellite Systems (GNSSs), including the utilization of advanced signal processing techniques and the incorporation of supplementary data sources. Additionally, we anticipate the future trajectory of GNSS applications in GIS-enabled climate studies. The Google Earth Engine (GEE) [2] provides a collection of strong tools and functions that considerably improve the exploitation of data from Global Navigation Satellite Systems (GNSSs) in the context of climate research.

One of the most significant benefits of GEE is its capacity to integrate geographic information produced from GNSSs with a large array of Earth observation datasets. These datasets include satellite imagery, terrain models, and environmental variables such as temperature and precipitation. Researchers can undertake thorough geographical and temporal assessments of climate-related phenomena as a result of this integration, which takes advantage of the synergies that exist between GNSS [3] location data and other environmental indicators. In addition, GEE offers sophisticated atmospheric correction tools that may be utilized to apply them to GNSS data. These tools help reduce mistakes brought about by atmospheric interference and enhance the precision of positional information for climate research.

To improve our understanding of climate dynamics, the scalable computer infrastructure that GEE provides enables the creation and deployment of machine learning algorithms specifically specialized to GNSS data processing. These algorithms make it possible to perform tasks such as pattern recognition, anomaly detection, and predictive modeling. Additionally, GEE makes real-time monitoring [4] and decision support easier to do by enabling near-instantaneous updates of information received from GNSSs within the context of dynamic environmental situations. This platform also encourages collaborative study and the exchange of data within the scientific community all over the world. It offers a centralized environment for the storage, access, and analysis of GNSS datasets and products created from them. In a nutshell, GEE gives academics the ability to make full use of the potential of GNSS data to advance climate science, assist decision-making based on evidence, and stimulate innovation in climate change adaptation and mitigation methods.

Although the satellite datasets accessible through the Google Earth Engine (GEE) play an important part in climate research, it is important to note that each dataset has its own set of limitations as well as chances for development. MODIS data [5], which has a moderate spatial resolution (250–1000 meters), is useful for monitoring the climate on a large scale. However, because of cloud cover and poor temporal resolution, it may not be able to capture phenomena that occur on a finer scale.

Errors have been reduced and interpretability has been improved thanks to advancements in MODIS data processing. These advancements include better atmospheric correction algorithms and fusion with higher-resolution datasets. Furthermore, these advancements enhance the value of MODIS for climate studies. The improved spatial resolution (10–60 meters) of Sentinel-2 makes it useful for conducting thorough analyses of land cover and vegetation, but cloud cover continues to be a barrier. Future developments, such as Sentinel-2B [6] and planned Sentinel missions, hold the potential of increased data availability and continuity, which will enable climate monitoring to be carried out more frequently and in more detail.

The data collected by Landsat [7, 8], which has a resolution of 30 meters, is useful for long-term climate research; nevertheless, it is susceptible to cloud cover and processing delays. The utility of Landsat for applications linked to climate is being improved by ongoing improvements in the data processing and cloud masking techniques that are being implemented. Although the GRACE data only provide a coarse spatial resolution, they do provide insights into the changes that occur in the water cycle of the Earth. When combined with other datasets, such as those about precipitation and soil moisture, these data can supplement hydrological and climate assessments.

We will continue to make progress in our understanding of climate-related processes with the completion of future missions such as GRACE-FO. The ERA5 reanalysis data provide comprehensive climate datasets at rather coarse resolutions. However, these datasets are continuously improving with better modeling and data assimilation techniques, which enables them to accommodate a wide variety of climate investigations. Using these datasets inside GEE to successfully address climate concerns and inform sustainable environmental management plans would require significant breakthroughs in satellite technology, data processing, and joint research activities. These advancements will be crucial in the future.

25.2 RELATED WORK

In the field of climate studies, previous research on the limitations, improvements, and future aspects of Global Navigation Satellite Systems (GNSSs) has produced significant findings that have the potential to inform the application and development of GNSS technology for environmental monitoring. Studies that investigated the limitations of GNSSs have contributed to the understanding of the influence that atmospheric impacts, signal blockage, and intrinsic accuracy concerns have on climate-related data produced via GNSSs. In light of these discoveries, it is apparent there is a pressing requirement for the development of sophisticated atmospheric correction models, signal processing techniques to reduce the impact of multipath effects, and the incorporation of several Global Navigation Satellite Systems to improve positioning accuracy and dependability in difficult settings. Improvements such as real-time data processing capabilities, which enable rapid responses to climate events, and integration with Geographic Information Systems (GIS) and Earth observation [9] data to give a full picture of climate dynamics, have also been stressed as a result of the research.

Furthermore, previous research has identified future aspects of Global Navigation Satellite Systems (GNSSs) in climate studies. These future aspects include the development of high-precision monitoring techniques for sea level rise and land subsidence, integration with satellite-based remote sensing for enhanced spatial resolution, and collaboration initiatives to standardize data formats and facilitate data sharing across research communities. The researchers can expand the use of GNSS technology in climate studies by using these past outputs, which contribute to more accurate and practical insights into the implications of climate change and adaptation measures.

The previous study that was conducted in the field of climate studies [10, 11] and focused on the Google Earth Engine (GEE) has yielded several significant findings regarding limitations, improvements, and potential future directions. Studies have revealed limitations in the effectiveness of GEE, notably in terms of efficiently managing large-scale climatic datasets and addressing concerns relating to the availability of data, the quality of the data, and the pace at which it processes. To increase the usability of satellite data for climate-related investigations, research has proposed changes such as the creation of improved algorithms for cloud detection, correction, and masking within GEE. These algorithms would mitigate the impact of cloud cover on the quality of the data. Additionally, developments in atmospheric correction techniques have been investigated within GEE to increase the accuracy of climate variables generated from satellite data.

Furthermore, previous research has highlighted the future potential of GEE in climate studies. These studies have highlighted opportunities for integrating diverse data sources (such as satellite imagery, climate models, and ground observations) to support comprehensive climate monitoring, impact assessment, and decision-making. The advancement of GEE's capabilities for real-time data processing, geographical analysis, and visualization has been made possible by the collaborative efforts of researchers and stakeholders. This has in turn paved the way for more informed strategies for climate change adaptation and mitigation.

Moving forward, the objective of research is to optimize and expand the features of GEE to handle complex climatic concerns and to promote environmentally friendly management practices on a worldwide scale. GEE also can be extensible in computing annual/monthly rainfall temperature and water presence over any study area. Much research has been done on these.

Significant scientific insights into the local climate dynamics and environmental variability have been presented as a result of the findings of research publications that focused on the computation of monthly rainfall, water temperature, and climatological conditions in study locations. For instance, [12] conducted research that proved the utilization of satellite-based rainfall estimates, such as the data from the Tropical Rainfall Measuring Mission (TRMM), to compute monthly precipitation patterns and discover seasonal variations in the distribution of rainfall across particular regions. According to [13, 14], this method makes it possible to obtain a characterization of the precipitation regimes and to identify periods of drought or wetness, both of which are significant for determining the availability of water resources and for agricultural planning.

Apart from this water temperature, research conducted by [15] has utilized remote sensing techniques and in situ data to estimate surface water temperatures across a variety of aquatic ecosystems. According to [16, 17], studies can evaluate the thermal conditions in lakes, rivers, and coastal areas by computing monthly fluctuations in water temperature. This provides insights into the implications of climate change on aquatic habitats and biodiversity. It is possible to analyze thermal conditions in these locations. In addition, the computation of climatological conditions requires the examination of long-term measures of temperature, humidity, wind, and other meteorological factors to provide a comprehensive description of the climate of the investigated regions. For instance, research conducted by [15] has utilized climate classification techniques to classify regions according to climatic characteristics. This has made it easier to comprehend the variability of climate across regions and the effects this variability has on ecosystems and human activities [18].

25.3 DISCUSSION

In the context of climate studies, research that investigated the outcomes associated with available satellite datasets within the framework of constraints, enhancements, and future issues has provided useful insights into improving the exploitation of satellite data. Certain satellite datasets have been shown to have certain limits, which include restrictions on geographical and temporal resolution, sensitivity to interference from the atmosphere, and difficulties in terms of data processing and interpretation. These limitations have been uncovered through research.

For instance, research has brought to light the limitations of moderate-resolution datasets like MODIS, which may not be able to capture fine-scale climate events. To increase the accuracy of the data, innovations such as improved atmospheric correction algorithms have been proposed. Utilizing upcoming satellite missions, such as Sentinel-2B, to enhance data resolution and coverage, integrating multi-source datasets (including GNSS and Earth observation data) to improve climate monitoring capabilities, and developing standardized workflows for data processing and analysis are some of the future aspects that will be emphasized. Academics, satellite agencies, and data suppliers need to work together to address these limitations and advance the utilization of satellite datasets to make educated decisions regarding climate change adaptation and mitigation methods.

The goal of ongoing research is to improve the integration of satellite data within systems such as the Google Earth Engine. This will make it possible to conduct climate studies that are more complete and trustworthy and will have wide-ranging effects on society. The Google Earth Engine (GEE) is a platform that integrates satellite imagery, remote sensing data, and computational capabilities inside a cloud-based [19, 20] environment. This integration makes it a revolutionary platform that can advance climate studies and environmental knowledge. Researchers are able to examine major environmental factors essential for climate studies [21–23] with the help of GEE.

These parameters include rainfall patterns, temperature fluctuations, and concentrations of greenhouse gases such as carbon dioxide and nitrogen dioxide. Researchers can monitor climate trends, evaluate the effects of climate change on ecosystems, and investigate the hazards of natural disasters by utilizing the tools and datasets provided by the Google Earth Engine (GEE). Our comprehension of environmental dynamics is improved by GEE because it makes it easier to analyze changes in land cover, the health of plants, and water supplies. This, in turn, enables us to make more informed decisions on managing natural resources and developing strategies for climate adaptation.

25.3.1 Climate Analysis

To comprehend climatic variability and the effects it has on ecosystems, agricultural practices, and human populations, it is vital to conduct rainfall and temperature study with tools such as the Google Earth Engine (GEE). To create insights into rainfall patterns and temperature variations at local, regional, and global scales, GEE provides powerful capabilities for processing and analyzing data that is acquired from satellites.

25.3.1.1 Rain Analysis

GEE can integrate precipitation data from satellites such as the Tropical Rainfall Measuring Mission (TRMM) and the Global Precipitation Measurement (GPM) to do rainfall analysis. This allows for the evaluation of rainfall patterns. Through the application of GEE, researchers can study temporal trends in rainfall, recognize seasonal fluctuations, and identify extreme precipitation occurrences such as high rainfall or drought conditions. It is possible to use the algorithms provided by GEE to forecast the quantity of rainfall that will fall over particular regions, which will make hydrological modeling, water resource management, and catastrophe risk assessment much easier. The computational output of rain analysis over the Google Earth Engine Code Editor is shown in Figure 25.1.

25.3.1.2 Temperature Analysis

GEE makes it possible to evaluate the dynamics of surface temperature by utilizing thermal infrared imagery obtained from satellites such as Landsat and MODIS (Moderate Resolution Imaging Spectroradiometer). The researchers are able to examine the fluctuations in temperatures that occur

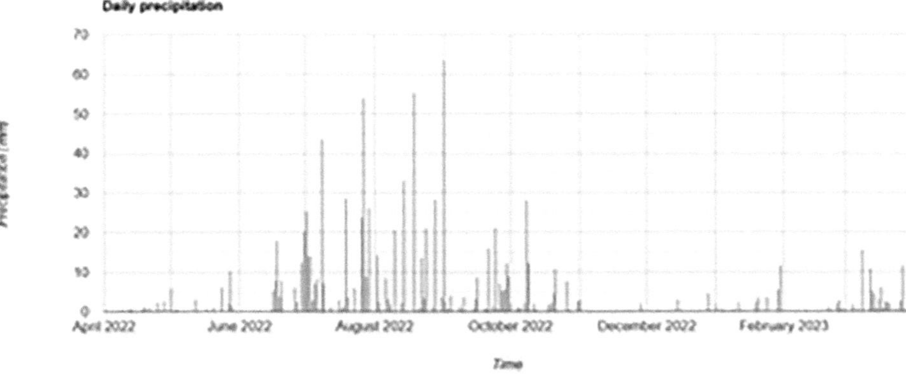

FIGURE 25.1 Monthly rain level over any selected study area in mm.

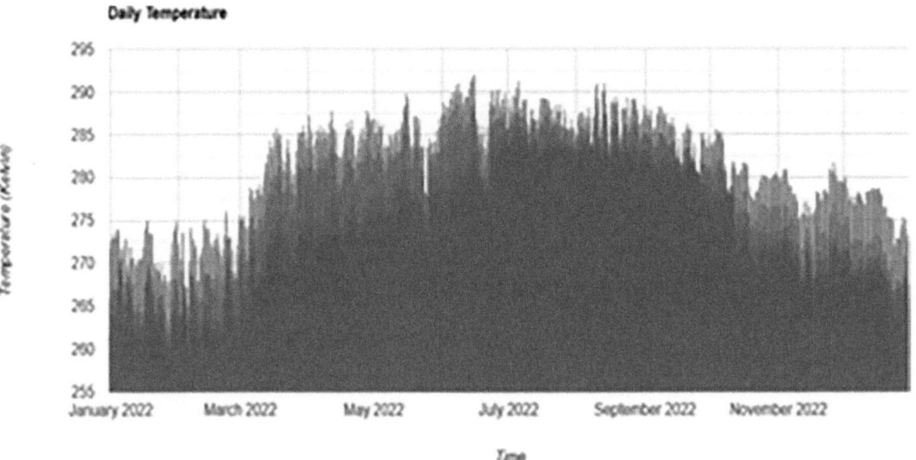

FIGURE 25.2 Daily temperature over any selected study area in Kelvin.

during the day and throughout the seasons, map urban heat islands, and evaluate the long-term temperature patterns connected with climate change. The capabilities of GEE's spatial analysis allow for the development of temperature anomaly maps, which are helpful in identifying locations susceptible to heat stress or temperature shifts caused by climate change. The computational output of temperature analysis over the Google Earth Engine platform is shown in Figure 25.2.

25.3.2 GREENHOUSE GAS MONITORING

When it comes to understanding the dynamics of atmospheric composition and evaluating the impacts of anthropogenic activities on climate change, greenhouse gas monitoring using tools such as the Google Earth Engine (GEE) plays a crucial role. In order to monitor the concentrations of critical greenhouse gases, such as carbon dioxide (CO_2) and nitrogen dioxide (NO_2), across a variety of spatial and temporal scales, GEE makes it easier to analyze data collected from satellites.

25.3.2.1 Carbon Dioxide (CO_2) Monitoring

GEE makes use of satellite data from sensors such as NASA's OCO-2 (Orbiting Carbon Observatory-2) and ESA's Sentinel-5P to monitor the concentrations of carbon dioxide in the atmosphere. Researchers analyze spatiotemporal fluctuations in CO_2 levels, identify sources and sinks of

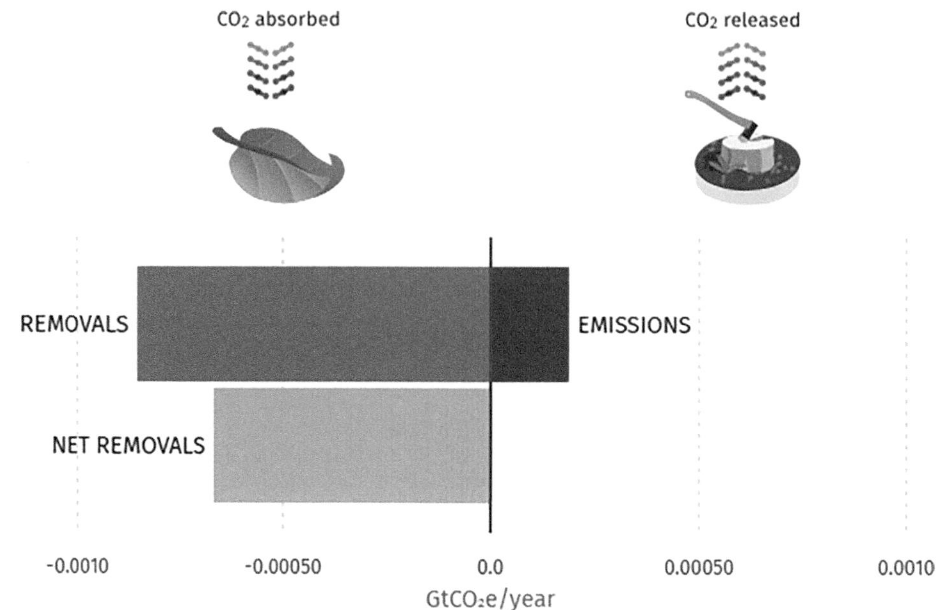

FIGURE 25.3 CO_2 absorbed and emitted computed over any geographical region.

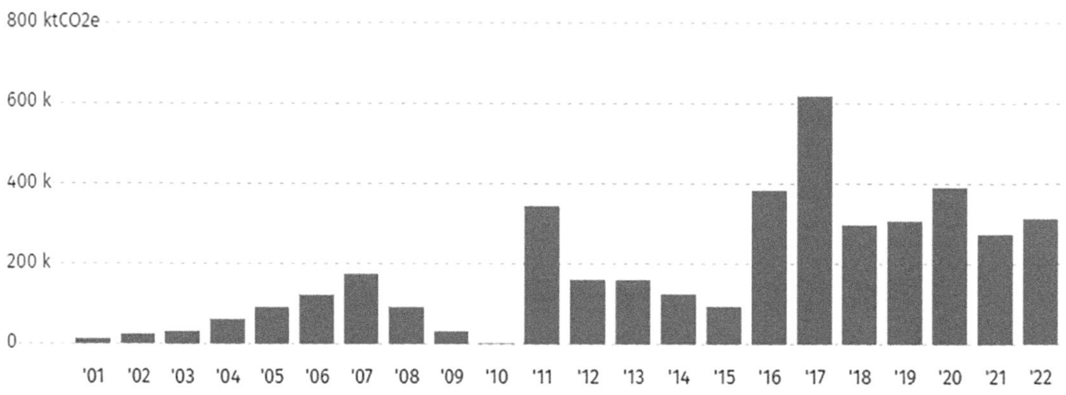

FIGURE 25.4 Total emitted greenhouse gases computed over any geographical region.

carbon emissions, and evaluate regional and global carbon budgets. The tools provided by GEE make it possible to record the dynamics of vegetation, deforestation, and changes in land use that affect CO_2 fluxes. This helps efforts to reduce carbon emissions and improve carbon sequestration strategies. The computational output of CO_2 emitted and absorbed over the Google Earth Engine Code Editor is shown in Figures 25.3 and 25.4.

25.3.2.2 Nitrogen Dioxide (NO₂) Monitoring

GEE processes satellite data from sensors such as Sentinel-5P TROPOMI to monitor NO_2 concentrations connected with industrial activity, traffic, and urban pollution. Researchers can utilize GEE to evaluate the effects of NO_2 emissions on public health and ecosystems, as well as to monitor the quality of the air in metropolitan areas, identify pollution hotspots, and detect pollution hotspots. The capabilities of GEE make it possible to create NO_2 emission inventories and provide support for air quality control activities that aim to lower nitrogen oxide emissions.

25.3.3 NATURAL DISASTER PREDICTION

One of the most important applications of geospatial data analysis is the identification of potential natural disasters utilizing tools such as the Google Earth Engine (GEE). This helps improve early warning systems and disaster preparedness initiatives. As a means of monitoring environmental conditions and identifying potential precursors to natural hazards such as earthquakes, floods, wildfires, and landslides, GEE makes use of satellite imagery and data collected through remote sensing operations.

25.3.3.1 Earthquake Prediction

For earthquake prediction, GEE can analyze seismic data, ground deformation measurements (InSAR), and characteristics of the terrain to identify seismic risk zones and evaluate the likelihood of earthquakes. GEE can contribute to seismic hazard mapping and early warning systems by merging historical earthquake data with real-time monitoring. This enables rapid responses and risk mitigation measures to be conducted.

25.3.3.2 Flood Prediction

Through the processing of satellite-derived data, GEE can monitor river discharge, patterns of precipitation, and variations in surface water extent to make predictions regarding flooding. Through hydrological models and real-time satellite measurements, GEE can anticipate flood events and evaluate the danger of flooding in areas susceptible to flooding. The flood prediction capabilities of GEE provide assistance with the preparation of emergency response tactics and evacuation procedures.

25.3.3.3 Wildfires Prediction

GEE conducts an analysis of vegetation indices, changes in land cover, and weather conditions to evaluate the risk of wildfires and monitor areas prone to fires. Through the integration of climate data with historical fire occurrence data, GEE can make predictions on the behavior of fires, map patterns of fire spread, and aid in the development of plans for fire management. It is possible to reduce the negative effects of wildfires on ecosystems and populations by using the wildfire prediction tools provided by GEE. These technologies help in early identification and rapid response. The computational output of Wildfires Prediction over the Google Earth Engine Code Editor is shown in Figure 25.5.

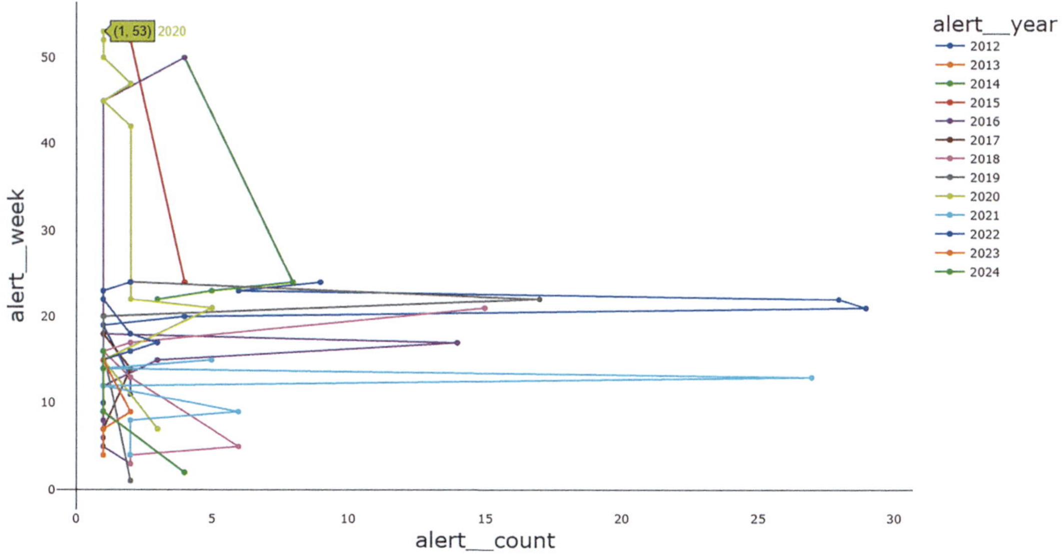

FIGURE 25.5 Number of wildfire computations over selected study areas.

25.3.3.4 Landslide and Debris Flow Prediction

To predict landslides and debris flows, GEE analyzes the characteristics of the topography, the stability of the slope, and the intensity of the rainfall to locate regions prone to landslides and debris flows. Through examination of topographic data and estimates of precipitation, GEE is able to evaluate the potential for landslides and send warnings to regions prone to mass movements. GEE's landslide prediction skills provide assistance for the mitigation of hazards and the planning of infrastructure in areas prone to landslides.

25.3.4 INTEGRATION WITH CLIMATE MODELS

The incorporation of the Google Earth Engine (GEE) into climate models holds much potential for enhancing our capacity for prediction and extending our understanding of the dynamics of climate. Researchers are able to improve climate models by including high-resolution observational data thanks to the capability of GEE to process large-scale geospatial datasets and to provide real-time data assimilation. This integration enables enhanced validation and calibration of climate models, which ultimately results in more accurate estimates of future climatic scenarios. Through the integration of GEE with dynamic Earth system models, scientists can simulate intricate interactions between the components of the atmosphere, land, ocean, and ice. This enables them to conduct in-depth analyses of the implications of climate change on specific regions and to devise strategies for adaptation.

By further facilitating the communication of climate model outputs to policymakers and stakeholders, the visualization tools provided by GEE contribute to the development of evidence-based decision-making for climate resilience and sustainable development efforts. In the end, the incorporation of GEE into climate models gives researchers and decision-makers the ability to more effectively handle the difficulties brought about by climate change. Climatological and soil conditions of any selected study area can be calculated as shown in Figure 25.6.

25.3.5 DATA VISUALIZATION AND DECISION SUPPORT

The purpose of the Google Earth Engine (GEE) is to provide a comprehensive platform for the visualization of data and the support of decision-making in the field of climate studies, particularly about Global Navigation Satellite Systems (GNSSs) and the computation of climatological conditions. Researchers must have the ability to study geographical and temporal patterns in climate-related parameters, such as temperature, precipitation, and vegetation dynamics, which are essential

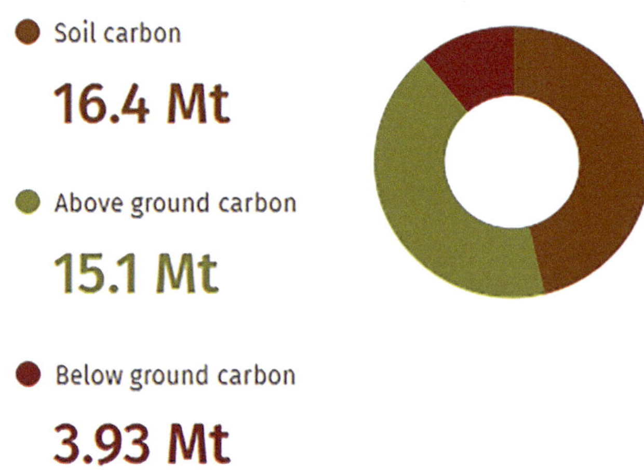

FIGURE 25.6 Climatological and soil conditions of the study area.

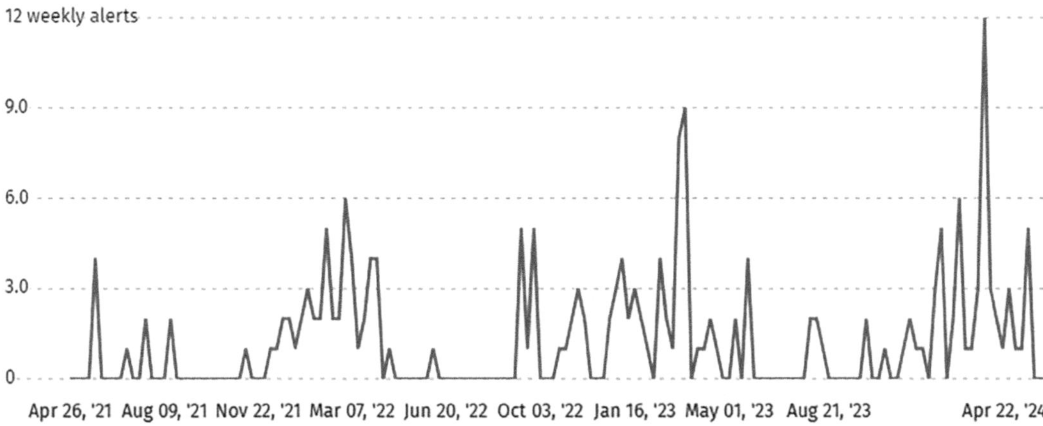

FIGURE 25.7 Weekly Alerts Computed by the Google Earth Engine.

for understanding environmental influences on GNSS performance. The data visualization tools provided by GEE make it possible for researchers to do so.

By combining a wide variety of datasets and determining the influence that climatological elements have on GNSS operations, decision support tools inside GEE give stakeholders the ability to make decisions based on accurate information. Through access to historical climate data and modern data processing tools, GEE enables the computation of climatological conditions, which in turn makes it easier to analyze long-term climate patterns and the linkages between climate and General Navigation Satellite Systems (GNSSs). Researchers can improve our understanding of how climate affects satellite-based positioning technologies and increase the reliability of GNSS data by utilizing the capabilities GEE provides. The output of weekly alerts computed by the GEE of any area is shown in Figure 25.7. The user can analyze any study area fire count by such computation.

25.4 CONCLUSION

The Google Earth Engine (GEE) platform is both powerful and versatile, capable of making substantial advancements in the field of climate studies and providing support for the optimization of applications that utilize Global Navigation Satellite Systems (GNSSs). Researchers can undertake in-depth analyses of complex climatic variables with high geographical and temporal resolution thanks to the capability of GEE to integrate a wide variety of geospatial information. These databases include satellite imagery, the outputs of climate models, and environmental factors. Our understanding of climate dynamics, changes in ecosystems, and the effects of climatic variability and change is improved as a result of this skill.

Researchers and practitioners are able to find appropriate deployment locations and techniques to mitigate environmental influences on GNSS operations with the help of GEE's key tools for assessing climatological variables that influence signal quality and accuracy. These tools are specifically designed for use in GNSS applications. In addition, the decision support functions of GEE make it easier to make decisions based on evidence by transforming complicated geographic data into insights that can be put into action, assisting with the design of climate resilience, and providing useful information for adaption plans. In the future, continued developments in GEE's capabilities—in conjunction with collaborations between other fields of study—hold the potential to stimulate innovation in the fields of climate science, geospatial technology, and sustainable development projects. This demonstrates the transformative role GEE plays in solving urgent climate concerns and generating a more resilient and sustainable future.

The potential uses of the Google Earth Engine (GEE) in the field of climate research and Global Navigation Satellite System (GNSS) applications are vast and will provide opportunities for enhanced climate modeling, real-time monitoring, impact assessment, and integration with developing technologies in the future. GEE has the potential to play a significant role in the development of climate models that are more accurate through the integration of satellite data with climate simulations and the utilization of machine learning techniques. In addition, the capabilities of GEE make it possible to construct early warning systems for climate-related dangers by merging General Navigation Satellite System (GNSS) data with meteorological observations. The use of GEE for vulnerability mapping may also be a potential future use. This would allow for the identification of regions at high risk of climate impacts and the support of targeted adaptation actions.

GEE's real-time monitoring capabilities can be further improved through integration with Internet of Things devices and sensor networks, which ultimately results in the provision of full insights into environmental dynamics. When everything is said and done, the future of GEE resides in its potential to drive creativity, collaboration, and ideas that can be put into action to address climate concerns and promote sustainability.

REFERENCES

[1] A. Wubalem, "Landslide susceptibility mapping using statistical methods in Uatzau catchment area, northwestern Ethiopia," *Geoenvironmental Disasters*, vol. 8, no. 1, pp. 1–21, 2021, doi: 10.1186/s40677-020-00170-y

[2] N. Gorelick, M. Hancher, M. Dixon, S. Ilyushchenko, D. Thau, and R. Moore, "Google Earth Engine: Planetary-scale geospatial analysis for everyone," *Remote Sens. Environ.*, vol. 202, pp. 18–27, 2017, doi: 10.1016/j.rse.2017.06.031

[3] H. Ullah et al., "Vegetation assessments under the influence of environmental variables from the Yakhtangay Hill of the Hindu-Himalayan range, North Western Pakistan," *Sci. Rep.*, vol. 12, no. 1, pp. 1–17, 2022, doi: 10.1038/s41598-022-21097-4

[4] P. S. Thenkabail et al., "Global Cropland-Extent Product at 30-m Resolution (GCEP30) Derived from Landsat Satellite Time-Series Data for the Year 2015 Using Multiple Machine-Learning Algorithms on Google Earth Engine Cloud," *US Geol. Surv. Prof. Pap.*, vol. 2021, no. 1868, pp. 1–63, 2021, doi: 10.3133/pp1868

[5] S. Bar, B. R. Parida, A. C. Pandey, and N. Kumar, "Pixel-Based Long-Term (2001–2020) Estimations of Forest Fire Emissions over the Himalaya," *Remote Sens.*, vol. 14, no. 21, 2022, doi: 10.3390/rs14215302

[6] A. Srivastava, S. Bharadwaj, R. Dubey, V. B. Sharma, and S. Biswas, "Mapping Vegetation and Measuring the Performance of Machine Learning Algorithm in Lulc Classification in the Large Area Using Sentinel-2 and Landsat-8 Datasets of Dehradun as a Test Case," *Int. Arch. Photogramm. Remote Sens. Spat. Inf. Sci. - ISPRS Arch.*, vol. 43, no. B3-2022, pp. 529–535, 2022, doi: 10.5194/isprs-archives-XLIII-B3-2022-529-2022

[7] A. Srivastava, R. Dubey, and S. Biswas, "Comparison of Sentinel and Landsat Data Sets over Lucknow Region Using Gradient Tree Boost Supervised Classifier," *Lect. Notes Networks Syst.*, vol. 730 LNNS, pp. 221–232, 2023, doi: 10.1007/978-981-99-3963-3_18

[8] R. Manandhar, I. O. A. Odeh, and T. Ancev, "Improving the Accuracy of Land Use and Land Cover Classification of Landsat Data Using Post-Classification Enhancement," *Remote Sens. 1*, pp. 330–344, 2009, doi: 10.3390/rs1030330

[9] A. Srivastava, S. Umrao, S. Biswas, R. Dubey, and M. I. Zafar, "FCCC: Forest Cover Change Calculator User Interface for Identifying Fire Incidents in Forest Region using Satellite Data," *Int. J. Adv. Comput. Sci. Appl.*, vol. 14, no. 7, pp. 948–959, 2023, doi: 10.14569/IJACSA.2023.01407103

[10] A. Srivastava, S. Umrao, and S. Biswas, "Exploring Forest Transformation by Analyzing Spatial-temporal Attributes of Vegetation using Vegetation Indices," *Int. J. Adv. Comput. Sci. Appl.*, vol. 14, no. 5, 2023, doi: 10.14569/IJACSA.2023.01405114

[11] A. Srivastava and S. Biswas, "Analyzing Land Cover Changes over Landsat-7 Data using Google Earth Engine," *Proc. 3rd Int. Conf. Artif. Intell. Smart Energy, ICAIS* 2023, pp. 1228–1233, 2023, doi: 10.1109/ICAIS56108.2023.10073795

[12] J. E. Hay, D. Easterling, K. L. Ebi, A. Kitoh, and M. Parry, "Introduction to the special issue: Observed and projected changes in weather and climate extremes," *Weather Clim. Extrem.*, vol. 11, pp. 1–3, 2016, doi: 10.1016/j.wace.2015.08.006

[13] F. Yang, J. Guo, C. Zhang, Y. Li, and J. Li, "A regional zenith tropospheric delay (Ztd) model based on gpt3 and ann," *Remote Sens.*, vol. 13, no. 5, pp. 1–19, 2021, doi: 10.3390/rs13050838

[14] A. Srivastava and P. Ahmad, "A Probabilistic Gossip-based Secure Protocol for Unstructured P2P Networks," *Procedia Comput. Sci.*, vol. 78, no. December 2015, pp. 595–602, 2016, doi: 10.1016/j. procs.2016.02.122

[15] Viana CM, Girão I, Rocha J. Long-Term Satellite Image Time-Series for Land Use/Land Cover Change Detection Using Refined Open Source Data in a Rural Region. *Remote Sensing.* 2019; 11(9):1104. doi: 10.3390/rs11091104

[16] R. Van Malderen et al., "Global Spatiotemporal Variability of Integrated Water Vapor Derived from GPS, GOME/SCIAMACHY and ERA-Interim: Annual Cycle, Frequency Distribution and Linear Trends," *Remote Sens.*, vol. 14, no. 4, 2022, doi: 10.3390/rs14041050

[17] W. Li, D. Zhao, Y. Shen, and K. Zhang, "Modeling Australian TEC maps using long-term observations of Australian regional GPS network by artificial neural network-aided spherical cap harmonic analysis approach," *Remote Sens.*, vol. 12, no. 23, pp. 1–20, 2020, doi: 10.3390/rs12233851

[18] Y. Liu, L. Liu, and Y. Yan, "Network Topology Change Detection Based on Statistical Process Control," *ACM Int. Conf. Proceeding Ser.*, pp. 145–151, 2020, doi: 10.1145/3409501.3409532

[19] A. Srivastava and P. Ahmad, "A Probabilistic Gossip-based Secure Protocol for Unstructured P2P Networks," *Phys. Procedia*, vol. 78, no. December 2015, pp. 595–602, 2016, doi: 10.1016/j. procs.2016.02.122

[20] A. Tassi and M. Vizzari, "Object-oriented lulc classification in google earth engine combining snic, glcm, and machine learning algorithms," *Remote Sens.*, vol. 12, no. 22, pp. 1–17, 2020, doi: 10.3390/ rs12223776

[21] R. T. Vilasan and V. S. Kapse, "Evaluation of the prediction capability of AHP and F-AHP methods in flood susceptibility mapping of Ernakulam district (India)," *Nat. Hazards*, 112, 1767–1793 (2022). doi: 10.1007/s11069-022-05248-4

[22] M. Vreugdenhil et al., "Microwave remote sensing for agricultural drought monitoring: Recent developments and challenges," *Front. Water*, vol. 4, 2022, doi: 10.3389/frwa.2022.1045451

[23] X. X. Yang et al., "A 30-m landsat-derived cropland extent product of Australia and China using random forest machine learning algorithm on Google Earth Engine cloud computing platform," *GIScience Remote Sens.*, vol. 57, no. February, pp. 85–91, 2020, doi: 10.1016/j.isprsjprs.2018.07.017

26 Advancements and Limitations of GNSS in Ionospheric Research

Raj Gusain, Anurag Vidyarthi, Rishi Prakash, and A. K. Shukla

26.1 INTRODUCTION

The ionosphere is the upper region of Earth's atmosphere extending in altitude about 50 km to 480 km into space, although this height is not fixed and varies with factors such as solar activity, time of day, and geographic location. The ionosphere plays a crucial role in long-distance radio communication and navigation because it can reflect and modify electromagnetic signals, making it possible to communicate over long distances and determine the user's location accurately. It is crucial to remember, too, that the signal traveling through the ionosphere encounters delay because of the high concentration of electrons and numerous charged particles. The signal's interaction with these charged particles, which modifies the electromagnetic waves' propagation, causes this delay.

The ionosphere's special qualities allow radio waves to bounce off and reach distant locations, even with this delay. This phenomenon is essential for accurate navigation and dependable long-distance communication. Besides navigational applications of the ionosphere, other major applications include ionospheric research, ionospheric remote sensing, Space weathering, astronomy and astrophysics, radio occultation, and earthquake and tsunami early warning.

The ionosphere is important because it is considered as the layer responsible for navigation, Space weather research, and characterizing the type of ionospheric irregularity. In order to understand and quantify these characteristics, scientists and researchers rely on a crucial parameter known as Total Electron Content (TEC), which is used to calculate the electron density in the ionosphere along the path of the satellite signal. It's commonly expressed in TEC units (TECU), where one TECU represents 10^{16} electrons per square meter along the signal path. By measuring TEC, scientists can gain valuable insights into the ionosphere's behavior, which is essential for space weather research, navigation, and understanding ionospheric irregularities that can impact various communication and navigation systems.

26.2 NAVIGATION THROUGH SATELLITES

Navigation means to move from one place to another by any person or any vehicle, which in earlier times was done with the help of stars, compass, maps, dead reckoning, pilotage, etc. Later, in medieval times the electronic navigation or radio navigation was done with the help of hyperbolic navigation systems such as Decca, OMEGA, and LORAN-C by determining the position based on the differences in the arrival times between radio signals transmitted from two or more synchronized radio transmitters; however, they had some drawbacks like limited coverage area, low accuracy, affected by obstructions to line of sight, vulnerable to interference, and maintenance and development of ground-based transmitters incurring huge infrastructure cost.

DOI: 10.1201/9781032712444-31

The use of satellite and radar systems has transformed navigation in the modern era. These days, receivers closely linked to satellite systems are essential for detecting and monitoring humans, vehicles, and spacecraft. A constellation of satellites is used in satellite navigation, using the latest technology, to determine the exact location of any point on Earth. This novel method is especially useful for monitoring and surveillance applications since it not only increases accuracy but also yields precise information about geographic coordinates (Ganguly and Brown 2001). Thus, a new era of navigation has begun with the incorporation of satellite technology, providing formerly unobtainable accuracy and dependability in locating and tracking locations worldwide, as shown in Figure 26.1.

Satellite navigation systems are classified into two main categories: GNSS (Global Navigation Satellite System) and RNSS (Regional Navigation Satellite System).

26.2.1 GNSS

GNSS represents the global navigational satellite system, consisting of a constellation of satellites used to provide precise location or geographic coordinates for users and vehicles across the Earth's surface. Some well-known GNSS systems include GPS (developed by the U.S.), GLONASS (developed by Russia), BeiDou (developed by China), and GALILEO (developed by Europe).

GNSS systems offer numerous advantages, including high accuracy, high-speed data delivery, global coverage, 26/7 availability, and versatility. However, there are also some drawbacks associated with GNSS, such as susceptibility to signal interference, dependency on satellites, and security concerns. GNSS systems can provide relatively accurate positioning, but they may not meet the stringent accuracy requirements of certain applications. Satellite augmentation systems can correct

FIGURE 26.1 Satellites at Work: Navigating the World from Above.

errors and enhance accuracy, making them necessary in certain fields like aviation (Moreno et al., 2011). Examples of such systems include India's GPS Aided Geo Augmented Navigation (GAGAN), the United States' Wide Area Augmented System (WAAS), and Europe's European Geostationary Navigation Overlay Service (EGNOS), all of which contribute to strengthening the reliability and accuracy of the navigation system.

26.2.2　RNSS

RNSS is the regional navigational satellite system that also has a constellation of satellites, but the area of coverage is limited to specific geographical regions. Some RNSS systems are NavIC (developed by India) and QZSS (developed by Japan). India's NavIC is a well-known example of RNSS. With a focus on specific geographic areas, NavIC also extends to neighboring countries including Afghanistan, the Maldives, Pakistan, Sri Lanka, Bangladesh, Nepal, and Bhutan. Japan's QZSS, another RNSS system, serves the navigation needs of its assigned region in a similar manner.

This illustrates the global trend of nations creating and implementing their own satellite constellations to improve navigation services within defined regions. A wide range of applications, including disaster relief, telecommunications, transportation, and agriculture, are supported by these RNSS systems, which together enhance worldwide navigation capabilities. Figure 26.2 shows satellite navigation and a satellite augmented system.

26.3　IONOSPHERIC STUDIES

Ionospheric studies are the scientific study of the ionosphere, a part of the Earth's upper atmosphere situated between 50 to 1,000 kilometers above the Earth's surface. The primary objectives of ionospheric studies are electron density profiling, ionospheric modeling, Space weather monitoring, radio wave propagation studies, and enhancement of accuracy and reliability of GNSS systems

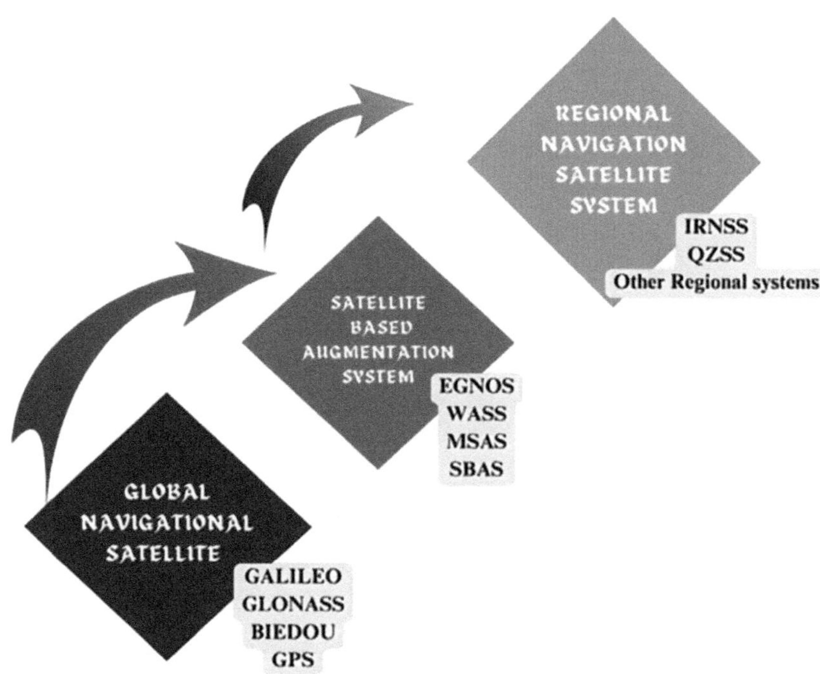

FIGURE 26.2　Navigating the Future: GNSS, RNSS, and SBAS Systems.

(Galkin et al., 2022; Pulinets et al., 2021; Huang, Du, and Wang, 2023). Our knowledge of the ionosphere and its complicated behavior has greatly improved because of GNSS systems.

Ionospheric research can benefit significantly from the amount of information provided by GNSS systems. These navigational satellite systems use signals that pass through the ionosphere to determine the accurate location of the user on the surface of the Earth. However, the charged particles present in the ionosphere, mainly electrons, cause delays and distortions in these signals as they travel through the ionosphere. Scientists take benefit from this phenomenon by measuring the Total Electron Content (TEC) of the ionosphere using GNSS data (Garner et al., 2008; Patari, Paul, and Guha, 2021; Jin and Song, 2023; Ghimire and Chapagain, 2022). Researchers may make precise ionospheric maps, examine variations, and gain insight into the effects of Space weather events by analyzing TEC data collected from various GNSS satellites. Scientists are able to make precise maps of the ionospheric layer, monitor changes, and analyze the effects of Space weather phenomena because of the quantity of information obtained from GNSS data, especially when total electron content is measured. As technology develops, ionospheric research and GNSS technology will surely work together to improve the precision and dependability of global navigation systems and deepen our understanding of the complex dynamics that exist within the ionosphere.

26.4　STUDY OF IONOSPHERE THROUGH NAVIGATION SATELLITES

The Global Positioning System (GPS) constellation and other navigational satellite systems are great resources for studying the ionosphere. As signals from these satellites travel through the ionospheric region, they are affected by variations in electron density, which result in signal delay and distortion. Scientists determine the composition of the ionosphere, variations in electron density, and responses to Space weather events by carefully analyzing the radio signals. This study is essential because it not only advances our knowledge of basic ionospheric processes, but it also significantly improves the precision and dependability of satellite-based navigation and communication systems. The study of the ionosphere through navigation satellites is conducted by analyzing TEC, Space weather monitoring and real-time data analysis. These key aspects are discussed in detail below.

26.4.1　TEC Analysis

GNSSs are greatly impacted by the ionosphere's dynamic behavior, which is characterized by a high concentration of ionized particles, mostly electrons. Total Electron Content (TEC), one of the primary ionospheric parameters, is frequently used by researchers and scientists when studying the ionosphere using navigation satellites. Understanding the electron density in the ionosphere is important for understanding ionospheric behavior, enhancing satellite-based navigation systems, and TEC measurements from signals sent by navigation satellites. The additional transmission delay in the ionosphere due to the high amount of total electron content is shown in Figure 26.3.

Total Electron Content (TEC) is measured and analyzed as part of the study of the ionosphere through navigation satellites employing TEC analysis to understand ionospheric conditions and their effects on satellite signals. The accuracy and reliability of satellite-based navigation and communication systems must be increased. The TEC is calculated using the following expression as given in equation (26.1).

$$TEC = -4.4192 \times \left(\phi_2 - \phi_1 \right) TECU \tag{26.1}$$

26.4.2　Ionospheric Scintillation

Radio transmissions are impacted by ionospheric scintillation as they travel through the ionosphere. It can be identified by rapid fluctuations in signal strength and phase that result from electron

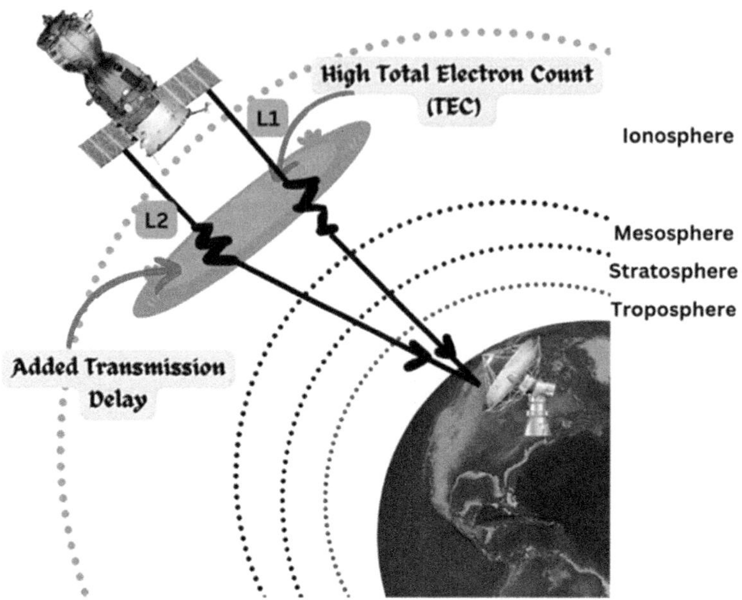

FIGURE 26.3 Analyzing TEC to understand Ionospheric behavior.

density irregularities in the ionosphere causing problems in the reception of radio signals and degrades the performance of GNSS receivers. Ionospheric scintillation is the result of radio waves being bent by ionospheric irregularities (Priyadarshi and Singh, 2012; Vani et al., 2019; Guo, Aquino, and Vadakke Veettil, 2019; Wernik, Alfonsi, and Materassi, 2004; Brassarote, de Souza, Monico, and Galera, 2018). When ionospheric scintillation is extreme, the signal intensity may become so weak that the receiver completely loses track of the signal. This may cause temporary signal loss, preventing interaction with the satellite or ground station, reducing the quality of recorded data and finally degrading the performance of any navigation system. Users may experience a loss of connection with GPS and other navigation and positioning systems. These interruptions can be particularly problematic in critical applications like aircraft and maritime navigation and emergency services, where continuous communication is essential. The effect of ionospheric scintillation and ionospheric irregularities in the form of amplitude and phase fluctuations is shown in Figure 26.4.

Also, ionospheric scintillation monitors are available to record the data during scintillation caused due to Space weather events. Scintillation predictive models have the potential to stop resource misallocation and encourage the development of new adaptive systems.

26.4.3 PLASMA BUBBLE

A localized area of increased, unpredictable, and varying electron density existing within the ionosphere is referred to as a plasma bubble, which have the potential to severely disrupt GNSS signals, satellite communication, and radio communication. Numerous applications, including navigation, weather monitoring, and military activities, may be impacted by these interruptions. The bubble's size extends from centimeters to hundreds of kilometers, and they can take on a variety of shapes. The amount of time they last can range from a few minutes to several hours. Plasma bubbles typically occur in the F-region at night or after sunset, mostly in the equatorial and low-latitude regions, reaching up to very high latitudinal heights and moving towards an eastward direction (Luo et al., 2021; Li et al., 2021; Shetti, Gurav, and Seemla, 2019; Magdaleno et al., 2017; Nishioka, Saito, and Tsugawa, 2008; Timoçin et al., 2020; Timoçin, Temuçin, and İnyurt, 2022; Yokoyama, 2017;

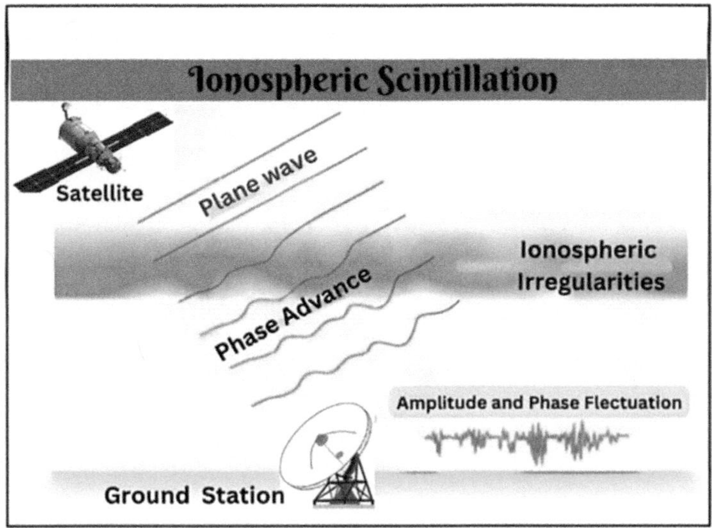

FIGURE 26.4 Ionospheric Scintillation.

Yamamoto et al., 2018). The F-region is mainly stable during the day; however, after sunset, the electron density of the ionosphere fluctuates. Plasma bubbles may form because of this instability, which is known as Rayleigh-Taylor Instability, causing significant TEC depletions.

The occurrence of EPBs depends on many factors such as local time, longitude, season, solar activity, Earth's magnetic field, and in particular, geomagnetic activity. The equatorial plasma bubble irregularities are significantly influenced by geomagnetic activity. With increasing geomagnetic activity, equatorial plasma bubble irregularities are more frequently seen during the hours after sunset, while the recovery phases of geomagnetic storms have an effect that suppresses equatorial plasma bubble irregularities. The plasma bubbles also increase during the September/October equinox as compared to the March/April equinox.

For accurate forecasting of Space weather, it is essential to understand plasma bubbles and how they behave in the ionosphere. Plasma bubbles have the potential to interfere with radio transmissions, GPS, and satellite communication systems, all of which are essential for different sectors such as aviation, telecommunications, and military activities. By monitoring and forecasting plasma bubbles, Space weather forecasters can provide alerts and warnings to the operators of these systems. By doing so, the adverse impacts of Space weather problems on technology and infrastructure are minimized.

The study of plasma bubble behavior enables the creation of more reliable navigational algorithms and correctional methods. For applications like aviation, maritime navigation, and precision agriculture—where precise positioning is essential for safety, efficiency, and productivity—improved navigation system performance is especially important. Scientists are interested in and conducting study on plasma bubbles in the ionosphere. They provide light on the complex structure of the ionosphere and how it reacts to numerous geophysical and space-related events. Various ionospheric disturbances are shown in Figure 26.5.

The study of plasma bubbles advances our knowledge of ionospheric science, geophysics, and space physics. Radio waves used for long-distance communication can be affected by plasma bubbles during their transmission. The performance of such communication systems can be enhanced by understanding and predicting the development of plasma bubbles. Applications including long-distance broadcasting, military operations, and emergency communications are particularly important in this respect.

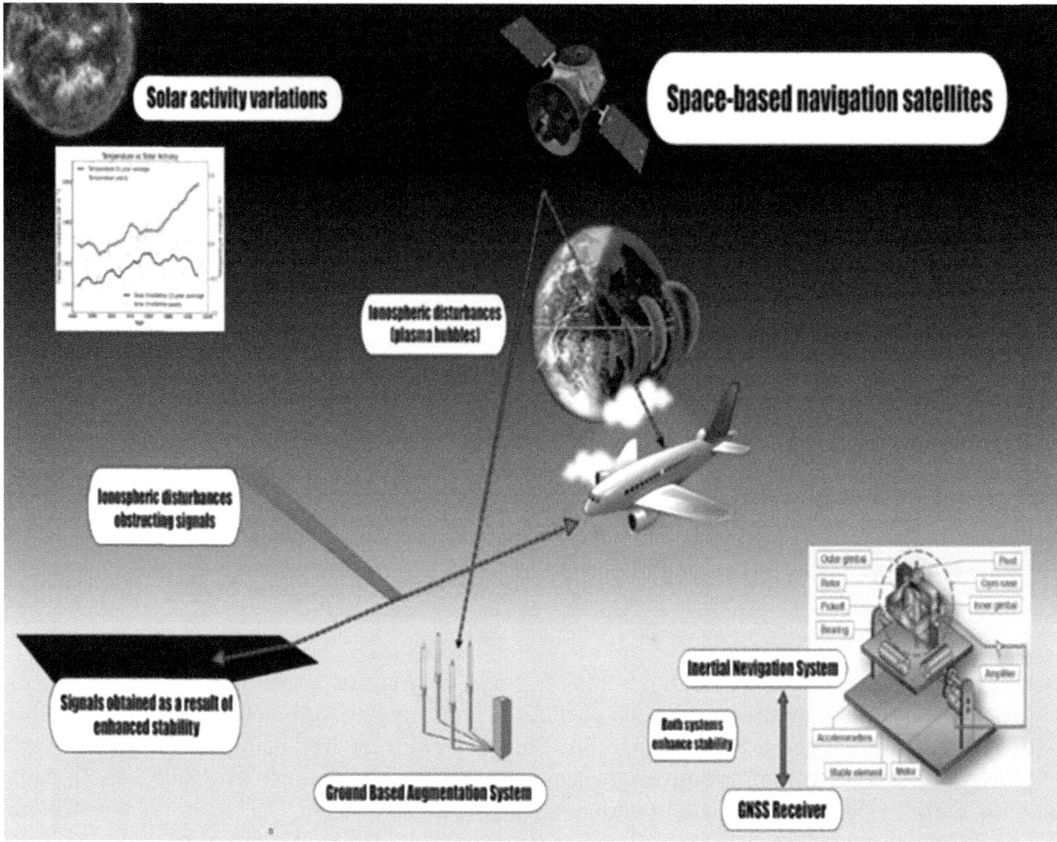

FIGURE 26.5 Ionospheric disturbances (plasma bubbles).

26.4.4 Equatorial Ionization Anomaly

A key characteristic of the ionosphere, especially in the equatorial area, is the Equatorial Ionization Anomaly (EIA), which is characterized by an asymmetric distribution of electron density. The F zone of the ionosphere, which is located between 150 and 500 kilometers above the Earth's surface, is where this phenomenon is most commonly observed. The EIA is more noticeable during the day and is more common in tropical areas than at higher latitudes.

A primary cause of the Equatorial Ionization Anomaly is the distinct geomagnetic and solar conditions present in the equatorial regions. Throughout the year, the equatorial region experiences high solar radiation, which is a major factor in ionizing the upper atmosphere. The Appleton anomaly layers are created, nevertheless, since the geomagnetic field in the equatorial zone is almost horizontal. By increasing the concentration of ionized particles, especially in the F2 layer, these layers contribute to the EIA, as shown in Figure 26.6.

The equator's strong solar radiation during the day causes the F area to become more ionized and electron dense. On the other hand, two crests that form on either side of the magnetic equator and are divided by a trough characterize the EIA in the equatorial zone. The migration of the Earth's ionosphere and the convergence of the geomagnetic field lines are two elements that combine to create this unusual structure. The propagation of radio waves and communication networks are significantly impacted by the Equatorial Ionization Anomaly. Radio communication disturbances, phase variations, and signal scintillation can be caused by the ionosphere's uneven distribution of

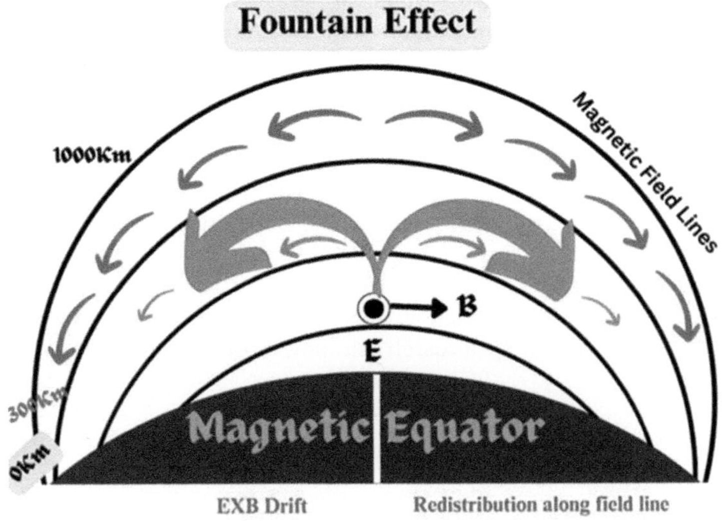

FIGURE 26.6 Understanding the Formation Mechanism of the Equatorial Anomaly.

electron density (M. A. Abdu, 2001; Balan, Liu, and Le, 2018; Sao et al., 1970; OHSHIO, 1971; Dabbakuti, Devanaboyina, and Kanchumarthi, 2016). Scientists hope to improve Space weather forecasts, increase the accuracy of ionospheric models, and lessen the effects of ionospheric anomalies on equatorial technology systems by learning more about the processes causing this anomaly.

26.4.5 SOLAR FLARE

A solar flare is a high energy burst of radiation emitted from the sun along with charged particles. Solar flares release energy in the X-ray, ultraviolet, and visible light ranges that extend the electromagnetic spectrum. X-rays are frequently used to track and categorize flare intensity. Solar flares typically last a few minutes to several hours, depending on their intensity. When a solar flare is active, it can shine in X-rays brighter than the rest of the Sun.

Based on their X-ray intensity, solar flares are divided into five groups, with X-class flares being the strongest and C-class flares being the least intense (Kiruthiga et al., 2022). Solar flares mostly occur during the day. Over the entire radio spectrum, the disturbances have a significant impact on radio navigation and communications. SIDs were generally identified as shortwave fadeouts, abrupt variations in frequency (or Doppler shifts), sudden absorption of cosmic disturbances, sudden enhancement/decrease of atmospherics and abrupt increment in TEC.

Solar flares can have major impacts on Earth's technologies and Space weather. Several of the outcomes include interference with radio transmission on Earth, especially at high frequencies. Military and emergency services, as well as maritime and aviation communication, may all be impacted by this interference. Solar flares may disrupt GPS accuracy and dependability, causing navigational errors (Moreno et al., 2011; Warnant and Pottiaux, 2000; Shagimuratov et al., 2020; Kolarski et al., 2023). Strong solar flares have the potential to trigger geomagnetic storms, which might harm power systems and result in blackouts. Solar flares may alter the orbits and lifetimes of low-Earth orbit satellites by increasing drag.

Solar flares are tracked and studied by scientists using a variety of tools and observatories, including solar telescopes and space-based observatories. The study of solar flares is essential for enhancing our capability to forecast and minimize their impact on the environment.

Several space agencies and organizations around the world, including NASA and the European Space Agency, continuously monitor the Sun's activity to predict when solar flares might occur. This early warning system helps protect astronauts in space, safeguard satellite operations, and mitigate potential disruptions on Earth (Dal'col et al., 2022; Makela, 2006; Ahmadzadeh et al., 2021).

Solar flares have a relationship with the Sun's 11-year solar cycle and are not just unpredictable occurrences. This cycle, commonly referred to as the solar activity cycle, exhibits an increasing and decreasing pattern of solar activity. Solar flares and other solar activity are more common and intense during solar maximum, which happens approximately every 11 years, whereas they are less frequent and less intense during solar minimum (Mendillo et al., 1974). In particular, during solar flare events, astronauts on the International Space Station (ISS) are exposed to more radiation, which may be damaging to their health. Continuing research into how Space weather affects human health is a priority for space agencies.

26.4.6 CORONAL MASS EJECTIONS

Solar flares are frequently accompanied by coronal mass ejections (CMEs), enormous blasts of solar wind, solar plasma, and magnetic fields. Geomagnetic storms can result from the interaction of CMEs with the Earth's magnetic field, which can carry billions of tonnes of material. These storms are a crucial component of Space weather studies because they can impair power networks, satellite operations, and communication systems. When a large amount of solar plasma and magnetic fields is released into space from the solar corona, the outermost region of the Sun, these magnificent and significant solar phenomena are known as coronal mass ejections. The Sun's corona, the outermost layer of the solar atmosphere, can suddenly and violently discharge charged particles, mostly protons and electrons, along with magnetic fields. With temperatures ranging from one to three million degrees Celsius, the corona is a delicate and extremely hot area of the Sun. CMEs arise in areas with highly twisted and complex magnetic fields and are frequently linked to solar flares (M. A. Abdu, 2001; Schmölter and Berdermann, 2021; Mosna et al., 2020; Moges et al., 2022).

When a CME hits Earth, it can interfere with the magnetic field, causing radio communication interruptions, damage to satellites and other space assets, geomagnetic storms, and even spectacular auroras (the Northern and Southern Lights). CMEs are of significant importance in the fields of solar and space physics for several reasons such as studying the behavior of the Sun, Space weather forecasting, magnetosphere and ionosphere dynamics, and planetary studies.

Ionized gases in the form of plasma, mostly hydrogen and helium, make up the majority of the corona's composition. Traces of other elements are also present. These atoms lose their electrons due to the intense heat of the corona, resulting in a plasma of charged particles. Understanding solar dynamics and Space weather requires an understanding of solar magnetic activity and the corona. This emission may be linked to sunspots, dynamic regions, or even solar flares, and it may be caused by phenomena such as magnetic connection. CME production and eruptions from the solar corona are caused by the interaction of these factors.

26.4.7 GEOMAGNETIC STORM

The disruption of the Earth's magnetosphere, which is primarily caused by the interaction of the solar wind with the planet's magnetic field, results in geomagnetic storms, a natural phenomenon (Bolaji, Adebiyi, and Fashae, 2019; Alfonsi et al., 2021; Olawepo et al., 2022). The results during the storm phase also reveal that over the low-latitude region, intensity of amplitude scintillation is increased whose magnitude enhances the effectiveness of the geomagnetic storm.

26.4.8 VTEC Profiling

The Vertical Total Electron Content (VTEC) is calculated by combining GNSS data with other instruments. These instruments are ionosondes, radiosondes, or radio occultation sensors. VTEC profiling offers important insights into the VTEC, which aids in the understanding of the ionosphere's complex structure. This method is essential for accurately tracking variations in electron density at various altitudes and for comprehending how Space weather events affect the ionosphere. Scientists can improve the accuracy of ionospheric models, the dependability of global navigation and communication systems, and our general understanding of the dynamic interactions between solar effects and Earth's upper atmosphere by analyzing VTEC data.

Essentially, VTEC profiling is an important instrument in the continuous effort to improve Space weather prediction and maximize the capabilities of technologies dependent on satellites. It was possible to monitor the ionospheric TEC, although in a one-dimensional way, by using satellite data. However, by combining one-dimensional satellite data over a certain region, it is possible to create ionospheric TEC maps. While other mapping techniques have been recognized, we have attempted to evaluate kriging interpolation in this study.

An algorithm for reconstructing Global Ionospheric Maps (GIMs) was created. We calculated the ideal number of stations by simulating 12 days in March 2015, which included times of high scintillation and an ionospheric storm caused by the 2015 St. Patrick's Day geomagnetic storm. During the day, there is a north-south gradient in the VTEC in this region. Depending on the ionospheric conditions, the VTEC in the south typically ranges from 60 to 90 TECU, whereas in the north it exceeds 100 TECU.

The VTEC can be further calculated by using equations (26.2) and (26.3):

$$\text{VTEC} = \frac{\text{STEC}}{\text{SF}} \tag{26.2}$$

$$\text{SF} = \frac{1}{\cos Z'} \tag{26.3}$$

and Z' is the zenith angle at the IPP.

The STEC using code-range is given as equation (26.4):

$$\text{STEC} = \frac{1}{40.3} \times \left(\frac{f_{S1}^2 - f_{L5}^2}{f_{S1}^2 f_{L5}^2} \right)\left(\pounds_{L5} - \pounds_{S1} \right) \tag{26.4}$$

where $f_{S1} = 2492.028$ MHz and $f_{L5} = 1176.45$ MHz are the two frequencies of the signals, and where $\pounds_i$ is the measured code-range (in *meters*) at dual-frequency ($i = $ L5 and S1).

The STEC using carrier ranges is given by equation (26.5):

$$\text{STEC} = \frac{1}{40.3} \times \left(\frac{f_{S1}^2 - f_{L5}^2}{f_{S1}^2 f_{L5}^2} \right)\left(\varphi_{L5} - \varphi_{S1} \right) \tag{26.5}$$

where φ_i carrier-phase range measurements (in *meters*) at dual-frequency ($i = $ L5 and S1).

26.4.9 Ionospheric Modeling

Ionospheric modeling is done when some disturbances occur in the ionosphere and the position of user or receiver is to be maintained, when the user should switch to some model for maintaining its position. Accurate modeling of the ionosphere for scientific and technological research is a challenging problem since many plasma instabilities are still undiscovered and need more research. The study employs NavIC-based measurements to examine the ionospheric TEC variability over the low-latitude Indian region. IRI-Plas 2017 model with additional TEC input applied to diurnal, seasonal, and storm time variations over the Indian low-latitude Hyderabad station is one of the major contributions in this work. Another major contribution is the comparison of regional NavIC-derived TEC with GIM TEC and popular global empirical model IRI-2016.

Finally, IRI 2016 models are better as compared to IRI-Plas models, but the IRI-Plas model predicts ionospheric as well as plasmaspheric part of the ionosphere. The suggested study can be modified to include significant data from numerous stations for various solar activity to draw further conclusions. The findings demonstrate that NavIC data, in addition to GPS data, can be utilized to evaluate the efficiency of global models. The results and evaluations reported in this study will be highly beneficial in developing regional error correction values. NavIC data for five distinct locations is used to assess the models' performance under the intense geomagnetic storm conditions of September 7–9. For five places, IRI Plas 2017 outperforms IRI 2016 model on the post-storm day.

In this study, the IRI Plas model overestimates TEC in comparison to the NavIC TEC regardless of the days of the year, and it performs poorly during the day over low-latitude regions. It might be because the model's representation of plasmaspheric TEC is poor. The model needs to be enhanced further. Otherwise, greater model data assimilation performance over low-latitude regions is required.

It was observed that ionospheric effects are related to the location, the local time, and the solar activity. Large higher-order ionospheric effects are found in the low-latitude area in the daytime as well as during strong solar activities. According to the authors' investigation, the assimilation process is crucial, and ionospheric modeling can be completed tag-free, highlighting the value of the data assimilation technique. Data is therefore taken into account for modeling and data assimilation requirements. As a result, data assimilation is crucial for real-time monitoring applications. In order to create real-time TEC maps and manage real-time conditions during extreme Space weather events, data assimilation is also carried out for the VTEC parameter. The contents of the TEC serve as the fundamental basis for assimilation.

26.4.10 REAL-TIME DATA ANALYSIS

Real-time data analysis is done for studying the ionosphere at the time of any Space weather events like plasma bubbles, geomagnetic storms, solar flares, coronal mass ejections, and solar storms. The forecasting of these events should be done in real time so that the position of the user or receiver is not affected. Various Space weather events occurring in the ionosphere are shown in Figure 26.7.

26.5 ADVANCEMENTS OF GNSS IN IONOSPHERIC RESEARCH

26.5.1 MULTI-GNSS SYSTEMS

Multi-GNSS systems are navigation and positioning systems that make use of numerous satellite constellations to determine exact locations and deliver precise timing data. The Global Positioning System (GPS), GLONASS (Russia), Galileo (European Union), and BeiDou (China) are examples of these constellations. Multi-GNSS basically refers to the concept of merging signals and data from different satellite systems to improve the precision, reliability, and accessibility of global navigation and positioning services. Global Navigation Satellite Systems (GNSS) have significantly enhanced ionospheric research by providing diverse and numerous data sources. The main benefit of multi-GNSS systems is increased positioning and navigational accuracy. By combining signals from many constellations, signal interference, atmospheric interference, and satellite geometry concerns reduce the likelihood of a problem, resulting in more accurate position data. While using numerous constellations, more satellites become visible, especially in difficult terrain with tall structures, dense

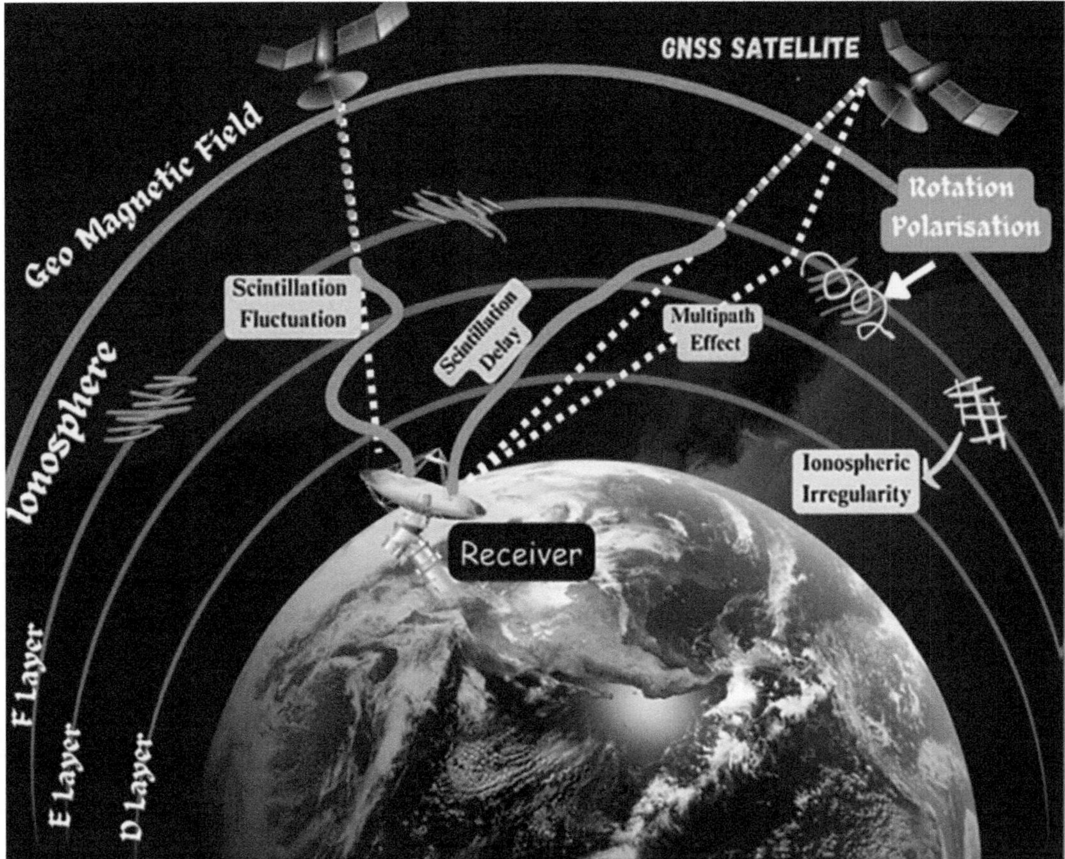

FIGURE 26.7   Real-time Space weather events.

vegetation, or other challenges. This results in improved signal reception and positioning data availability.

There is a higher level of redundancy when there are many constellations. Other constellations continue to provide data if one system's signals are obstructed or disrupted, ensuring more accurate navigational and timing data. Satellite coverage by different GNSS systems may vary throughout various regions of the globe. Multi-GNSS ensures global coverage by merging different systems, making it especially useful for applications that require navigation across various geographical locations.

The geometry of satellite constellations can be unfavorable for a single GNSS system at high-latitude areas close to the Earth's poles. By offering improved coverage in these regions, multi-GNSS helps to reduce this problem. In order to deliver more reliable and precise location, navigation, and timing services, the idea of numerous GNSS constellations cooperating involves combining signals and data from several satellite systems. Multi-GNSS-enabled receivers simultaneously gather signals from several constellations, including GPS, GLONASS, Galileo, and BeiDou. In order to determine the most precise position iond timing data, the receiver processes and merges the data from these constellations. The receiver can reduce errors and increase the estimated position's overall accuracy by integrating signals from several systems.

Through the combination of signals from many GNSS constellations, inaccuracies caused due to multipath interference, satellite geometry, and signal delay caused by the ionosphere and tropo-sphere are reduced. As a result, location determination for users is significantly more accurate. Even in difficult situations, such as urban canyons, dense forests, and areas with restricted line-of-sight to satellites, multi-GNSS ensures that signals are accessible. Receivers can switch to another constellation that is accessible if one GNSS system has signal disturbances or interference, ensuring uninterrupted operation and improved reliability. Ionospheric anomalies can be more efficiently detected and tracked by using multiple GNSS systems. When data from several systems are combined, sudden variations in electron density, such as ionospheric storms or scintillation occurrences, can be better understood and tracked. This knowledge is helpful for detecting and minimizing expected communications and navigational failures.

26.5.2 DUAL-FREQUENCY MEASUREMENTS

Dual frequency measurements like L5 and S1 are used for calculating TEC at dual frequencies to study ionospheric irregularities and characterize the ionosphere. The TEC is the only parameter utilized in the climatology of ionospheric irregularities and ionospheric studies.

26.5.3 NETWORKS OF GNSS RECEIVERS

The IGS (International GNSS systems) receivers are used for getting the TEC information at different stations. But in the case of the Indian region, there are limited IGS stations, which again raises the problem of sparse networks of receivers.

26.5.4 IONOSPHERIC TEC MAPS

The TEC maps are used for knowing the coordinates (altitude, latitude, and longitude) at the time of Space weather events (Olawepo et al., 2022; Gubenko et al., 2021; Danilov and Konstantinova, 2019; Wan et al., 2021; Candido et al., 2018; Cheng et al., 2021; Alfonsi et al., 2021). There are three type of GIMs: final, rapid, and predicted GIMs. The final GIMs provide highly accurate and reliable TEC data. There are a limited number of TEC maps than the IGS ROTI developed in 2018 that is used to map the novel ionospheric product created for ionospheric irregularity analysis (Gogie, 2021; Ayyagari et al., 2020). The most recent IGS ionospheric element determines the GPS signal phase fluctuation behavior as well as the amount and location of huge irregular structures in the composition of the ionospheric plasma. Positioning errors in the vicinity of strong GPS signal variations, as shown by the ROTI map, have an impact on ground-based GNSS systems, as given in equations (26.6) and (26.7).

$$ROT = \frac{TEC_{ph}(i) - TEC_{ph}(i-1)}{\Delta t} \tag{26.6}$$

$$ROTI = \sqrt{\frac{1}{N-1} \sum_{j=i}^{N} \left(ROT(i) - \overline{ROT}\right)^2} \tag{26.7}$$

26.5.5 IONOSPHERIC MONITORING SERVICES

Ionospheric monitoring services are essential for tracking and evaluating the dynamic behavior of the ionosphere, an area important for satellite operations, worldwide navigation, and communication, among other technical uses. These services gather data on important ionospheric characteristics using a variety of satellite- and ground-based devices. They provide historical and real-time data critical to comprehending the dynamics of Space weather.

Ionospheric Monitoring Services measure Total Electron Content (TEC) in various geographic places as one of its main priorities. TEC data contributes to the creation of ionospheric maps, which are useful for determining electron density fluctuations and forecasting possible ionospheric irregularity-induced disturbances to radio wave transmission. Additionally, these services play a major role in Space weather monitoring. They offer early warnings and forecasts for possible disruptions to radio transmissions and satellite communications by continuously monitoring ionospheric conditions. This is especially crucial for reducing the negative effects of Space weather events on satellite operations, navigation systems, and other essential technologies that depend on ionospheric conditions.

A network of strategically placed ground-based sensors, including magnetometers, GPS receivers, and ionosondes, is frequently used by ionospheric monitoring services. Furthermore, information from multiple satellite missions adds to a thorough knowledge of global ionospheric behavior. A wide spectrum of users, including researchers, space agencies, telecommunications corporations, and sectors dependent on satellite-based technologies, are served by these services. Ionospheric Monitoring Services' data is essential for maximizing the dependability and performance of systems that operate in or through the ionosphere.

Ionospheric Monitoring Services are expected to develop with time, bringing with them more advanced sensors, better data assimilation methods, and enhanced modeling methodologies. This development will improve our capacity to anticipate and lessen the effects of ionospheric variability on the infrastructure of contemporary technology. To sum up, in our globalized society, Ionospheric Monitoring Services are essential for promoting a better knowledge of the ionosphere and guaranteeing the uninterrupted functioning of several important technologies.

ESA (European Space Agency) and IGS are space monitoring services that continuously track and monitor the ionosphere.

26.5.6 Real-time Monitoring of Ionosphere

It is critical for assessing impact of solar events on the ionosphere and ensuring the safety of satellite-based systems. Real-time data is significant for early warning of natural hazards such as earthquakes, tsunamis, Space weather monitoring, etc. For a wide range of applications, from navigation to the detection of Space weather events, monitoring the vertical total electron content (VTEC) in the ionosphere is important. As a result, numerous analysis centers have been actively working to develop various approaches for real-time VTEC estimation. Due to its extensive network of receivers, high temporal resolution, and capacity for rapid data dissemination with minimal latency, Global Navigation Satellite Systems (GNSSs) stand as a key technology for ionosphere modeling. The careful selection of a suitable technique to extract ionospheric data from GNSSs and the creation of VTEC representation through an appropriate mathematical model are essential for providing quick and accurate ionosphere products.

A variety of internal and external forces, such as gravity, seismic activity, precipitation, and human actions, can all contribute to the movement of rock-soil masses on slopes. It is crucial to monitor and issue early warnings for landslides since they constitute a serious concern to many nations and can result in casualties and property destruction. Due to its many benefits, such as high precision, all-weather functionality, continuous three-dimensional tracking, and the absence of communication requirements, the Global Navigation Satellite System (GNSS) has been a highly effective method for landslide monitoring since the 1990s.

In Space weather research and ionospheric studies, real-time data analysis is becoming more and more crucial. Researchers are now able to monitor and react quickly to changes in the ionosphere due to the availability of additional instruments that can offer real-time data. This is particularly critical for applications that depend on precise ionospheric data for reliable operation such as communication systems and GPS navigation. The use of differential GNSS in real-time scenarios reduces the chances of inaccuracy while navigating, as shown in Figure 26.8.

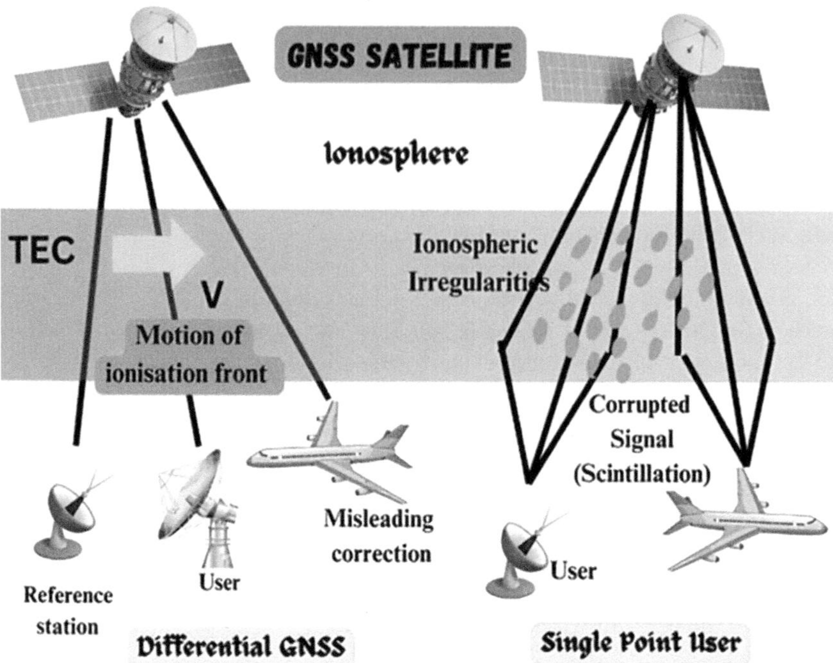

FIGURE 26.8   Requirement of real-time monitoring of Ionosphere.

26.5.7 Integration with Other Instruments

Ionospheric researchers now rely heavily on global navigation satellite systems (GNSSs) like the Global Positioning System (GPS). Because of their sensitivity to fluctuations in electron density, these devices offer a great chance to research the ionosphere. GNSSs improve our understanding of the complex ionospheric environment when combined with other tools and methods. The following are some developments in GNSS-based ionospheric research as a result of instrument integration:

1. **Total Electron Content (TEC) Measurements**:
   - **Integration with Ground-Based Receivers**: The total electron content is determined by ground-based GNSS receivers following the signal path. TEC maps can be created more easily when numerous strategically positioned receivers offer spatial distribution information.
   - **Combining with Ionosondes**: The electron density profile can be better understood by combining TEC readings from GNSS with ionosonde data. This integration contributes to the improvement and validation of TEC measurement accuracy.
2. **Ionospheric Tomography**:
   - **Combining GNSS and Tomography Techniques**: Ionospheric tomography systems benefit from the use of GNSS receivers, which supply important information for reconstructing three-dimensional electron density profiles. Ionospheric tomography's accuracy and spatial resolution are improved by this integration.
3. **Scintillation Monitoring**:
   - **GNSS Scintillation Monitors**: Ionospheric anomalies give rise to scintillations, which are abrupt variations in the amplitude and phase of GNSS signals. GNSS scintillation monitors are useful for characterizing the nature and effects of ionospheric abnormalities on signal propagation when combined with other equipment such as radar and satellites (Beach and Kintner, 2001; Mohanty et al., 2018).

4. **Space Weather Monitoring**:
   - **Integration with Space Weather Instruments**: For tracking Space weather events like solar flares and geomagnetic storms, GNSS data is essential. Researchers can examine the ionospheric reaction to outside influences and enhance Space weather prediction models by integration with Space weather instrumentation.
5. **Combining with Satellite Observations**:
   - **Correlation with Satellite Payload Data**: Data from satellite-borne scientific devices can be linked with GNSS observations. This integration makes it possible to investigate ionospheric dynamics more thoroughly and improves our understanding of the interaction between various ionosphere regions.
6. **Improved Modeling and Prediction**:
   - **Assimilation into Ionospheric Models**: To increase the precision and dependability of ionospheric models, GNSS data can be incorporated. This integration aids in the development of more accurate models by researchers to forecast the behavior of the ionospheric layer under various circumstances (Jin and Song, 2023; Tang et al., 2022; Solomon et al., 2012; Cherniak, Zakharenkova, and Krankowski, 2014; Tariku, 2015; Kumar, Tan, and Murti, 2015).
7. **Validation of Remote Sensing Techniques**:
   - **Cross-Validation with Lidar and Radar Data**: Remote sensing techniques can be validated by combining lidar and radar data with GNSS measurements. By using cross-validation, measurements of the ionosphere become more accurate and strengthen our understanding of the ionosphere.
8. **Real-time Monitoring and Early Warning Systems**:
   - **Integration for Space Weather Forecasting**: When GNSS data is integrated into real-time monitoring systems, early warning systems for Space weather phenomena can be developed. In order to decrease the influence of ionospheric disturbances on communication and navigation systems, this integration is essential.

26.5.8 High-data Rate Receivers

Receivers such as ISM (Ionospheric Scintillation Monitor) are utilized for studying fast-changing ionospheric phenomena like ionospheric scintillation. A key component of contemporary ionospheric monitoring systems is integration with high-data rate receivers, which improves the accuracy and productivity of data collecting in ionospheric research. Specialized receivers designed to manage massive amounts of data from Global Navigation Satellite System (GNSS) signals are known as high-data rate receivers. Because they allow researchers to gather high-frequency and high-resolution data on the Total Electron Content (TEC) and other ionospheric characteristics, these receivers are essential to ionospheric investigations. Large-scale data sets can be collected simultaneously when High-Data Rate Receivers are integrated with ionospheric monitoring devices like ionosondes and dual-frequency GNSS receivers. More temporal and spatial resolution in the study of the ionosphere's dynamic behavior is made possible by this combination. Because of these receivers' high data rates, scientists can record abrupt changes in ionospheric conditions, which advances our knowledge of Space weather phenomena. Working together with High-Data Rate Receivers in ionospheric monitoring networks helps enhance the forecasting of ionospheric disturbances and to construct accurate ionospheric models.

Applications like satellite communication, navigation, and surveillance systems depend on the ability to detect and analyze ionospheric abnormalities, which is further enhanced by the real-time data these receivers provide. The use of High-Data Rate Receivers into ionospheric research is evidence of the continuous efforts to utilize cutting-edge technology for enhanced and comprehensive Space weather surveillance. The cooperation with ionospheric monitoring instruments will further advance our understanding of the ionosphere and its effects on satellite-based technologies and radio wave propagation as these receivers continue to evolve, adding features like multi-frequency and multi-constellation capabilities.

26.5.9 ADVANCED SIGNAL PROCESSING TECHNIQUES

Algorithms are used to extract valuable ionospheric information from GNSS data, including detection of ionospheric irregularities and scintillation.

26.6 LIMITATIONS OF GNSS IN IONOSPHERIC RESEARCH

26.5.10 TEC DENSITY

The total electron density is an important parameter for measuring the density of electrons and the study of ionospheric irregularities, Space weather monitoring studies, etc. In spite of that, the ionosphere height is not fixed, so ionospheric study is crucial for knowing the uppermost layer dimensions and height of the ionosphere.

26.5.11 SIGNAL BLOCKAGE

Signal blockage or multipath is a problem because of tall buildings or obstructions in line-of-sight. Thus, measures should be taken to prevent signal blockage.

26.5.12 NEED OF ADDITIONAL INSTRUMENTS

The GNSS alone is insufficient to conduct ionospheric research; additional instruments such as radiosondes, ionosondes, radar, etc., should support research as well. These instruments are available for real-time data analysis to better support research in real-time scenarios.

The following tools and methods are frequently used by researchers in addition to GNSS to bypass GNSS's limitations and gain deeper knowledge of the ionosphere:

1. **Ionosondes** – Ionosondes are fixed devices that measure the time required for radio waves to return from the ionosphere by sending them vertically into the atmosphere. With this data, scientists may produce electron density profiles that provide insight into the structure of the ionosphere at various elevations. Information regarding the critical frequency of the ionospheric layers can also be obtained using ionosondes. Every instrument contributes distinct data that enhances the information derived by GNSS. Understanding the ionospheric structure and its fluctuations is crucial for forecasting how electromagnetic signals will behave as they travel through it, and these profiles aid in this process. Ionosondes can also measure the critical frequency of ionospheric layers, which helps determine the conditions for radio signal transmission. This method improves Space weather prediction models and enables researchers to better investigate ionospheric anomalies and disturbances due to continuous technological developments. Utilizing additional instruments like radiosonde, ionosonde, and radar minimizes the inaccuracy in ionospheric research data or Space weather studies, as shown in Figure 26.9.
2. **Radar Systems** – Ionospheric electron density, temperature, and composition can all be obtained using radar systems, especially incoherent scatter radars. To investigate distinct ionospheric layers, these radar systems can function at a range of frequencies and angles.
3. **Digisondes** – Digisondes are advanced ionosondes that offer comprehensive data regarding the ionosphere's electron density profile. They can measure ionospheric characteristics at several altitudes using frequency-swept signals.

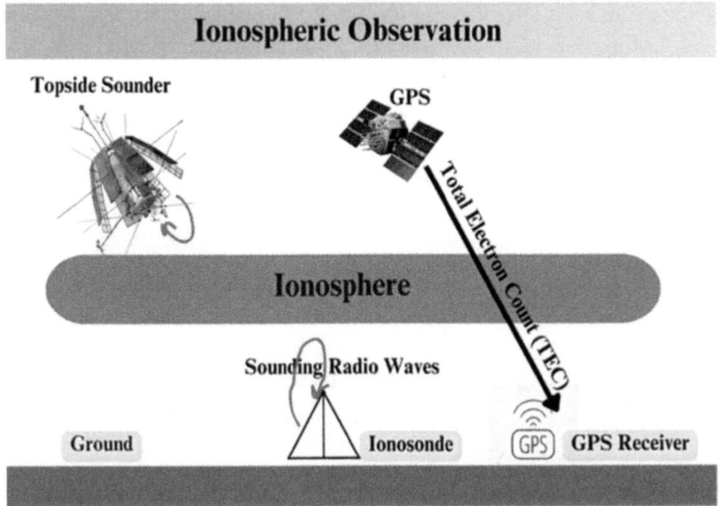

FIGURE 26.9 Ionospheric observation using additional instruments.

4. **Ionospheric Tomography Systems** – These devices generate a three-dimensional picture of the ionosphere using transmitters and receivers. Through the measurement of signal variations during ionosphere advancement, scientists can reconstruct profiles of electron density.
5. **Langmuir Probes** – For in-situ measurements in space, Langmuir probes are utilized. They take direct measurements of the temperature, spacecraft potential, and electron density in the plasma. Ionosphere-focused satellite missions frequently employ Langmuir probes.
6. **Satellites** – Important data for ionospheric study is provided by satellites fitted with ionospheric equipment. Instruments on board scientific satellites, communication satellites, and specialized ionospheric research satellites are some examples.
7. **Fabry-Perot Interferometers** – These devices monitor the Doppler shift of specific ionosphere-derived emission lines. Researchers can determine temperature and ionospheric plasma drift by examining the shift.
8. **LIDAR (Light Detection and Ranging)** – By monitoring backscattered light from the atmosphere, laser-based LIDAR systems provide data on the density and altitude of ionospheric layers.
9. **Ionospheric Sounders** – These devices use a set frequency to broadcast radio waves and track the frequency at which the waves reappear after being reflected by the ionosphere. This data aids in calculating the reflecting layer's virtual height.

These instruments enable multi-instrument methods, real-time monitoring, and a variety of observational techniques in addition to enhancing the data provided by GNSS. They serve a vital role in expanding our knowledge of the ionosphere and enhancing the precision and dependability of systems for communication, navigation, and Space weather forecasting.

26.5.13 RECEIVER LOCATION

Sparse receiver networks or uneven receiver distribution can lead to data gaps in remote regions to examine the total electron content (TEC) variation at low, mid, and high latitudes in the Northern Hemisphere on a daily, weekly, yearly, and seasonal scale. In this chapter, TEC fluctuations are compared to solar proxies (sunspot number, DST, Kp index, and F10.7) at three different latitudes during the 24th solar cycle's low activity period (January to December 2016). Several distinct features were noted, including periodic seasonal TEC oscillation and diurnal TEC sinusoidal patterns. At low-latitude locations, maximum diurnal and monthly TEC results were computed. At low, mid, and high latitudes, the maximum seasonal TEC was recorded during the equinox. The study demonstrated that low-latitude regions display greater TEC magnitude and more TEC fluctuations than mid- or high-latitude regions (Bolaji, Adebiyi, and Fashae, 2019).

The observer mostly observes low elevation angles of satellites because the northern portion of India is situated at low latitudes. The primary issue with the location of the receiver is that the low-latitude elevation angle persists, a significant concern in the hilly northern region. It is a major problem in positioning; therefore, we should design a specific algorithm for receivers located at low latitudes.

26.6 LACK OF DATA ON OTHER IONOSPHERIC PARAMETERS

Although GNSS is a great tool for measuring TEC, it lacks direct information about other important ionospheric characteristics, such as:

Profiles of Electron Density: The electron density of the various layers that make up the ionosphere vary depending on altitude. The vertical distribution of electron density is not provided by GNSS. For a thorough understanding of the structure and behavior of the ionosphere, it is crucial to comprehend these profiles.

Electron Temperature: The behavior of the ionosphere, including the propagation of radio waves, can be influenced by the temperature of the electrons within it. Data on electron temperature cannot be obtained solely from GNSS.

Ion Composition: A mixture of positively charged electrons and ions makes up the ionosphere. Understanding the ionosphere's interactions with GNSS signals and other electromagnetic waves requires knowledge of its composition.

26.6.1 DATA ACCESS AND PROCESSING

In order to examine how GNSS signals are impacted as they travel through the ionosphere, ionospheric research employs sophisticated data processing techniques, and specialized software is required for this process. To assess ionospheric parameters like total electron content (TEC) or ionospheric scintillation, researchers may have to develop complex algorithms or employ ones that already exist. In some cases, researchers may need real-time or near-real-time access to GNSS data for ionospheric monitoring and forecasting.

Data latency can be a limitation, as it may take time for data to be collected, transmitted, and processed. VTECs in data availability can be problematic in applications such as Space weather monitoring or geomagnetic storm prediction (Erdogan et al., 2021). For ionospheric monitoring and forecasting, researchers sometimes need access to GNSS data in real time or almost real time. The distribution of GNSS data is not uniform on Earth, and insufficient coverage may exist, especially in isolated or marine areas. This may restrict the geographic scope of ionospheric research or make it more challenging to investigate ionospheric irregularity events in particular regions.

Several elements can impact the precision and accuracy of GNSS data, such as receiver characteristics, atmospheric conditions, and satellite geometry. High-quality data are necessary for ionospheric research to make insightful observations and develop accurate conclusions. Research might become inaccurate due to biases and errors in data.

26.7 CONCLUSION

This chapter has shown how GNSS technology has significantly improved our capacity to gather and process ionospheric data, which has refined ionospheric models and increased our understanding of Space weather events.

The developments covered in this chapter demonstrate how GNSS can help a number of sectors, including communication, navigation, and Space weather forecasting. Research on ionospheric irregularities, the impacts of solar and geomagnetic activity, and the development of more precise location and timing solutions have all been made possible by the high precision and worldwide coverage of GNSS systems. Nevertheless, it is necessary to recognize the limitations and challenges that accompany GNSS-based ionospheric research. Among the limitations are the sensitivity of GNSS signals to ionospheric disturbances and the requirement for precise data processing. Future studies in this area should concentrate on resolving these issues, advancing methods for the assimilation of data, and strengthening the adaptability of ionospheric models. GNSS technology will enable the scientific community and technological sectors to work together to advance the subject of ionospheric research. GNSS-based ionospheric research will be crucial in reducing the effects of Space weather events and deepening our knowledge of this rapidly changing and complex area of Earth's atmosphere as these events become more significant in a modern connected world.

In summary, the development and limitations of GNSSs in ionospheric research indicate a dynamic and developing field that continues influencing our understanding of the ionosphere and its useful applications. This study not only serves as evidence of the advancements made so far but also as motivation for more research in this important and significant field of study.

REFERENCES

Abdu, M. A. 2001. "Outstanding Problems in the Equatorial Ionosphere-Thermosphere Electrodynamics Relevant to Spread F." *Journal of Atmospheric and Solar-Terrestrial Physics* 63 (9): 869–84. https://doi.org/10.1016/S1364-6826(00)00201-7

Ahmadzadeh, Azim, Berkay Aydin, Manolis K. Georgoulis, Dustin J. Kempton, Sushant S. Mahajan, and Rafal A. Angryk. 2021. "How to Train Your Flare Prediction Model: Revisiting Robust Sampling of Rare Events."*TheAstrophysicalJournalSupplementSeries*254(2):23.https://doi.org/10.3847/1538-4365/abec88

Alfonsi, L., C. Cesaroni, L. Spogli, M. Regi, A. Paul, S. Ray, S. Lepidi, et al. 2021. "Ionospheric Disturbances Over the Indian Sector During 8 September 2017 Geomagnetic Storm: Plasma Structuring and Propagation." *Space Weather* 19 (3): 1–16. https://doi.org/10.1029/2020SW002607

Ayyagari, Deepthi, Sumanjit Chakraborty, Saurabh Das, Ashish Shukla, Ashik Paul, and Abhirup Datta. 2020. "Performance of NavIC for Studying the Ionosphere at an EIA Region in India." *Advances in Space Research* 65 (6): 1544–58. https://doi.org/10.1016/j.asr.2019.12.019

Balan, Nanan, Li Bo Liu, and Hui Jun Le. 2018. "A Brief Review of Equatorial Ionization Anomaly and Ionospheric Irregularities." *Earth and Planetary Physics* 2 (4): 257–75. https://doi.org/10.26464/epp2018025

Beach, Theodore L., and Paul M. Kintner. 2001. "Development and Use of a GPS Ionospheric Scintillation Monitor." *IEEE Transactions on Geoscience and Remote Sensing* 39 (5): 918–28. https://doi.org/10.1109/36.921409

Bolaji, O. S., S. J. Adebiyi, and J. B. Fashae. 2019. "Characterization of Ionospheric Irregularities at Different Longitudes during Quiet and Disturbed Geomagnetic Conditions." *Journal of Atmospheric and Solar-Terrestrial Physics* 182: 93–100. https://doi.org/10.1016/j.jastp.2018.11.007

Brassarote, de Oliveira Nascimento, Gabriela de Souza, Eniuce Menezes Monico, and João Francisco Galera. 2018. "S4index: Does It Only Measure Ionospheric Scintillation?" *GPS Solutions* 22 (1). https://doi.org/10.1007/s10291-017-0666-x

Candido, Claudia M.N., Inez S. Batista, Virginia Klausner, Patricia M. de Siqueira Negreti, Fabio Becker-Guedes, Eurico R. de Paula, Jiankui Shi, and Emilia S. Correia. 2018. "Response of the Total Electron Content at Brazilian Low Latitudes to Corotating Interaction Region and High-Speed Streams during Solar Minimum 2008." *Earth, Planets and Space* 70 (1). https://doi.org/10.1186/s40623-018-0875-8

Cheng, Pin Hsuan, Charles Lin, Yuichi Otsuka, Hanli Liu, Panthalingal Krishanunni Rajesh, Chia Hung Chen, Jia Ting Lin, and M T Chang. 2021. "Statistical Study of Medium - Scale Traveling Ionospheric Disturbances in Low - Latitude Ionosphere Using an Automatic Algorithm." *Earth, Planets and Space*, 1–17. https://doi.org/10.1186/s40623-021-01432-1

Cherniak, Iurii, Irina Zakharenkova, and Andrzej Krankowski. 2014. "Approaches for Modeling Ionosphere Irregularities Based on the TEC Rate Index." *Earth, Planets and Space* 66 (1): 165. https://doi.org/10.1186/preaccept-1949710399128347

Dabbakuti, J. R. K. K., Venkata Ratnam Devanaboyina, and S. Ramesh Kanchumarthi. 2016. "Analysis of Local Ionospheric Variability Based on SVD and MDS at Low-Latitude GNSS Stations 6. Geodesy." *Earth, Planets and Space* 68 (1): 1–11. https://doi.org/10.1186/s40623-016-0478-1

Dal'col, Claudio, Levi W, Resende Filho, Rosa Correa Pabon, Ralf Moura, and Gustavo Pessin. 2022. "The Impact of Ionospheric Irregularities in Autonomous Systems Operations." *International Geoscience and Remote Sensing Symposium (IGARSS)* July: 7681–84. https://doi.org/10.1109/IGARSS46834.2022.9883702

Danilov, A. D., and A. V. Konstantinova. 2019. "Ionospheric Precursors of Geomagnetic Storms. 1. A Review of the Problem." *Geomagnetism and Aeronomy* 59 (5): 554–66. https://doi.org/10.1134/S0016793219050025

Erdinç Timoçin, Hüseyin Temuçin, and Samed İnyurt. 2022. "The Seasonal Characteristics of Equatorial Plasma Bubbles in Conjugate Hemispheres During 2015." *Geomagnetism and Aeronomy* 62 (3): 309–23. https://doi.org/10.1134/S0016793222030203

Erdogan, Eren, Michael Schmidt, Andreas Goss, Barbara Görres, and Florian Seitz. 2021. "Real-Time Monitoring of Ionosphere VTEC Using Multi-GNSS Carrier-Phase Observations and B-Splines." *Space Weather* 19 (10). https://doi.org/10.1029/2021SW002858

Galkin, Ivan, Adam Froń, Bodo Reinisch, Manuel Hernández-Pajares, Andrzej Krankowski, Bruno Nava, Dieter Bilitza, et al. 2022. "Global Monitoring of Ionospheric Weather by GIRO and GNSS Data Fusion." *Atmosphere* 13 (3). https://doi.org/10.3390/atmos13030371

Ganguly, S., and A. Brown. 2001. "Real-Time Characterization of the Ionosphere Using Diverse Data and Models." *Radio Science* 36 (5): 1181–97. https://doi.org/10.1029/1999RS002412

Garner, T. W., T. L. Gaussiran, B. W. Tolman, R. B. Harris, R. S. Calfas, and H. Gallagher. 2008. "Total Electron Content Measurements in Ionospheric Physics." *Advances in Space Research* 42 (4): 720–26. https://doi.org/10.1016/j.asr.2008.02.025

Ghimire, Basu Dev, and Narayan P. Chapagain. 2022. "Ionospheric Anomalies Due to Nepal Earthquake-2015 as Observed from GPS-TEC Data." *Geomagnetism and Aeronomy* 62 (4): 460–73. https://doi.org/10.1134/S0016793222040041

Gogie, T. K. 2021. "The Rate of Ionospheric Total Electron Content Index (ROTI) as a Proxy for Nighttime Ionospheric Irregularity Using Ethiopian Low-Latitude GPS Data." *Geomagnetism and Aeronomy* 61 (3): 464–75. https://doi.org/10.1134/S0016793221030051

Gubenko, V. N., V. E. Andreev, I. A. Kirillovich, T. V. Gubenko, A. A. Pavelyev, and D. V. Gubenko. 2021. "Radio Occultation Studies of Disturbances in the Earth's Ionosphere During a Magnetic Storm on June 22–23, 2015." *Geomagnetism and Aeronomy* 61 (6): 810–18. https://doi.org/10.1134/S0016793221060050

Guo, Kai, Marcio Aquino, and Sreeja Vadakke Veettil. 2019. "Ionospheric Scintillation Intensity Fading Characteristics and GPS Receiver Tracking Performance at Low Latitudes." *GPS Solutions* 23 (2): 1–12. https://doi.org/10.1007/s10291-019-0835-1

Huang, Guanwen, Shi Du, and Duo Wang. 2023. "GNSS Techniques for Real-Time Monitoring of Landslides: A Review." *Satellite Navigation* 4 (1). https://doi.org/10.1186/s43020-023-00095-5

Jin, Xulei, and Shuli Song. 2023. "Near Real-Time Global Ionospheric Total Electron Content Modeling and Nowcasting Based on GNSS Observations." *Journal of Geodesy* 97 (3): 1–18. https://doi.org/10.1007/s00190-023-01715-3

Kiruthiga, S., S. Mythili, M. Vijay, and R. Mukesh. 2022. "Analysis of Ionospheric TEC Variations Due to X, M & C Class Solar Flares during the Years 2003 to 2018 and Comparison with IRI Models." *Geomagnetism and Aeronomy* 62 (8): 971–91. https://doi.org/10.1134/S0016793222080138

Kolarski, Aleksandra, Nikola Veselinović, Vladimir A. Srećković, Zoran Mijić, Mihailo Savić, and Aleksandar Dragić. 2023. "Impacts of Extreme Space Weather Events on September 6th, 2017 on Ionosphere and Primary Cosmic Rays." *Remote Sensing* 15 (5). https://doi.org/10.3390/rs15051403

Kumar, Sanjay, Eng Leong Tan, and Dhimas Sentanu Murti. 2015. "Impacts of Solar Activity on Performance of the IRI-2012 Model Predictions from Low to Mid Latitudes." *Earth, Planets and Space* 67 (1). https://doi.org/10.1186/s40623-015-0205-3

Li, Guozhu, Baiqi Ning, Yuichi Otsuka, Mangalathayil Ali Abdu, Prayitno Abadi, Zhizhao Liu, Luca Spogli, and Weixing Wan. 2021. *Challenges to Equatorial Plasma Bubble and Ionospheric Scintillation Short-Term Forecasting and Future Aspects in East and Southeast Asia. Surveys in Geophysics.* Vol. 42. https://doi.org/10.1007/s10712-020-09613-5

Luo, Xiaomin, Yidong Lou, Shengfeng Gu, Guozhu Li, Chao Xiong, Weiwei Song, and Zhengyu Zhao. 2021. "Local Ionospheric Plasma Bubble Revealed by BDS Geostationary Earth Orbit Satellite Observations." *GPS Solutions* 25 (3): 1–10. https://doi.org/10.1007/s10291-021-01155-6

Magdaleno, Sergio, Miguel Herraiz, David Altadill, and Benito A. De La Morena. 2017. "Climatology Characterization of Equatorial Plasma Bubbles Using GPS Data." *Journal of Space Weather and Space Climate* 7. https://doi.org/10.1051/swsc/2016039

Makela, Jonathan J. 2006. "A Review of Imaging Low-Latitude Ionospheric Irregularity Processes." *Journal of Atmospheric and Solar-Terrestrial Physics* 68 (13): 1441–58. https://doi.org/10.1016/j.jastp.2005.04.014

Mendillo, M., J. A. Klobuchar, R. B. Fritz, A. V. da Rosa, L. Kersley, K. C. Yeh, B. J. Flaherty, et al. 1974. " Behavior of the Ionospheric F Region during the Great Solar Flare of August 7, 1972." *Journal of Geophysical Research* 79 (4): 665–72. https://doi.org/10.1029/ja079i004p00665

Moges, Samson T., Nigussie Mezgebe Giday, Daniel Atnafu Chekole, Thomas Ulich, and R. O. Sherstyukov. 2022. "Storm-Time Observations of Traveling Ionospheric Disturbances and Ionospheric Irregularities in East Africa." *Radio Science* 57 (8): 1–17. https://doi.org/10.1029/2022RS007426

Mohanty, S., G. Singh, C. S. Carrano, and S. Sripathi. 2018. "Ionospheric Scintillation Observation Using Space-Borne Synthetic Aperture Radar Data." *Radio Science* 53 (10): 1187–1202. https://doi.org/10.1029/2017RS006424

Moreno, B., S. Radicella, M. C. de Lacy, M. Herraiz, and G. Rodriguez-Caderot. 2011. "On the Effects of the Ionospheric Disturbances on Precise Point Positioning at Equatorial Latitudes." *GPS Solutions* 15 (4): 381–90. https://doi.org/10.1007/s10291-010-0197-1

Mosna, Zbysek, Daniel Kouba, Petra Koucka Knizova, Dalia Buresova, Jaroslav Chum, Tereza Sindelarova, Jaroslav Urbar, Josef Boska, and Dana Saxonbergova-Jankovicova. 2020. "Ionospheric Storm of September 2017 Observed at Ionospheric Station Pruhonice, the Czech Republic." *Advances in Space Research* 65 (1): 115–28. https://doi.org/10.1016/j.asr.2019.09.024

Nishioka, M., A. Saito, and T. Tsugawa. 2008. "Occurrence Characteristics of Plasma Bubble Derived from Global Ground-Based GPS Receiver Networks." *Journal of Geophysical Research: Space Physics* 113 (5): 1–12. https://doi.org/10.1029/2007JA012605

OHSHIO, MITSUO. 1971. "Negative Sudden Phase Anomaly." *Nature Physical Science* 229 (8): 239–40. https://doi.org/10.1038/physci229239a0

Olawepo, A. O., O. A. Oladipo, S. Sodiq, O. Odeyemi, B. W. Joshua, S. J. Adebiyi, J. O. Adeniyi, and I. A. Adimula. 2022. "Geomagnetic Storms Impact on Post-Sunset Equatorial Ionosphere at Ilorin During 2010." *Geomagnetism and Aeronomy* 62 (5): 632–51. https://doi.org/10.1134/S0016793222050097

Patari, Arup, Bapan Paul, and Anirban Guha. 2021. "Statistics of GPS TEC at the Northern EIA Crest Region of the Indian Subcontinent during the Solar Cycle 24 (2013-2018): Comparison with IRI-2016 and IRI-2012 Models." *Astrophysics and Space Science* 366 (5). https://doi.org/10.1007/s10509-021-03950-6

Priyadarshi, S., and A. K. Singh. 2012. "Statistics of Night Time Ionospheric Scintillation Using GPS Data at Low Latitude Ground Station Varanasi," no. May 2014. http://arxiv.org/abs/1208.1914

Pulinets, Sergey, Andrzej Krankowski, Manuel Hernandez-Pajares, Sergio Marra, Iurii Cherniak, Irina Zakharenkova, Hanna Rothkaehl, et al. 2021. "Ionosphere Sounding for Pre-Seismic Anomalies Identification (INSPIRE): Results of the Project and Perspectives for the Short-Term Earthquake Forecast." *Frontiers in Earth Science* 9 (April): 1–16. https://doi.org/10.3389/feart.2021.610193

Sao, K., M. Yamashita, S. Tanahashi, H. Jindoh, and K. Ohta. 1970. "Sudden Enhancements (SEA) and Decreases (SDA) of Atmospherics." *Journal of Atmospheric and Terrestrial Physics* 32 (9): 1567–76. https://doi.org/10.1016/0021-9169(70)90072-3

Schmölter, Erik, and Jens Berdermann. 2021. "Predicting the Effects of Solar Storms on the Ionosphere Based on a Comparison of Real-Time Solar Wind Data with the Best-Fitting Historical Storm Event." *Atmosphere* 12 (12). https://doi.org/10.3390/atmos12121684

Shagimuratov, I. I., I. E. Zakharenkova, N. Yu Tepenitsina, G. A. Yakimova, and I. I. Efishov. 2020. "Ionospheric Effects of Solar Flares in September 2017 and an Evaluation of Their Influence on Errors in Navigation Measurements." *Geomagnetism and Aeronomy* 60 (5): 597–605. https://doi.org/10.1134/S0016793220050138

Shetti, D. J., O. B. Gurav, and Gopi K. Seemla. 2019. "Occurrence Characteristics of Equatorial Plasma Bubbles and Total Electron Content during Solar Cycle Peak 23rd to Peak 24th over Bangalore (13.02° N, 77.57° E)." *Astrophysics and Space Science* 364 (9). https://doi.org/10.1007/s10509-019-3643-8

Solomon, Stanley C., Alan G. Burns, Barbara A. Emery, Martin G. Mlynczak, Liying Qian, Wenbin Wang, Daniel R. Weimer, and Michael Wiltberger. 2012. "Modeling Studies of the Impact of High-Speed Streams and Co-Rotating Interaction Regions on the Thermosphere-Ionosphere." *Journal of Geophysical Research: Space Physics* 117 (8): 1–14. https://doi.org/10.1029/2011JA017417

Tang, Jun, Shimeng Zhang, Xingliang Huo, and Xuequn Wu. 2022. "Ionospheric Assimilation of GNSS TEC into IRI Model Using a Local Ensemble Kalman Filter." *Remote Sensing* 14 (14): 1–15. https://doi.org/10.3390/rs14143267

Tariku, Yekoye Asmare. 2015. "TEC Prediction Performance of the IRI-2012 Model over Ethiopia during the Rising Phase of Solar Cycle 24 (2009-2011)." *Earth, Planets and Space* 67 (1): 1–10. https://doi.org/10.1186/s40623-015-0312-1

Timoçin, Erdinç, Samed Inyurt, Hüseyin Temuçin, Kutubuddin Ansari, and Punyawi Jamjareegulgarn. 2020. "Investigation of Equatorial Plasma Bubble Irregularities under Different Geomagnetic Conditions during the Equinoxes and the Occurrence of Plasma Bubble Suppression." *Acta Astronautica* 177 (June): 341–50. https://doi.org/10.1016/j.actaastro.2020.08.007

Vani, Bruno Cesar, Biagio Forte, Joao Francisco Galera Monico, Susan Skone, Milton Hirokazu Shimabukuro, Alison De Oliveira Moraes, Igor Ponte Portella, and Haroldo Antonio Marques. 2019. "A Novel Approach to Improve GNSS Precise Point Positioning during Strong Ionospheric Scintillation: Theory and Demonstration." *IEEE Transactions on Vehicular Technology* 68 (5): 4391–4403. https://doi.org/10.1109/TVT.2019.2903988

Wan, Xin, Chao Xiong, Shunzu Gao, Fuqing Huang, Yiwen Liu, Ercha Aa, Fan Yin, and Hongtao Cai. 2021. "The Nighttime Ionospheric Response and Occurrence of Equatorial Plasma Irregularities during Geomagnetic Storms: A Case Study." *Satellite Navigation* 2 (1). https://doi.org/10.1186/s43020-021-00055-x

Warnant, René, and Eric Pottiaux. 2000. "The Increase of the Ionospheric Activity as Measured by GPS." *Earth, Planets and Space* 52 (11): 1055–60. https://doi.org/10.1186/BF03352330

Wernik, Andrzej W., Lucilla Alfonsi, and Massimo Materassi. 2004. "Ionospheric Irregularities, Scintillation and Its Effect on Systems." *Acta Geophysica Polonica* 52 (2): 237–49.

Yamamoto, Mamoru, Yuichi Otsuka, Hidekatsu Jin, and Yasunobu Miyoshi. 2018. "Relationship between Day-to-Day Variability of Equatorial Plasma Bubble Activity from GPS Scintillation and Atmospheric Properties from Ground-to-Topside Model of Atmosphere and Ionosphere for Aeronomy (GAIA) Assimilation." *Progress in Earth and Planetary Science* 5 (1). https://doi.org/10.1186/s40645-018-0184-7

Yokoyama, Tatsuhiro. 2017. "A Review on the Numerical Simulation of Equatorial Plasma Bubbles toward Scintillation Evaluation and Forecasting." *Progress in Earth and Planetary Science* 4 (1). https://doi.org/10.1186/s40645-017-0153-6

27 Enhancing GNSS Applications in Climate Studies
A Path Forward

Nitin Arora, Sartajvir Singh, and Apoorva Sharma

27.1 INTRODUCTION

Since the inception of the technical era, everything has been carefully marked and measured before practical application in the real world. These markings and assessments must be well-suited and precisely tailored for the specific tasks they perform. The Global Navigation Satellite System (GNSS) is a technology excellently suited for precise marking and measurements. GNSS constitutes a vast domain encompassing satellite monitoring and location using specific parameters, providing a precise 3-axis location parameter for the observed object. The object of observation can range from an airplane to a small smartphone.

GNSS comprises Europe's Galileo, the USA's NAVSTAR Global Positioning System (GPS), Russia's Global'naya Navigatsionnaya Sputnikovaya Sistema (GLONASS), and China's BeiDou Navigation Satellite System [1]. In this chapter, we will discuss the core concept of GNSS and its role in climate studies. We will cover the limitations of GNSS technology and scenarios where this technology fails to examine and provide real-time coordinates of a location, thus leading to improper examination of climatic changes [2]. To obtain accurate information about a geographic location, it is crucial to comprehend its coordinates. Neglecting this factor when recording information about an area can result in conflicts between regions due to inadequate coordination in examination.

27.2 GNSS

GNSS or Global Navigation Satellite System is a comprehensive term that includes the techniques used for navigation and precise positioning of locomotive vehicles such as aircraft and ships, and electronic devices like smartphones. GNSS itself consists of a vast variety of terms, encompassing several technologies responsible for precise navigation, e.g., GPS, GLONAS, Galileo, BDS, etc. They all collectively make GNSS. GNSS uses the principle of trilateration for positioning and has three major components: Satellite Constellation, Ground Control Stations, and User Receivers (see Figure 27.1) [3–5]. All these work in a sequence to provide the precise location of a region or an object (see Figure 27.1).

Satellite Constellation refers to the group of satellites that revolves around the Earth. These satellite broadcast signals are further received by GNSS receivers on the ground, whereas GCS or Ground Control Stations are specially made stations that are responsible for monitoring and controlling the satellite in the GNSS constellation, ensuring their proper functioning and accurate signal transmission. GCS tracks a constant record of the satellite orbit path, makes necessary adjustments and calculates corrections to ensure the precision of positioning information received by GNSS users. User receivers are responsible for calculating the user's positioning, velocity, and time-based information received from multiple satellites and then transmitted to GCS.

All these systems work together to provide real-time coordinates of the receiver. First, the location coordinates are depicted by the satellite constellation, and the collected data—along with

DOI: 10.1201/9781032712444-32

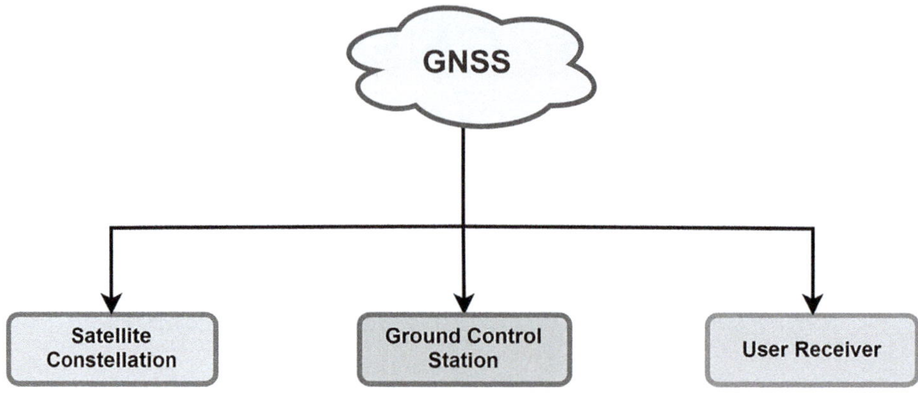

FIGURE 27.1   Components of GNSS technology.

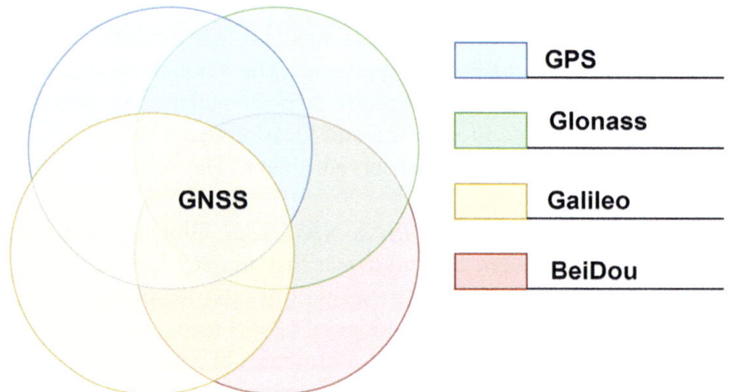

FIGURE 27.2   These four GNSS systems are among the most prominent and widely used globally, each contributing to the availability of accurate and reliable positioning services across different regions.

real-time information—is sent to the ground control station after processing with certain algorithms and removing any imperfection due to noise; this is then sent to the user (receiver) [6]. The ground control station, along with serving as a bridge for the satellite and the receiver, serves a much important purpose, i.e., it keeps the satellite aligned and is responsible for correcting any time-based imperfection causing the satellite to go off track [7].

Figure 27.2 provides an added view of the overlap and interaction between the several GNSSs, which include GPS, GLONASS, Galileo, and BeiDou. To clarify that all these systems are a part of the GNSS and provide global satellite navigation services, a Venn diagram has been prepared (refer to Table 27.1).

GNSS works on two basic principles: one is trilateration, and the other is the mathematical relationship between the rate speed and time given as:

$$D = R \times T \tag{27.1}$$

Where:

$'R'$ Rate = Given based on satellite transmission

$'T'$ Time = How long until it reaches the receiver

$'D'$ Distance = With the known rate and time, we can solve for the distance between the sender and the receiver

TABLE 27.1

Representation of Various GNSS Systems

Name of Technology	Introductory Year	Place of Origin	Use Case
GPS	1978	United States Department of Defense	Global Positioning and Navigation
GLONASS	1982	Russia	Global Navigation and Positioning
Galileo	2016	European Union	Global Navigation, Positioning, and Timing
BDS	2020	China	Global Navigation, Positioning, and Timing

Trilateration is a geometric technique used in navigation and positioning systems to determine the location of a point in space by measuring its distance from three known points. Unlike triangulation, which involves measuring angles between known points, trilateration relies on the precise measurement of distances or ranges. In the context of navigation and satellite systems, trilateration is commonly employed by receiving signals from at least three satellites to calculate the exact position of a receiver on or near the Earth's surface. The process involves determining the intersection point of three spheres, each representing the possible locations based on the measured distances. Trilateration is a fundamental principle in various applications, including GPS (Global Positioning System) and other global navigation satellite systems, providing accurate and reliable positioning information for a wide range of purposes [8–10]. Trilateration is the fundamental principle underlying all GNSS, with variations in data representation being the primary distinction among these systems [11, 12].

According to the principle, there are four satellites (see Figure 27.3) that are a must for navigation and precise positioning. In the case of GPS, 24 satellites revolve in the Earth's orbit and are responsible for precision in location. Satellite-1 is responsible for sending the first frequency from the satellite to the receiver. Satellite-2 is responsible for sending other/second signals in correspondence with the first satellite to find the exact location of the receiver. Satellite-3 is placed closer to the horizon and is responsible for third signal modulation; hence, as the name suggests, it becomes signal triangulation. The third satellite in conjunction with the first two satellites is responsible for locating the position. The common point of the three satellites tells the exact location of the receiver with the most precision. Satellite-4 is responsible for proper timing, since timing has a crucial role in positioning.

27.3 ROLE OF GNSS IN CLIMATE STUDIES

GNSSs have been used in navigation technology for the last two decades and provide a strong base for observation and studying climate studies and environmental processes. GNSSs provide a precise and more accurate model for remote sensing as compared to other present models [13, 14]. They can be used to monitor many important aspects necessary for climatic studies, some of which are listed below:

27.3.1 ATMOSPHERIC SOUNDING

Earth's ionosphere harms the transmission of GNSS signals, and that delay in the signal receiving allows researchers to study the varying environment. It also enables them to know the dynamically changing components of the environment such as temperature, humidity, and many more.

27.3.2 SEA LEVEL RISE MONITORING

Monitoring sea-level patterns is an essential aspect in climate studies. Studying the changing pattern allows the researchers to understand the shift in sea levels within the past few years and helps them

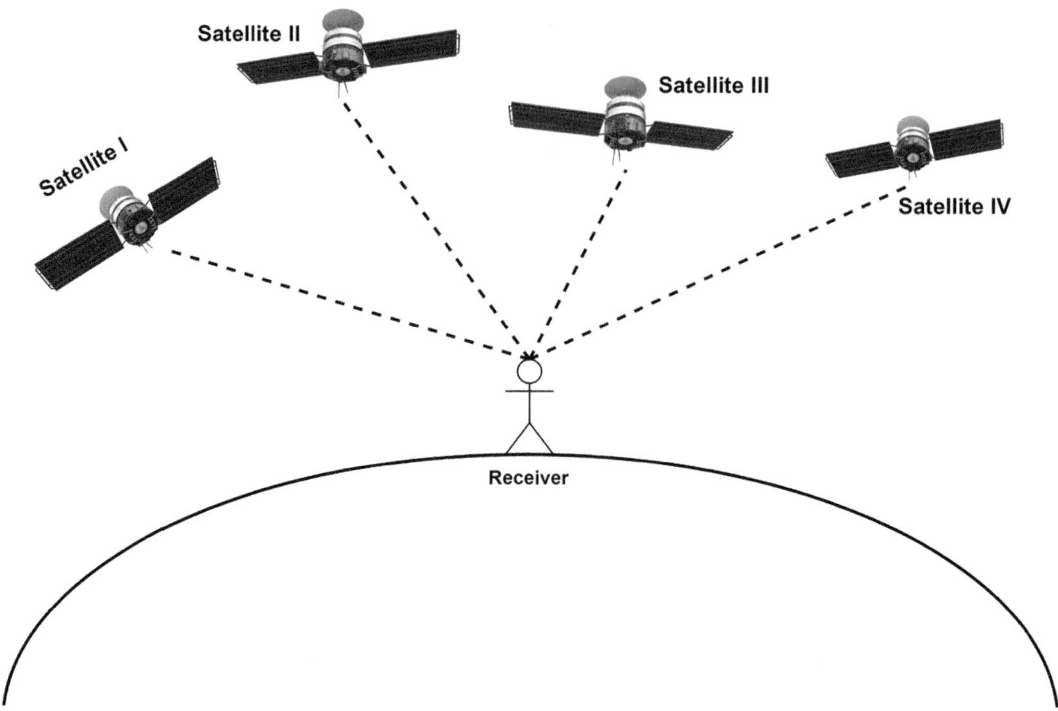

FIGURE 27.3 Representation of the four satellites (minimum) required for recording precise location and further sending the information to the GCS located on the ground state.

to make predictions for the future. This study also enables weather forecasters to analyze any upcoming natural calamity or disaster near sea coasts. GNSSs can monitor any dip or rise in sea level with high precision and accuracy. The GNSS model can be trained for future events, enabling GNSS to evaluate the probability of an event happening before it occurs.

27.3.3 EXTREME WEATHER EVENT ANALYSIS

The GNSS system has proven reliable when there is a requirement for in-depth analysis of weather conditions. It can predict the occurrence of an event well before it happens, thanks to machine learning models included inside its system. This is useful in situations of natural calamity such as hurricanes or storms [15]. Information collected by the GNSS system serves as a valuable source of knowledge, helping scholars with their research.

27.3.4 CLIMATE MODEL VALIDATION

As discussed earlier, GNSSs can monitor the climate and make an accurate real-time map. This property allows researchers for in-depth examination and monitoring of any unintended patterns visible in the environment [16].

27.4 LIMITATIONS OF GNSS

Every system has its limitations, and GNSS is no exception—but there have been constant improvements in the system over time. Currently, the system is overall more reliable and less prone to environmental factors such as ionospheric effects and spoofing. Earlier sections of this chapter covered the importance of GNSS in climate studies and how it is involved in understanding the dynamic patterns of important parameters of the environment.

The following sections will explain the limitations faced by the GNSS system and how these limitations can be eradicated with improvements. Lastly, some suggestions, if implemented correctly, will ultimately benefit the system, thus increasing its precision and accuracy more than anticipated [17].

Despite having many advantages, GNSSs also have limitations. Some instances where GNSSs fail to provide precise monitoring and accurate navigation are cited below:

27.4.1　Signal Blocking

GNSS signals are sensitive to the line of sight—any hindrance caused by high-rise buildings or dense forests covered with dense vegetation would compromise the precision and accuracy of the monitored data; thus making it almost impossible to achieve 100% accuracy in such regions [17, 18].

27.4.2　Limited Vertical Resolution

GNSSs are reliable when it comes to horizontal positioning, but their vertical accuracy is not as precise. Considering the case of sea level monitoring, GNSSs are proven to be the most reliable means of monitoring, whereas if the system is used for monitoring a heavily dense forest or any place with high-rise buildings, GNSS fails to provide such high accuracy and is considered a less reliable means of mapping.

27.4.3　Polar Coverage Limitations

Polar coverage refers to the accessibility of the GNSS system in regions near the poles. It is noted that GNSSs often face challenges associated with signal availability and reliability near the Earth's polar regions. This is because satellites are inclined at a certain angle to provide reliable measurements, and this inclination is by the equator. However, as one goes towards the polar regions, the inclination angle is lower. As a result, satellites appear to be closer to the horizon, leading to a hindrance in signal visibility. This reduced visibility usually results in weaker signals, giving rise to signal blockage and hence weaker signal strength, overall lowering the accuracy in positioning data in high-latitude regions near the poles.

27.4.4　Weather-induced Impact

Bad weather can hinder the signal strength of the GNSS system, thus making the signals less accurate, which directly impact the quality of information from GNSS during the task of climate monitoring. Rain and stormy conditions act as pebbles between the path of the signal to be shared between the GNSS satellite and research stations at ground level. These weaker signals are a result of less reliable observations, thus making it difficult to get exact and trustworthy location information.

27.4.5　Interference and Spoofing Threats

Spoofing is the act of producing misleading signals to disrupt the signals emitted by the satellite, which were ultimately about to be received by the receiver. This leads the receiver to calculate the wrong monitoring, making the system deceptive [19]. There are many instances where interference in signal transfer occurs, leading to data manipulation and subsequently reducing the accuracy and precision of monitoring. This also introduces other risks, making it challenging to trust the accuracy and reliability of climate data collected through GNSS technology.

27.4.6 Temporal Resolution Constraints

Good precision is all about real-time observation and locating any change in the stats from the past. Comparing both provides a base for the research of the changes observed in the climate and the environmental elements. While GNSSs are excellent in capturing real-time stats and data, they lack in capturing the rapid change in climate over a short period; this limitation of GNSSs is termed a temporal resolution constraint [20, 21]. Due to this constraint, the GNSS itself is not enough to capture any sudden change. Additional technologies with better temporal resolution are necessary for obtaining a detailed and accurate analysis of the sudden changes in climatic conditions. Essentially, GNSS data may not be enough to update the sudden climate changes, and other tools are needed for a closer look at rapid climate shifts.

27.4.7 Dependency on Satellite Constellation

Satellite constellations form the backbone of the GNSS system, and dependence on these satellites is crucial for accurate navigation. Therefore, it is essential for satellites to be placed in proper coordinates and to provide consistent real-time responses [22]. Any malfunction or degradation in even one satellite has a direct impact on the functioning of other satellites, ultimately leading to a loss of precision. In simpler terms, the smooth operation of GNSS technology depends on the health and functionality of the satellite network. Any issues with the satellites can lead to challenges in providing accurate and reliable positioning information through GNSS [23].

27.5 IMPROVEMENT AND FUTURE PROSPECTS OF GNSS IN CLIMATE STUDIES

Every system requires improvements, and GNSSs are no exception. There is a need for significant enhancements in their current state to address the previously cited limitations in Climate Studies. Some improvement strategies and amendments are outlined below:

27.5.1 Introduction of Wide-Area Augmentation System (WAAS)

WAAS, which stands for Wide Area Augmentation System, is an augmentation system designed to enhance the accuracy and reliability of the GNSS (see Figure 27.4). The concept is similar to that of GPS, but with a slight difference: A ground control stage before the receiver receives the signals [24, 25].

27.5.2 Reference Stations (Ref Stations)

Positioned strategically nationwide, these ground-based stations systematically gather data on errors within GPS signals arising from atmospheric conditions and satellite clock discrepancies.

27.5.3 Master Station

The accumulated data is relayed to a central Master Station, where meticulous corrections are computed.

27.5.4 WAAS Satellite

These precise corrections are then transmitted to a geostationary satellite exclusively designated for the WAAS system, as shown in Figure 27.5.

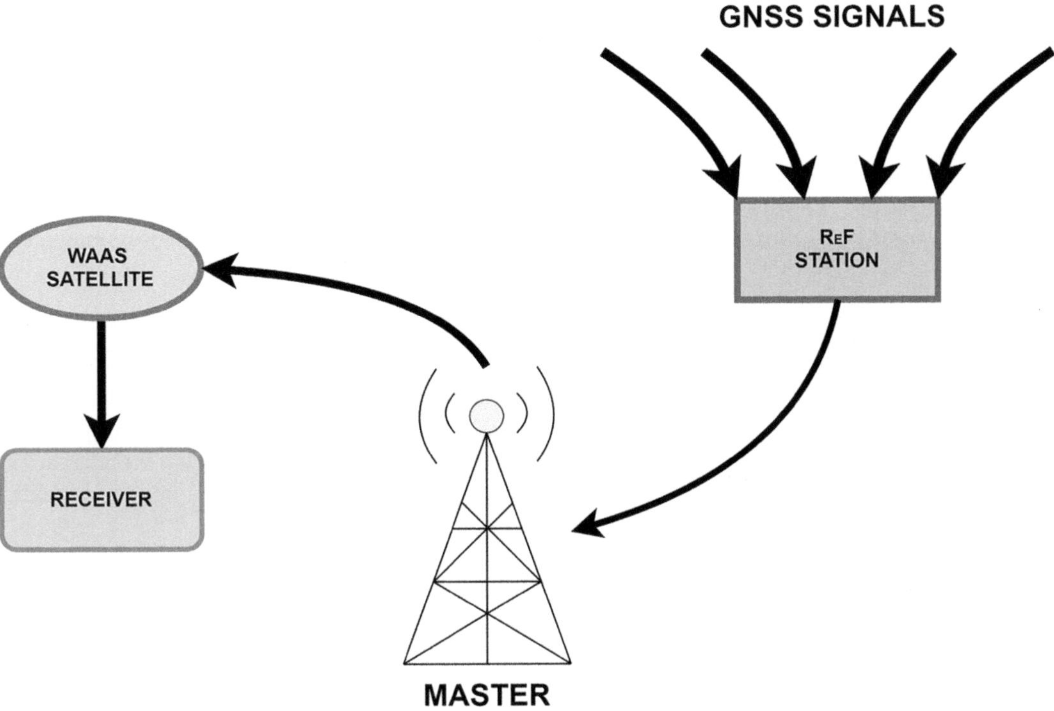

FIGURE 27.4 Illustration of the architecture of the WAAS (Wide Area Augmentation System) with key components, including Reference Stations (Ref Stations), a Master Station, a WAAS satellite, and the ultimate receiver.

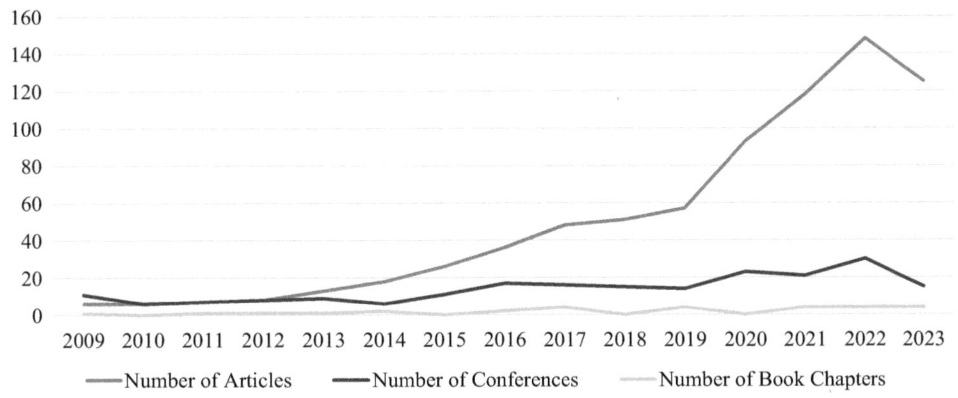

FIGURE 27.5 The stacked bar graph of publications related to GNSS and Climate (a) article, (b) conferences, (c) book chapter (Note: Year with no publication has been omitted).

27.5.5 BROADCASTING CORRECTED SIGNALS

The WAAS satellite disseminates the refined signals extensively, enveloping a broad geographic expanse, including airspace.

27.5.6 RECEIVER

The end user, equipped with a GPS receiver, captures these enhanced signals, resulting in an elevated level of accuracy and reliability for navigation and positioning.

WAAS is currently used in airplanes, but it can also be used for climate studies. WAAS systems have a close resemblance with the GPS, sharing a similar working principle [26, 27]. WAAS also includes a ground-level monitoring station that accesses all the signal data and recalibrates it to remove any uninvited noise or imperfection. WAAS has an advantage over GPS (a component of the GNSS system), i.e., it holds more accuracy since there is a ground-level-cited body for removing noise and any damaged signal that might have entered the system as a result of spoofing or any other unavoidable reason. In conclusion, the implementation of the WAAS system could be a better approach towards providing accuracy and more precision to the system, along with providing a barrier to prevent spoofing threats.

27.5.7 Enhanced Security Measures Towards Spoofing:

As discussed in the earlier sections of the chapter, spoofing is the act of implanting unwanted/fake signals into the navigation system, thus reducing the accuracy and making the system practically unreliable [28, 29]. GNSSs unfortunately face a similar type of issue, hence reducing the accuracy of climate monitoring. These signals could be intentional or unintentional, but they need to be eliminated from the system as much as possible [30]. Spoofing can be removed from the system with defensive measures such as developing algorithms to eliminate unexpected noise that may have entered the system intentionally or unintentionally, and extracting the essential signals originally needed could be a better solution for ensuring uninterrupted system operation. This approach would lead to much more accurate measurements and monitoring. Time synchronization is an essential key component for GNSS to work flawlessly and thus act as the backbone of the whole system. A satellite works on time coordination, and making the time synchronization more secure will ultimately lead to better results. Using encryption to secure transmitted signals can prevent spoofing since the harmful signals cannot blend with the encrypted signals, thus making the system less vulnerable to data loss.

27.5.8 Research in Ionospheric Effect

Earth's atmosphere consists of several layers stacked on one another, namely the Troposphere, Stratosphere, Mesosphere, Thermosphere, and Exosphere; their characteristics are shown in Table 27.2. Every layer has its importance in making up the Earth's atmosphere as a whole. Refer to Table 27.2 to see the extension of layers and its effect on the GNSS system [31–33]. The ionosphere is not one single layer but consists of multiple layers, mainly the mesosphere and thermosphere. The ionosphere got its name because it contains a higher concentration of ions and free electrons, the result of the ionization of neutral particles by solar ultraviolet (UV) radiation [34–36].

TABLE 27.2

Different Atmospheric Layers and Their Effects on GNSS

Layer	Altitude Range	Key Characteristics	Effects on GNSS Systems
Troposphere	0–8 to 15 km	Weather events, clouds, and precipitation	Signal delay and refraction due to atmospheric gases
Stratosphere	15–50 km	Ozone layer absorbs UV radiation	Minimal impact on GNSS systems
Mesosphere	50–85 km	Meteorites burn up here	Minimal impact on GNSS systems
Ionosphere	85 km–~600 km	Ionization due to solar radiation, multiple layers	Causes signal scintillation and delay in GNSS systems
Thermosphere	~600 km and beyond	High temperatures, low air density	Significant impact on GNSS signals due to ionization
Exosphere	Beyond 600 km	Gradual transition into space	Negligible impact on GNSS systems

Understanding the role of the ionosphere and its influence in the transmission and reception of signals is crucial for applications such as radio communication, navigation systems, and satellite communication. The ionosphere plays an important role in the propagation of radio waves, including those used by the GNSS. The number of ionized particles in the atmosphere directly affects the navigation systems and satellite constellations such as the GNSS system. In conclusion, despite having many disadvantages, the ionosphere is an essential aspect of GNSS, so researchers should learn more about the ionosphere and how to increase the efficiency of the GNSS system [36].

The objective is to design a system capable of eliminating the overall imperfections of the GNSS system, including alterations to the signals caused by spoofing or ionospheric effects. While WAAS appears to be a better approach, it also faces similar limitations as those mentioned above. Enabling the GNSS system to adapt to the changing environment ensures that it can consistently provide trustworthy and precise data transfer, even in the presence of extreme atmospheric disruptions.

27.6 META-ANALYSIS

A meta-analysis has also been presented to understand the research contribution made by different authors in the field of GNSS and climate studies. To compute the meta-analysis, the SCOPUS database has been utilized, and the terms "GNSS" and "Climate" were searched on 27-Dec-2023 at 12:45 p.m. (Indian Standard Time – IST). The search was performed within the "Article title, Abstract and Keywords." The search was performed for the past two decades, i.e., 2009 to 2023 (till 27-Dec-2023). The terms, i.e., "GNSS" and "Agriculture," have been cross-verified concerning their relevance and eliminated redundant publications. Moreover, the Scopus-indexed publications from articles, reviews, conferences, and book chapters have been considered, as shown in Figure 27.5.

27.7 SUMMARY AND FUTURE SCOPE

This chapter sheds light on the GNSS system and the role of GNSS in climate studies. GNSS itself is a combined term used for all of the different navigation systems utilized by countries across the globe. The GNSS system is not limited to the navigation of aircraft, ships, or smartphones but also has an important role in other sectors too, including disaster management systems, flood monitoring and mapping, and, most importantly, agriculture. In this chapter, we discussed the working principle of GNSS systems and to what extent they can provide high precision and accuracy.

The GNSS has been used in climate studies for two decades, but it has not yet been replaced by any other system. No system is perfect, and the same applies to the GNSS system, thus requiring consistent amendments and improvements. The second phase of the chapter included the limitations of the GNSS system, showing where improvement is needed. Going deep into the studies, it became clear that there are two major drawbacks of the GNSS system: improper signal strength because of spoofing issues and any error caused by the coordinates of the satellite constellation.

Moving ahead, the third phase of the chapter discussed the improvements needed that could be helpful in increasing the overall efficiency of the GNSS. Since it is important for the system to be precise, any delay in receiving signals might affect the overall performance. The chapter discusses several measures that, if implemented correctly, would ultimately be beneficial for GNSS going forward. Having some special emphasis on the study of the ionosphere and its effect on the signal transmission of GNSS could ultimately enhance the signal quality and coverage area for GNSS satellites. Since being an essential element in radio transmission over longer distances, this layer also has some adverse effects on the normal operation of the system, thus reducing the precision of the overall system. In conclusion, one can say that using GNSS can be considered as a perfect approach, but it also has some limitations just like any other system. However, with some future amendments, it could be predicted that the GNSS would incorporate much more precise monitoring than it provides today.

REFERENCES

[1] S. Severi, H. Wymeersch, J. Harri, M. Ulmschneider, B. Denis, and M. Bartels, "Beyond GNSS: Highly accurate localization for cooperative-intelligent transport systems," in *2018 IEEE Wireless Communications and Networking Conference (WCNC)*, IEEE, Apr. 2018, pp. 1–6. doi: 10.1109/WCNC.2018.8377457

[2] N. Shen et al., "A review of Global Navigation Satellite System (GNSS)-based dynamic monitoring technologies for structural health monitoring," *Remote Sensing*, vol. 11, no. 9. MDPI AG, May 01, 2019. doi: 10.3390/rs11091001

[3] N. Shen et al., "A Review of Global Navigation Satellite System (GNSS)-based Dynamic Monitoring Technologies for Structural Health Monitoring," *Remote Sens (Basel)*, vol. 11, no. 9, p. 1001, Apr. 2019, doi: 10.3390/rs11091001

[4] "Global Navigation Satellite System (GNSS)."

[5] M. T. Tran, T. H. D. Dao, H. Quoc, V. Rd, C. Giay, and V. Hanoi, "Global Navigation Satellite Systems (GNSS): An Overview."

[6] C. J. Hegarty and E. Chatre, "Evolution of the Global Navigation SatelliteSystem (GNSS)," *Proceedings of the IEEE*, vol. 96, no. 12, pp. 1902–1917, Dec. 2008, doi: 10.1109/JPROC.2008.2006090

[7] E. Ochin, "Fundamentals of structural and functional organization of GNSS," in *GPS and GNSS Technology in Geosciences*, Elsevier, 2021, pp. 21–49. doi: 10.1016/B978-0-12-818617-6.00010-X

[8] P. Zabalegui, G. De Miguel, A. Perez, J. Mendizabal, J. Goya, and I. Adin, "A Review of the Evolution of the Integrity Methods Applied in GNSS," *IEEE Access*, vol. 8, pp. 45813–45824, 2020, doi: 10.1109/ACCESS.2020.2977455

[9] X. He et al., "Multilevel-teaching/training practice on GNSS principle and application for undergraduate educations: A case study in China," *Advances in Space Research*, vol. 69, no. 1, pp. 778–793, Jan. 2022, doi: 10.1016/j.asr.2021.11.021

[10] Z. Yao and M. Lu, "Signal Multiplexing Techniques for GNSS: The Principle, Progress, and Challenges Within a Uniform Framework," *IEEE Signal Process Mag*, vol. 34, no. 5, pp. 16–26, Sep. 2017, doi: 10.1109/MSP.2017.2713882

[11] J. Armstrong, Y. Sekercioglu, and A. Neild, "Visible light positioning: a roadmap for international standardization," *IEEE Communications Magazine*, vol. 51, no. 12, pp. 68–73, Dec. 2013, doi: 10.1109/MCOM.2013.6685759

[12] F. Vincent, E. Chaumette, C. Charbonnieras, J. Israel, M. Aubault, and F. Barbiero, "Asymptotically efficient GNSS trilateration," *Signal Processing*, vol. 133, pp. 270–277, Apr. 2017, doi: 10.1016/j.sigpro.2016.11.027

[13] K. Yu, C. Rizos, D. Burrage, A. G. Dempster, K. Zhang, and M. Markgraf, "An overview of GNSS remote sensing," *EURASIP J Adv Signal Process*, vol. 2014, no. 1, p. 134, Dec. 2014, doi: 10.1186/1687-6180-2014-134

[14] G. Guerova et al., "Review of the state of the art and future prospects of the ground-based GNSS meteorology in Europe," *Atmos Meas Tech*, vol. 9, no. 11, pp. 5385–5406, Nov. 2016, doi: 10.5194/amt-9-5385-2016

[15] S. Jin, G. P. Feng, and S. Gleason, "Remote sensing using GNSS signals: Current status and future directions," *Advances in Space Research*, vol. 47, no. 10, pp. 1645–1653, May 2011, doi: 10.1016/j.asr.2011.01.036

[16] S. Bonafoni and R. Biondi, "The usefulness of the Global Navigation Satellite Systems (GNSS) in the analysis of precipitation events," *Atmos Res*, vol. 167, pp. 15–23, Jan. 2016, doi: 10.1016/j.atmosres.2015.07.011

[17] A. Hussain, F. Akhtar, Z. H. Khand, A. Rajput, and Z. Shaukat, "Complexity and Limitations of GNSS Signal Reception in Highly Obstructed Enviroments," *Engineering, Technology & Applied Science Research*, vol. 11, no. 2, pp. 6864–6868, Apr. 2021, doi: 10.48084/etasr.3908

[18] P. Puricer and P. Kovar, "Technical Limitations of GNSS Receivers in Indoor Positioning," in *2007 17th International Conference Radioelektronika*, IEEE, Apr. 2007, pp. 1–5. doi: 10.1109/RADIOELEK.2007.371487

[19] P. Dabove, A. M. Manzino, and C. Taglioretti, "GNSS network products for post-processing positioning: limitations and peculiarities," *Applied Geomatics*, vol. 6, no. 1, pp. 27–36, Mar. 2014, doi: 10.1007/s12518-014-0122-3

[20] C. D. Bussy-Virat, C. S. Ruf, and A. J. Ridley, "Relationship between Temporal and Spatial Resolution for a Constellation of GNSS-R Satellites," *IEEE J Sel Top Appl Earth Obs Remote Sens*, vol. 12, no. 1, pp. 16–25, Jan. 2019, doi: 10.1109/JSTARS.2018.2833426

[21] T. Ning and G. Elgered, "High-temporal-resolution wet delay gradients estimated from multi-GNSS and microwave radiometer observations," *Atmos Meas Tech*, vol. 14, no. 8, pp. 5593–5605, Aug. 2021, doi: 10.5194/amt-14-5593-2021

[22] M. Martin-Neira, W. Li, A. Andres-Beivide, and X. Ballesteros-Sels, "'Cookie': A Satellite Concept for GNSS Remote Sensing Constellations," *IEEE J Sel Top Appl Earth Obs Remote Sens*, vol. 9, no. 10, pp. 4593–4610, Oct. 2016, doi: 10.1109/JSTARS.2016.2585620

[23] L. Xing, Y. Wen, D. W. P. Thomas, J. Zhang, D. Zhang, and J. Xiao, "A Joint Time-Frequency Analytical Method for Electromagnetic Interference in Railway GNSS System," in *2020 International Symposium on Electromagnetic Compatibility - EMC EUROPE*, IEEE, Sep. 2020, pp. 1–4. doi: 10.1109/EMCEUROPE48519.2020.9245779

[24] R. Loh, V. Wullschleger, B. Elrod, M. Lage, and F. Haas, "The U.S. Wide-Area Augmentation System (WAAS)," *Navigation*, vol. 42, no. 3, pp. 435–465, Sep. 1995, doi: 10.1002/j.2161-4296.1995.tb01900.x

[25] P. Enge, "WAAS Messaging System: Data Rate, Capacity, and Forward Error Correction*," *Navigation*, vol. 44, no. 1, pp. 63–76, Mar. 1997, doi: 10.1002/j.2161-4296.1997.tb01940.x

[26] S. Pullen and T. Walter, "5 System Overview, Recent Developments, and Future Outlook for WAAS and LAAS," 2010. [Online]. Available: https://www.researchgate.net/publication/268372652

[27] K. Shallberg and Fang Sheng, "WAAS measurement processing; current design and potential improvements," in *2008 IEEE/ION Position, Location and Navigation Symposium, IEEE*, 2008, pp. 253–262. doi: 10.1109/PLANS.2008.4570014

[28] L. Meng, L. Yang, W. Yang, and L. Zhang, "A Survey of GNSS Spoofing and Anti-Spoofing Technology," *Remote Sens (Basel)*, vol. 14, no. 19, p. 4826, Sep. 2022, doi: 10.3390/rs14194826

[29] L. Junzhi, L. Wanqing, F. Qixiang, and L. Beidian, "Research Progress of GNSS Spoofing and Spoofing Detection Technology," in *2019 IEEE 19th International Conference on Communication Technology (ICCT)*, IEEE, Oct. 2019, pp. 1360–1369. doi: 10.1109/ICCT46805.2019.8947107

[30] M. L. Psiaki and T. E. Humphreys, "GNSS Spoofing and Detection," *Proceedings of the IEEE*, vol. 104, no. 6, pp. 1258–1270, Jun. 2016, doi: 10.1109/JPROC.2016.2526658

[31] O. P. Verkhoglyadova, A. J. Mannucci, C. O. Ao, B. A. Iijima, and E. R. Kursinski, "Effect of small-scale ionospheric variability on GNSS radio occultation data quality," *J Geophys Res Space Phys*, vol. 120, no. 9, pp. 7937–7951, Sep. 2015, doi: 10.1002/2015JA021055

[32] A. J. Knisely and D. J. Emmons, "Impacts of Kelvin-Helmholtz billow formation on GNSS radio occultation measurements of sporadic-E," *Frontiers in Astronomy and Space Sciences*, vol. 10, Dec. 2023, doi: 10.3389/fspas.2023.1280228

[33] M. Zhran and A. Mousa, "Planetary boundary layer height retrieval using GNSS Radio Occultation over Egypt," *The Egyptian Journal of Remote Sensing and Space Science*, vol. 25, no. 2, pp. 551–559, Aug. 2022, doi: 10.1016/j.ejrs.2022.03.013

[34] M. M. Hoque and N. Jakowski, "Higher order ionospheric effects in precise GNSS positioning," *J Geod*, vol. 81, no. 4, pp. 259–268, Mar. 2007, doi: 10.1007/s00190-006-0106-0

[35] T. Hadas et al., "Impact and Implementation of Higher-Order Ionospheric Effects on Precise GNSS Applications," *J Geophys Res Solid Earth*, vol. 122, no. 11, pp. 9420–9436, Nov. 2017, doi: 10.1002/2017JB014750

[36] Z. G. Elmas, M. Aquino, H. A. Marques, and J. F. G. Monico, "Higher order ionospheric effects in GNSS positioning in the European region," *Ann Geophys*, vol. 29, no. 8, pp. 1383–1399, 2011.

Index

Note: Pages in *italics* refer to figures and pages in **bold** refer to tables.